T0312150

Modeling and Analysis of Stochastic Systems

Third Edition

CHAPMAN & HALL/CRC
Texts in Statistical Science Series

Series Editors

Francesca Dominici, *Harvard School of Public Health, USA*
Julian J. Faraway, *University of Bath, UK*
Martin Tanner, *Northwestern University, USA*
Jim Zidek, *University of British Columbia, Canada*

Texts in Statistical Science

Modeling and Analysis of Stochastic Systems

Third Edition

Vidyadhar G. Kulkarni

Department of Statistics and Operations Research
University of North Carolina at Chapel Hill, USA

CRC Press
Taylor & Francis Group
Boca Raton London New York

CRC Press is an imprint of the
Taylor & Francis Group, an **informa** business
A CHAPMAN & HALL BOOK

CRC Press
Taylor & Francis Group
6000 Broken Sound Parkway NW, Suite 300
Boca Raton, FL 33487-2742

First issued in paperback 2020

© 2017 by Taylor & Francis Group, LLC
CRC Press is an imprint of Taylor & Francis Group, an Informa business

No claim to original U.S. Government works

ISBN-13: 978-1-4987-5661-7 (hbk)
ISBN-13: 978-0-367-73679-8 (pbk)

Library of Congress Cataloging-in-Publication Data

Names: Kulkarni, Vidyadhar G.
Title: Modeling and analysis of stochastic systems / Vidyadhar G. Kulkarni.
Description: Third edition. | Boca Raton : CRC Press, 2017. | Series: Chapman & hall/CRC texts in statistical science series | Includes bibliographical references and index.
Identifiers: LCCN 2016021331 | ISBN 9781498756617 (alk. paper)
Subjects: LCSH: Stochastic processes. | Stochastic systems.
Classification: LCC QA274 .K844 2017 | DDC 519.2/3--dc23
LC record available at https://lccn.loc.gov/2016021331

Visit the Taylor & Francis Web site at
http://www.taylorandfrancis.com

and the CRC Press Web site at
http://www.crcpress.com

To

my sons

Milind, Ashwin, and Arvind

Jack and Harry were lost over a vast farmland while on their balloon ride. When they spotted a bicyclist on a trail going through the farmland below, they lowered their balloon and yelled, "Good day, sir! Could you tell us where we are?"

The bicyclist looked up and said, "Sure! You are in a balloon!"

Jack turned to Harry and said, "This guy must be a mathematician!"

"What makes you think so?" asked Harry.

"Well, his answer is correct, but totally useless!"

The author sincerely hopes that a student mastering this book will be able to use stochastic models to obtain correct as well as useful answers.

Contents

Preface

Preface to the second edition.

Probabilistic methodology has now become a routine part of graduate education in operations research, statistics, computer science, economics, business, public policy, bioinformatics, engineering, etc. The following three aspects of the methodology are most vital for the students in these disciplines:

1. Modeling a "real-life" situation with stochastic or random elements
2. Analysis of the resulting stochastic model
3. Implementation of the results of the analysis

Of course, if the results of Step 2 show that the model does not "fit" the real-life situation, then one needs to modify the model and repeat Steps 1 and 2 until a satisfactory solution emerges. Then one proceeds to Step 3. As the title of the book suggests, we emphasize the first two steps. The selection, the organization, and the treatment of topics in this book are dictated by the emphasis on modeling and analysis.

Based on my teaching experience of over 25 years, I have come to the conclusion that it is better (from the students' points of view) to introduce Markov chains before renewal theory. This enables the students to start building interesting stochastic models right away in diverse areas such as manufacturing, supply chains, genetics, communications, biology, queueing, and inventory systems, etc. This gives them a feel for the modeling aspect of the subject early in the course. Furthermore, the analysis of Markov chain models uses tools from matrix algebra. The students feel comfortable with these tools since they can use the matrix-oriented packages, such as MATLAB®, to do numerical experimentation. Nothing gives them better confidence in the subject than seeing the analysis produce actual numbers that quantify their intuition. We have also developed a collection of MATLAB®-based programs that can be downloaded from:

1. www.unc.edu/~vkulkarn/Maxim/maxim.zip
2. www.unc.edu/~vkulkarn/Maxim/maximgui.zip

The instructions for using them are included in the readme files in these two zip files.

After students have developed familiarity with Markov chains, they are ready for renewal theory. They can now appreciate it because they now have a lot of renewal,

renewal-reward, or regenerative processes models. Also, they are more ready to use the tools of Laplace transforms.

I am aware that this sequence is contrary to the more prevalent approach that starts with renewal theory. Although it is intellectually appealing to start with renewal theory, I found that it confuses and frustrates students, and it does not give them a feel for the modeling aspect of the subject early on. In this new edition, I have also changed the sequence of topics within Markov chains; I now cover the first passage times before the limiting behavior. This seems more natural since the concepts of transience and recurrence depend upon the first passage times.

The emphasis on the analysis of the stochastic models requires careful development of the major useful classes of stochastic processes: discrete and continuous time Markov chains, renewal processes, regenerative processes, and Markov regenerative processes. In the new edition, I have included a chapter on diffusion processes. In order to keep the length of the book under control, some topics from the earlier edition have been deleted: discussion of numerical methods, stochastic ordering, and some details from the Markov renewal theory. We follow a common plan of study for each class: characterization, transient analysis, first passage times, limiting behavior, and cost/reward models. The main aim of the theory is to enable the students to "solve" or "analyze" the stochastic models, to give them general tools to do this, rather than show special tricks that work in specific problems.

The third aspect, the implementation, involves actually using the results of Steps 1 and 2 to manage the "real-life" situation that we are interested in managing. This requires knowledge of statistics (for estimating the parameters of the model) and organizational science (how to persuade the members of an organization to follow the new solution, and how to set up an organizational structure to facilitate it), and hence is beyond the scope of this book, although, admittedly, it is a very important part of the process.

The book is designed for a two-course sequence in stochastic models. The first six chapters can form the first course, and the last four chapters, the second course. The book assumes that the students have had a course in probability theory (measure theoretic probability is not needed), advanced calculus (familiarity with differential and difference equations, transforms, etc.), and matrix algebra, and a general level of mathematical maturity. The appendix contains a brief review of relevant topics. In the second edition, I have removed the appendix devoted to stochastic ordering, since the corresponding material is deleted from the chapters on discrete and continuous time Markov chains. I have added two appendices: one collects relevant results from analysis, and the other from differential and difference equations. I find that these results are used often in the text, and hence it is useful to have them readily accessible.

The book uses a large number of examples to illustrate the concepts as well as computational tools and typical applications. Each chapter also has a large number of exercises collected at the end. The best way to learn the material of this course is by doing the exercises. Where applicable, the exercises have been separated into three classes: modeling, computational, and conceptual. Modeling exercises do not involve analysis, but may involve computations to derive the parameters of the problem. A computational exercise may ask for a numerical or algebraic answer. Some computational exercises may involve model building as well as analysis. A conceptual exercise generally involves proving some theorem, or fine tuning the understanding of some concepts introduced in the chapter, or it may introduce new concepts. Computational exercises are not necessarily easy, and conceptual exercises are not necessarily hard. I have deleted many exercises from the earlier edition, especially those that I found I never assigned in my classes. Many new exercises have been added. I found it useful to assign a model building exercise and then the corresponding analysis exercise. The students should be encouraged to use computers to obtain the solutions numerically.

It is my belief that a student, after mastering the material in this book, will be well equipped to build and analyze useful stochastic models of situations that he or she will face in his or her area of interest. It is my fond hope that the students will see a stochastic model lurking in every corner of their world as a result of studying this book.

What's new in the third edition?

I have added several new applications in the third edition, for example, Google search Algorithm in discrete time Markov chains, several health care and finance related applications in the continuous time Markov chains, etc. I have also added over fifty new exercises throughout the book. These new exercises were developed over the last ten years for my own exams in the course and the qualifying exams for our PhD candidates. I have heeded the request from the instructors using this as a textbook not to change the exercise numbers, so the new exercises are added at the end in appropriate sections. However this means that the exercise sequence no longer follows the sequence in which topics are developed in the chapter. To make space for these additions, I have deleted some material from the second edition: most notably the section on extended key Markov renewal theorem and the related conceptual exercises have been deleted. The material in Chapter 10 on diffusion processes has been rewritten in several places so as to make it more precise and clearer. Many graduate students have helped me find and correct the embarrassingly many typos that were present in the second edition. In particular, Jie Huang has helped proofread the third edition especially carefully, and I thank her for her help. However, there are bound to be a few new typos in the third edition. I will appreciate it if the readers are kind enough to send me an email when they find any.

Vidyadhar Kulkarni

vkulkarn@email.unc.edu

Department of Statistics and Operations Research

University of North Carolina

Chapel Hill, NC

CHAPTER 1

Introduction

The discipline of operations research was born out of the need to solve military problems during World War II. In one story, the air force was using the bullet holes on the airplanes used in combat duty to decide where to put extra armor plating. They thought they were approaching the problem in a scientific way until someone pointed out that they were collecting the bullet hole data from the planes that returned safely from their sorties.

1.1 What in the World Is a Stochastic Process?

Consider a system that evolves randomly in time, for example, the stock market index, the inventory in a warehouse, the queue of customers at a service station, water level in a reservoir, the state of a machines in a factory, etc.

Suppose we observe this system at discrete time points $n = 0, 1, 2, \cdots$, say, every hour, every day, every week, etc. Let X_n be the state of the system at time n. For example, X_n can be the Dow-Jones index at the end of the n-th working day; the number of unsold cars on a dealer's lot at the beginning of day n; the intensity of the n-th earthquake (measured on the Richter scale) to hit the continental United States in this century; or the number of robberies in a city on day n, to name a few. We say that $\{X_n, n \geq 0\}$ is a *discrete-time* stochastic process describing the system.

If the system is observed continuously in time, with $X(t)$ being its state at time t, then it is described by a *continuous time* stochastic process $\{X(t), t \geq 0\}$. For example, $X(t)$ may represent the number of failed machines in a machine shop at time t, the position of a hurricane at time t, or the amount of money in a bank account at time t, etc.

More formally, a stochastic process is a collection of random variables $\{X(\tau), \ \tau \in T\}$, indexed by the parameter τ taking values in the *parameter set T*. The random variables take values in the set S, called the *state-space* of the stochastic process. In many applications the parameter τ represents time, but it can represent any index. Throughout this book we shall encounter two cases:

1. $T = \{0, 1, 2, \cdots\}$. In this case we write $\{X_n, n \geq 0\}$ instead of $\{X(\tau),\ \tau \in T\}$.
2. $T = [0, \infty)$. In this case we write $\{X(t),\ t \geq 0\}$ instead of $\{X(\tau),\ \tau \in T\}$.

Also, we shall almost always encounter $S \subseteq \{0, 1, 2, \cdots\}$ or $S \subseteq (-\infty, \infty)$. We shall refer to the former case as the discrete state-space case, and the latter case as the continuous state-space case.

Let $\{X(\tau), \tau \in T\}$ be a stochastic process with state-space S, and let $x : T \to S$ be a function. One can think of $\{x(\tau), \tau \in T\}$ as a possible evolution (trajectory) of $\{X(\tau), \tau \in T\}$. The functions x are called the sample paths of the stochastic process. Figure 1.1 shows typical sample paths of stochastic processes. Since the stochastic process follows one of the sample paths in a random fashion, it is sometimes called a random function. In general, the set of all possible sample paths, called the *sample space* of the stochastic process, is uncountable. This can be true even in the case of a discrete time stochastic process with finite state-space. One of the aims of the study of the stochastic processes is to understand the behavior of the random sample paths that the system follows, with the ultimate aim of prediction and control of the future of the system.

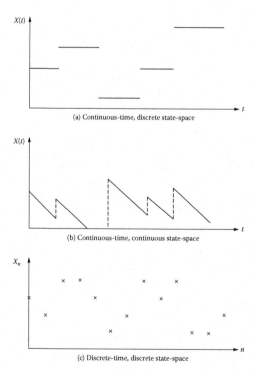

(a) Continuous-time, discrete state-space

(b) Continuous-time, continuous state-space

(c) Discrete-time, discrete state-space

Figure 1.1 *Typical sample paths of stochastic processes.*

Stochastic processes are used in epidemiology, biology, demography, health care systems, polymer science, physics, telecommunication networks, economics, finance, marketing, and social networks, to name a few areas. A vast literature exists in each of these areas. Our applications will generally come from queueing theory, inventory systems, supply chains, manufacturing, health care systems, computer and communication networks, reliability, warranty management, mathematical finance, and statistics. We illustrate a few such applications in the example below. Although most of the applications seem to involve continuous time stochastic process, they can easily be converted into discrete time stochastic processes by simply assuming that the system in question is observed at a discrete set of points, such as each hour, or each day, etc.

Example 1.1 Examples of Stochastic Processes in Real Life

Queues. Let $X(t)$ be the number of customers waiting for service in a service facility such as an outpatient clinic. $\{X(t), t \geq 0\}$ is a continuous time stochastic process with state-space $S = \{0, 1, 2, \cdots\}$.

Inventories. Let $X(t)$ be the number of automobiles in the parking lot of a dealership available for sale at time t, and $Y(t)$ be the number of automobiles on order (the customers have paid a deposit for them and are now waiting for delivery) at the dealership at time t. Both $\{X(t), t \geq 0\}$ and $\{Y(t), t \geq 0\}$ are continuous time stochastic processes with state-space $S = \{0, 1, 2, \cdots\}$.

Supply Chains. Consider a supply chain of computer printers with three levels: the manufacturer (level 1), the regional warehouse (level 2), and the retail store (level 3). The printers are stored at all three levels. Let $X_i(t) = $ the number of printers at level i, $(1 \leq i \leq 3)$. Then $\{X(t) = (X_1(t), X_2(t), X_3(t)), t \geq 0\}$ is a vector-valued continuous time stochastic process with state-space $S = \{(x_1, x_2, x_3) : x_1, x_2, x_3 = 0, 1, 2, \cdots\}$.

Manufacturing. Consider a machine shop with N machines. Each machine can be either working or under repair. Let $X_i(t) = 1$ if the i-th machine is working at time t, and 0 otherwise. Then $\{X(t) = (X_1(t), X_2(t), \cdots, X_N(t)), t \geq 0\}$ is a vector valued continuous time stochastic process with state-space $S = \{0, 1\}^N$.

Communication Networks. A telecommunication network is essentially a network of buffers (say N) connected by communication links. Let $X_i(t)$ be the number of packets in the i-th buffer ($i = 1, 2, \cdots, N$). Then the state of the system at time t is given by $X(t) = (X_1(t), X_2(t), \cdots, X_N(t))$. Then $\{X(t), t \geq 0\}$ is a vector valued continuous time stochastic process with state-space $S = \{0, 1, \cdots\}^N$.

Reliability. A complex system consists of many (say N) parts, each of which can be up or down. Let $X_i(t) = 1$ if the i-th component is working at time t, and 0 otherwise. The state of the system at time t is given by the vector $X(t) = (X_1(t), X_2(t), \cdots, X_N(t))$. $\{X(t), t \geq 0\}$ is a continuous time stochastic process with state-space $S = \{0, 1\}^N$. Finally, the functionality of the system is described by a structure function $\Phi : S \rightarrow \{0, 1\}$, with the following interpretation: The system is up at time t if $\Phi(X(t)) = 1$, and down otherwise.

Warranty Management. Consider a company that makes and sells hard drives for personal computers. Each hard drive is sold under a one-year warranty, which stipulates that the company will replace free of charge any hard drive that fails while under warranty by a new one, which is under warranty for the remaining time period. Such a warranty is called a free-replacement non-renewable warranty. In order to keep track of warranty liabilities the company typically keeps track of $X(t)$, the number of hard drives under warranty at time t. In addition, it keeps track of the remaining warranty period for each of those hard drives. This produces a rather complicated stochastic process.

Mathematical Finance. Let $X(t)$ be the stock price of IBM stock at time t. Then $\{X(t), t \geq 0\}$ is a continuous time stochastic process with state space $[0, \infty)$. A European call option with strike price K, and expiry date T, gives the owner the option (but not the obligation) of buying one share of IBM stock at price K at time T, regardless of the actual price of the stock at that time. Clearly, the holder of the option stands to make $\max(0, X(T) - K)$ amount of money at time T. Since there is no downside risk (no loss) in this option, how much money should the owner of this option be willing to pay to buy this option at time 0? Clearly that will depend on how the stock process $\{X(t), t \geq 0\}$ evolves, what the price of the stock is at time of purchase of the option, and what other investment opportunities are available.

Quality Control. A sequential quality control plan is used to continuously monitor the quality of a product. One such plan is described by two non-negative integers k and r, and operates in two phases as follows: in phase one we inspect each item and mark it as defective or non-defective. If we encounter k non-defective items in a row, we switch to phase two, in which we inspect every r-th item. We continue in phase two until we hit a defective item, at which time we switch back to phase 1. Let X_n be the phase the plan is in after inspecting the n-th item. Then $\{X_n, n \geq 0\}$ is a discrete time stochastic process on state-space $\{1, 2\}$.

Insurance. Let $X(t)$ be the number of active policies at time t that are covered by a life insurance company. Each active policy pays premiums at a fixed rate r per unit time. When a policy expires, the company pays out a random payoff. Let $R(t)$ be the total funds at the company at time t. Then $\{X(t), t \geq 0\}$ is a continuous time stochastic process with state-space $S = \{0, 1, 2, \cdots\}$. It goes up by one every time a new policy is sold, and goes down by one every time an active policy expires. $\{R(t), t \geq 0\}$ is a continuous time stochastic process with state-space $S = [0, \infty)$. It increases at rate $rX(t)$ at time t and jumps down by a random amount whenever a policy expires. If this process goes negative, we say that the company has gone bankrupt!

DNA Analysis. DNA (deoxyribonucleic acid) is a long molecule that consists of a string of four basic molecules called bases, represented by A, T, G, C. Let X_n be the n-th base in the DNA. Then $\{X_n, n \geq 1\}$ is a discrete time stochastic process with state-space $S = \{A, T, G, C\}$.

1.2 How to Characterize a Stochastic Process

In this book we develop the tools for the analysis of stochastic processes and study several special classes of stochastic processes in great detail. To do this we need a mathematically precise method to describe a stochastic process unambiguously. Since a stochastic process is a collection of random variables, it makes sense to start by reviewing how one "describes" a single random variable.

From elementary probability theory (see Appendices A and B) we see that a single random variable X is completely described by its cumulative distribution function (cdf)

$$F(x) = \mathsf{P}(X \le x), \quad -\infty < x < \infty. \tag{1.1}$$

A multivariate random variable $(X_1, X_2, \cdots, X_n)$ is completely described by its joint cdf

$$F(x_1, x_2, \cdots, x_n) = \mathsf{P}(X_1 \le x_1, X_2 \le x_2, \cdots, X_n \le x_n), \tag{1.2}$$

for all $-\infty < x_i < \infty$ and $i = 1, 2, \cdots, n$. Thus if the parameter set T is finite, the stochastic process $\{X(\tau), \ \tau \in T\}$ is a multivariate random variable, and hence is completely described by the joint cdf. But what about the case when T is not finite?

Consider the case $T = \{0, 1, 2, \cdots\}$ first. One could naively look for a direct extension of the finite dimensional joint cdf to an infinite dimensional case as follows:

$$F(x_0, x_1, \cdots) = \mathsf{P}(X_0 \le x_0, X_1 \le x_1, \cdots), \quad -\infty < x_i < \infty, \ i = 0, 1, \cdots. \tag{1.3}$$

However, the probability on the right-hand side of the above equation is likely to be zero or one in most of the cases. Thus such a function is not likely to give much information regarding the stochastic process $\{X_n, n \ge 0\}$. Hence we need to look for an alternative. Suppose we are given a family of finite dimensional joint cdfs $\{F_n, n \ge 0\}$ such that

$$F_n(x_0, x_1, \cdots, x_n) = \mathsf{P}(X_0 \le x_0, X_1 \le x_1, \cdots, X_n \le x_n), \tag{1.4}$$

for all $-\infty < x_i < \infty$, and $i = 0, 1, \cdots, n$. Such a family is called consistent if it satisfies

$$F_n(x_0, x_1, \cdots, x_n) = F_{n+1}(x_0, x_1, \cdots, x_n, \infty), \tag{1.5}$$

for all $-\infty < x_i < \infty$, and $i = 0, 1, \cdots, n, \ n \ge 0$. A discrete time stochastic process is completely described by a consistent family of finite dimensional joint cdfs, that is, any probabilistic question about $\{X_n, n \ge 0\}$ can be answered in terms of $\{F_n, \ n \ge 0\}$. Technically speaking, what one means by "completely describe" a stochastic process is to construct a probability space $(\Omega, \mathcal{F}, \mathcal{P})$ on which the process is defined. The more curious reader is referred to more advanced texts on stochastic processes to answer further questions.

Next we turn to the case of $T = [0, \infty)$. Unfortunately the matter of completely describing a continuous time stochastic process $\{X(t), t \ge 0\}$ is not so simple, since this case deals with an uncountable number of random variables. The situation can be simplified if we can make certain assumptions about the continuity of the sample paths of the process. We shall not deal with the details here, but shall give the main result:

Suppose the sample paths of $\{X(t), t \geq 0\}$ are, with probability 1, right continuous with left limits, i.e.,

$$\lim_{s \downarrow t} X(s) = X(t), \tag{1.6}$$

and $\lim_{s \uparrow t} X(s)$ exists for each t. Furthermore, suppose the sample paths have a finite number of discontinuities in a finite interval of time with probability one. Then $\{X(t), \ t \geq 0\}$ is completely described by a consistent family of finite dimensional joint cdfs

$$F_{t_1, t_2, \cdots, t_n}(x_1, x_2, \cdots, x_n) = P(X(t_1) \leq x_1, X(t_2) \leq x_2, \cdots, X(t_n) \leq x_n), \tag{1.7}$$

for all $-\infty < x_i < \infty$, $i = 1, \cdots, n$, $n \geq 1$ and all $0 \leq t_1 < t_2 < \cdots < t_n$.

Example 1.2   Independent and Identically Distributed Random Variables. Let $\{X_n, n \geq 1\}$ be a sequence of independent and identically distributed (iid) random variables with common distribution $F(\cdot)$. This stochastic process is completely described by $F(\cdot)$ as we can create the following consistent family of joint cdfs:

$$F_n(x_1, x_2, \cdots, x_n) = \prod_{i=1}^{n} F(x_i), \tag{1.8}$$

for all $-\infty < x_i < \infty$, $i = 1, 2, \cdots, n$, and $n \geq 1$.

In a sense a sequence of iid random variables is the simplest kind of stochastic process, but it does not have any interesting structure. However, one can construct more complex and interesting stochastic processes from it as shown in the next example.

Example 1.3   Random Walk. Let $\{X_n, n \geq 1\}$ be as in Example 1.2. Define

$$S_0 = 0, \ S_n = X_1 + \cdots + X_n, \ n \geq 1. \tag{1.9}$$

The stochastic process $\{S_n, \ n \geq 0\}$ is called a random walk. It is also completely characterized by $F(\cdot)$, since the joint distribution of $(S_0, S_1, \cdots, S_n)$ is completely determined by that of $(X_1, X_2, \cdots, X_n)$, which, from Example 1.2, is determined by $F(\cdot)$.

1.3 What Do We Do with a Stochastic Process?

Now that we know what a stochastic process is and how to describe it precisely, the next question is: what do we with it?

In Section 1.1 we have seen several situations where stochastic processes appear. For each situation we have a specific goal for studying it. The study of stochastic processes will be useful if it somehow helps us achieve that goal. In this book we shall develop a set of tools to help us achieve those goals. This will lead us to the study of special classes of stochastic processes: Discrete time Markov chains (Chapters 2, 3, 4), Poisson processes (Chapter 5), continuous time Markov chains (Chapter 6), renewal and regenerative processes (Chapter 8), Markov-regenerative processes (Chapter 9) and Diffusion processes (Chapter 10).

For each of these classes, we follow a more or less standard format of study, as described below.

1.3.1 Characterization

The first step is to define the class of stochastic processes under consideration. Then we look for ways of uniquely characterizing the stochastic process. In many cases we find that the consistent family of finite dimensional joint probability distributions can be described by a rather compact set of parameters. Identifying these parameters is part of this step.

1.3.2 Transient Behavior

The second step in the study of the stochastic process is to study its transient behavior. We concentrate on two aspects of transient behavior. First, we study *the marginal distribution*, that is, the distribution of X_n or $X(t)$ for a fixed n or t. We develop methods of computing this distribution. In many cases, the distribution is too hard to compute, in which case we satisfy ourselves with the moments or transforms of the random variable. (See Appendices B through F.) Mostly we shall find that the computation of transient behavior is quite difficult. It may involve computing matrix powers, or solving sets of simultaneous differential equations, or inverting transforms. Very few processes have closed form expressions for transient distributions, for example, the Poisson process. Second, we study *the occupancy times*, that is, the expected total time the process spends in different states up to time n or t.

1.3.3 First Passage Times

Let $\{X_n, n \geq 0\}$ be a stochastic process with state-space S. Let $B \subset S$ be a given subset of states. The first passage time to B is a random variable defined as

$$T = \min\{n \geq 0 : X_n \in B\}. \tag{1.10}$$

Thus the stochastic process "passes into" B for the first time at time T. We can define similar first passage time in a continuous time stochastic process as follows:

$$T = \min\{t \geq 0 : X(t) \in B\}. \tag{1.11}$$

We study the random variable T: the probability that it is finite, its distribution, moments, transforms, etc. The first passage times occur naturally when we use stochastic processes to model a real-life system. For example it may represent time until failure, or time until bankruptcy, or time until a production process goes out of control and requires external intervention, etc.

1.3.4 Limiting Distribution

Since computing the transient distribution is intractable in most cases, we next turn our attention to the limiting behavior, viz., studying the convergence of X_n or $X(t)$ as n or t tends to infinity. Now, there are many different modes of convergence of a sequence of random variables: convergence in distribution, convergence in moments, convergence in probability, and convergence with probability one (or almost sure convergence). These are described in Appendix G. Different stochastic processes exhibit different types of convergence. For example, we generally study convergence in distribution in Markov chains. For renewal processes, we obtain almost sure convergence results.

The first question is if the convergence occurs at all, and if the limit is unique when it does occur. This is generally a theoretical exploration, and proceeds in a theorem-proof fashion. The second question is how to compute the limit if the convergence does occur and the limit is unique. Here we may need mathematical tools like matrix algebra, systems of difference and differential equations, Laplace and Laplace–Stieltjes transforms, generating functions, and, of course, numerical methods. The study of the limiting distributions forms a major part of the study of stochastic processes.

1.3.5 Costs and Rewards

Stochastic processes in this book are intended to be used to model systems evolving in time. These systems typically incur costs or earn rewards depending on their evolution. In practice, the system designer has a specific set of operating policies in mind. Under each policy the system evolution is modeled by a specific stochastic process which generates specific costs or rewards. Thus the analysis of these costs and rewards (we shall describe the details in later chapters) is critical in evaluating comparative worth of the policies. Thus we develop methods of computing different cost and reward criteria for the stochastic processes.

If the reader keeps in mind these five main aspects that we study for each class of stochastic processes, the organization of the rest of the book will be relatively transparent. The main aim of the book is always to describe general methods of studying the above aspects, and not to go into special methods, or "tricks," that work (elegantly) for specialized problems, but fail in general. For this reason, sometimes our analysis may seem a bit long-winded for those who are already familiar with the tricks. However, it is the general philosophy of this book that the knowledge of general methods is superior to that of the tricks. Finally, the general methods can be adopted for implementation on computers, while the tricks cannot. This is important, since computers are a great tool for solving practical problems.

CHAPTER 2

Discrete-Time Markov Chains: Transient Behavior

"Come, let us hasten to a higher plane
Where dyads tread the fairy fields of Venn,
Their indices bedecked from one to n
commingled in an endless Markov Chain!"
– Stanislaw Lem, 'Cyberiad'

2.1 Definition and Characterization

Consider a system that is modeled by a discrete-time stochastic process $\{X_n, n \geq 0\}$ with a countable state-space S, say $\{0, 1, 2, \cdots\}$. Consider a fixed value of n that we shall call "the present time" or just the "present." Then X_n is called the present (state) of the system, $\{X_0, X_1, \cdots, X_{n-1}\}$ is called the past of the system, and $\{X_{n+1}, X_{n+2}, \cdots\}$ is called the future of the system. If $X_n = i$ and $X_{n+1} = j$, we say that the system has jumped (or made a transition) from state i to state j from time n to $n + 1$.

In this chapter we shall restrict attention to the systems having the following property: *if the present state of the system is known, the future of the system is independent of its past.* This is called the Markov property and this seemingly innocuous property has surprisingly far-reaching consequences. For a system having the Markov property, the past affects the future only through the present; or, stated in yet another way, the present state of the system contains all the relevant information needed to predict the future of the system in a probabilistic sense. The stochastic process $\{X_n, n \geq 0\}$ with state-space S used to describe a system with Markov property is called a discrete-time Markov chain, or DTMC for short. We give the formal definition below.

Definition 2.1 Discrete-Time Markov Chain. *A stochastic process* $\{X_n, n \geq 0\}$ *with countable state-space S is called a DTMC if*

(i). for all $n \geq 0$, $X_n \in S$,

(ii). for all $n \geq 0$, and $i, j \in S$

$$\mathsf{P}(X_{n+1} = j | X_n = i, X_{n-1}, X_{n-2}, \cdots, X_0) = \mathsf{P}(X_{n+1} = j | X_n = i). \quad (2.1)$$

Equation 2.1 is a formal way of stating the Markov property for discrete-time stochastic processes with countable state space. We next define an important subclass of DTMCs called the time-homogeneous DTMCs.

Definition 2.2 Time-Homogeneous DTMC. *A DTMC $\{X_n, n \geq 0\}$ with countable state-space S is said to be time-homogeneous if*

$$\mathsf{P}(X_{n+1} = j | X_n = i) = p_{i,j} \quad \textit{for all } n \geq 0, \ i, j \in S. \quad (2.2)$$

Thus if a time-homogeneous DTMC is in state i at time n, it jumps to state j at time $n + 1$ with probability $p_{i,j}$, for all values of n. From now on we assume that the DTMCs under consideration are time-homogeneous, unless otherwise specified. Let

$$P = [p_{i,j}]$$

denote the matrix of the conditional probabilities $p_{i,j}$. We call $p_{i,j}$ the transition probability from state i to state j. The matrix P is called the *one-step transition probability matrix* or just the transition probability matrix. When S is finite, say $S = \{1, 2, \cdots, m\}$, one can display P as a matrix as follows:

$$P = \begin{bmatrix} p_{1,1} & p_{1,2} & \cdots & p_{1,m-1} & p_{1,m} \\ p_{2,1} & p_{2,2} & \cdots & p_{2,m-1} & p_{2,m} \\ \vdots & \vdots & \ddots & \vdots & \vdots \\ p_{m-1,1} & p_{m-1,2} & \cdots & p_{m-1,m-1} & p_{m-1,m} \\ p_{m,1} & p_{m,2} & \cdots & p_{m,m-1} & p_{m,m} \end{bmatrix}. \quad (2.3)$$

Notice that a transition probability matrix of a DTMC is a square matrix. Next we define an important property of a square matrix.

Definition 2.3 Stochastic Matrix. *A square matrix $P = [p_{i,j}]$ is called stochastic if*

$$(i). \quad p_{i,j} \geq 0 \quad \textit{for all } i, j \in S, \quad (2.4)$$

$$(ii). \quad \textstyle\sum_{j \in S} p_{i,j} = 1 \quad \textit{for all } i \in S. \quad (2.5)$$

In short, the elements on each row of a stochastic matrix are non-negative and add up to one. The relevance of this definition is seen from the next theorem.

Theorem 2.1 Transition Probability Matrix. *The one-step transition probability matrix of a DTMC is stochastic.*

Proof. Statement (i) is obvious since $p_{i,j}$ is a conditional probability. Statement (ii) can be seen as follows:

$$\sum_{j \in S} p_{i,j} = \sum_{j \in S} P(X_{n+1} = j | X_n = i),$$

$$= P(X_{n+1} \in S | X_n = i) = 1,$$

since, according to the definition of DTMC, $X_{n+1} \in S$ with probability 1. ∎

Next, following the general road map laid out in Chapter 1, we turn our attention to the question of characterization of a DTMC. Clearly, any stochastic matrix can be thought of as a transition matrix of a DTMC. This generates a natural question: is a DTMC completely characterized by its transition probability matrix? In other words, are the finite dimensional distributions of a DTMC completely specified by its transition probability matrix? The answer is no, since we cannot derive the distribution of X_0 from the transition probability matrix, since its elements are conditional probabilities. So suppose we specify the distribution of X_0 externally. Let

$$a_i = P(X_0 = i), \quad i \in S, \tag{2.6}$$

and

$$a = [a_i]_{i \in S} \tag{2.7}$$

be a row vector representing the probability mass function (pmf) of X_0. We say that a is the *initial distribution* of the DTMC.

Next we ask: is a DTMC completely described by its transition probability matrix and its initial distribution? The following theorem answers this question in the affirmative. The reader is urged to read the proof, since it clarifies the role played by the Markov property and the time-homogeneity of the DTMC.

Theorem 2.2 Characterization of a DTMC. *A DTMC $\{X_n, n \geq 0\}$ is completely described by its initial distribution a and the transition probability matrix P.*

Proof. We shall prove the theorem by showing how we can compute the finite dimensional joint probability mass function $P(X_0 = i_0, X_1 = i_1, \cdots, X_n = i_n)$ in terms of a and P. Using Equation 2.6 we get

$$a_{i_0} = P(X_0 = i_0), \quad i_0 \in S.$$

Next we have

$$P(X_0 = i_0, X_1 = i_1) = P(X_1 = i_1 | X_0 = i_0)P(X_0 = i_0)$$

$$= a_{i_0} p_{i_0, i_1}$$

by using the definition of P. Now, as an induction hypothesis, assume that

$$P(X_0 = i_0, X_1 = i_1, \cdots, X_k = i_k) = a_{i_0} p_{i_0, i_1} p_{i_1, i_2} \cdots p_{i_{k-1}, i_k} \tag{2.8}$$

for $k = 1, 2, \cdots, n - 1$. We shall show that it is true for $k = n$. We have

$$
\begin{aligned}
&P(X_0 = i_0, X_1 = i_1, \cdots, X_n = i_n) \\
&= \quad P(X_n = i_n | X_{n-1} = i_{n-1}, \cdots, X_1 = i_1, X_0 = i_0) \cdot \\
&\qquad P(X_0 = i_0, X_1 = i_1, \cdots, X_{n-1} = i_{n-1}) \\
&= \quad P(X_n = i_n | X_{n-1} = i_{n-1}) P(X_0 = i_0, X_1 = i_1, \cdots, X_{n-1} = i_{n-1}) \\
&\qquad \text{(by Markov Property)} \\
&= \quad p_{i_{n-1}, i_n} P(X_0 = i_0, X_1 = i_1, \cdots, X_{n-1} = i_{n-1}) \\
&\qquad \text{(by time-homogeneity)} \\
&= \quad p_{i_{n-1}, i_n} a_{i_0} p_{i_0, i_1} p_{i_1, i_2} \cdots p_{i_{n-2}, i_{n-1}} \\
&\qquad \text{(by induction hypothesis),}
\end{aligned}
$$

which can be rearranged to show that the induction hypothesis holds for $k = n$. Hence the result follows. ∎

The transition probability matrix of a DTMC can be represented graphically by its *transition diagram*, which is a directed graph with as many nodes as there are states in the state-space, and a directed arc from node i to node j if $p_{i,j} > 0$. In particular, if $p_{i,i} > 0$, there is a self loop from node i to itself. The dynamic behavior of a DTMC is best visualized via its transition diagram as follows: imagine a particle that moves from node to node by choosing the outgoing arcs from the current node with the corresponding probabilities. In many cases it is easier to describe the DTMC by its transition diagram, rather than displaying the transition matrix.

Example 2.1 Transition Diagram. Consider a DTMC $\{X_n, n \geq 0\}$ on state-space $\{1, 2, 3\}$ with the following transition probability matrix:

$$
P = \begin{bmatrix}
.1 & .2 & .7 \\
.6 & 0 & .4 \\
.4 & 0 & .6
\end{bmatrix}.
$$

The transition diagram corresponding to this DTMC is shown in Figure 2.1. ∎

Example 2.2 Joint Distributions. Let $\{X_n, n \geq 0\}$ be a DTMC on state-space $\{1, 2, 3, 4\}$ with the transition probability matrix given below:

$$
P = \begin{bmatrix}
0.1 & 0.2 & 0.3 & 0.4 \\
0.2 & 0.2 & 0.3 & 0.3 \\
0.5 & 0.0 & 0.5 & 0.0 \\
0.6 & 0.2 & 0.1 & 0.1
\end{bmatrix}.
$$

The initial distribution is

$$
a = [0.25 \ 0.25 \ 0.25 \ 0.25].
$$

1. Compute $P(X_3 = 4, X_2 = 1, X_1 = 3, X_0 = 1)$.

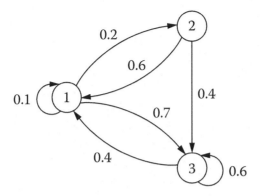

Figure 2.1 *Transition diagram of a DTMC.*

From Equation 2.8, we get

$$P(X_3 = 4, X_2 = 1, X_1 = 3, X_0 = 1)$$
$$= a_1 p_{1,3} p_{3,1} p_{1,4} = 0.25 \times 0.3 \times 0.5 \times 0.4 = 0.015.$$

2. Compute $P(X_3 = 4, X_2 = 1, X_1 = 3)$.

We have

$$P(X_3 = 4, X_2 = 1, X_1 = 3)$$
$$= \sum_{i=1}^{4} P(X_3 = 4, X_2 = 1, X_1 = 3 | X_0 = i) P(X_0 = i)$$
$$= \sum_{i=1}^{4} a_i p_{i,3} p_{3,1} p_{1,4} = 0.06. \qquad \blacksquare$$

2.2 Examples

Now we describe several examples of time homogeneous DTMCs. In many cases we omit the specification of the initial distribution, since it can be arbitrary. Also, we specify the transition probability matrix, or the transition probabilities, or the transition diagram, depending on what is more convenient. As we saw before, they carry the same information, and are thus equivalent.

Example 2.3   Two-State DTMC. One of the simplest DTMCs is one with two states, labeled 1 and 2. Thus $S = \{1, 2\}$. Such a DTMC has a transition matrix as follows:

$$\begin{bmatrix} \alpha & 1 - \alpha \\ 1 - \beta & \beta \end{bmatrix},$$

where $0 \leq \alpha, \beta \leq 1$. The transition diagram is shown in Figure 2.2. Thus if the

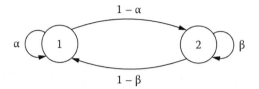

Figure 2.2 *Transition diagram of a two-state DTMC.*

DTMC is in state 1, it jumps to state 2 with probability $1 - \alpha$; if it is in state 2, it jumps to state 1 with probability $1 - \beta$, independent of everything else. ∎

Next we give several examples of two-state DTMCs.

Example 2.4   Two-State Weather Model. Consider a simple weather model in which we classify the day's weather as either "sunny" or "rainy." On the basis of previous data we have determined that if it is sunny today, there is an 80% chance that it will be sunny tomorrow regardless of the past weather; whereas, if it is rainy today, there is a 30% chance that it will be rainy tomorrow, regardless of the past. Let X_n be the weather on day n. We shall label sunny as state 1, and rainy as state 2. Then $\{X_n, n \geq 0\}$ is a DTMC on $S = \{1, 2\}$ with transition probability matrix

$$\begin{bmatrix} .8 & .2 \\ .7 & .3 \end{bmatrix}.$$

Clearly, the DTMC has a higher tendency to move to state 1, thus implying that this is a model of the weather at a sunny place! ∎

Example 2.5   Clinical Trials. Suppose two drugs are available to treat a particular disease, and we need to determine which of the two drugs is more effective. This is generally accomplished by conducting clinical trials of the two drugs on actual patients. Here we describe a clinical trial setup that is useful if the response of a patient to the administered drug is sufficiently quick, and can be classified as "effective" or "ineffective." Suppose drug i is effective with probability p_i, $i = 1, 2$. In practice the values of p_1 and p_2 are unknown, and the aim is to determine if $p_1 \geq p_2$ or $p_2 \geq p_1$. Ethical reasons compel us to use the better drug on more patients. This is achieved by using the *play the winner rule* as follows.

The initial patient (indexed as patient 0) is given either drug 1 or 2 at random. If the nth patient is given drug i $(i = 1, 2)$ and it is observed to be effective for that patient, then the same drug is given to the $(n + 1)$-st patient; if it is observed to be ineffective then the $(n + 1)$-st patient is given the other drug. Thus we stick with a drug as long as its results are good; when we get a bad result, we switch to the other drug – hence the name "play the winner." Let X_n be the drug (1 or 2) administered to the n-th patient. If the successive patients are chosen from a completely randomized pool, then we see that

$$P(X_{n+1} = 1 | X_n = 1, X_{n-1}, \cdots, X_0)$$

$$= \quad \text{P(drug 1 is effective on the } n\text{-th patient)} = p_1.$$

We can similarly derive $P(X_{n+1} = j | X_n = i; \text{history})$ for all other (i, j) combinations, thus showing that $\{X_n, n \geq 0\}$ is a DTMC. Its transition probability matrix is given by

$$\begin{bmatrix} p_1 & 1 - p_1 \\ 1 - p_2 & p_2 \end{bmatrix}.$$

If $p_1 > p_2$, the DTMC has a higher tendency to move to state 1, thus drug 1 (the better drug) is used more often. Thus the ethical purpose is served by the play the winner rule. ∎

Example 2.6 Two-State Machine. Consider a machine that can be either up (working) or down (failed). If it is up on day n, it is up on day $n + 1$ with probability p_u, independent of the past. If it is down on day n, it is down on day $n + 1$ with probability p_d, also independent of the past. Let X_n be the state of the machine (0 if is down, and 1 if it is up) on day n. We see that the Markov property is given as an assumption. Thus $\{X_n, n \geq 0\}$ is a DTMC on state-space $S = \{0, 1\}$ with transition probability matrix given by

$$\begin{bmatrix} p_d & 1 - p_d \\ 1 - p_u & p_u \end{bmatrix}.$$

Note that the above transition probability matrix implies that the up and down times are geometrically distributed. ∎

Example 2.7 Two-Machine Workshop. Suppose a workshop has two identical machines as described in Example 2.6. The two machines behave independently of each other. Let X_n be the number of working machines on day n. Is $\{X_n, n \geq 0\}$ a DTMC?

From the definition of X_n we see that the state-space is $S = \{0, 1, 2\}$. Next we verify the Markov property given by Equation 2.1. For example, we have

$$P(X_{n+1} = 0 | X_n = 0, X_{n-1}, \cdots, X_0)$$
$$= P(X_{n+1} = 0 | \text{Both machines are down on day } n, X_{n-1}, \cdots, X_0)$$
$$= P(\text{Both machines are down on day } n + 1 | \text{ Both machines are down on day } n)$$
$$= p_d p_d.$$

Similarly we can verify that $P(X_{n+1} = j | X_n = i, X_{n-1}, \cdots, X_0)$ depends only on i and j for all $i, j \in S$. Thus $\{X_n, n \geq 0\}$ is a DTMC on state-space $S = \{0, 1, 2\}$ with transition probability matrix given by

$$P = \begin{bmatrix} p_d^2 & 2p_d(1 - p_d) & (1 - p_d)^2 \\ p_d(1 - p_u) & p_d p_u + (1 - p_d)(1 - p_u) & p_u(1 - p_d) \\ (1 - p_u)^2 & 2p_u(1 - p_u) & p_u^2 \end{bmatrix}.$$

This example can be extended to $r \geq 2$ machines in a similar fashion. ∎

Example 2.8 Independent and Identically Distributed Random Variables. Let $\{X_n, n \geq 0\}$ be a sequence of iid random variables with common probability mass function (pmf)

$$\alpha_j = \mathsf{P}(X_n = j), \quad j \geq 0.$$

Then $\{X_n, n \geq 0\}$ is a DTMC on $S = \{0, 1, 2, \cdots\}$ with transition probabilities given by

$$p_{i,j} = \mathsf{P}(X_{n+1} = j | X_n = i, X_{n-1}, \cdots, X_0) = \mathsf{P}(X_{n+1} = j) = \alpha_j,$$

for all $i \in S$. Thus the transition probability matrix P of this DTMC has identical rows, each row being the row vector $\alpha = [\alpha_0, \alpha_1, \cdots]$. ∎

Example 2.9   Random Walk. Let $\{Z_n, n \geq 1\}$ be a sequence of iid random variables with common pmf

$$\alpha_k = \mathsf{P}(Z_n = k), \quad k = 0, \pm 1, \pm 2, \cdots.$$

Define

$$X_0 = 0, \quad X_n = \sum_{k=1}^{n} Z_k, \quad n \geq 1.$$

The process $\{X_n, n \geq 0\}$ has state-space $S = \{0, \pm 1, \pm 2, \cdots\}$. To verify that it is a DTMC, we see that, for all $i, j \in S$,

$$\mathsf{P}(X_{n+1} = j | X_n = i, X_{n-1}, \cdots, X_0)$$
$$= \mathsf{P}(\sum_{k=1}^{n+1} Z_k = j | \sum_{k=1}^{n} Z_k = i, X_{n-1}, \cdots, X_0)$$
$$= \mathsf{P}(Z_{n+1} = j - i) = \alpha_{j-i}.$$

Thus $\{X_n, n \geq 0\}$ is a DTMC with transition probabilities

$$p_{i,j} = \alpha_{j-i}, \quad i, j \in S.$$

This random walk is called space-homogeneous, or state-independent, since, Z_n, the size of the n-th step, does not depend on the position of the random walk at time n. ∎

Example 2.10   State-Dependent Random Walk.

Consider a particle that moves randomly on a doubly infinite one-dimensional lattice where the lattice points are labeled $\cdots, -2, -1, 0, 1, 2, \cdots$. The particle moves by taking steps of size 0 or 1 or -1 as follows: if it is on site i at time n, then at time $n+1$ it moves to site $i+1$ with probability p_i, or to site $i-1$ with probability q_i, or stays at site i with probability $r_i = 1 - p_i - q_i$, independent of its motion up to time n. Let X_n be the position of the particle (the label of the site occupied by the particle) at time n. Thus $\{X_n, n \geq 0\}$ is a DTMC on state-space $S = \{0, \pm 1, \pm 2, \cdots\}$ with transition probabilities given by

$$p_{i,i+1} = p_i, \quad p_{i,i-1} = q_i, \quad p_{i,i} = r_i, \quad i \in S.$$

This random walk is not space-homogeneous, since the step size distribution depends upon where the particle is. When $r_i = 0$, $p_i = p$, $q_i = q$ for all $i \in S$, the random walk is indeed space-homogeneous, and we call it a simple random walk. Further, if $p = q = 1/2$, we call it a simple symmetric random walk.

If $q_0 = 0$ and $P(X_0 \geq 0) = 1$, the random walk $\{X_n, n \geq 0\}$ has state-space $S = \{0, 1, 2, \cdots\}$. If $q_0 = r_0 = 0$ (and hence $p_0 = 1$), the random walk is said to have a reflecting barrier at 0, since whenever the particle hits site zero, it "bounces back" to site 1. If $q_0 = p_0 = 0$ (and hence $r_0 = 1$), the random walk is said to have an absorbing barrier at 0, since in this case when the particle hits site zero, it stays there forever. If, for some integer $N \geq 1$, $q_0 = p_N = 0$, and $P(0 \leq X_0 \leq N) = 1$, the random walk has state-space $S = \{0, 1, 2, \cdots, N - 1, N\}$. The barrier at N can be reflecting or absorbing according to whether r_n is zero or one. Simple random walks with appropriate barriers have many applications. We describe a few in the next three examples. ■

Example 2.11   Gambler's Ruin. Consider two gamblers, A and B, who have a combined fortune of N dollars. They bet one dollar each on the toss of a coin. If the coin turns up heads, A wins a dollar from B, and if the coin turns up tails, B wins a dollar from A. Suppose the successive coin tosses are independent, and the coin turns up heads with probability p and tails with probability $q = 1 - p$. The game ends when either A or B is broke (or ruined).

Let X_n denote the fortune of gambler A after the n-th toss. We shall assume the coin tossing continues after the game ends, but no money changes hands. With this convention we can analyze the stochastic process $\{X_n, n \geq 0\}$. Obviously, if $X_n = 0$ (A is ruined) or $X_n = N$ (B is ruined), $X_{n+1} = X_n$. If $0 < X_n < N$, we have

$$X_{n+1} = \begin{cases} X_n + 1 & \text{with probability } p \\ X_n - 1 & \text{with probability } q. \end{cases}$$

This shows that $\{X_n, n \geq 0\}$ is a DTMC on state-space $S = \{0, 1, 2, \cdots, N\}$. Its

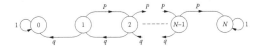

Figure 2.3 *Transition diagram of the gambler's ruin DTMC.*

transition diagram is shown in Figure 2.3. Thus $\{X_n, n \geq 0\}$ is a simple random walk on $\{0, 1, 2, \cdots, N\}$ with absorbing barriers at 0 and N.

A somewhat more colorful situation giving rise to the same Markov chain involves a suitably intoxicated person taking a step towards home (at site 0) with probability q or towards a bar (placed at site N) with probability p, in a random manner. As soon as he reaches the home or the bar he stays there forever. The random walk executed by such a person is called the "drunkard's walk." ■

Example 2.12   Discrete-Time Queue: Bernoulli Arrivals and Departures. Consider a service system where customers arrive at times $n = 0, 1, 2, \cdots$ and form a queue for service. The departures from this queue also occur at times $n = 0, 1, 2, \cdots$. At each time n, a customer is removed (a departure occurs) from the head of the queue (if it is not empty) with probability q, then a customer is added (an arrival occurs) at the tail end of the queue with probability p $(0 \leq p, q \leq 1)$. All the arrivals and departures are independent of everything else. Thus the number of arrivals and (potential) departures at each time are Bernoulli random variables. Let X_n be the number of customers in the system at time n, after the departures and arrivals at n are accounted for. Here we study the queue-length process $\{X_n, n \geq 0\}$. Its state-space is $S = \{0, 1, 2, \cdots\}$. Now, for $i > 0$, we have

$$P(X_{n+1} = i + 1 | X_n = i, X_{n-1}, \cdots, X_0)$$
$$= P(\text{one arrival and no departure at time } n + 1 | X_n = i)$$
$$= p(1 - q).$$

Carrying out similar calculations, one can show that $\{X_n, n \geq 0\}$ is a simple random walk on $\{0, 1, 2, \cdots\}$ with the following parameters:

$$p_0 = p, \quad q_0 = 0, \quad p_i = p(1 - q), \quad q_i = q(1 - p), \quad i \geq 1.$$

We also have $r_i = 1 - p_i - q_i$. Note that the barrier at 0 is neither absorbing, nor reflecting. ∎

Example 2.13   Urn Model. Consider two urns labeled A and B, containing a total of N white balls and N red balls among them. An experiment consists of picking one ball at random from each urn and interchanging them. This experiment is repeated in an independent fashion. Let X_n be the number of white balls in urn A after n repetitions of the experiment. Assume that initially urn A contains all the white balls, and urn B contains all the red balls. Thus $X_0 = N$. Note that X_n tells us precisely the contents of the two urns after n experiments: if $X_n = i$, urn A contains i white balls and $N - i$ red balls; and urn B contains $N - i$ white balls and i red balls. That $\{X_n, n \geq 0\}$ is a DTMC on state-space $S = \{0, 1, \cdots, N\}$ can be seen from the following calculation. For $0 < i < N$

$$P(X_{n+1} = i + 1 | X_n = i, X_{n-1}, \cdots, X_0)$$
$$= \quad P(\text{A red ball from urn } A \text{ and a white ball from urn } B$$
$$\text{are picked on the } n\text{-th experiment})$$
$$= \quad \frac{N - i}{N} \cdot \frac{N - i}{N} = p_{i,i+1}.$$

The other transition probabilities can be computed similarly to see that $\{X_n, n \geq 0\}$ is a random walk on $S = \{0, 1, \cdots, N\}$ with the following parameters:

$$r_0 = 0, \quad p_0 = 1,$$

$$q_i = \left(\frac{i}{N}\right)^2, \quad r_i = 2\left(\frac{i}{N}\right) \cdot \left(\frac{N - i}{N}\right), \quad p_i = \left(\frac{N - i}{N}\right)^2, \quad 0 < i < N,$$

$$r_N = 0, \quad q_N = 1.$$

Thus the random walk has reflecting barriers at 0 and N. This urn model was used initially by Ehrenfest to model diffusion of molecules across a permeable membrane. One can think of the white balls and red balls as the molecules of two different gases and the switching mechanism as the model for the diffusion across the membrane. It also appears as Moran model in genetics. ∎

Example 2.14   Brand Switching. A customer chooses among three brands of beer, say A, B, and C, every week when he buys a six-pack. Let X_n be the brand he purchases in week n. From his buying record so far, it has been determined that $\{X_n, n \geq 0\}$ is a DTMC with state-space $S = \{A, B, C\}$ and transition probability matrix given below:

$$P = \begin{bmatrix} 0.1 & 0.2 & 0.7 \\ 0.2 & 0.4 & 0.4 \\ 0.1 & 0.3 & 0.6 \end{bmatrix}. \tag{2.9}$$

Thus if he purchases brand A beer in week n, he will purchase brand C beer in the next week with probability .7, regardless of the purchasing history and the value of n. Such brand switching models are used quite often in practice to analyze and influence consumer behavior. A major effort is involved in collecting and analyzing data to estimate the transition probabilities. This involves observing the buying habits of a large number of customers. ∎

Example 2.15   Success Runs. Consider a game where a coin is tossed repeatedly in an independent fashion. Whenever the coin turns up heads, which happens with probability p, the player wins a dollar. Whenever the coin turns up tails, which happens with probability $q = 1 - p$, the player loses all his winnings so far. Let X_n denote the player's fortune after the n-th toss. We have

$$X_{n+1} = \begin{cases} 0 & \text{with probability } q \\ X_n + 1 & \text{with probability } p. \end{cases}$$

This shows that $\{X_n, n \geq 0\}$ is a DTMC on state-space $S = \{0, 1, 2, \cdots\}$ with transition probabilities

$$p_{i,0} = q, \quad p_{i,i+1} = p, \quad i \in S.$$

A slightly more general version of this DTMC can be considered with transition

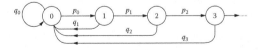

Figure 2.4 *Transition diagram for the success-runs DTMC.*

probabilities

$$p_{i,0} = q_i, \quad p_{i,i+1} = p_i = 1 - q_i, \quad i \in S.$$

Such a DTMC is called a success runs Markov chain. Its transition diagram is shown in Figure 2.4. ∎

Example 2.16   Production-Inventory System: Batch Production. Consider a production inventory system in discrete time where production occurs in integer valued random batches and demands occur one at a time at times $n = 1, 2, \cdots$. The inventory of unsold items is maintained in a warehouse. We assume that demand at time n occurs before the production at time n, so the production at time n is not available to satisfy the demand at time n. Thus, if the warehouse is empty when a demand occurs, the demand is lost. Let X_0 be the initial number of items in the inventory and X_n be the number of items in the warehouse at time n, after accounting for the demand and production at time n, $n \geq 1$. Let Y_n be the size of the batch produced at time n (after the demand is satisfied at time n). This implies that

$$X_{n+1} = \begin{cases} Y_{n+1} & \text{if } X_n = 0 \\ X_n - 1 + Y_{n+1} & \text{if } X_n > 0. \end{cases} \tag{2.10}$$

This can also be written as

$$X_{n+1} = \max\{X_n - 1 + Y_{n+1}, Y_{n+1}\}.$$

Now suppose $\{Y_n, n \geq 1\}$ is a sequence of iid random variables, with common pmf given by

$$\alpha_k = \mathsf{P}(Y_n = k), \quad k = 0, 1, 2, \cdots.$$

We shall show that under this assumption $\{X_n, n \geq 0\}$ is a DTMC on state-space $S = \{0, 1, 2, \cdots\}$. First, we have

$$\mathsf{P}(X_{n+1} = j | X_n = 0, X_{n-1}, \cdots, X_0)$$
$$= \mathsf{P}(Y_{n+1} = j | X_n = 0, X_{n-1}, \cdots, X_0) = \alpha_j,$$

since Y_{n+1} is independent of $X_0, X_1, \cdots, X_n$. Similarly, for $i > 0$, we have

$$\mathsf{P}(X_{n+1} = j | X_n = i, X_{n-1}, \cdots, X_0)$$
$$= \mathsf{P}(X_n - 1 + Y_{n+1} = j | X_n = i, X_{n-1}, \cdots, X_0)$$
$$= \mathsf{P}(Y_{n+1} = j - i + 1 | X_n = i, X_{n-1}, \cdots, X_0) = \alpha_{j-i+1},$$

assuming $j \geq i - 1$. Otherwise the above transition probability is zero. Thus $\{X_n, n \geq 0\}$ is a DTMC with transition probability matrix given by

$$P = \begin{bmatrix} \alpha_0 & \alpha_1 & \alpha_2 & \alpha_3 & \cdots \\ \alpha_0 & \alpha_1 & \alpha_2 & \alpha_3 & \cdots \\ 0 & \alpha_0 & \alpha_1 & \alpha_2 & \cdots \\ 0 & 0 & \alpha_0 & \alpha_1 & \cdots \\ 0 & 0 & 0 & \alpha_0 & \cdots \\ \vdots & \vdots & \vdots & \vdots & \ddots \end{bmatrix}. \tag{2.11}$$

Note that $\{X_n, n \geq 0\}$ can jump up by any integral amount in one step, but can decrease by at most one. Matrices of the form above are known as the upper Hessenberg matrices. Markov chains with this type of transition probability matrix arise in many applications, especially in queueing theory. ∎

Example 2.17   Production-Inventory System: Batch Demands. Now consider a production-inventory system where the demands occur in integer valued random batches and the production occurs one at a time at times $n = 1, 2, \cdots$. The inventory of unsold items is maintained in a warehouse. Unlike in the previous example, here we assume that the production occurs before demand. Thus the production (of one unit) at time n available to satisfy the demand at time n. Let X_0 be the initial inventory and X_n be the number of items in the warehouse at time n, after the production and demand at time n is accounted for. Let Y_n be the size of the batch demand at time $n \geq 1$. Any part of the demand that cannot be satisfied is lost. This implies that

$$X_{n+1} = \begin{cases} 0 & \text{if } Y_{n+1} \geq X_n + 1 \\ X_n + 1 - Y_{n+1} & \text{if } Y_{n+1} < X_n + 1. \end{cases}$$

This can also be written as

$$X_{n+1} = \max\{X_n + 1 - Y_{n+1}, 0\}. \tag{2.12}$$

Now suppose $\{Y_n, n \geq 1\}$ is a sequence of iid random variables, with common pmf given by

$$\alpha_k = \mathsf{P}(Y_n = k), \quad k = 0, 1, 2, \cdots.$$

We shall show that under this assumption $\{X_n, n \geq 0\}$ is a DTMC on state-space $S = \{0, 1, 2, \cdots\}$. We have, for $0 < j \leq i + 1$

$$\begin{aligned} \mathsf{P}(X_{n+1} = j | X_n &= i, X_{n-1}, \cdots, X_0) \\ &= \mathsf{P}(\max\{X_n + 1 - Y_{n+1}, 0\} = j | X_n = i, X_{n-1}, \cdots, X_0) \\ &= \mathsf{P}(X_n + 1 - Y_{n+1} = j | X_n = i, X_{n-1}, \cdots, X_0) \\ &= \mathsf{P}(Y_{n+1} = i - j + 1 | X_n = i, X_{n-1}, \cdots, X_0) = \alpha_{i-j+1}, \end{aligned}$$

since Y_{n+1} is independent of $X_0, X_1, \cdots, X_n$. For $j = 0$, we have

$$\begin{aligned} \mathsf{P}(X_{n+1} = 0 | X_n &= i, X_{n-1}, \cdots, X_0) \\ &= \mathsf{P}(\max\{X_n + 1 - Y_{n+1}, 0\} = 0 | X_n = i, X_{n-1}, \cdots, X_0) \\ &= \mathsf{P}(Y_{n+1} \geq i + 1 | X_n = i, X_{n-1}, \cdots, X_0) \\ &= \sum_{k=i+1}^{\infty} \alpha_k = \beta_i \text{ (say)}. \end{aligned}$$

All other transition probabilities are zero. Thus $\{X_n, n \geq 0\}$ is a DTMC with transition probability matrix given by

$$P = \begin{bmatrix} \beta_0 & \alpha_0 & 0 & 0 & 0 & \cdots \\ \beta_1 & \alpha_1 & \alpha_0 & 0 & 0 & \cdots \\ \beta_2 & \alpha_2 & \alpha_1 & \alpha_0 & 0 & \cdots \\ \beta_3 & \alpha_3 & \alpha_2 & \alpha_1 & \alpha_0 & \cdots \\ \vdots & \vdots & \vdots & \vdots & \vdots & \ddots \end{bmatrix}. \tag{2.13}$$

Note that $\{X_n, n \geq 0\}$ can increase by at most one. Matrices of the form above are known as the lower Hessenberg matrices. Markov chains with this type of transition probability matrix also arise in many applications, again in queueing theory. ∎

The last two examples illustrate a general class of DTMCs $\{X_n, n \geq 0\}$ that are generated by the following recursion

$$X_{n+1} = f(X_n, Y_{n+1}), \quad n \geq 0,$$

where $\{Y_n, n \geq 1\}$ is a sequence of iid random variables. The reader is urged to construct more DTMCs of this structure for further understanding of the DTMCs.

2.3 DTMCs in Other Fields

In this section we present several examples where the DTMCs have been used to model real-life situations.

2.3.1 Genomics

In 1953 Francis Crick and James Watson, building on the experimental research of Rosalind Franklin, first proposed the double helix model of DNA (deoxyribonu-cleic acid), the key molecule that contains the genetic instructions for the making of a living organism. It can be thought of as a long sequence of four basic nu-cleotides (or bases): adenine, cytosine, guanine, and thymine, abbreviated as $A, C, G,$ and T, respectively. The human DNA consists of roughly 3 billion of these bases. The sequence of these bases is very important. A typical sequence may read as: $CTTCTCAAATAACTGTGCCTC\cdots$.

Let X_n be the n-th base in the sequence. It is clear that $\{X_n, n \geq 1\}$ is a discrete-time stochastic process with state-space $S = \{A, C, G, T\}$. Clearly, whether it is a DTMC or not needs to be established by statistical analysis. We will not get into the mechanics of doing so in this book. By studying a section of the DNA molecule one might conclude that the $\{X_n, n \geq 1\}$ is a DTMC with transition probability matrix given below (rows and column are ordered as A, C, G, T)

$$P(1) = \begin{bmatrix} 0.180 & 0.274 & 0.426 & 0.120 \\ 0.170 & 0.368 & 0.274 & 0.188 \\ 0.161 & 0.339 & 0.375 & 0.135 \\ 0.079 & 0.355 & 0.384 & 0.182 \end{bmatrix}.$$

However, by studying another section of the molecule one may conclude that $\{X_n, n \geq 1\}$ is a DTMC with transition probability matrix given below:

$$P(2) = \begin{bmatrix} 0.300 & 0.205 & 0.285 & 0.210 \\ 0.322 & 0.298 & 0.078 & 0.302 \\ 0.248 & 0.246 & 0.298 & 0.208 \\ 0.177 & 0.239 & 0.292 & 0.292 \end{bmatrix}.$$

Clearly the sequence behaves differently in the two sections. In the first section, with transition probability matrix $P(1)$, the base G is much more likely to follow A, than in the second section with transition probability matrix $P(2)$.

The researchers have come up with two ways to model this situation. We describe them briefly below.

Hidden Markov Models. We assume that a DNA molecule is segmented into non-overlapping sections of random lengths, some of which are described by the matrix $P(1)$ (called type 1 segments) and the others by the matrix $P(2)$ (called type 2 segments). We define Y_n to be the section (1 or 2) the n-th base falls in. This cannot be observed. We assume that $\{Y_n, n \geq 1\}$ is itself a DTMC on state-space $\{1, 2\}$, with the following transition probability matrix (estimated by the statistical analysis of the entire sequence):

$$Q = \begin{bmatrix} .1 & .9 \\ .3 & .7 \end{bmatrix}.$$

The above transition probability matrix says that the sequence is more likely to switch from section type 1 to 2, than from 2 to 1. Now we create a bivariate stochastic process $\{(X_n, Y_n), n \geq 1\}$. It is clear that this is a DTMC. For example,

$$P(X_{n+1} = b, Y_{n+1} = j | X_n = a, Y_n = i) = Q_{i,j} P(i)_{a,b},$$

for $i, j \in \{1, 2\}$ and $a, b \in \{A, C, G, T\}$. Such a DTMC model is called a hidden Markov model, because we can observe X_n, but not Y_n. The second component is only a modeling tool to describe the behavior of the DNA molecule.

Higher-Order Markov Models. We illustrate with an example of a second-order DTMCs. A stochastic process $\{X_n, n \geq 0\}$ is called a second-order DTMC on state-space S if, for $n \geq 1$,

$$P(X_{n+1} = k | X_n = j, X_{n-1} = i, X_{n-2}, \cdots, X_0)$$
$$= P(X_{n+1} = k | X_n = j, X_{n-1} = i) = p_{(i,j);k},$$

for all $i, j, k \in S$. Note that we do not need new theory to study such processes. We simply define $Z_n = (X_{n-1}, X_n)$. Then $\{Z_n, n \geq 1\}$ is a DTMC on $S \times S$. Such higher-order Markov models have been found useful in DNA analysis since it has been empirically observed that, for example, the frequency with which A is observed to follow CG, is different than the frequency with which it follows AG. This should not be the case if the simple 4-state DTMC model is valid. For our DNA sequence, the second-order Markov model produces a DTMC with a state-space having 16 elements $\{AA, AC, \cdots, TG, TT\}$, and needs a 16 by 16 transition probability matrix to describe it. The matrix has only $16 \cdot 4 = 64$ non-zero entries since the Z_n process can jump from a state to only four other states. The higher-order Markov models have more parameters to estimate (the k-th order Markov model has a state-space of cardinality 4^k), but generally produce a better fit, and a better description of the sequence.

One can try to use non-homogeneous DTMCs to model a DNA molecule, but this tends to have too many parameters (one transition probability matrix for each n), and cannot be estimated from a single sequence. The reader is referred to *Biological Sequence Analysis* by Durbin, Eddy, Krogh, and Mitchison (2001) for more details.

2.3.2 Genetics

The study of genetics originated with Gregor Mendel's pioneering experiments on the interbreeding of pea plants. Since then the subject of genetics, based on solid experimental work, has given rise to sophisticated stochastic models. We shall illustrate here the simplest models from mathematical genetics. First we introduce some basic background information about population genetics.

It is well known that the physical characteristics like skin color, eye color, height, etc., are passed from parents to the children through genes. A human DNA contains about 20,000 distinct genes.

Many characteristics, such as eye color or presence of freckles, are determined by a single gene. The genes that control such characteristics can have two variants, called alleles. These alleles can be classified as *dominant* (d) or *recessive* (r). Each cell in the parent contains two alleles for each characteristic. During conception, a cell from each parent contributes one of its alleles to create a complete new cell of the offspring. An individual who carries two dominant alleles for this characteristic is denoted by dd and is called *dominant*, an individual carrying two recessive alleles is called *recessive* and is denoted by rr, and an individual carrying one recessive and one dominant allele is called *hybrid*, denoted by dr. The physical characteristic of a dd individual is indistinguishable from that of a dr individual, i.e., they have different genotypes, but the same phenotype.

The Genotype Evolution Model. Consider a population of two individuals. We represent the state of the population by specifying the genotype of each individual. Thus a state (dd, dd) means both individuals are dominant. There are six possible states of the population: $S = \{(dd, dd), (dd, dr), (dd, rr), (dr, dr), (dr, rr), (rr, rr)\}$. For example, suppose the initial state of the population is (dd, dr). When these two individuals produce an offspring, there are four possibilities—depending upon which allele from each parent the offspring inherits—namely, dd, dr, dd, dr. Thus the offspring will be dominant with probability .5, and hybrid with probability .5. Suppose the initial two individuals produce two independent offsprings. Thus the next generation again consists of two individuals. This process is repeated indefinitely. Let X_n be the state of the population in the n-th generation. Then it can be seen that $\{X_n, n \geq 0\}$ is a DTMC on state-space S with transition probability matrix given below (where the rows and columns are indexed in the same order as in S):

$$P = \begin{bmatrix} 1 & 0 & 0 & 0 & 0 & 0 \\ 1/4 & 1/2 & 0 & 1/4 & 0 & 0 \\ 0 & 0 & 0 & 1 & 0 & 0 \\ 1/16 & 1/4 & 1/8 & 1/4 & 1/4 & 1/16 \\ 0 & 0 & 0 & 1/4 & 1/2 & 1/4 \\ 0 & 0 & 0 & 0 & 0 & 1 \end{bmatrix}. \qquad (2.14)$$

Thus once the population state reaches (dd, dd) or (rr, rr), it stays that way forever. This model can be extended to populations of size more than two, however, the state-space gets increasingly large. The next two models attempt to alleviate this difficulty.

The Wright–Fisher Model. Consider a population of a fixed number, N, of alleles. If i of these are dominant and $N - i$ are recessive, we say that the state of the population is i. The composition of the N alleles in the next generation is determined by randomly sampling N genes from this population with replacement in an independent fashion. Let X_n be the state of the population in the n-th generation.

Since a dominant allele is picked with probability X_n/N and a recessive allele is picked with probability $1 - X_n/N$, it can be seen that $\{X_n, n \geq 0\}$ is a DTMC on state-space $S = \{0, 1, 2, \cdots, N\}$ with transition probabilities:

$$p_{i,j} = \binom{N}{j} \left(\frac{i}{N}\right)^j \left(1 - \frac{i}{N}\right)^{N-j}, \quad i, j \in S.$$

In other words,

$$X_{n+1} \sim \text{Bin}(N, X_n/N).$$

Once all the alleles in the population are dominant (state N), or recessive (state 0) they stay that way forever.

The Moran Model. This model also considers a population with a fixed number, N, of alleles. If i of these are dominant and $N - i$ are recessive, we say that the state of the population is i. Unlike in the Wright–Fisher model, the composition of genes in the population changes by sampling one allele from the existing population, and replacing a randomly chosen allele with it. Let X_n be the state of the population after the n-th step. Thus the randomly deleted allele is dominant with probability X_n/N and the randomly added allele is independent of the deleted allele, and is dominant with probability X_n/N. Thus $\{X_n, n \geq 0\}$ is a simple random walk on $S = \{0, 1, 2, \cdots, N\}$ with parameters:

$$p_0 = 0, \quad r_0 = 1,$$

$$p_i = q_i = \frac{i}{N} \cdot \left(1 - \frac{i}{N}\right), \quad r_i = 1 - p_i - q_i, \quad 1 \leq i \leq N - 1$$

$$q_N = 0 \quad r_N = 1.$$

States 0 and N are absorbing.

The above two models assume that there are no mutations, migrations, or selections involved in the process. It is possible to modify the models to account for these factors. For more details see *Statistical Processes of Evolutionary Theory* by P. A. P. Moran.

2.3.3 Genealogy

Genealogy is the study of family trees. In 1874 Francis Galton and Henry Watson wrote a paper on extinction probabilities of family names, a question that was originally posed by Galton about the likelihood of famous family names disappearing due to lack of male heirs. It is observed that ethnic populations such as the Koreans and Chinese have been using family names that are passed on via male children for several thousand years, and hence are left with very few family names. While in other societies where people assume new family names more easily, or where the tradition of family names is more recent, there are many family names.

In its simplest form, we consider a patriarchal society where the family name is carried on by the male heirs only. Consider the Smith family tree as shown in Figure 2.5.

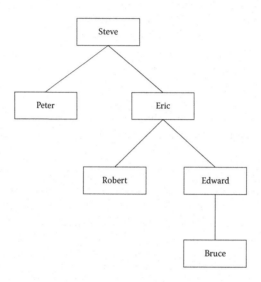

Figure 2.5 *The Smith family tree.*

The initiator is Steve Smith, who we shall say constitutes the zeroth generation. Steve Smith has two sons: Peter and Eric Smith, who constitute the first generation. Peter does not have any male offspring, whereas Eric has two: Robert and Edward Smith. Thus the second generation has two males. The third generation has only one male: Bruce Smith, who dies without a male heir, and hence the family name Smith initiated by Steve dies out. We say that the family name of Steve Smith became extinct after the third generation.

To model this situation let X_n be the number of individuals in the n-th generation, starting with $X_0 = 1$. We index the individuals of the n-th generation by integers $1, 2, \cdots, X_n$. Let $Y_{r,n}$ be the number of male offspring to the r-th individual of the n-th generation. Then we have

$$X_{n+1} = \sum_{r=1}^{X_n} Y_{r,n}. \qquad (2.15)$$

Now suppose $\{Y_{r,n}\}$ are iid random variables. Then $\{X_n, n \geq 0\}$ is a DTMC on state-space $S = \{0, 1, 2, \cdots\}$ with transition probabilities

$$
\begin{aligned}
p_{i,j} &= \mathsf{P}(X_{n+1} = j | X_n = i, X_{n-1}, \cdots, X_0) \\
&= \mathsf{P}\left(\sum_{r=1}^{X_n} Y_{r,n} = j \,\Big|\, X_n = i, X_{n-1}, \cdots, X_0\right) \\
&= \mathsf{P}\left(\sum_{r=1}^{i} Y_{r,n} = j\right),
\end{aligned}
$$

where the last probability can be (in theory) computed as a function of i and j. The $\{X_n, n \geq 0\}$ process is called a branching process.

Typical questions of interest are: What is the probability that the family name eventually becomes extinct? How many generations does it take before it becomes extinct, given that it does become extinct? How many total males are produced in the family? What is the size of the n-th generation?

Although we have introduced the branching process as a model of propagation of family names, the same model arises in other areas, such as nuclear physics, spread of diseases, or rumors or chain-letters or internet jokes in very large populations. We give an example of a nuclear reaction: A neutron (zeroth generation) is introduced from outside in a fissionable material. The neutron may pass through the material without hitting any nucleus. If it does hit a nucleus, it will cause a fission resulting in a random number of new neutrons (first generation). These new neutrons themselves behave like the original neutron, and each will produce its own random number of new neutrons through a possible collision. Thus X_n, the number of neutrons after the n-th generation, can be modeled as a branching process. In nuclear reactors the evolution of this branching process is controlled by inserting moderator rods in the fissionable material to absorb some of the neutrons.

2.3.4 Finance

DTMCs have become an important tool in mathematical finance. This is a very rich and broad subject, hence we concentrate on a relatively narrow and basic model of stock fluctuations. Let X_n be the value of a stock at time n (this could be a day, or a minute). We assume that $\{X_n, n \geq 0\}$ is a stochastic process with state-space $(0, \infty)$, not necessarily discrete, although stock values are reported in integer cents. Define the *return* in period n as

$$
R_n = \frac{X_n - X_{n-1}}{X_{n-1}}, \quad n \geq 1.
$$

Thus $R_3 = .1$ implies that the stock value increased by 10% from period two to three, $R_2 = -.05$ is equivalent to saying that the stock value decreased by 5% from period one to two. From this definition it follows that

$$
X_n = X_0 \prod_{i=1}^{n} (1 + R_i), \quad n \geq 1. \tag{2.16}
$$

Thus specifying the stochastic process of returns $\{R_n, n \geq 1\}$ along with X_0 is equivalent to specifying the stochastic process $\{X_n, n \geq 0\}$. The simplest model for the return process is to assume that it is a sequence of iid random variables taking values in $(-1, \infty)$. Note that this ensures that $X_n > 0$ for all $n \geq 1$ if $X_0 > 0$. Also, since

$$X_{n+1} = X_n(1 + R_{n+1}), \quad n \geq 0,$$

it follows that $\{X_n, n \geq 0\}$ is a DTMC with state-space $(0, \infty)$, which may not be discrete.

A further simplification occurs if we assume that $\{R_n, n \geq 1\}$ is a sequence of iid random variables with common pmf

$$P(R_n = u) = p, \quad P(R_n = -d) = q = 1 - p,$$

where $0 \leq p \leq 1$, $u \geq 0$, and $0 \leq d < 1$ are fixed real numbers. Thus if $u = .1$ and $d = .05$, the stock goes up by 10% with probability p and goes down by 5% with probability $q = 1 - p$ in each period. This is a very useful model of discrete time stock movements and is called the *binomial model*. It is used frequently in financial literature. To see why it is called a binomial model, let $\{Y_n, n \geq 1\}$ be a sequence of iid Ber(p) (Bernoulli with parameter p) random variables. Then $Z_n = Y_1 + \cdots + Y_n$ is a binomial random variable with parameters n and p. With this notation we can write Equation 2.16 as

$$X_n = X_0(1 + u)^{Z_n}(1 - d)^{n - Z_n}.$$

Note that for each value of X_0, X_n can take $n + 1$ discrete values.

Since stock values can go up or down, investors take a risk when they invest in stocks, rather than putting their money in a risk-free money market account that gives a fixed positive rate of return. Hence the financial industry has created several financial instruments that mitigate or bound such risks. A simple instrument of this type is the European call option. It gives the holder of the option a right, but not an obligation, to buy the stock on a specified day, say T, at a specified price, say K. Clearly, if X_T, the value of the stock on day T, is less than K, the holder will not exercise the option, and the option will expire with no profit and no loss to the holder. If $X_T > K$, the holder will exercise the option and buy the stock at price K, and immediately sell it at price X_T and make a tidy profit of $X_T - K$. In general the holder gets $\max\{0, X_T - K\}$ at time T, which is never negative. Thus there is no risk of loss in this financial instrument. Clearly, the buyer of this option must be willing to pay a price to buy this risk-less instrument. How much should she pay? What is the fair value of such an option? There are many other such options. For example, a put option gives a right to sell. American versions of these options can be exercised anytime until T, and not just at time T, as is the case in the European options. Valuation of these options is a highly technical area, and DTMCs play a major role in the discrete time versions of these problems. See *Options, Futures, and Other Derivatives* by J. C. Hull for more details at an elementary level.

2.3.5 *Manpower Planning*

Large organizations, such as government, military, or multinational corporations, employ a large number of people who are generally categorized into several grades and are promoted from one grade to another depending upon performance, seniority, and need. The employees also leave the organization from any grade, and new employees are hired into different grades. Thus the state of the organization, described by the employee histories, changes dynamically over time, and it is of interest to model this evolution by a stochastic process.

Here we present a simple stochastic model of an organization consisting of M employees distributed in N grades. We assume that the number of employees is kept constant by the hiring of a new employee whenever an existing employee leaves the organization. We assume that all employees behave independently and identically. We assume that the grade changes and hiring and departures occur at times $n = 1, 2, \cdots$. If an employee is in grade i at time n, he or she moves to grade j at time $n+1$ with probability $r_{i,j}$, or leaves the organization with probability r_i. Clearly, we must have

$$r_i + \sum_{j=1}^{N} r_{i,j} = 1, \quad \text{for all } i = 1, 2, \cdots, N.$$

When an employee leaves the organization, he or she is replaced instantaneously by a new employee who starts in grade j with probability α_j. Note that this is an extremely simplified version of reality: we have ignored the issues of seniority, constraints on the number of employees in a given grade, changing number of employees, etc.

Let $X_n(k)$ be the grade of the k-th employee ($1 \leq k \leq M$) at time n. Think of k as an employee identification number, and when the k-th employee leaves the system and is replaced by a new employee, that new employee gets the identification number k. From the above description we see that $\{X_n(k), n \geq 0\}$ is a DTMC on state-space $S = \{1, 2, \cdots, N\}$ with transition probabilities given by

$$
\begin{aligned}
p_{i,j} &= \mathsf{P}(X_{n+1}(k) = j | X_n(k) = i) \\
&= \mathsf{P}(k\text{-th employee moves from grade } i \text{ to } j) \\
&\quad + \mathsf{P}(k\text{-th employee leaves and is replaced by a new one in grade } j) \\
&= r_{i,j} + r_i \alpha_j.
\end{aligned}
$$

Now let $X_n = [X_n(1), X_n(2), \cdots, X_n(M)]$. Since $X_n(k), 1 \leq k \leq M$, are independent, we see that $\{X_n, n \geq 0\}$ is a DTMC on state-space S^M, with the following transition probabilities:

$$
\begin{aligned}
\mathsf{P}(X_{n+1} &= [j_1, j_2, \cdots, j_M] | X_n = [i_1, i_2, \cdots, i_M]) \\
&= \prod_{k=1}^{M} \mathsf{P}(X_{n+1}(k) = j_k | X_n(k) = i_k) \\
&= \prod_{k=1}^{M} p_{i_k, j_k}.
\end{aligned}
$$

An alternate method of modeling the system is to define $Y_n(j)$ to be the number of employees in grade j at time n. We leave it to the reader to show that $Y_n = [Y_n(1), Y_n(2), \cdots, Y_n(N)]$ is a DTMC and compute its transition probabilities.

Typically we are interested in the following: what is the composition of the workforce at time n? What is the expected number of employees in grade j at time n? What is the expected amount of time an employee spends in the organization before leaving? How can we control the distribution of workers in various grades? We shall see how these questions can be answered with the help of the tools we shall develop in this and the following two chapters.

2.3.6 Telecommunications

Wireless communication via cell phones has become commonly available. In such a system a wireless device such as a cell phone communicates with a cell tower in a bi-directional fashion, that is, it receives data from the cell tower and sends data to it. The rate at which data can be transmitted changes randomly with time due to many factors, for example, changing position of the user, weather, topology of the terrain, to name a few. The cell tower knows at all times how many users are registered with it, and what data-rate is available to each user. We consider a technology in which the time is slotted into short intervals of length δ, say a millisecond long, and the cell tower can communicate with exactly one user during each time slot.

Let $R_n(u)$ be the data rate (in kilobits per second) available to user u in the n-th slot. It is commonly assumed that $\{R_n(u), n \geq 0\}$ is a DTMC with a finite state-space, for example, $\{38.4, 76.8, 102.6, 153.6, 204.8, 307.2, 614.4, 921.6, 1228.8, 1843.2, 2457.6\}$, and that the data-rates available to different users are independent. Now let $X_n(u)$ be the amount of data (in kilobits) waiting for transmission at user u at the beginning of the n-th time slot, and $A_n(u)$ be the new data that arrives for the user in the n-th slot. Thus if user u is served during the n-th time slot, $X_n(u) + A_n(u)$ amount of data is available for transmission, out of which $\delta R_n(u)$ is actually transmitted. Using $v(n)$ to denote the user that is served in the n-th slot, we see that the following recursion holds:

$$X_{n+1}(u) = \begin{cases} \max\{X_n(u) + A_n(u) - \delta R_n(u), 0\} & \text{if } u = v(n) \\ X_n(u) + A_n(u) & \text{if } u \neq v(n). \end{cases}$$

Now suppose there are a fixed number N of users in the reception area of the cell tower. Let $R_n = [R_n(1), R_n(2), \cdots, R_n(N)]$, $X_n = [X_n(1), X_n(2), \cdots, X_n(N)]$, and $A_n = [A_n(1), A_n(2), \cdots, A_n(N)]$. Suppose the cell tower knows the state of the system (R_n, X_n, A_n) at the beginning of the n-th time slot. It decides which user to serve next based solely on this information. If we assume that $\{A_n(u), n \geq 0\}$ are independent (for different u's) sequences of iid random variables, it follows that $\{(R_n, X_n), n \geq 0\}$ is a DTMC with a rather large state-space and complicated transition probabilities.

The cell tower has to decide which user to serve in each time slot so that the data is transferred at the highest possible rate (maximize throughput) and at the same time no user is starved for too long (ensure fairness). This is the main scheduling problem in wireless communications. For example, the cell tower may decide to serve that user in the n-th time slot who has the highest data rate available to it. This may maximize the throughput, but may be unfair for those users who are stuck with low data rate environment. On the other hand, the cell tower may decide to serve the user u with the highest backlog $X_n(u) + A_n(u)$. This may also be unfair, since the user with largest data requirement will be served most of the time. One rule that attempts to strike a balance between these two conflicting objectives serves the user u that has the largest value of $\min(\delta R_n(u), X_n(u) + A_n(u)) \cdot (X_n(u) + A_n(u))$.

DTMCs have been used in a variety of problems arising in telecommunications. The above model is but one example. DTMCs have been used in the analysis of radio communication networks using protocols like ALOHA, performance of the ethernet protocol and the TCP protocol in the Internet, to name a few other applications. The reader is referred to the classic book *Data Networks* by Bertsekas and Gallager for a simple introduction to this area.

2.3.7 *Google Search*

The Internet has become an integral part of modern living. The majority of the information on the Internet is kept in the form of web pages, which can be accessed by using their URLs (user resource locaters), a kind of address in the cyberspace. The Internet itself would be quite hard to use unless there was an easy way to find the URLs for the pages containing the information users are looking for. This service is provided by the web search engines. Google is the most popular among such search engines. It handles several billion user search queries a day. A typical query asks Google to find web pages containing a given text, such as "tropical flowers." The Google search engine responds with a list of web pages (that is, their URLs). A typical search can result in several thousand or sometimes millions of web pages that contain the desired phrase. Such a list will be overwhelming without some indication as to the relative importance of each web page. Google uses a special algorithm, called the PageRank algorithm, to generate a rank for each page. The higher the page rank, the more important the page. Google then lists the web pages starting with the highest ranked page first.

The PageRank algorithm was first developed by Larry Page and Sergey Brin in 1997, who then proceeded to found the company Google. The algorithm is based on a simple mathematical model of how a user surfs the web, that is, how a user moves from one page to another on the web. Suppose the Internet contains N web pages. We think of the internet as a directed network with N nodes, with one node for each page. There is an arc going from node i to node j if there is at least one hyperlink in page i that leads the user to page j if she clicks on it. Suppose a user starts on page X_0, and then clicks on one of the hyperlinks in it to go to the next web page, say X_1. In general suppose X_n is the page the user lands on after clicking n times.

A simple model of websurfing is to assume that $\{X_n, n \geq 0\}$ is a DTMC, and that a user clicks on any of the hyperlinks on the current webpage in a uniform random fashion. That is, if there are k hyperlinks on the current page, she will click on any one of them with probability $1/k$, regardless of her web surfing history. Let us call this Model 1.

One possible problem with Model 1 arises if the user reaches a webpage with no outgoing hyperlinks. Then the user is stuck on that page with nowhere to go. We modify the user behavior model to avoid this problem as follows: when a user visits a page with no outgoing links, the user simply moves to any of the N pages with probability $1/N$ regardless of history. We call this the Model 2. Under Model 2, $\{X_n, n \geq 0\}$ continues to have the Markov property.

Model 2 can also be problematic if, for example, page 1 only links to page 2, and page 2 only links to page 1. Thus once the user visits page 1 or 2, she will never visit any other pages in the Internet under Model 2. To avoid this problem, one can further modify the user behavior model by introducing a damping factor d and create Model 3. Under this modification, with probability d the user chooses the next webpage according to the mechanism of Model 2, and with probability $1 - d$, the user simply moves to any one of the N pages with probability $1/N$. Model 3 ensures that the user has positive probability of visiting any of the pages in the Internet.

The PageRank algorithm estimates the rank of a page as the long run frequency with which the user visits that page. We shall develop methods of computing such quantities in Chapter 4. See Examples 4.27 and 4.28 for other interpretations of Google page rank. We also refer the reader to *Google's PageRank and Beyond: The Science of Search Engine*, by Langville and Meyer for further details.

2.4 Marginal Distributions

The examples in Sections 2.2 and 2.3 show that many real-world systems can be modeled by DTMCs and provide the motivation to study them. Section 2.1 tells us how to characterize a DTMC. Now we follow the road map laid out in Chapter 1 and study the transient behavior of the DTMCs. In this section we first study the marginal distributions of DTMCs.

Let $\{X_n, n \geq 0\}$ be a DTMC on state-space $S = \{0, 1, 2, \cdots\}$ with transition probability matrix P and initial distribution a. In this section we shall study the distribution of X_n. Let the pmf of X_n be denoted by

$$a_j^{(n)} = \mathsf{P}(X_n = j), \quad j \in S, \ n \geq 0. \tag{2.17}$$

Clearly $a_j^{(0)} = a_j$ is the initial distribution. By using the law of total probability we get

$$\mathsf{P}(X_n = j) = \sum_{i \in S} \mathsf{P}(X_n = j | X_0 = i)\mathsf{P}(X_0 = i)$$

$$= \sum_{i \in S} P(X_n = j | X_0 = i) a_i$$

$$= \sum_{i \in S} a_i p_{i,j}^{(n)}, \qquad (2.18)$$

where

$$p_{i,j}^{(n)} = P(X_n = j | X_0 = i), \quad i, j \in S, \ n \geq 0. \qquad (2.19)$$

It is called the *n-step transition probability*, since it is the probability of going from state i to state j in n transitions. We have

$$p_{i,j}^{(0)} = P(X_0 = j | X_0 = i) = \delta_{i,j}, \quad i, j \in S, \qquad (2.20)$$

where $\delta_{i,j}$ is one if $i = j$ and zero otherwise, and

$$p_{i,j}^{(1)} = P(X_1 = j | X_0 = i) = p_{i,j}, \quad i, j \in S. \qquad (2.21)$$

If we can compute the n-step transition probabilities $p_{i,j}^{(n)}$, we can compute the marginal distribution of X_n. Intuitively, the event of going from state i to state j involves going from state i to some intermediate state r at time $k \leq n$, followed by a trajectory from state r to state j in the remaining $n - k$ steps. This intuition is used to derive a method of computing $p_{i,j}^{(n)}$ in the next theorem. The proof of the theorem is enlightening in itself since it shows the critical role played by the assumptions of Markov property and time homogeneity.

Theorem 2.3 Chapman–Kolmogorov Equations. *The n-step transition probabilities satisfy the following equations:*

$$p_{i,j}^{(n)} = \sum_{r \in S} p_{i,r}^{(k)} p_{r,j}^{(n-k)}, \quad i, j \in S, \qquad (2.22)$$

where k is a fixed integer such that $0 \leq k \leq n$.

Proof: Fix an integer k such that $0 \leq k \leq n$. Then

$$
\begin{aligned}
p_{i,j}^{(n)} &= P(X_n = j | X_0 = i) \\
&= \sum_{r \in S} P(X_n = j, X_k = r | X_0 = i) \\
&= \sum_{r \in S} P(X_n = j | X_k = r, X_0 = i) P(X_k = r | X_0 = i) \\
&= \sum_{r \in S} P(X_n = j | X_k = r) P(X_k = r | X_0 = i) \quad \text{(Markov property)} \\
&= \sum_{r \in S} P(X_{n-k} = j | X_0 = r) P(X_k = r | X_0 = i) \quad \text{(time homogeneity)} \\
&= \sum_{r \in S} p_{r,j}^{(n-k)} p_{i,r}^{(k)}
\end{aligned}
$$

which can be rearranged to get Equation 2.22. This proves the theorem. ∎

Equations 2.22 are called the Chapman–Kolmogorov equations and can be written more succinctly in matrix notation. Define the n-step transition probability matrix as

$$P^{(n)} = [p_{i,j}^{(n)}].$$

The Chapman–Kolmogorov equations can be written in matrix form as

$$P^{(n)} = P^{(k)} P^{(n-k)}, \quad 0 \le k \le n. \tag{2.23}$$

The next theorem gives an important implication of the Chapman–Kolmogorov equations.

Theorem 2.4   The n-Step Transition Probability Matrix.

$$P^{(n)} = P^n, \tag{2.24}$$

where P^n is the n-th power of P.

Proof: We prove the result by induction on n. Let I be the identity matrix of the same size as P. Equations 2.20 and 2.21 imply that

$$P^{(0)} = I = P^0$$

and

$$P^{(1)} = P = P^1.$$

Thus the result is valid for $n = 0$ and 1. Suppose it is valid for some $k \ge 1$. From Equation 2.23 we get

$$
\begin{aligned}
P^{(k+1)} &= P^{(k)} P^{(1)} \\
&= P^k P \quad \text{(From the induction hypothesis)} \\
&= P^{k+1}.
\end{aligned}
$$

The theorem then follows by induction. ∎

Example 2.18   Two-State DTMC. Let P be the transition probability matrix of the two-state DTMC of Example 2.3 on page 13. If $\alpha + \beta = 2$, we must have $\alpha = \beta = 1$ and hence $P = I$. In that case $P^n = I$ for all $n \ge 0$. If $\alpha + \beta < 2$, it can be shown by induction that, for $n \ge 0$,

$$P^n = \frac{1}{2 - \alpha - \beta} \begin{bmatrix} 1 - \beta & 1 - \alpha \\ 1 - \beta & 1 - \alpha \end{bmatrix} + \frac{(\alpha + \beta - 1)^n}{2 - \alpha - \beta} \begin{bmatrix} 1 - \alpha & \alpha - 1 \\ \beta - 1 & 1 - \beta \end{bmatrix}.$$

In general such a closed form expression for P^n is not available for DTMCs with larger state-spaces. ∎

Example 2.19   Simple Random Walk. Consider a simple random walk on all integers with the following transition probabilities

$$p_{i,i+1} = p, \quad p_{i,i-1} = q = 1 - p, \quad -\infty < i < \infty,$$

where $0 < p < 1$. Compute

$$p_{0,0}^{(n)} = \mathsf{P}(X_n = 0 | X_0 = 0), \quad n \ge 0.$$

Starting from state 0, the random walk can return to state 0 only in an even number of steps. Hence we must have

$$p_{0,0}^{(n)} = 0, \quad \text{for all odd } n.$$

Now let $n = 2k$ be an even integer. To return to state 0 in $2k$ steps starting from state 0, the random walk must take a total of k steps to the right, and k steps to the left, in any order. There are $(2k)!/(k!k!)$ distinct sequences of length $2k$ made up of k right and k left steps, and the probability of each sequence is $p^k q^k$. Hence we get

$$p_{0,0}^{(2k)} = \frac{(2k)!}{k!k!} p^k q^k, \quad k = 0, 1, 2, \cdots. \tag{2.25}$$

In a similar manner one can show that

$$p_{i,j}^{(n)} = \binom{n}{b} p^a q^b, \quad \text{if } n + j - i \text{ is even}, \tag{2.26}$$

where $a = (n + j - i)/2$ and $b = (n + i - j)/2$. If $n + j - i$ is odd the above probability is zero. ∎

It is not always possible to get a closed form expression for the n-step transition probabilities, and one must do so numerically. We will study this in more detail in Section 2.6.

Now let $a^{(n)} = [a_j^{(n)}]$ be the pmf of X_n. The next theorem gives a simple expression for $a^{(n)}$.

Theorem 2.5   Probability Mass Function of X_n.

$$a^{(n)} = aP^n, \quad n \geq 0. \tag{2.27}$$

Proof: Equation 2.18 can be written in matrix form as

$$a^{(n)} = a^{(0)} P^{(n)}.$$

The result follows from the observation that $a^{(0)} = a$, and Equation 2.24. ∎

It follows that the i-th row of P^n gives the pmf of X_n conditioned on $X_0 = i$. The above theorem gives a numerically straightforward method of computing the marginal distribution of X_n. We illustrate these concepts with several examples below.

Example 2.20   A Numerical Example. Let $\{X_n, n \geq 0\}$ be the DTMC of Example 2.2 on page 12. Compute the pmf of X_4.

From Theorem 2.5 we get the pmf of X_4 as

$$a^{(4)} = aP^4 = [0.25 \ 0.25 \ 0.25 \ 0.25] \cdot \begin{bmatrix} 0.1 & 0.2 & 0.3 & 0.4 \\ 0.2 & 0.2 & 0.3 & 0.3 \\ 0.5 & 0.0 & 0.5 & 0.0 \\ 0.6 & 0.2 & 0.1 & 0.1 \end{bmatrix}^4$$

$$= [0.25\ 0.25\ 0.25\ 0.25] \cdot \begin{bmatrix} 0.3616 & 0.1344 & 0.3192 & 0.1848 \\ 0.3519 & 0.1348 & 0.3222 & 0.1911 \\ 0.3330 & 0.1320 & 0.3340 & 0.2010 \\ 0.3177 & 0.1404 & 0.3258 & 0.2161 \end{bmatrix}$$

$$= [0.34105\ 0.13540\ 0.32530\ 0.19825]. \blacksquare$$

Example 2.21   Urn Model Continued. Let $\{X_n, n \geq 0\}$ be the stochastic process of the urn model of Example 2.13 on page 18 with $N = 10$. Compute $E(X_n)$ for $n = 0, 5, 10, 15,$ and 20, starting with $X_0 = 10$. Using the transition matrix P from Example 2.13 we get

$$e_n = E(X_n | X_0 = 10) = \sum_{i=1}^{10} iP(X_n = i | X_0 = 10) = aP^n b,$$

where $a = (0, 0, \cdots, 0, 1)$ and $b = (0, 1, \cdots, 9, 10)^\top$. Numerical computations yield: $e_0 = 10$, $e_5 = 6.6384$, $e_{10} = 5.5369$, $e_{15} = 5.1759$, and $e_{20} = 5.0576$. We can see numerically that as $n \to \infty$, e_n converges to 5. $\blacksquare$

Example 2.22   The Branching Process. Consider the branching process $\{X_n, n \geq 0\}$ introduced in Section 2.3.3. Here we compute $\mu_n = E(X_n)$ and $\sigma_n^2 = Var(X_n)$ as functions of n. Since $X_0 = 1$, clearly

$$\mu_0 = 1, \quad \sigma_0^2 = 0. \tag{2.28}$$

Let μ and σ^2 be the mean and the variance of the number of offspring to a single individual. Since X_1 is the number of offspring to the single individual in generation zero, we have

$$\mu_1 = \mu, \quad \sigma_1^2 = \sigma^2. \tag{2.29}$$

Furthermore, one can show that

$$E(X_n) = E\left(\sum_{r=1}^{X_{n-1}} Y_{r,n-1}\right) = \mu E(X_{n-1}),$$

$$Var(X_n) = Var\left(\sum_{r=1}^{X_{n-1}} Y_{r,n-1}\right) = \sigma^2 E(X_{n-1}) + \mu^2 Var(X_{n-1}).$$

Hence we have the following recursive equations:

$$\mu_n = \mu\mu_{n-1},$$
$$\sigma_n^2 = \sigma^2\mu_{n-1} + \mu^2\sigma_{n-1}^2.$$

Using the initial condition from Equation 2.28, we can solve these equations to get

$$\mu_n = \mu^n$$

$$\sigma_n^2 = \begin{cases} n\sigma^2 & \text{if } \mu = 1 \\ \sigma^2\mu^{n-1}\frac{\mu^n - 1}{\mu - 1} & \text{if } \mu \neq 1. \end{cases}$$

Thus, if $\mu > 1$, $\mu_n \to \infty$, as $n \to \infty$, that is, the branching process grows without bounds. If $\mu < 1$, $\mu_n \to 0$, as $n \to \infty$, that is, the branching process becomes extinct. If $\mu = 1$, we see that $\mu_n = 1$ for all n. We shall later show a rather counterintuitive result that the branching process becomes extinct with probability 1 in this case.

In the context of nuclear reaction, if $\mu > 1$, the reaction grows without bounds and we get an explosion or a reactor meltdown. If $\mu < 1$, the reaction is not self sustaining, and dies out naturally. In a nuclear power station, in the initial firing stage, one maintains $\mu > 1$ to get the reactor to a "hot" stage. Once the reactor is hot, moderator rods are inserted to absorb some neutrons, essentially reducing μ, and the reaction is controlled. The moderator rods are moved in and out constantly in response to the reactor temperature. If $\mu < 1$, the reaction starts dying out, the reactor starts cooling, and the moderator rods are pulled out. This raises μ to above 1, the reaction starts exploding, the reactor starts heating up, and the moderator rods are pushed back in. This control keeps the reactor producing heat (which is converted to electricity) at a constant rate.

In the context of propagation of family names, royal (or wealthy) males in historical times have been known to marry repeatedly until they produce a male heir. This is an obvious attempt to maintain $\mu \geq 1$. ∎

Example 2.23   Manpower Planning. Consider the manpower planning model of Section 2.3.5. Suppose the organization classifies the employees into four grades $\{1, 2, 3, 4\}$. Every year an employee in grades 1, 2, and 3 gets promoted from the current grade to the next grade with probability .2, or leaves with probability .2, or stays in the same grade. An employee in the fourth grade leaves with probability .2 or stays in the same grade. A departing employee is immediately replaced by a new one starting in grade 1. Suppose the organization has 100 employees distributed evenly in the four grades. What is the expected number of employees in each grade in the eighth year?

Suppose we pick an employee at random and track his evolution. Let X_n be the grade of this employee in year n. Thus the initial grade of this employee is i with probability .25 for $i = 1, 2, 3, 4$. Following the analysis in Section 2.3.5 it follows that $\{X_n, n \geq 0\}$ is a DTMC on $\{1, 2, 3, 4\}$ with initial distribution $[0.25,\ 0.25,\ 0.25,\ 0.25]$ and the following transition probability matrix:

$$
\begin{bmatrix}
0.8 & 0.2 & 0 & 0 \\
0.2 & 0.6 & 0.2 & 0 \\
0.2 & 0 & 0.6 & 0.2 \\
0.2 & 0 & 0 & 0.8
\end{bmatrix}.
$$

Then the pmf of the grade of this employee in year 8 is given by

$$[0.25\ \ 0.25\ \ 0.25\ \ 0.25] \cdot P^8 = [0.4958,\ 0.2388,\ 0.1140,\ 0.1514].$$

We can interpret this to mean that 49.58% of the employees are in grade 1 in year 8, etc. Thus the expected number of employees in the four grades are $[49.58,\ 23.88,\ 11.40,\ 15.14]$. One can numerically see that after several years (24 in this example), the vector of the expected number of employees in the four grades stabilizes at $[50, \ 25, \ 12.5, \ 12.5]$. ∎

2.5 Occupancy Times

Let $\{X_n, n \geq 0\}$ be a DTMC on state-space S with transition probability matrix P. In this section we compute the expected time spent by the DTMC in various states. Let $V_j^{(n)}$ be the number of visits to state j by the DTMC over $\{0, 1, 2, \cdots, n\}$. Note that we count the visit at time 0, that is, $V_j^{(0)} = 1$ if $X_0 = j$, and zero otherwise. Define

$$M_{i,j}^{(n)} = \mathsf{E}(V_j^{(n)} | X_0 = i), \quad i, j \in S, \ n \geq 0. \tag{2.30}$$

$M_{i,j}^{(n)}$ is called the *occupancy time* of state j up to time n starting from state i. Define the occupancy times matrix as

$$M^{(n)} = [M_{i,j}^{(n)}].$$

The next theorem shows how to compute the occupancy times matrix $M^{(n)}$.

Theorem 2.6   Occupancy Times Matrix. *We have*

$$M^{(n)} = \sum_{r=0}^{n} P^r, \quad n \geq 0, \tag{2.31}$$

where $P^0 = I$, the identity matrix.

Proof: Fix a $j \in S$. Let $Z_r = 1$ if $X_r = j$, and zero otherwise. Then

$$V_j^{(n)} = \sum_{r=0}^{n} Z_r.$$

Hence we get

$$
\begin{aligned}
M_{i,j}^{(n)} &= \mathsf{E}(V_j^{(n)} | X_0 = i) \\
&= \mathsf{E}\left(\sum_{r=0}^{n} Z_r \Big| X_0 = i \right) \\
&= \sum_{r=0}^{n} \mathsf{E}(Z_r | X_0 = i) \\
&= \sum_{r=0}^{n} \mathsf{P}(X_r = j | X_0 = i) \\
&= \sum_{r=0}^{n} p_{i,j}^{(r)} = \sum_{r=0}^{n} [P^{(r)}]_{i,j} = \sum_{r=0}^{n} [P^r]_{i,j},
\end{aligned}
$$

where the last equality follows from Theorem 2.4. Writing the above equation in matrix form yields Equation 2.31. This proves the theorem. ∎

We illustrate with two examples.

Example 2.24 Two-State DTMC. Consider the two-state DTMC of Example 2.3 on page 13. Assume $\alpha + \beta < 2$. The n-step transition probability matrix of the DTMC was given in Example 2.18 on page 34. Using that, and a bit of algebra, we see that the occupancy matrix for the two-state DTMC is given by

$$M^{(n)} = \frac{n+1}{2-\alpha-\beta} \begin{bmatrix} 1-\beta & 1-\alpha \\ 1-\beta & 1-\alpha \end{bmatrix} + \frac{1-(\alpha+\beta-1)^{(n+1)}}{(2-\alpha-\beta)^2} \begin{bmatrix} 1-\alpha & \alpha-1 \\ \beta-1 & 1-\beta \end{bmatrix}.$$

Thus, if the DTMC starts in state 1, the expected number of times it visits state 2 up to time n is given by $M_{1,2}^{(n)}$. ∎

It is rarely possible to compute the occupancy times matrix analytically. In most cases we have to do the calculations numerically, as shown in the next example.

Example 2.25 Brand Switching Model Continued. Consider the brand switching model of Example 2.14 on page 19. Compute the expected number of each brand sold in the first ten weeks.

Let P be the transition matrix given by Equation 2.9. Since we are interested in purchases over the weeks 0 through 9, we need to compute $M^{(9)}$. Using Theorem 2.6 we get

$$M^{(9)} = \sum_{n=0}^{9} P^n = \begin{bmatrix} 2.1423 & 2.7412 & 5.1165 \\ 1.2631 & 3.9500 & 4.7869 \\ 1.1532 & 2.8511 & 5.9957 \end{bmatrix}.$$

Thus if a customer chooses brand A in the initial week, his expected number of purchases of brand A over the weeks 0 through 9 is 2.1423, of brand B is 2.7412, and of brand C is 5.1165. ∎

2.6 Computation of Matrix Powers

If P is a finite matrix with numerical entries, computing P^n is relatively straightforward, especially with the help of matrix oriented languages like MATLAB® or Mathematica®. Here we describe two other methods of computing P^n.

2.6.1 Method of Diagonalization

We start with some preliminaries from matrix algebra. The reader is referred to texts by Fuller (1962) and Gantmacher (1960) for proofs and other details. An $m \times m$ square matrix A is called *diagonalizable* if there exist an invertible matrix X and a diagonal matrix

$$D = \text{diag}[\lambda_1, \lambda_2, \cdots, \lambda_m],$$

such that

$$A = XDX^{-1}. \tag{2.32}$$

Elements of X and D maybe complex even if A is real. It is known that the λ's are the eigenvalues of A, the j-th column x_j of X is the right eigenvector of λ_j, and the j-th row y_j of X^{-1} is the left eigenvector of λ_j. That is, the λ's are the m roots of

$$\det(\lambda I - A) = 0,$$

where det is short for determinant, x_j satisfies

$$Ax_j = \lambda_j x_j, \tag{2.33}$$

and y_j satisfies

$$y_j A = \lambda_j y_j.$$

If all the eigenvalues are distinct, then A is diagonalizable. This is a sufficient condition, but not necessary. With this notation we get the following theorem:

Theorem 2.7 Powers of a Square Matrix. *Suppose A is diagonalizable and Equation 2.32 holds. Then*

$$A^n = XD^nX^{-1} = \sum_{j=1}^{m} \lambda_j^n x_j y_j. \tag{2.34}$$

Proof: Since A is diagonalizable, and Equation 2.32 holds, we have

$$A^n = [XDX^{-1}][XDX^{-1}] \cdots [XDX^{-1}] = XD^nX^{-1}.$$

Using

$$D^n = \text{diag}[\lambda_1^n, \lambda_2^n, \cdots, \lambda_m^n]$$

we get the second equality in Equation 2.34. ∎

Computation of eigenvalues and eigenvectors is done easily using a matrix oriented language like MATLAB®. The following theorem gives important information about the eigenvalues of a stochastic matrix P.

Theorem 2.8 Eigenvalues of P. *Let P be an $m \times m$ transition probability matrix, with m eigenvalues λ_i, $1 \le i \le m$. Then*

1. At least one of the eigenvalues is one.

2. $|\lambda_i| \le 1$ for all $1 \le i \le m$.

Proof: The first result follows from the fact that P is stochastic and hence the column vector e with all coordinates equal to one satisfies

$$Pe = e,$$

thus proving that 1 is an eigenvalue of P.

To derive the second results, define the norm of an m-vector x as

$$\|x\| = \max\{|x_i| : 1 \le i \le m\},$$

and the norm of P as

$$\|P\| = \sup_{x:\|x\|=1} \|Px\|.$$

Then one can show that

$$\|P\| = \max_i\{\sum_{j=1}^{m} p_{i,j}\} = 1,$$

since P is stochastic. Furthermore,

$$\|Px\| \le \|P\|\|x\| = \|x\|.$$

Now, using Equation 2.33, we get

$$|\lambda_i|\|x_i\| = \|\lambda_i x_i\| = \|Px_i\| \le \|x_i\|.$$

Since $0 < \|x_i\| < \infty$, the above inequality implies

$$|\lambda_i| \le 1$$

for all $1 \le i \le m$. This proves the Theorem. ∎

We illustrate with two examples.

Example 2.26 Three-state DTMC. Consider a three state DTMC with the following transition probability matrix:

$$P = \begin{bmatrix} 0 & 1 & 0 \\ q & 0 & p \\ 0 & 1 & 0 \end{bmatrix},$$

where $0 < p < 1$ and $q = 1 - p$. Simple matrix multiplications show that

$$P^{2n} = \begin{bmatrix} q & 0 & p \\ 0 & 1 & 0 \\ q & 0 & p \end{bmatrix}, \quad n \ge 1, \tag{2.35}$$

$$P^{2n+1} = \begin{bmatrix} 0 & 1 & 0 \\ q & 0 & p \\ 0 & 1 & 0 \end{bmatrix}, \quad n \ge 0. \tag{2.36}$$

Derive these formulas using the method of diagonalization.

Simple calculations show that P has three eigenvalues $1, 0$, and -1, consistent with Theorem 2.8. Thus P is diagonalizable with

$$D = \begin{bmatrix} 1 & 0 & 0 \\ 0 & 0 & 0 \\ 0 & 0 & -1 \end{bmatrix},$$

$$X = \begin{bmatrix} 1 & p & 1 \\ 1 & 0 & -1 \\ 1 & -q & 1 \end{bmatrix},$$

and

$$X^{-1} = \frac{1}{2} \begin{bmatrix} q & 1 & p \\ 2 & 0 & -2 \\ q & 1 & -p \end{bmatrix}.$$

Thus

$$P^n = XD^nX^{-1} = \frac{1}{2} \begin{bmatrix} q(1+(-1)^n) & 1-(-1)^n & p(1+(-1)^n) \\ q(1-(-1)^n) & 1+(-1)^n & p(1-(-1)^n) \\ q(1+(-1)^n) & 1-(-1)^n & p(1+(-1)^n) \end{bmatrix}, \quad n \geq 1.$$

The above equation reduces to Equation 2.35 when n is even, and Equation 2.36 when n is odd. Thus the powers of P show an oscillatory behavior as a function of n. ■

Example 2.27   Genotype Evolution. Consider the six-state DTMC of the Genotype Evolution Model described on page 24, with the transition matrix P given by Equation 2.14 on page 24. Compute P^n.

A tedious calculation shows that the six eigenvalues in decreasing order are $\lambda_1 = 1, \lambda_2 = 1, \lambda_3 = (1+\sqrt{5})/4 = 0.8090, \lambda_4 = 0.5, \lambda_5 = 0.25, \lambda_6 = (1-\sqrt{5})/4 = -0.3090$. The matrix of right eigenvectors is given by

$$X = \begin{bmatrix} 4 & 0 & 0 & 0 & 0 & 0 \\ 3 & 1 & 0 & 0 & 0 & 0 \\ 2 & 2 & \lambda_3^2 & -1 & -1 & \lambda_6^2 \\ 2 & 2 & 1 & 0 & 4 & 1 \\ 1 & 3 & \lambda_3 & 0 & 1 & \lambda_6 \\ 0 & 4 & \lambda_3^2 & 1 & -1 & \lambda_6^2 \end{bmatrix}.$$

Note that the eigenvalues are not distinct, since the eigenvalue 1 is repeated twice. However, the matrix X is invertible. Hence P is diagonalizable, and the representation in Equation 2.32 holds with

$$D = \text{diag}[1, \ 1, \ 0.809, \ 0.5, \ 0.25, \ -0.3090].$$

Thus we get $P^n = XD^nX^{-1}$. ■

2.6.2 Method of Generating Functions

We assume that the reader is familiar with generating functions; see Appendix D for relevant details. Let P be an $m \times m$ matrix of transition probabilities and define

$$P(z) = \sum_{n=0}^{\infty} z^n P^n, \tag{2.37}$$

where z is a complex number. The above series converges absolutely, element by element, if $|z| < 1$. From Theorem 2.8 it follows that the eigenvalues of zP are all strictly less than one in absolute value if $|z| < 1$, and hence $I - zP$ is invertible. Now,

$$P(z) = I + \sum_{n=1}^{\infty} z^n P^n = I + zP(z)P.$$

Hence

$$P(z) = (I - zP)^{-1}.$$

Thus the method of generating function works as follows:

1. Compute $P(z) = [p_{i,j}(z)] = (I - zP)^{-1}$.
2. The function $p_{i,j}(z)$ is a rational functions of z, that is, it is a ratio of two polynomials in z. Expand it in power series of z to get

$$p_{i,j}(z) = \sum_{n=0}^{\infty} p_{i,j}^{(n)} z^n.$$

3. $P^n = [p_{i,j}^{(n)}]$ is the required power of P.

Generally, this method is numerically inferior to the method of diagonalization.

Example 2.28   Manpower Planning. Compute $[P^n]_{1,1}$ for the transition probability matrix P on page 37 of Example 2.23 using the generating function method.

Direct calculation, or the use of a symbolic calculation tool on a computer, yields (after cancelling common factors from the numerator and denominator)

$$[P(z)]_{1,1} = \sum_{n=0}^{\infty} [P^n]_{1,1} z^n = [(I - zP)^{-1}]_{1,1} = \frac{5 - 4z}{3z^2 - 8z + 5}.$$

The denominator has two roots:

$$z_1 = 5/3, \quad z_2 = 1.$$

Using partial fractions, and simplifying, we get

$$[P(z)]_{1,1} = \frac{1}{2} \left(\frac{1}{1 - z} + \frac{1}{1 - .6z} \right).$$

Expanding in power series in z, we get

$$\sum_{n=0}^{\infty} [P^n]_{1,1} z^n = \frac{1}{2} \left(\sum_{n=0}^{\infty} (1 + .6^n) z^n \right).$$

Equating the coefficients of z^n we get

$$[P^n]_{1,1} = .5(1 + .6^n).$$

One can compute the expressions for other elements of P^n in a similar fashion. ∎

2.7 Modeling Exercises

2.1 We have an infinite supply of light bulbs, and Z_i is the lifetime of the i-th light bulb. $\{Z_i, i \geq 1\}$ is a sequence of iid discrete random variables with common pmf

$$\mathsf{P}(Z_i = k) = p_k, \quad k = 1, 2, 3, \cdots,$$

with $\sum_{k=1}^{\infty} p_k = 1$. At time zero, the first light bulb is turned on. It fails at time Z_1, when it is replaced by the second light bulb, which fails at time $Z_1 + Z_2$, and so on. Let X_n be the age of the light bulb that is on at time n. Note that $X_n = 0$, if a new light bulb was installed at time n. Show that $\{X_n, n \geq 0\}$ is a DTMC and compute its transition probability matrix.

2.2 In the above exercise, let Y_n be the remaining life of the bulb that is in place at time n. For example, $Y_0 = Z_1$. Show that $\{Y_n, n \geq 0\}$ is a DTMC and compute its transition probability matrix.

2.3 An urn contains w white balls and b black balls initially. At each stage a ball is picked from the urn at random and is replaced by k balls of similar color ($k \geq 1$). Let X_n be the number of black balls in the urn after n stages. Is $\{X_n, n \geq 0\}$ a DTMC? If yes, give its transition probability matrix.

2.4 Consider a completely connected network of N nodes. At time 0 a cat resides on node N and a mouse on node 1. During one time unit, the cat chooses a random node uniformly from the remaining $N - 1$ nodes and moves to it. The mouse moves in a similar way, independently of the cat. If the cat and the mouse occupy the same node, the cat promptly eats the mouse. Model this as a Markov chain.

2.5 Consider the following modification of the two-state weather model of Example 2.4: given the weather condition on day $n - 1$ and n, the weather condition on day $n + 1$ is independent of the weather on earlier days. Historical data suggests that if it rained yesterday and today, it will rain tomorrow with probability 0.6; if it was sunny yesterday and today, it will rain tomorrow with probability 0.2; if it was sunny yesterday but rained today, it will rain tomorrow with probability .5; and if it rained yesterday but is sunny today, it will rain tomorrow with probability .25. Model this as a four state DTMC.

2.6 A machine consists of K components in series, i.e., all the components must be in working condition for the machine to be functional. When the machine is functional at the beginning of the nth day, each component has a probability p of failing at the beginning of the next day, independent of the other components. (More than one component can fail at the same time.) When the machine fails, a single repair person repairs the failed components one by one. It takes exactly one day to repair one failed component. When all the failed components are repaired the machine is functional again, and behaves as before. When the machine is down, the working components do not fail. Let X_n be the number of failed components at the beginning of the nth day, after all the failure and repair events at that time are accounted for.

Show that $\{X_n, n \geq 0\}$ is a DTMC. Display its transition matrix or the transition diagram.

2.7 Two coins are tossed simultaneously and repeatedly in an independent fashion. Coin i ($i = 1, 2$) shows heads with probability p_i. Let $Y_n(i)$ be the number of heads observed during the first n tosses of the i-th coin. Let $X_n = Y_n(1) - Y_n(2)$. Show that $\{X_n, n \geq 0\}$ is a DTMC. Compute its transition probabilities.

2.8 Consider the following weather forecasting model: if today is sunny (rainy) and it is the k-th day of the current sunny (rainy) spell, then it will be sunny (rainy) tomorrow with probability p_k (q_k) regardless of what happened before the current sunny (rainy) spell started ($k \geq 1$). Model this as a DTMC. What is the state-space? What are the transition probabilities?

2.9 Let $Y_n \in \{1, 2, 3, 4, 5, 6\}$ be the outcome of the n-th toss of a fair six-sided die. Let $S_n = Y_1 + \cdots + Y_n$, and $X_n = S_n(\text{mod } 7)$, the remainder when S_n is divided by 7. Assume that the successive tosses are independent. Show that $\{X_n, n \geq 1\}$ is a DTMC, and display its transition probability matrix.

2.10 This is a generalization of the above problem. Let $\{Y_n, n \geq 1\}$ be a sequence of iid random variables with common pmf $P(Y_n = k) = \alpha_k$, $k = 1, 2, 3, \cdots$. Let r be a given positive integer and define $S_n = Y_1 + \cdots + Y_n$, and $X_n = S_n(\text{mod } r)$. Show that $\{X_n, n \geq 1\}$ is a DTMC, and display its transition probability matrix.

2.11 A boy and a girl move into a two-bar town on the same day. Each night, the boy visits one or the other of the two bars, starting in bar 1, according to a DTMC with transition probability matrix

$$\begin{bmatrix} a & 1-a \\ 1-b & b \end{bmatrix}.$$

Likewise the girl visits one or the other of the same two bars according to a DTMC with transition probability matrix

$$\begin{bmatrix} c & 1-c \\ 1-d & d \end{bmatrix},$$

but starting in bar 2. Here, $0 < a, b, c, d < 1$. Assume that the Markov chains are independent. Naturally, the story ends when the boy meets the girl, i.e., when they both go to the same bar. Model the situation by a DTMC that will help answer when and where the story ends.

2.12 Suppose that $\{Y_n, n \geq 0\}$ is a DTMC on $S = \{0, 1, \cdots, 10\}$ with transition probabilities as follows:

$$p_{0,0} = .98, \quad p_{0,1} = .02,$$
$$p_{i,i-1} = .03, \quad p_{i,i} = .95, \quad p_{i,i+1} = .02, \quad 1 \leq i \leq 9,$$
$$p_{10,9} = .03, \quad p_{10,10} = .97.$$

Let X_n be the price (in dollars) of a gallon of unleaded gas on day n. We model the price fluctuation by assuming that there is an increasing function $f : S \rightarrow (0, \infty)$ such that $X_n = f(Y_n)$. For example, $f(i) = 4 + .05i$ implies that the gas price fluctuates between \$4 and \$4.50, with increments of \$.05. A student uses exactly one gallon of gas per day. In order to control gas expenditure, the student purchases gas only when the gas tank is empty (don't ask how he does it!). If he needs gas on day n, he purchases $11 - Y_n$ gallons of gas. Let Z_m be the price per gallon of gas paid by the student on his m-th purchase. Show that $\{Z_m, m \geq 0\}$ is a DTMC. Compute its transition probability matrix.

2.13 Consider the following extension of Example 2.5 on page 14. Suppose we follow the play the winner rule with k drugs ($k \geq 2$) as follows. The initial player is given drug 1. If the drug is effective with the current patient, we give it to the next patient. If the result is negative, we switch to drug 2. We continue this way until we reach drug k. When we observe a failure of drug k, we switch back to drug 1 and continue. Suppose the successive patients are independent, and that drug i is effective with probability p_i. Let X_n be i if the n-th patient is given drug i. Show that $\{X_n, n \geq 1\}$ is a DTMC. Derive its transition probability matrix.

2.14 A machine with two components is subject to a series of shocks that occur deterministically one per day. When the machine is working a shock can cause failure of component 1 alone with probability α_1, or of component 2 alone with probability α_2 or of both the components with probability α_{12} or no failures with probability α_0. (Obviously $\alpha_0 + \alpha_1 + \alpha_2 + \alpha_{12} = 1$.) When a failure occurs, the machine is shut down and no more failures occur until the machine is repaired. The repair time (in days) of component i ($i = 1, 2$) is a geometric random variable with parameter r_i, $0 < r_i < 1$. Assume that there is a single repair person and all repair times are independent. If both components fail simultaneously, the repair person repairs component one first, followed by component two. Give the state-space and the transition probability matrix of an appropriate DTMC that can be used to model the state-evolution of the machine.

2.15 Suppose three players — 1,2,3 — play an infinite tournament as follows: Initially player 1 plays against player 2. The winner of the n-th game plays against the player who was not involved in the n-th game. Suppose $b_{i,j}$ is the probability that in a game between players i and j, player i will win. Obviously, $b_{i,j} + b_{j,i} = 1$. Suppose the outcomes of the successive games are independent. Let X_n be the pair that plays the n-th game. Show that $\{X_n, n \geq 0\}$ is a DTMC. Display its transition probability matrix or the transition diagram.

2.16 Mr. Al Anon drinks one six-pack of beer every evening! Let Y_n be the price of the six-pack on day n. Assume that the price is either L (low) or H (high), and that $\{Y_n, n \geq 0\}$ is a DTMC on state-space $\{H, L\}$ with transition probability matrix

$$\begin{bmatrix} \alpha & 1 - \alpha \\ 1 - \beta & \beta \end{bmatrix}.$$

Mr. Al Anon visits the beer store each day in the afternoon. If the price is high and he has no beer at home, he buys one six pack, which he consumes in the evening. If the price is high and he has at least one six pack at home, he does not buy any beer. If the price is low, he buys enough six packs so that he will have a total of five six packs in the house when he reaches home. Model this system by a DTMC. Describe its state-space and compute the transition probability matrix.

2.17 Let $\{Y_n, n \geq 1\}$ be a sequence of iid random variables with common pmf

$$\mathsf{P}(Y_n = k) = \alpha_k, \quad k = 0, 1, 2, 3, \cdots, M.$$

Define $X_0 = 0$ and

$$X_n = \max\{Y_1, Y_2, \cdots, Y_n\}, \quad n = 1, 2, 3, \cdots.$$

Show that $\{X_n, n \geq 0\}$ is a DTMC. Display its transition probability matrix.

2.18 Consider a machine that alternates between two states: up and down. The successive up and down times are independent of each other. The successive up times are iid positive integer valued random variables with common pmf

$$\mathsf{P}(\text{up time} = i) = u_i, \quad i = 1, 2, 3, \cdots,$$

and the successive down times are iid positive integer valued random variables with common pmf

$$\mathsf{P}(\text{down time} = i) = d_i, \quad i = 1, 2, 3, \cdots.$$

Assume that

$$\sum_{i=1}^{\infty} u_i = 1, \quad \sum_{i=1}^{\infty} d_i = 1.$$

Model this system by a DTMC. Describe its state-space and the transition probability matrix.

2.19 Ms. Friendly keeps in touch with her friends via email. Every day she checks her email inbox at 8:00 am. She processes each message in the inbox at 8:00 am independently in the following fashion: she answers it with probability $p > 0$ and deletes it, or she leaves it in the inbox, to be visited again the next day. Let Y_n be the number of messages that arrive during 24 hours on day n. Assume that $\{Y_n, n \geq 0\}$ is a sequence of iid random variables with common pmf

$$\alpha_k = \mathsf{P}(Y_n = k), \quad k = 0, 1, 2, \cdots.$$

Let X_n be the number of messages in the inbox at 8:00 am on day n. Assume that no new messages arrive while she is processing the messages. Show that $\{X_n, n \geq 0\}$ is a DTMC.

2.20 A buffer of size B bytes is used to store and play a streaming audio file. Suppose the time is slotted so that one byte is played (and hence removed from the buffer) at the end of each time slot. Let A_n be the number of bytes streaming into the buffer from the Internet during the nth time slot. Suppose that $\{A_n, n \geq 1\}$ is a sequence of iid random variables with common pmf

$$\alpha_k = \mathsf{P}(A_n = k), \quad k = 0, 1, 2.$$

Let X_n be the number of bytes in the buffer at the end of the nth time slot, after the input during that slot followed by the output during that slot. If the buffer is empty no sound is played, and if the buffer becomes full, some of the incoming bytes may be lost if there is no space for them. Both these events create a loss of quality. Model $\{X_n, n \geq 0\}$ as a DTMC. What is its state-space and transition probability matrix?

2.21 A shuttle bus with finite capacity B stops at bus stops numbered $0, 1, 2, \cdots$ on an infinite route. Let Y_n be the number of riders waiting to ride the bus at stop n. Assume that $\{Y_n, n \geq 0\}$ is a sequence of iid random variables with common pmf

$$\alpha_k = \mathsf{P}(Y_n = k), \quad k = 0, 1, 2, \cdots.$$

Every passenger who is on the bus alights at a given bus stop with probability p. The passengers behave independently of each other. After the passengers alight, as many of the waiting passengers board the bus as there is room on the bus. Show that $\{X_n, n \geq 0\}$ is a DTMC. Compute the transition probabilities.

2.22 A production facility produces one item per hour. Each item is defective with probability p, the quality of successive items being independent. Consider the following quality control policy parameterized by two positive integers k and r. In the beginning the policy calls for 100% inspections, the expensive mode of operation. As soon as k consecutive non-defective items are encountered, it switches to economy mode and calls for inspecting each item with probability $1/r$. It reverts back to the expensive mode as soon as an inspected item is found to be defective. The process alternates this way forever. Model the inspection policy as a DTMC. Define the state-space, and show the transition probability matrix or the transition diagram.

2.23 Let D_n be the demand for an item at a store on day n. Suppose $\{D_n, n \geq 0\}$ is a sequence of iid random variables with common pmf

$$\alpha_k = \mathsf{P}(D_n = k), \quad k = 0, 1, 2, \cdots.$$

Suppose the store follows the following inventory management policy, called the (s, S) policy: If the inventory at the end of the n-th day (after satisfying the demands for that day) is s or more, (here $s \geq 0$ is a fixed integer) the store manager does nothing. If it is less than s, the manager orders enough items to bring the inventory at the beginning of the next day up to S. Here $S \geq s$ is another fixed integer. Assume the delivery to the store is instantaneous. Let X_n be the number of items in the inventory in the store at the beginning of the n-th day, before satisfying that day's demand, but after the inventory is replenished. Show that $\{X_n, n \geq 0\}$ is a DTMC, and compute its transition probabilities.

2.24 A machine requires a particular component in functioning order in order to operate. This component is provided by two vendors. The maintenance policy is as follows: whenever the component fails, it is instantaneously replaced by a new component from vendor i with probability v_i. ($v_1 > 0$, $v_2 > 0$, $v_1 + v_2 = 1$.) The lifetimes of successive components from vendor i are iid random variables with common distribution

$$\alpha_k^i = P(\text{Life Time} = k), \quad k = 1, 2, 3, \ldots$$

with

$$\sum_{k=1}^{\infty} \alpha_k^i = 1, \quad (i = 1, 2).$$

Let X_n be the age of the component in use on day n. (If there is a replacement at time n, $X_n = 0$.) Let $Y_n = i$ if the component in place at time n (after replacement, if any) is from vendor i. Show that $\{(X_n, Y_n), n \geq 0\}$ is a DTMC. Display its transition diagram.

2.25 Consider the following simple model of a software development process: the software is tested at times $n = 0, 1, 2 \cdots$. If the software has k bugs, the test at time n will reveal a bug with probability β_k independent of the history. Assume $\beta_0 = 0$. If a bug is revealed, the software is updated so that the bug is fixed. However, in the process of fixing the bug, additional i bugs are introduced with probability α_i ($i = 0, 1, 2$) independent of the history. Let X_n be the number of bugs in the software just before it is tested at time n. Show that $\{X_n, n \geq 0\}$ is a DTMC and display its transition diagram.

2.26 Consider a closed society of N individuals. At time 0 one of these N individuals hears a rumor (from outside the society). At time 1 he tells it to one of the remaining $N - 1$ individuals, chosen uniformly at random. At each time n, every individual who has heard the rumor (and has not stopped spreading it) picks a person from the remaining $N - 1$ individuals at random and spreads the rumor. If he picks a person who has already heard the rumor, he stops spreading the rumor any more; else, he continues. If $k \geq 2$ persons tell the rumor to the same person who has not heard the rumor before, all $k + 1$ will continue to spread the rumor in the next period. Model this rumor spreading phenomenon as a DTMC.

2.27 Consider the following genetic experiment. Initially we cross an rr individual with a dr individual, thus producing the next generation individual. We generate the $(n + 1)$-st generation individual by crossing the n-th generation individual with a dr individual. Let X_n be the genotype of the n-th generation individual. What is the transition probability matrix of the DTMC $\{X_n, n \geq 0\}$?

2.28 Redo the above problem by assuming that we begin with an rr individual but always cross with a dd individual.

2.29 An electronic chain letter scheme works as follows. The initiator emails K persons exhorting each to email it to K of their own friends. It mentions that complying with the request will bring mighty good fortunes, while ignoring the request would bring dire supernatural consequences. Suppose a recipient complies with the request with probability α and ignores it with probability $1 - \alpha$, independently of other recipients. Assume the population is large enough so that this process continues forever. Show that we can model this situation by a branching process where the zeroth generation consists of all the individuals who get the letter from the initiator, that is, with $X_0 = K$. Why can we not consider the initiator as the zeroth generation and use $X_0 = 1$?

2.30 Slotted ALOHA is a protocol used to transmit packets by radio communications. Under this protocol time is slotted and each user can transmit its message at the beginning of the n-th slot ($n = 0, 1, 2, \cdots$). Each message takes exactly one time slot to transmit. When a new user arrives to the system, it transmits its message at the beginning of the next time slot. If only one user transmits during a time slot, the message is received by its intended recipient successfully. If no one transmits during a slot, that slot is wasted. If two or more users transmit during a slot, a collision occurs and the messages are garbled, and have to be retransmitted. If a user has experienced a collision while transmitting its message, it is said to be backlogged. A backlogged user attempts retransmission at the beginning of the next time slot with probability p, independent of its history. All users behave independently of each other. Once a user's message is transmitted successfully, the user leaves the system. Let Y_n be the number of new users that arrive during the n-th time slot. Assume that $\{Y_n, n \geq 0\}$ is a sequence of iid random variables with common pmf

$$\alpha_k = \mathsf{P}(Y_n = k), \quad k = 0, 1, 2, \cdots.$$

Let X_n be the number of backlogged users at time n (before any transmissions have occurred). Show that $\{X_n, n \geq 0\}$ is a DTMC and compute its transmission probabilities.

2.31 Under TDM (time division multiplexing) protocol, a user is allowed to transmit packets one at a time at times $n = 0, 1, 2, \cdots$. If the user has no packets to transmit at time n, he must wait until time $n + 1$ for the next opportunity to transmit, even though new packets may arrive between time n and $n + 1$. Let X_n be the number of packets before transmission is completed at time n and before any arrivals, say Y_n, between time n and $n + 1$. Assume that $\{Y_n, n \geq 0\}$ is a sequence of iid random variables with common pmf

$$\alpha_k = \mathsf{P}(Y_n = k), \quad k = 0, 1, 2, \cdots.$$

Show that $\{X_n, n \geq 0\}$ is a DTMC and compute its transition probabilities.

2.32 A manufacturing setup consists of two distinct machines, each producing one component per hour. Each component is tested instantly and is identified as defective or non-defective. Let α_i be the probability that a component produced by machine i is non-defective, $i = 1, 2$. The defective components are discarded and the non-defective components are stored in two separate bins, one for each machine. When a component is present in each bin, the two are instantly assembled together and shipped out. Bin i can hold at most B_i components, $i = 1, 2$. (Here B_1 and B_2 are fixed positive integers.) When a bin is full the corresponding machine is turned off. It is turned on again when the bin has space for at least one component. Assume that successive components are independent. Model this system by a DTMC.

2.33 Consider the following variation of Modeling Exercise 2.1: We replace the light bulb upon failure, or upon reaching age K, where $K > 0$ is a fixed integer. Assume that replacement occurs before failure if there is a tie. Let X_n be as in Modeling Exercise 2.1. Show that $\{X_n, n \geq 0\}$ is a DTMC and compute its transition probability matrix.

2.34 A citation model. Consider a set of N journal articles on the topic of Markov chains. Each article has a list of citations at the end, referring to earlier published relevant articles. (Clearly there cannot be any references to future articles.) Develop methods of ranking the N articles in the order of their importance by developing the three models of reader behavior as in Section 2.3.7.

2.8 Computational Exercises

2.1 Redo Example 2.21 on page 36 for the following initial conditions

$$1.\ X_0 = 8, \quad 2.\ X_0 = 5, \quad 3.\ X_0 = 3.$$

2.2 Let $\{X_n, n \geq 0\}$ be a DTMC with state-space $\{1, 2, 3, 4, 5\}$ and the following transition probability matrix:

$$\begin{bmatrix} 0.1 & 0.0 & 0.2 & 0.3 & 0.4 \\ 0.0 & 0.6 & 0.0 & 0.4 & 0.0 \\ 0.2 & 0.0 & 0.0 & 0.4 & 0.4 \\ 0.0 & 0.4 & 0.0 & 0.5 & 0.1 \\ 0.6 & 0.0 & 0.3 & 0.1 & 0.0 \end{bmatrix}.$$

Suppose the initial distribution is $a = [0.5,\ 0,\ 0,\ 0,\ 0.5]$. Compute the following:

1. The pmf of X_2,
2. $P(X_2 = 2, X_4 = 5)$,
3. $P(X_7 = 3 | X_3 = 4)$,
4. $P(X_1 \in \{1, 2, 3\}, X_2 \in \{4, 5\})$.

2.3 Prove the result in Example 2.18 on page 34 by using (a) induction, and (b) the method of diagonalization of Theorem 2.7 on page 40.

2.4 Compute the expected fraction of the patients who get drug 1 among the first n patients in the clinical trial of Example 2.5 on page 14. Hint: Use the results of Example 2.24 on page 39.

2.5 Suppose a market consists of k independent customers who switch between the three brands A, B, and C according to the DTMC of Example 2.14 on page 19. Suppose in week 0 each customer chooses brand A with probability 0.3 and brand B with probability 0.3 in an independent fashion. Compute the probability distribution of the number of customers who choose brand B in week 3.

2.6 Consider the two machine workshop of Example 2.7 on page 15. Suppose each machine produces a revenue of $\$r$ per day when it is up, and no revenue when it is down. Compute the total expected revenue over the first n days assuming both machines are up initially. Hint: Use independence and the results of Example 2.24 on page 39.

2.7 Consider the binomial model of stock fluctuation as described in Section 2.3.4 on page 27. Suppose $X_0 = 1$. Compute $E(X_n)$ and $\text{Var}(X_n)$ for $n \geq 0$.

2.8 Let $\{X_n, n \geq 0\}$ be a DTMC with state-space $\{1, 2, 3, 4\}$ and transition probability matrix given below:

$$P = \begin{bmatrix} 0.4 & 0.3 & 0.2 & 0.1 \\ 0.5 & 0.0 & 0.0 & 0.5 \\ 0.5 & 0.0 & 0.0 & 0.5 \\ 0.1 & 0.2 & 0.3 & 0.4 \end{bmatrix}.$$

Suppose $X_0 = 1$ with probability 1. Compute

1. $P(X_2 = 4)$,
2. $P(X_1 = 2, X_2 = 4, X_3 = 1)$,
3. $P(X_7 = 4 | X_5 = 2)$,
4. $E(X_3)$.

2.9 Consider Modeling Exercise 2.23 with the following parameters: $s = 10, S = 20, \alpha_0 = 0.1, \alpha_1 = 0.2, \alpha_2 = 0.3$, and $\alpha_3 = 0.4$. Suppose $X_0 = 20$ with probability 1. Compute $E(X_n)$ for $n = 1, 2, \cdots, 10$.

2.10 Consider the discrete time queue of Example 2.12 on page 17. Compute $P(X_2 = 0 | X_0 = 0)$.

2.11 Derive the n-step transition probabilities in Equation 2.26 on page 35.

2.12 Consider the weather forecasting model of Modeling Exercise 2.5. What is the probability distribution of the length of the rainy spell predicted by this model? Do the same for the length of the sunny spell.

2.13 Consider Modeling Exercise 2.21 with a bus of capacity 20. Suppose $\{Y_n, n \geq 0\}$ are iid Poisson random variables with mean 10, and assume $p = .4$. Compute $E(X_n | X_0 = 0)$ for $n = 0, 1, \cdots, 20$.

2.14 Let P be a $m \times m$ transition probability matrix given below

$$P = \begin{bmatrix} \alpha_1 & \alpha_2 & \alpha_3 & \cdots & \alpha_m \\ \alpha_m & \alpha_1 & \alpha_2 & \cdots & \alpha_{m-1} \\ \alpha_{m-1} & \alpha_m & \alpha_1 & \cdots & \alpha_{m-2} \\ \vdots & \vdots & \vdots & \ddots & \vdots \\ \alpha_2 & \alpha_3 & \alpha_4 & \cdots & \alpha_1 \end{bmatrix}.$$

Such a matrix is called the circulant matrix. Let $\iota = \sqrt{-1}$, and

$$e_k = \exp(\frac{\iota 2\pi}{m} k), \; 1 \leq k \leq m$$

be the m-th roots of unity. Define

$$\lambda_k = \sum_{i=1}^{m} \alpha_i e_k^{i-1}, \; 1 \leq k \leq m,$$

$$x_{j,k} = \exp(\frac{\iota 2\pi}{m} kj), \; 1 \leq j, k \leq m,$$

$$y_{k,j} = \frac{1}{m} \exp(-\frac{\iota 2\pi}{m} kj), \; 1 \leq j, k \leq m.$$

Show that λ_k is the k-th eigenvalue of P, with right eigenvector $[x_{1,k} \; x_{2,k} \; \cdots \; x_{m,k}]^\top$ and the left eigenvector $[y_{k,1} \; y_{k,2} \; \cdots \; y_{k,m}]$. Hence, using $D = \text{diag}(\lambda_1 \; \lambda_2 \; \cdots \lambda_m)$, $X = [x_{j,k}]$ and $Y = [y_{k,j}]$, show that

$$P = XDY.$$

Thus the powers of a circulant transition probability matrix can be written down analytically.

2.15 Let $\{X_n, n \geq 0\}$ be a success runs DTMC on $\{0, 1, 2, \cdots\}$ with

$$p_{i,0} = q = 1 - p_{i,i+1}, \; i \geq 0.$$

Show that $p_{0,0}^n = q$ for all $n \geq 1$.

2.16 Four points are arranged in a circle in a clockwise order. A particle moves on these four points by taking a clockwise step with probability p and a counterclockwise step with probability $q = 1 - p$, at times $n = 0, 1, 2, \cdots$. Let X_n be the position of the particle at time n. Thus $\{X_n, n \geq 0\}$ is a DTMC on $\{1, 2, 3, 4\}$. Display its transition probability matrix P. Compute P^n by using the diagonalization method of Section 2.6. Hint: Use the results of Computational Exercise 2.14 above.

2.17 Consider the urn model of Example 2.13 on page 18. Show that the transition probability matrix has the following eigenvalues: $\lambda_k = k(k+1)/N^2 - 1/N, 0 \leq k \leq N$. Show that the transition probability matrix is diagonalizable. In general, finding the eigenvalues is the hard part. Finding the corresponding eigenvectors is the easy part.

2.18 Compute $p_{i,j}^{(n)}$ for the success runs Markov chain of Computational Exercise 2.15.

2.19 The transition probability matrix of a three-state DTMC is as given below:

$$P = \begin{bmatrix} 0.3 & 0.4 & 0.3 \\ 0.4 & 0.5 & 0.1 \\ 0.6 & 0.2 & 0.2 \end{bmatrix}.$$

Find $p_{1,1}^{(n)}$ using the method of generating functions.

2.20 Consider a DTMC on $S = \{0, 1, 2, \cdots\}$ with the following transition probabilities:

$$p_{i,j} = \begin{cases} \frac{1}{i+1} & \text{for } 0 \le j \le i \\ 0 & \text{otherwise.} \end{cases}$$

Show that

$$p_{i,j}(z) = \sum_{n=0}^{\infty} p_{i,j}^{(n)} z^n = \begin{cases} \frac{z}{i+1} \prod_{k=j}^{i} \frac{k+1}{k+1-z} & \text{if } 0 \le j < i \\ \frac{i+1}{i+1-z} & \text{if } i = j. \end{cases}$$

2.21 Use the method of diagonalization to compute the probability distribution of the genotype of the n-th individual in the model described in Modeling Exercise 2.28.

2.22 Use the method of diagonalization to compute the probability distribution of the genotype of the n-th individual in the model described in Modeling Exercise 2.27.

2.23 Compute the mean of variance of the number of individuals in the n-th generation in the branching process of Section 2.3.3 on page 25, assuming the initial generation consists of i individuals. Hence compute the mean and variance of the number of letters in the n-th generation in Modeling Exercise 2.29. Hint: Imagine i independent branching processes, each initiated by a single individual, and use the results of Example 2.22 on page 36.

2.24 Use the method of diagonalization to compute the probability distribution of the gambler A's fortune after n games in the gambler's ruin model of Example 2.11 on page 17, with $N = 3$.

2.25 Consider the Wright–Fisher model described on page 25. Let $i \in \{0, 1, \cdots, N\}$ be a given integer and suppose $X_0 = i$ with probability 1. Compute $E(X_n)$ and $\text{Var}(X_n)$ for $n \ge 0$.

2.26 Consider the Moran model described on page 25. Let $i \in \{0, 1, \cdots, N\}$ be a given integer and suppose $X_0 = i$ with probability 1. Compute $E(X_n)$ and $\text{Var}(X_n)$ for $n \ge 0$.

2.9 Conceptual Exercises

2.1 Suppose $\{X_n, n \geq 0\}$ is a time homogeneous DTMC on $S = \{0, 1, 2, \cdots\}$ with transition probability matrix P. Show that

$$\mathsf{P}(X_{n+2} \in B, X_{n+1} \in A | X_n = i, X_{n-1}, \cdots, X_0) = \mathsf{P}(X_2 \in B, X_1 \in A | X_0 = i),$$

where A and B are subsets of S.

2.2 Suppose $\{X_n, n \geq 0\}$ and $\{Y_n, n \geq 0\}$ are two independent DTMCs with state-space $S = \{0, 1, 2, \cdots\}$. Prove or give a counterexample to the following statements:

1. $\{X_n + Y_n, n \geq 0\}$ a DTMC.
2. $\{(X_n, Y_n), n \geq 0\}$ is a DTMC.

2.3 Suppose $\{X_n, n \geq 0\}$ is a time homogeneous DTMC on $S = \{0, 1, 2, \cdots\}$. Prove or give a counterexample to the following statements:

1. $\mathsf{P}(X_{n+1} = j | X_n \in A, X_{n-1}, \cdots, X_0) = \mathsf{P}(X_{n+1} = j | X_n \in A)$, where $A \subset S$ has more than one element.
2. $\mathsf{P}(X_n = j_0 | X_{n+1} = j_1, X_{n+2} = j_2, \cdots, X_{n+k} = j_k) = \mathsf{P}(X_n = j_0 | X_{n+1} = j_1)$, where $j_0, j_1, \cdots, j_k \in S$ and $n \geq 0$.
3. $\mathsf{P}(X_n = j_0, X_{n+1} = j_1, X_{n+2} = j_2, \cdots, X_{n+k} = j_k) = \mathsf{P}(X_0 = j_0, X_1 = j_1, X_2 = j_2, \cdots, X_k = j_k)$.

2.4 Suppose $\{X_n, n \geq 0\}$ is a time homogeneous DTMC on $S = \{0, 1, 2, \cdots\}$. Prove or give a counterexample to the following statements:

1. Let $b_j = \mathsf{P}(X_k = j)$, for $j \in S$, and a given $k > 0$. Then $\{X_n, n \geq 0\}$ is completely described by $[b_j, j \in S]$ and the transition probability matrix.
2. Let $f : S \to S$ be any function. Then $\{f(X_n), n \geq 0\}$ is a DTMC.

2.5 Suppose $\{X_n, n \geq 0\}$ and $\{Y_n, n \geq 0\}$ are two independent DTMCs with state-space $S = \{0, 1, 2, \cdots\}$. Let $\{Z_n, n \geq 0\}$ be a sequence iid Ber(p) random variables. Define

$$W_n = \begin{cases} X_n & \text{if } Z_n = 0 \\ Y_n & \text{if } Z_n = 1. \end{cases}$$

Is $\{W_n, n \geq 0\}$ a DTMC (not necessarily time homogeneous)?

2.6 Let $\{Y_n, n \geq 0\}$ be a sequence of iid random variables with common pmf

$$\alpha_k = \mathsf{P}(Y_n = k), \quad k = 0, 1, 2, \cdots.$$

We say that Y_n is a record if $Y_n > Y_r$, $0 \leq r \leq n - 1$. Let $X_0 = Y_0$, and X_n be the value of the n-th record, $n \geq 1$. Show that $\{X_n, n \geq 0\}$ is a DTMC and compute its transition probability matrix.

2.7 Suppose $\{X_n, n \geq 0\}$ is a time homogeneous DTMC on $S = \{0, 1, 2, \cdots\}$ and transition probability matrix P. Define

$$N = \min\{n > 0 : X_n \neq X_0\}.$$

Thus N is the first time the DTMC leaves the initial state. Compute

$$\mathsf{P}(N = k|X_0 = i), \quad k \geq 1.$$

2.8 Suppose $\{X_n, n \geq 0\}$ is a time homogeneous DTMC on $S = \{0, 1, 2, \cdots\}$ and transition probability matrix P. Let A be a strict non-empty subset of S. Assume $X_0 \in A$ with probability 1. Define $N_0 = 0$ and

$$N_{r+1} = \min\{n > N_r : X_n \in A\}, \quad r \geq 0.$$

Thus N_r is the time of the r-th visit by the DTMC to the set A. Define

$$Y_r = X_{N_r}, \quad r \geq 0.$$

Is $\{Y_r, r \geq 0\}$ a DTMC? Prove or give a counterexample.

2.9 Suppose $\{X_n, n \geq 0\}$ is a time homogeneous DTMC with the following property: there is a $j \in S$ such that $p_{i,j} = p$ for all $i \in S$. Show that $\mathsf{P}(X_n = j) = p$ for all $n \geq 1$, no matter what the initial distribution is.

2.10 Suppose $\{X_n, n \geq 0\}$ is a time non-homogeneous DTMC with the following transition probabilities:

$$\mathsf{P}(X_{n+1} = j|X_n = i) = \begin{cases} a_{i,j} & \text{if } n \text{ is even} \\ b_{i,j} & \text{if } n \text{ is odd,} \end{cases}$$

where $A = [a_{i,j}]$ and $B = [b_{i,j}]$ are two given stochastic matrices. Construct a time homogeneous DTMC $\{Y_n, n \geq 0\}$ that is equivalent to $\{X_n, n \geq 0\}$, i.e., a sample path of $\{X_n, n \geq 0\}$ uniquely determines that of $\{Y_n, n \geq 0\}$ and vice versa.

2.11 Let $\{X_n, n \geq 0\}$ be a simple random walk of Example 2.19 on page 34. Show that $\{|X_n|, n \geq 0\}$ is a DTMC. Compute its transition probability matrix.

2.12 Suppose $\{X_n, n \geq 0\}$ is a time homogeneous DTMC on $S = \{0, 1, 2, \cdots\}$ and transition probability matrix P. Let $f : S \to \{1, 2, \cdots, M\}$ be a given on-to function, that is, $f^{-1}(i)$ is non-empty for all $1 \leq i \leq M$. Give the necessary and sufficient condition under which $\{f(X_n), n \geq 0\}$ is a DTMC.

CHAPTER 3

Discrete-Time Markov Chains: First Passage Times

A frequent flyer business traveler is concerned about the risk of encountering a terrorist bomb on one of his flights. He consults his statistician friend to get an estimate of the risk. After studying the data the statistician estimates that the probability of finding a bomb on a random flight is one in a thousand. Alarmed by such a large risk, the businessman asks his friend if there is any way to reduce the risk. The statistician offers, "Carry a bomb with you on the plane, since the probability of two bombs on the same flight is one in a million."

3.1 Definitions

Let $\{X_n, n \geq 0\}$ be a DTMC on $S = \{0, 1, 2, \cdots\}$, with transition probability matrix P, and initial distribution a. Let

$$T = \min\{n \geq 0 : X_n = 0\}. \tag{3.1}$$

The random variable T is called the first passage time into state 0, since T is the first time the DTMC "passes into" state 0. Although we shall specifically concentrate on this first passage time, the same techniques can be used to study the first passage time into any set $A \subset S$. (See Conceptual Exercises 3.1 and 3.2.) In this chapter we shall study the following aspects of the first passage time T:

(1) Complementary cdf of T: $\mathsf{P}(T > n)$, $n \geq 0$,

(2) Probability of eventually visiting state 0: $\mathsf{P}(T < \infty)$,

(3) Factorial Moments of T: $\mathsf{E}(T^{(k)})$, $k \geq 1$,
    where $T^{(k)} = T(T-1)\cdots(T-k+1)$,

(4) Generating function of T: $\mathsf{E}(z^T)$.

There are two reasons for studying the first passage times. First, they appear naturally in applications when we are interested in the time until a given event occurs in a stochastic system modeled by a DTMC. Second, we shall see in the next chapter that the quantities u (probability of eventually visiting a state) and $m(1)$ (the expected time to visit a state) play an important part in the study of the limiting behavior of the DTMC. Thus the study of the first passage times has practical as well as theoretical motivation.

Next we introduce the conditional quantities for $i \in S$:

$$
\begin{aligned}
v_i(n) &= \mathsf{P}(T > n | X_0 = i), \\
u_i &= \mathsf{P}(T < \infty | X_0 = i), \\
m_i(k) &= \mathsf{E}(T^{(k)} | X_0 = i), \\
\phi_i(z) &= \mathsf{E}(z^T | X_0 = i).
\end{aligned}
$$

Using the initial distribution we see that

$$
\begin{aligned}
\mathsf{P}(T > n) &= \sum_{i \in S} a_i v_i(n), \\
\mathsf{P}(T < \infty) &= \sum_{i \in S} a_i u_i, \\
\mathsf{E}(T^{(k)}) &= \sum_{i \in S} a_i m_i(k), \\
\mathsf{E}(z^T) &= \sum_{i \in S} a_i \phi_i(z).
\end{aligned}
$$

Thus we can compute the unconditional quantities from the conditional ones. In this chapter we develop a method called the *first-step analysis* to compute the conditional quantities. The method involves computing the conditional quantities by further conditioning on the value of X_1 (i.e., the first step), and then using time-homogeneity and Markov property to derive a set of linear equations for the conditional quantities. These equations can then be solved numerically or algebraically depending on the problem at hand.

We shall see later that sometimes we need to study an alternate first passage time as defined below:

$$
\tilde{T} = \min\{n > 0 : X_n = 0\}. \tag{3.2}
$$

We leave it to the reader to establish the relationship between T and $\tilde{T}$. (See Conceptual Exercises 3.3 and 3.4.)

3.2 Cumulative Distribution Function of T

The following theorem illustrates how the first-step analysis produces a recursive method of computing the cumulative distribution of T. We first introduce the following matrix notation:

$$
v(n) = [v_1(n), \ v_2(n), \ \cdots]^\top, \quad n \geq 0
$$

$$B = [p_{i,j} : i, j \geq 1]. \tag{3.3}$$

Thus B is a submatrix of P obtained by deleting the row and column corresponding to the state 0.

Theorem 3.1

$$v(n) = B^n e, \quad n \geq 0, \tag{3.4}$$

where e is column vector of all ones.

Proof: We prove the result by using the first-step analysis. For $n \geq 1$ and $i \geq 1$ we have

$$
\begin{aligned}
v_i(n) &= \mathsf{P}(T > n | X_0 = i) \\
&= \sum_{j=0}^{\infty} \mathsf{P}(T > n | X_1 = j, X_0 = i) \mathsf{P}(X_1 = j | X_0 = i) \\
&= \sum_{j=0}^{\infty} p_{i,j} \mathsf{P}(T > n | X_1 = j, X_0 = i) \\
&= p_{i,0} \mathsf{P}(T > n | X_1 = 0, X_0 = i) + \sum_{j=1}^{\infty} p_{i,j} \mathsf{P}(T > n | X_1 = j, X_0 = i) \\
&= \sum_{j=1}^{\infty} p_{i,j} \mathsf{P}(T > n | X_1 = j) \\
&= \sum_{j=1}^{\infty} p_{i,j} \mathsf{P}(T > n - 1 | X_0 = j) \\
&= \sum_{j=1}^{\infty} p_{i,j} v_j(n - 1).
\end{aligned}
$$

Here we have used the fact that $X_1 = 0$ implies that $T = 1$ and hence $\mathsf{P}(T > n | X_1 = 0, X_0 = i) = 0$, and the Markov property and time homogeneity implies that the probability of $T > n$ given $X_1 = j, X_0 = i$ is the same as the probability that $T > n - 1$ given $X_0 = j$. Writing the final equation in matrix form yields

$$v(n) = Bv(n - 1), \quad n \geq 1. \tag{3.5}$$

Solving this equation recursively yields

$$v(n) = B^n v(0).$$

Finally, $X_0 = i \geq 1$ implies that $T \geq 1$. Hence $v_i(0) = 1$. Thus

$$v(0) = e. \tag{3.6}$$

This yields Equation 3.4. Note that it is valid for $n = 0$ as well, since $B^0 = I$, the identity matrix, by definition. ∎

Since we have studied the computation of matrix powers in Section 2.6, we have an easy way of computing the complementary cdf of T. We illustrate with several examples.

Example 3.1   Two State DTMC. Consider the two state DTMC of Example 2.3 on page 13. The state-space is $\{1, 2\}$. Let T be the first passage time to state 1. One can use Theorem 3.1 with $B = [\beta]$, or direct probabilistic reasoning, to see that

$$v_2(n) = \mathsf{P}(T > n | X_0 = 2) = \beta^n, \quad n \geq 0.$$

Hence we get

$$\mathsf{P}(T = n | X_0 = 2) = v_2(n-1) - v_2(n) = \beta^{n-1}(1 - \beta), \quad n \geq 1,$$

which shows that T is a geometric random variable with parameter $1 - \beta$. ∎

Example 3.2   Genotype Evolution. Consider the genotype evolution model on page 24 involving a DTMC with state-space $\{1, 2, 3, 4, 5, 6\}$ and transition probability matrix given below:

$$P = \begin{bmatrix} 1 & 0 & 0 & 0 & 0 & 0 \\ 1/4 & 1/2 & 0 & 1/4 & 0 & 0 \\ 0 & 0 & 0 & 1 & 0 & 0 \\ 1/16 & 1/4 & 1/8 & 1/4 & 1/4 & 1/16 \\ 0 & 0 & 0 & 1/4 & 1/2 & 1/4 \\ 0 & 0 & 0 & 0 & 0 & 1 \end{bmatrix}. \tag{3.7}$$

Let

$$T = \min\{n \geq 0 : X_n = 1\}$$

be the first passage time to state 1. Let B be the submatrix of P obtained by deleting rows and columns corresponding to state 1. Then, using $v(n) = [v_2(n), \cdots, v_6(n)]^\top$, Theorem 3.1 yields

$$v(n) = B^n e, \quad n \geq 0.$$

Direct numerical calculation yields, for example,

$$v(5) = B^5 e = [0.4133, \ 0.7373, \ 0.6921, \ 0.8977, \ 1]^\top,$$

and

$$\lim_{n \to \infty} v(n) = [0.25, \ 0.50, \ 0.50, \ 0.75, \ 1]^\top.$$

Thus T is a defective random variable, and the probability that the DTMC will never visit state 1 starting from state 2 is .25, which is the same as saying that the probability of eventually visiting state 1 starting from state 2 is .75. ∎

Example 3.3   Success Runs. Consider the success runs DTMC of Example 2.15 on page 19 with transition probabilities

$$p_{i,0} = q_i, \quad p_{i,i+1} = p_i, \quad i = 0, 1, 2, \cdots.$$

Let T be the first passage time to state 0. Compute the complementary cdf of T starting from state 1.

In this case it is easier to use the special structure of the DTMC. We have $v_1(0) = 1$ and, for $n \geq 1$,

$$\begin{aligned} v_1(n) &= P(T > n | X_0 = 1) \\ &= P(X_1 = 2, X_2 = 3, \cdots, X_n = n+1 | X_0 = 1) \\ &= p_1 p_2 \cdots p_n, \end{aligned}$$

since the only way $T > n$ starting from $X_0 = 1$ is if the DTMC increases by one for each of the next n steps. Since the p_i's can take any value in $[0, 1]$, it follows that we can construct a success runs DTMC such that T has any pre-specified distribution on $\{0, 1, 2, 3, \cdots\}$. ∎

Example 3.4 Consider a particle performing simple random walk on three positions labeled 0,1,2 that are arranged on a circle in a clockwise fashion. At each time $n = 1, 2, 3, \cdots$ the particle moves one step in the clockwise fashion with probability p and one step in the counter-clockwise fashion with probability $q = 1 - p$, starting at position 1 at time 0. Compute the distribution of the time it takes for the particle to visit position 0.

Let X_n be the position of the particle at time n. Then we see that $\{X_n, n \geq 0\}$ is a DTMC on $S = \{0, 1, 2\}$ with transition probability matrix

$$P = \begin{bmatrix} 0 & p & q \\ q & 0 & p \\ p & q & 0 \end{bmatrix}.$$

Let $T = \min\{n \geq 0 : X_n = 0\}$. We are asked to compute the distribution of the first passage time T. Using Theorem 3.1 we get $v(n) = B^n e$, where $v(n) = [v_1(n), v_2(n)]^\top$ and

$$B = \begin{bmatrix} 0 & p \\ q & 0 \end{bmatrix}.$$

Direct calculations show that

$$v_1(2n) = P(T > 2n | X_0 = 1) = p^n q^n, \quad n \geq 0,$$

$$v_1(2n+1) = P(T > 2n + 1 | X_1 = 1) = p^{n+1} q^n, \quad n \geq 0. \quad ∎$$

Example 3.5 Coin Tosses. Suppose a coin is tossed repeatedly and independently. The probability of heads is p and the probability of tails is $q = 1 - p$ on any given toss. Compute the distribution of the number of tosses needed to get two heads in row.

Let $\{X_n, n \geq 0\}$ be a success runs DTMC with transition probabilities

$$p_{i,0} = q, \quad p_{i,i+1} = p, \quad i = 0, 1, 2, \cdots.$$

One can think of X_n as the length of the current run of heads after the nth toss. Let

$$T = \min\{n \geq 0 : X_n = 2\}.$$

Then starting with $X_0 = 0$, T is the number of tosses needed to obtain two heads in a row. Let

$$v(n) = [v_0(n),\ v_1(n)]^\top,\ n \geq 0,$$

and

$$B = \begin{bmatrix} q & p \\ q & 0 \end{bmatrix}.$$

Then we can see that

$$v(n) = B^n e,\ n \geq 0.$$

Using methods of eigenvalues of Section 2.6 we get

$$v_0(n) = \frac{1 - \lambda_2}{\lambda_1 - \lambda_2}\lambda_1^n - \frac{1 - \lambda_1}{\lambda_1 - \lambda_2}\lambda_2^n,\ n \geq 0,$$

where

$$\lambda_1 = \frac{1}{2}(q + \sqrt{q^2 + 4pq}),\ \ \lambda_2 = \frac{1}{2}(q - \sqrt{q^2 + 4pq}).$$

We can obtain the required distribution as follows:

$$\begin{aligned}
&\text{P}(n \text{ tosses are needed to get two heads in a row}) \\
&= \ \text{P}(T = n | X_0 = 0) \\
&= \ v_0(n - 1) - v_0(n) \\
&= \ \frac{(1 - \lambda_1)(1 - \lambda_2)}{\lambda_1 - \lambda_2}[\lambda_1^{n-1} - \lambda_2^{n-1}],\ n \geq 1. \quad \blacksquare
\end{aligned}$$

3.3 Absorption Probabilities

Since $v_i(n)$ is a monotone bounded sequence we know that

$$v_i = \lim_{n \to \infty} v_i(n)$$

exists. The quantity

$$u_i = 1 - v_i = \text{P}(T < \infty | X_0 = i)$$

is the probability that the DTMC eventually visits state 0 starting from state i, also called the *absorption probability* into state 0 starting from state i when $p_{0,0} = 1$. In this section we develop methods of computing u_i or v_i. The main result is given by the next theorem.

Theorem 3.2 *The vector $v = \lim_{n \to \infty} v(n)$ is given by the largest solution to*

$$v = Bv \tag{3.8}$$

such that $v \leq e$, where B is as defined in Equation 3.3, and e is a vector of all ones.

Proof: Equation 3.8 follows by letting $n \to \infty$ on both sides of Equation 3.5 since we know the limit exists. To show that it is the largest solution bounded above by

e, suppose $w \leq e$ is another solution to Equation 3.8. Then, from Equation 3.6, we have

$$v(0) = e \geq w.$$

As an induction hypothesis assume

$$v(k) \geq w$$

for $k = 0, 1, 2, \cdots, n$. Using the fact that $w = Bw$, and that B is a non-negative matrix, we can use Equation 3.5 to get

$$v(n+1) = Bv(n) \geq Bw = w.$$

Thus

$$v(n) \geq w$$

for all $n \geq 0$. Thus letting $n \to \infty$

$$v = \lim_{n \to \infty} v(n) \geq w.$$

This proves the theorem. ∎

Notice that $v = 0$ is always a solution to Equation 3.8. Thus the largest solution is bound to be non-negative. What is more surprising is that a vector v that is componentwise maximum exists at all! We can also derive Equation 3.8 by the first-step analysis. However, that analysis will not show that the required solution is the largest one bounded above by one. How does one find the largest solution in practice? The general method is to identify all the solutions, and then pick the one that is the maximum. In practice, this is not as impractical as it sounds. We explain by several examples below.

Example 3.6 Genotype Evolution. Suppose that the population in the genotype evolution model of Example 3.2 initially consists of one dominant and one hybrid individual. What is the probability that eventually the population will contain only dominant individuals?

Let T be the first passage time as defined in Example 3.2. We are asked to compute $u_2 = P(T < \infty | X_0 = 2) = 1 - v_2$. From Theorem 3.2 we see that

$$v = Bv$$

where $v = [v_2, v_3, \cdots, v_6]^\top$ and

$$B = \begin{bmatrix} 1/2 & 0 & 1/4 & 0 & 0 \\ 0 & 0 & 1 & 0 & 0 \\ 1/4 & 1/8 & 1/4 & 1/4 & 1/16 \\ 0 & 0 & 1/4 & 1/2 & 1/4 \\ 0 & 0 & 0 & 0 & 1 \end{bmatrix}. \tag{3.9}$$

The above equation implies that $v_6 = v_6$, thus there are an infinite number of solutions to $v = Bv$. Since we are looking for the largest solution bounded above by 1, we must choose $v_6 = 1$. Notice that we could have concluded this by observing that $X_0 = 6$ implies that the DTMC can never visit state 1 and hence $T = \infty$. Thus we must have $u_6 = 0$ or $v_6 = 1$. Once v_6 is chosen to be 1, we see that $v = Bv$ has a unique solution given by

$$v = [0.25,\ 0.50,\ 0.50,\ 0.75,\ 1]^\top.$$

Thus the required answer is given by $u_2 = 1 - v_2 = 0.75$. ∎

Example 3.7   Gambler's Ruin. Consider the Gambler's ruin model of Example 2.11 on page 17. The DTMC in that example is a simple random walk on $\{0, 1, 2, \cdots, N\}$ with transition probabilities given by

$$
\begin{aligned}
p_{i,i+1} &= p \text{ for } 1 \le i \le N - 1, \\
p_{i,i-1} &= q = 1 - p \text{ for } 1 \le i \le N - 1, \\
p_{0,0} &= p_{N,N} = 1.
\end{aligned}
$$

Compute the probability that the DTMC eventually visits state 0 starting from state i.

Let $T = \min\{n \ge 0 : X_n = 0\}$. We can write the Equation 3.8 as follows

$$v_i = pv_{i+1} + qv_{i-1}, \quad 1 \le i \le N - 1. \tag{3.10}$$

The above equation can also be derived from first-step analysis, or argued as follows: If the DTMC starts in state i ($1 \le i \le N - 1$), it will jump to state $i + 1$ in one step with probability p, and the probability of never visiting state 0 from then on is v_{i+1}; and it will jump to state $i - 1$ in one step with probability q and the probability of never visiting state 0 from then on will be v_{i-1}. Hence we get Equation 3.10. To complete the argument we must decide appropriate values for v_0 and v_N. Clearly, we must have $v_0 = 0$, since the probability of never visiting state 0 will be 0 if the DTMC starts in state 0. We can also argue that $v_N = 1$, since the DTMC can never visit state 0 if it starts in state N. We also see that Equation 3.8 implies that $v_N = v_N$, and hence we set $v_N = 1$ in order to get the largest solution bounded above by 1.

Equation 3.10 is a difference equation with constant coefficients (with boundary conditions $v_0 = 0$ and $v_N = 1$), and hence using the standard methods of solving such equations (See Appendix I) we get

$$
v_i = \begin{cases} \frac{1 - (q/p)^i}{1 - (q/p)^N} & \text{if } q \ne p \\ \frac{i}{N} & \text{if } q = p. \end{cases} \tag{3.11}
$$

The required probability is $u_i = 1 - v_i$. ∎

Example 3.8   Success Runs. Consider the success runs DTMC of Example 3.3. Compute the probability that the DTMC never visits state 0 starting from state i. Using the derivation in Example 3.3, we get

$$v_i(n) = p_i p_{i+1} \cdots p_{i+n-1}, \quad i \ge 1, \ n \ge 1.$$

Hence

$$v_i = \lim_{n \to \infty} p_i p_{i+1} \cdots p_{i+n-1}, \quad i \ge 1.$$

The above limit is zero if $p_n = 0$ for some $n \geq i$. To avoid this triviality, assume that $p_i > 0$ for all $i \geq 1$. Now it is known that

$$\lim_{n \to \infty} p_1 p_2 \cdots p_n = 0 \Leftrightarrow \sum_{n=1}^{\infty} q_n = \infty,$$

and

$$\lim_{n \to \infty} p_1 p_2 \cdots p_n > 0 \Leftrightarrow \sum_{n=1}^{\infty} q_n < \infty.$$

Hence we see that

$$v_i = 0, \quad \text{i.e., } u_i = 1 \text{ for all } i \Leftrightarrow \sum_{n=1}^{\infty} q_n = \infty,$$

and

$$v_i > 0, \quad \text{i.e., } u_i < 1 \text{ for all } i \Leftrightarrow \sum_{n=1}^{\infty} q_n < \infty.$$

The reader should try to obtain the same results by using Theorem 3.2. ∎

Example 3.9 Simple Space Homogeneous Random Walk. Consider a simple random walk on $\{0, 1, 2, \cdots\}$ with an absorbing barrier at zero. That is, we have

$$\begin{aligned} p_{i,i+1} &= p \text{ for } i \geq 1, \\ p_{i,i-1} &= q = 1 - p \text{ for } i \geq 1, \\ p_{0,0} &= 1. \end{aligned}$$

Let T be the first passage time into state 0. Equation 3.8 for this system, written in scalar form, is as follows:

$$v_i = q v_{i-1} + p v_{i+1}, \quad i \geq 1, \tag{3.12}$$

with $v_0 = 0$. This is a difference equation with constant coefficients. The general solution is given by

$$v_i = \alpha + \beta \left(\frac{q}{p} \right)^i, \quad i \geq 0.$$

Using the condition $v_0 = 0$ we get $\beta = -\alpha$, and thus

$$v_i = \alpha \left(1 - \left(\frac{q}{p} \right)^i \right), \quad i \geq 0. \tag{3.13}$$

Now, if $q \geq p$, the largest solution bounded above by one is obtained by setting $\alpha = 0$; if $q < p$, such a solution is obtained by setting $\alpha = 1$. Thus we get

$$v_i = \begin{cases} 0 & \text{if } q \geq p \\ 1 - \left(\frac{q}{p} \right)^i & \text{if } q < p. \end{cases}$$

If $q \geq p$, the random walk has a drift toward zero (since it goes toward zero with higher probability than away from it), and it makes intuitive sense that the random walk will eventually hit zero with probability 1, and hence $v_i = 0$. On the other hand, if $q < p$, there is a drift away from zero, and there is a positive probability that the random walk will never visit zero. Note that without Theorem 3.2 we would not be able to choose any particular value of α in Equation 3.13. ■

Example 3.10   General Simple Random Walk. Now consider the space-nonhomogeneous simple random walk with the following transition probabilities:

$$
\begin{aligned}
p_{i,i+1} &= p_i \ \text{ for } i \geq 1, \\
p_{i,i-1} &= q_i = 1 - p_i \ \text{ for } i \geq 1, \\
p_{0,0} &= 1.
\end{aligned}
$$

Let T be the first passage time into state 0. Equation 3.8 for this system, written in scalar form, is as follows:

$$
v_i = q_i v_{i-1} + p_i v_{i+1}, \quad i \geq 1, \tag{3.14}
$$

with a single boundary condition $v_0 = 0$. Now let

$$
x_i = v_i - v_{i-1}, \quad i \geq 1.
$$

Using $p_i + q_i = 1$, we can write Equation 3.14 as

$$
(p_i + q_i)v_i = q_i v_{i-1} + p_i v_{i+1}
$$

which yields

$$
q_i x_i = p_i x_{i+1},
$$

or, assuming $p_i > 0$ for all $i \geq 1$,

$$
x_{i+1} = \frac{q_i}{p_i} x_i, \quad i \geq 1.
$$

Solving recursively, we get

$$
x_{i+1} = \alpha_i x_1, \quad i \geq 0, \tag{3.15}
$$

where $\alpha_0 = 1$ and

$$
\alpha_i = \frac{q_1 q_2 \cdots q_i}{p_1 p_2 \cdots p_i}, \quad i \geq 1. \tag{3.16}
$$

Summing Equation 3.15 for $i = 0$ to j, and rearranging, we get

$$
v_{j+1} = \left(\sum_{i=0}^{j} \alpha_i \right) v_1.
$$

Now, Theorem 3.2 says that v_1 must be the largest value such that $0 \leq v_j \leq 1$ for all $j \geq 1$. Hence we must choose

$$
v_1 = \begin{cases} 1/\sum_{i=0}^{\infty} \alpha_i & \text{if } \sum_{i=0}^{\infty} \alpha_i < \infty, \\ 0 & \text{if } \sum_{i=0}^{\infty} \alpha_i = \infty. \end{cases}
$$

Thus the final solution is

$$
v_i = \begin{cases} \sum_{j=0}^{i-1} \alpha_j / \sum_{j=0}^{\infty} \alpha_j & \text{if } \sum_{j=0}^{\infty} \alpha_j < \infty, \\ 0 & \text{if } \sum_{j=0}^{\infty} \alpha_j = \infty. \end{cases} \tag{3.17}
$$

We can compute $u_i = 1 - v_i$ from the above. Notice that visiting state 0 is certain from any state i (that is, $u_i = 1$) if the sum $\sum \alpha_j$ diverges. The reader is urged to verify that this is consistent with the results of Example 3.9. ∎

Example 3.11 Branching Processes. Consider the branching process $\{X_n, n \geq 0\}$ described in Section 2.3.3. Let α_k be the probability that an individual produces k offsprings, and let μ be the expected number of offsprings produced by an individual. In Example 2.22 on page 36 we saw that $E(X_n) = \mu^n$. Here we shall compute the probability that the process eventually becomes extinct, i.e., it hits state 0.

Let T be the first passage time into the state 0, and let $u_i = P(T < \infty | X_0 = i)$ and $u = u_1$. We first argue that $u_i = u^i$, $i \geq 0$. Note that a branching process starting with i individuals can be thought of as i independent branching processes, each starting with one individual. Thus a branching process starting with i individuals will become extinct if and only if each of these i separate processes become extinct, the probability of which is u^i. Hence $u_i = u^i$. Note that this relationship is valid even for $i = 0$. With this we apply the first-step analysis:

$$
\begin{aligned}
u &= P(T < \infty | X_0 = 1) \\
&= \sum_{j=0}^{\infty} P(T < \infty | X_0 = 1, X_1 = j) P(X_1 = j | X_0 = 1) \\
&= \sum_{j=0}^{\infty} \alpha_j P(T < \infty | X_1 = j) \\
&= \sum_{j=0}^{\infty} \alpha_j P(T < \infty | X_0 = j) \\
&= \sum_{j=0}^{\infty} \alpha_j u_j = \sum_{j=0}^{\infty} \alpha_j u^j.
\end{aligned}
$$

Now let

$$
\psi(u) = \sum_{j=0}^{\infty} \alpha_j u^j \tag{3.18}
$$

be the generating function of the offspring size. The absorption probability u satisfies

$$
u = \psi(u). \tag{3.19}
$$

Since $\psi'(u) \geq 0$ and $\psi''(u) \geq 0$ for all $u \geq 0$, ψ is a non-decreasing convex function of $u \in [0, 1]$. Also, $\psi(0) = \alpha_0$ and $\psi(1) = 1$. Clearly, if $\alpha_0 = 0$, each individual produces at least one offspring, and hence the branching process never goes extinct, i.e., $u = 0$. Hence assume that $\alpha_0 > 0$. Figures 3.1 and 3.2 show two possible cases that can arise.

The situation in Figure 3.1 arises if

$$
\psi'(1) = \sum_{j=0}^{\infty} j\alpha_j = \mu > 1,
$$

while that in Figure 3.2 arises if $\mu \leq 1$. In the first case there is a unique value

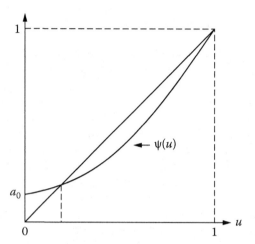

Figure 3.1 *Case 1:* $\mu > 1$.

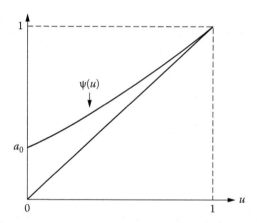

Figure 3.2 *Case 2:* $\mu \leq 1$.

of $u \in (0, 1)$ satisfying $\psi(u) = u$, in the second case the only value of $u \in [0, 1]$ that satisfies $u = \psi(u)$ is $u = 1$. Thus we conclude that extinction is certain if $\mu \leq 1$, while there is a positive probability of the branching process growing without bounds if $\mu > 1$. This is intuitive, except for the critical case of $\mu = 1$. In this case our previous analysis shows that $\mathsf{E}(X_n) = 1$ for all $n \geq 0$, however the branching process eventually becomes extinct with probability one. This seemingly inconsistent result is a manifestation of the fact that convergence in distribution or with probability one does not imply convergence of the means. ∎

We will encounter Equation 3.19 multiple times during the study of DTMCs. Hence it is important to know how to solve it. Except in very special cases (see Computational Exercise 3.34) we need to resort to numerical methods to solve this equation. We describe one such method here.

Let ψ be as defined in Equation 3.18, and define

$$\rho_0 = 0, \ \rho_{n+1} = \psi(\rho_n), \quad n \geq 0.$$

We shall show that if $\mu > 1$, $\rho_n \to u$, the desired solution in $(0,1)$ to Equation 3.19. We have

$$\rho_1 = a_0 > 0 = \rho_0.$$

Now, since ψ is an increasing function, it is clear that

$$\rho_{n+1} \geq \rho_n, \quad n \geq 0.$$

Also, $\rho_n \leq 1$ for all $n \geq 0$. Hence $\{\rho_n, n \geq 0\}$ is a bounded monotone increasing sequence. Hence it has a limit u, and this u satisfies Equation 3.19. The sequence $\{\rho_n\}$ is geometrically illustrated in Figure 3.3.

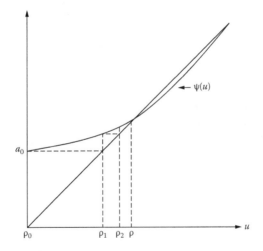

Figure 3.3 *The iteration $\rho_{n+1} = \psi(\rho_n)$.*

Similarly, if $\mu > 1$, one can use the bisection method to find a $\rho_0^* < 1$ such that $\psi(\rho_0^*) < \rho_0^*$. Then recursively define

$$\rho_{n+1}^* = \psi(\rho_n^*), \quad n \geq 0.$$

One can show that

$$\rho_{n+1}^* \leq \rho_n^*, \quad n \geq 0$$

so that the sequence $\{\rho_n^*\}$ monotonically decreases to u. Using these two monotonic sequences we see that if, for a given $\epsilon > 0$, we stop the iterations when $|\rho_n^* - \rho_n| < \epsilon$, then we are guaranteed that $0 \leq u - \rho_n < \epsilon$. Thus we have a simple numerical method of solving Equation 3.19 to any degree of accuracy.

Example 3.12   Production-Inventory System: Batch Production. Consider the DTMC of Example 2.16 on page 20. Compute u_i, the probability that the DTMC eventually visits state 0 starting from state i.

The DTMC has state-space $\{0, 1, 2, \cdots\}$ and transition probability matrix as given in Equation 2.11. First step analysis yields

$$u_i = \sum_{j=0}^{\infty} \alpha_j u_{i+j-1}, \quad i \geq 1, \tag{3.20}$$

with the boundary condition $u_0 = 1$. Since the above equation is a difference equation with constant coefficients, we try a geometric solution $u_i = u^i$, $i \geq 0$. Substituting in the above equation and canceling u^{i-1} from both sides we get

$$u = \sum_{j=0}^{\infty} \alpha_j u^j, \quad i \geq 1.$$

Using the notation $\psi(u)$ of Equation 3.18 we see that this is the same as Equation 3.19. Using the notation $\mu = \sum i \alpha_i$ and the results of Example 3.11, we see that $u = 1$ if and only if $\mu \leq 1$, and $u < 1$ if and only if $\mu > 1$. ∎

Example 3.13   Production-Inventory System: Batch Demands. Consider the DTMC of Example 2.17 on page 21. Compute v_i, the probability that the DTMC never visits state 0 starting from state i.

The DTMC has state-space $\{0, 1, 2, \cdots\}$ and transition probability matrix as given in Equation 2.13. First step analysis yields

$$v_i = \sum_{j=0}^{i} \alpha_j v_{i-j+1}, \quad i \geq 1. \tag{3.21}$$

It is intuitively clear that v_i is an increasing function of i. It can be formally proved by induction, see Conceptual Exercise 3.12. There is no closed form expression for v_i in terms of the α's. Hence we compute the generating function of the v_i's defined as

$$\phi(z) = \sum_{i=1}^{\infty} z^i v_i$$

in terms of the generating function of the α's as defined below

$$\psi(z) = \sum_{i=0}^{\infty} z^i \alpha_i.$$

Multiplying Equation 3.21 by z^i on both sides and summing up from $i = 1$ to ∞ and rearranging, we get

$$
\begin{aligned}
\phi(z) &= \sum_{i=1}^{\infty} z^i v_i \\
&= \sum_{i=1}^{\infty} z^i \sum_{j=0}^{i} \alpha_j v_{i-j+1} \\
&= -\alpha_0 v_1 + \frac{1}{z} \psi(z)\phi(z).
\end{aligned}
$$

This yields

$$
\phi(z) = \frac{\alpha_0 z}{\psi(z) - z} v_1.
$$

Since $\{v_i, i \geq 1\}$ is a monotone bounded sequence, it has a limit. Using the properties of the generating functions (see Appendix D) we get

$$
\begin{aligned}
\lim_{i \to \infty} v_i &= \lim_{z \to 1}(1 - z)\phi(z) \\
&= \lim_{z \to 1} \frac{(1 - z)\alpha_0 z}{\psi(z) - z} v_1.
\end{aligned}
$$

Using L'Hopital's rule we get

$$
\lim_{z \to 1} \frac{(1 - z)\alpha_0 z}{\psi(z) - z} v_1 = \frac{\alpha_0}{1 - \psi'(1)} v_1.
$$

Since we want to find the largest solution to Equation 3.21 that is bounded above by 1, the above limit must be 1. Hence, using

$$
\psi'(1) = \sum_{k=0}^{\infty} k\alpha_k = \mu,
$$

we see that we must choose

$$
v_1 = \begin{cases} \frac{1-\mu}{\alpha_0} & \text{if } \mu < 1,\ \alpha_0 > 0,\ \alpha_0 + \alpha_1 < 1, \\ 0 & \text{if } \mu \geq 1. \end{cases}
$$

The conditions $\mu < 1$, $\alpha_0 > 0$, $\alpha_0 + \alpha_1 < 1$ are necessary and sufficient to ensure that $0 < \frac{1-\mu}{\alpha_0} < 1$. ∎

3.4 Expectation of T

Let $\{X_n, n \geq 0\}$ and T be as defined in Section 3.1. In this section we assume that $u_i = \mathsf{P}(T < \infty | X_0 = i) = 1$ for all $i \geq 1$ and compute the expected time to reach state 0 starting from state i as

$$
m_i = \mathsf{E}(T | X_0 = i), \quad i \geq 1.
$$

Notice that $u_i < 1$ would imply that $m_i = \infty$. Let

$$
m = [m_1,\ m_2,\ m_3,\ \cdots]^{\top},
$$

and B be as defined in Equation 3.3. The main result is given by the following theorem.

Theorem 3.3 *Suppose $u_i = 1$ for all $i \geq 1$. Then m is given by the smallest nonnegative solution to*

$$m = e + Bm, \tag{3.22}$$

where e is a column vector of all ones.

Proof: We shall follow the first-step analysis:

$$
\begin{aligned}
m_i &= E(T|X_0 = i) \\
&= \sum_{j=0}^{\infty} E(T|X_0 = i, X_1 = j)P(X_1 = j|X_0 = i) \\
&= \sum_{j=0}^{\infty} p_{i,j}E(T|X_0 = i, X_1 = j).
\end{aligned}
$$

Now, using

$$E(T|X_0 = i, X_1 = 0) = 1,$$

and

$$E(T|X_0 = i, X_1 = j) = 1 + E(T|X_0 = j) = 1 + m_j, \quad j \geq 1$$

and simplifying, we get

$$m_i = 1 + \sum_{j=1}^{\infty} p_{i,j}m_j, \quad i \geq 1. \tag{3.23}$$

This yields Equation 3.22 in matrix form. The proof that it is the smallest nonnegative solution follows along the same lines as the proof of Theorem 3.2. ∎

Notice that Equation 3.23 can be intuitively understood as follows: If the DTMC starts in state $i \geq 1$, it has to take at least one step before hitting state 0. If after the first step it is in state 0, it does not need anymore steps to reach state zero. On the other hand, if it is in state $j \geq 1$, it will need an additional m_j steps on average to reach state zero. Weighing all these possibilities we get Equation 3.23. We illustrate with several examples below.

Example 3.14 Genotype Evolution. Consider the six-state DTMC in the Genotype Evolution model of Example 3.2 with transition probability matrix given in Equation 3.7. Suppose initially the population consists of one dominant and one hybrid individual. Compute the expected time until the population becomes entirely dominant or entirely recessive.

Let

$$T = \min\{n \geq 0 : X_n = 1 \text{ or } 6\}.$$

We are asked to compute $m_2 = E(T|X_0 = 2)$. It is easy to show that eventually the entire population will be either dominant or recessive, $u_i = 1$ for all i. Now let

$m = [m_2, m_3, m_3, m_4]^\top$, and B be the 4 by 4 submatrix of P obtained by deleting rows and columns corresponding to states 1 and 6. Then, from Theorem 3.3 we get

$$m = e + Bm.$$

Solving numerically, we get

$$m = [4\frac{5}{6}, \ 6\frac{2}{3}, \ 5\frac{2}{3}, \ 4\frac{5}{6}]^\top.$$

Thus, on the average, the population genotype gets fixed in $4\frac{5}{6}$ steps if it starts with one dominant and one hybrid individual. ∎

The next example shows that some questions can be answered by using the first passage times even if the problem does not explicitly involve a DTMC!

Example 3.15 Coin Tossing. Suppose a coin is tossed repeatedly in an independent fashion until k consecutive heads are obtained. Compute the expected number of tosses needed if probability of obtaining a head on any toss is p.

Consider the DTMC $\{X_n, n \geq 0\}$ defined in Example 3.5. Since X_n can be thought of as the length of the current run of heads, it is clear that the answer is given by $m_0 = \mathsf{E}(T|X_0 = 0)$, where

$$T = \min\{n \geq 0 : X_n = k\}.$$

Using the first-step analysis we get

$$m_i = 1 + qm_0 + pm_{i+1}, \ \ 0 \leq i \leq k - 1$$

with $m_k = 0$. Solving recursively we get

$$m_i = \left(\frac{1}{q} + m_0\right)(1 - p^{k-i}), \ \ 0 \leq i \leq k - 1.$$

Setting $i = 0$ we get an equation for m_0, which can be solved to get

$$m_0 = \frac{1}{q}\left(\frac{1}{p^k} - 1\right).$$

The reader should verify that this produces the correct result if we set $k = 1$. ∎

Example 3.16 General Simple Random Walk. Compute the expected time to hit state 0 starting from state $i \geq 1$ in the general simple random walk of Example 3.10.

Let α_i be as given in Equation 3.16. We shall assume that

$$\sum_{r=1}^{\infty} \alpha_r = \infty,$$

so that, from Example 3.10, $u_i = 1$ for all i. The first-step analysis yields

$$m_i = 1 + q_i m_{i-1} + p_i m_{i+1}, \ \ i \geq 1 \tag{3.24}$$

with boundary condition $m_0 = 0$. Now let

$$x_i = m_i - m_{i-1}, \quad i \geq 1.$$

Then Equation 3.24 can be written as

$$q_i x_i = 1 + p_i x_{i+1}, \quad i \geq 1.$$

Solving recursively, we get

$$x_{i+1} = -\alpha_i b_i + \alpha_i m_1, \quad i \geq 0 \qquad (3.25)$$

where

$$b_i = \sum_{j=1}^{i} \frac{1}{p_j \alpha_j},$$

with $b_0 = 0$. Summing Equation 3.25 we get

$$\begin{aligned} m_{i+1} &= x_{i+1} + x_i + \cdots + x_1 \\ &= -\sum_{j=1}^{i} \alpha_j b_j + m_1 \sum_{j=0}^{i} \alpha_j, \quad i \geq 0. \end{aligned}$$

Since m_1 is the smallest non-negative solution we must have

$$m_1 \geq \frac{\sum_{j=1}^{i} \alpha_j b_j}{\sum_{j=0}^{i} \alpha_j} = \sum_{j=1}^{i} \frac{1}{p_j \alpha_j} \frac{\sum_{r=j}^{i} \alpha_r}{\sum_{r=0}^{i} \alpha_r}, \quad i \geq 1. \qquad (3.26)$$

It can be shown that the right-hand side of the above equation is an increasing function of i, and its limit as $i \to \infty$ is $\sum_{j=1}^{\infty} 1/p_j \alpha_j$, since we have assumed $\sum_{r=1}^{\infty} \alpha_r = \infty$. Hence the smallest m_1 to satisfy Equation 3.26 is given by

$$m_1 = \sum_{j=1}^{\infty} \frac{1}{p_j \alpha_j}.$$

Note that m_1 may be finite or infinite. Using the above expression for m_1 we get

$$m_i = \sum_{k=0}^{i-1} \alpha_k \left(\sum_{j=k+1}^{\infty} \frac{1}{p_j \alpha_j} \right). \quad \blacksquare$$

Example 3.17 Production-Inventory System: Batch Production. Consider the DTMC of Example 2.16 on page 20. Compute m_i, the expected time when the DTMC reaches state 0 starting from state i.

Let μ be the mean production batch size as defined in Example 3.12. Assume that $\mu \leq 1$, so that, from the results of Example 3.12, the DTMC reaches state 0 with probability 1 starting from any state. The first-step analysis yields

$$m_i = 1 + \sum_{j=0}^{\infty} \alpha_j m_{i+j-1}, \quad i \geq 1,$$

with the boundary condition $m_0 = 0$. It can be shown that the solution to this set of equations is given by $m_i = im$ for some positive m. Substituting in the above equation we get

$$im = 1 + \sum_{j=0}^{\infty} \alpha_j (i + j - 1)m, \quad i \geq 1.$$

Simplifying, we get

$$(1 - \mu)m = 1.$$

Since m_i is the smallest non-negative solution, we must have

$$m = \begin{cases} \frac{1}{1-\mu} & \text{if } \mu < 1 \\ \infty & \text{if } \mu = 1. \end{cases}$$

Hence

$$m_i = \begin{cases} \frac{i}{1-\mu} & \text{if } \mu < 1 \\ \infty & \text{if } \mu \geq 1. \end{cases} \quad \blacksquare$$

Example 3.18 Production-Inventory System: Batch Demands. Consider the DTMC of Example 2.17 on page 21. Compute m_i, the expected time when the DTMC reaches state 0 starting from state i.

Let μ be the mean demand batch size as defined in Example 3.13. Assume that $\mu \geq 1$, so that, from the results of Example 3.13, the DTMC reaches state 0 with probability 1 starting from any state. First step analysis yields

$$m_i = 1 + \sum_{j=0}^{i} \alpha_j m_{i-j+1}, \quad i \geq 1. \tag{3.27}$$

It is intuitively clear that m_i is an increasing function of i. There is no closed form expression for m_i in terms of the α's. Hence we compute the generating function of the m_i's defined as

$$\phi(z) = \sum_{i=1}^{\infty} z^i m_i$$

in terms of the generating function of the α's as defined below

$$\psi(z) = \sum_{i=0}^{\infty} z^i \alpha_i.$$

Multiplying Equation 3.27 by z^i on both sides and summing up from $i = 1$ to ∞ and rearranging, we get

$$\phi(z) = \sum_{i=1}^{\infty} z^i m_i$$

$$= \sum_{i=1}^{\infty} z^i \left(1 + \sum_{j=0}^{i} \alpha_j m_{i-j+1} \right)$$

$$= \frac{z}{1-z} - \alpha_0 m_1 + \frac{1}{z}\psi(z)\phi(z).$$

This yields

$$\phi(z) = \frac{\alpha_0 m_1 z(1-z) - z^2}{(1-z)(\psi(z) - z)}.$$

One can show that $\phi(z) < \infty$ for $|z| < 1$. Hence if there is a z with $|z| < 1$ for which the denominator on the above equation becomes zero, the numerator must also become zero. From the results of Example 3.11, we see that $\psi(z) - z$ has a solution $z = \alpha$ with $0 < \alpha < 1$ if and only if $\mu = \sum k\alpha_k > 1$. Thus, in this case we must have

$$\alpha_0 m_1 \alpha(1-\alpha) = \alpha^2$$

or

$$m_1 = \frac{\alpha}{\alpha_0(1-\alpha)}.$$

Substituting in $\phi(z)$ we see that

$$\phi(z) = \frac{\frac{\alpha}{1-\alpha}z(1-z) - z^2}{(1-z)(\psi(z) - z)},$$

if $\alpha_0 > 0$ and $\mu > 1$. Clearly, if $\alpha_0 = 0$ or if $\mu \leq 1$ we get $m_1 = \infty$. ∎

3.5 Generating Function and Higher Moments of T

Let $\{X_n, n \geq 0\}$ and T be as defined in Section 3.1. We begin this section with the study of the generating function of T defined as

$$\phi_i(z) = \mathsf{E}(z^T | X_0 = i) = \sum_{n=0}^{\infty} z^n \mathsf{P}(T = n | X_0 = i), \quad i \geq 1.$$

The above generating function is well defined for all complex z with $|z| \leq 1$. Let B be as defined in Equation 3.3 and let

$$b = [p_{1,0}, \ p_{2,0}, \ \cdots]^{\mathsf{T}}.$$

The next theorem gives the main result concerning

$$\phi(z) = [\phi_1(z), \ \phi_2(z), \ \phi_3(z), \ \cdots]^{\mathsf{T}}.$$

Theorem 3.4 *The vector $\phi(z)$ is the smallest solution (for $z \in [0,1]$) to*

$$\phi(z) = zb + zB\phi(z). \tag{3.28}$$

Proof: Using the first-step analysis, we get, for $i \geq 1$,

$$\phi_i(z) = \mathsf{E}(z^T | X_0 = i) = \sum_{j=0}^{\infty} p_{i,j} \mathsf{E}(z^T | X_0 = i, X_1 = j).$$

Now, $X_0 = i, X_1 = 0 \Rightarrow T = 1$. Hence

$$\mathsf{E}(z^T|X_0 = i, X_1 = 0) = z.$$

Also, for $j \geq 1$, $X_0 = i, X_1 = j$ implies that T has the same distribution as $1 + T$ starting from $X_0 = j$. Hence

$$\mathsf{E}(z^T|X_0 = i, X_1 = j) = \mathsf{E}(z^{1+T}|X_0 = j) = z\phi_j(z).$$

Using these two observations we get

$$\phi_i(z) = zp_{i,0} + z\sum_{j=1}^{\infty} p_{i,j}\phi_j(z).$$

The above equation in matrix form yields Equation 3.28. The proof that it is the smallest solution (for $z \in [0, 1]$) follows along the same lines as that of the proof of Theorem 3.2. ∎

Note that

$$\phi_i(1) = \sum_{n=0}^{\infty} \mathsf{P}(T = n|X_0 = i) = \mathsf{P}(T < \infty|X_0 = i) = u_i = 1 - v_i. \qquad (3.29)$$

Using $u = \phi(1)$ in Equation 3.28 we get

$$u = b + Bu, \qquad (3.30)$$

or using $v = e - u$ in the above equation, we get

$$v = Bv$$

which is Theorem 3.2. We have

$$\phi'_i(1) = \sum_{n=0}^{\infty} n\mathsf{P}(T = n|X_0 = i).$$

Thus, if $u_i = 1$, $\phi'_i(1) = \mathsf{E}(T|X_0 = i) = m_i$. Now, taking the derivatives of both sides of Equation 3.28 we get

$$\phi'(z) = b + B\phi(z) + zB\phi'(z).$$

Setting $z = 1$ and recognizing $\phi(1) = u$ the above equation reduces to

$$\phi'(1) = b + Bu + B\phi'(1).$$

Using Equation 3.30 this yields

$$\phi'(1) = u + B\phi'(1).$$

Now assume $u = e$, that is absorption in state 0 is certain from all states. Then, using $\phi'(1) = m$, we get

$$m = e + Bm,$$

which is Equation 3.22. We can similarly compute the higher moments by taking higher derivatives. In particular, if we assume that $u = e$, and denoting the k-th derivative of $\phi(z)$ by $\phi^{(k)}(z)$, we see that

$$\phi^{(k)}(1) = E(T^{(k)}|X_0 = i) = m_i(k), \qquad (3.31)$$

where $T^{(k)} = T(T - 1) \cdots (T - k + 1)$ is the k-th factorial moment of T. Note that $T^{(1)} = T$ and $m(1) = m$. The next theorem gives the equations satisfied by

$$m(k) = [m_1(k), \ m_2(k), \ \cdots]^\top.$$

Theorem 3.5 *Suppose $u = e$. Then $m(k)$ is the smallest positive solution to*

$$m(k) = kBm(k - 1) + Bm(k), \quad k \geq 2.$$

Proof: Follows by successively differentiating Equation 3.28 and using Equation 3.31. The proof of the minimality is similar to the proof of Theorem 3.2. ∎

Example 3.19 Genotype Evolution. Consider the six-state DTMC in the Genotype Evolution model of Example 3.2 with transition probability matrix given in Equation 3.7. Suppose initially the population consists of one dominant and one hybrid individual. Compute the variance of the time until the population becomes entirely dominant or entirely recessive.

Let

$$T = \min\{n \geq 0 : X_n = 1 \text{ or } 6\}.$$

We will first compute $m_i(2) = E(T(T - 1)|X_0 = i)$, $i = 2, 3, 4, 5$. Since we know that eventually the entire population will be either dominant or recessive, $u_i = 1$ for all i. Now let $m(2) = [m_2(2), \ m_3(2), \ m_3(2), \ m_4(2)]^\top$, and B be the 4 by 4 submatrix of P obtained by deleting rows and columns corresponding to states 1 and 6. From Example 3.14, we have

$$m = m(1) = [4\frac{5}{6}, \ 6\frac{2}{3}, \ 5\frac{2}{3}, \ 4\frac{5}{6}]^\top.$$

Theorem 3.5 implies that

$$m(2) = 2Bm(1) + Bm(2).$$

Numerical calculations yield

$$m(2) = [39\frac{8}{9}, \ 60\frac{4}{9}, \ 49\frac{1}{9}, \ 39\frac{8}{9}]^\top.$$

Hence the required answer is

$$\text{Var}(T|X_0 = 2) = m_2(2) + m_2(1) - m_2(1)^2 = 22\frac{2}{3}. \ \blacksquare$$

3.6 Computational Exercises

3.1 Let $\{X_n, n \geq 0\}$ be a DTMC on state space $\{0, 1, 2, 3\}$ with the following transition probability matrix:

$$\begin{bmatrix} 0.2 & 0.1 & 0.0 & 0.7 \\ 0.1 & 0.3 & 0.6 & 0.0 \\ 0.0 & 0.4 & 0.2 & 0.4 \\ 0.7 & 0.0 & 0.1 & 0.2 \end{bmatrix}.$$

Let $T = \min\{n \geq 0 : X_n = 0\}$. Compute

1. $P(T \geq 3 | X_0 = 1)$,
2. $E(T | X_0 = 1)$,
3. $Var(T | X_0 = 1)$.

3.2 Do the above problem with the following transition probability matrix:

$$\begin{bmatrix} 0.0 & 1.0 & 0.0 & 0.0 \\ 0.2 & 0.0 & 0.8 & 0.0 \\ 0.0 & 0.8 & 0.0 & 0.2 \\ 0.0 & 0.0 & 1.0 & 0.0 \end{bmatrix}.$$

3.3 Consider a 4 by 4 chutes and ladders game as shown in Figure 3.4. A player

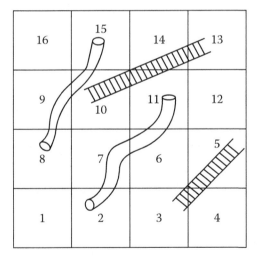

Figure 3.4 *The 4 by 4 chutes and ladders.*

starts on square one. He tosses a six-sided fair die and moves 1 square if the die shows 1 or 2, 2 squares if it shows 3 or 4, and 3 squares if it shows 5 or 6. Toward the end of the play, when the player is near square 16, he has to land on square 16 exactly in order to finish the game. If he overshoots 16, he has to toss again. (In a two person game he loses a turn.) Compute the expected number of tosses needed to finish the game. (You may need to use a computer.)

3.4 If two players are playing the game in Computational Exercise 3.3, compute the probability distribution of the time until the game ends (that is, when one of the two players lands on 16.) Assume that there is no interaction among the two players, except that they take turns. Compute the mean time until the game terminates.

3.5 Compute the expected number of times the ladder from 3 to 5 is used by a player during one game of chutes and ladders as described in Computational Exercise 3.3.

3.6 Develop a computer program that analyzes a general chutes and ladders game. The program accepts the following input: n, size of the board (n by n); k, the number of players; the placements of the chutes and the ladders, and the distribution of the number of squares moved in one turn. The program produces the following output:

1. Distribution of the number of tosses needed for one player to finish the game.
2. Distribution of the time to complete the game played by k players,
3. Expected length of the game played by k players.

3.7 Consider a maze of nine cells as shown in Figure 3.5. At time 0 a rat is placed

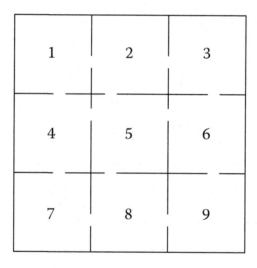

Figure 3.5 *The nine cell maze.*

in cell one, and food in cell 9. The rat stays in the current cell for one unit of time and then chooses one of the doors in the cell at random and moves to an adjacent cell. Its successive moves are independent and completely uninfluenced by the food. Compute the expected time required for the rat to reach the food.

3.8 Suppose the food in the above problem is replaced by a cat that moves like the rat, but in an independent fashion. Of course, when the cat and the rat occupy the same cell, the cat promptly eats the rat. Compute the expected time when the cat gets its meal.

3.9 Let $\{X_n, n \geq 0\}$ be a DTMC on state-space $\{0, 1, 2, \cdots\}$ with the following transition probabilities

$$p_{i,j} = \begin{cases} \frac{1}{i+2} & 0 \leq j \leq i+1, \ i \geq 0 \\ 0 & \text{otherwise.} \end{cases}$$

Let $T = \min\{n \geq 0 : X_n = 0\}$. Compute $E(T|X_0 = 1)$.

3.10 Consider the Moran model on page 25 with N genes in the population, i of which are initially dominant ($1 \leq i \leq N$). Compute the expected time until all the genes are dominant or all the genes are recessive.

3.11 Suppose a coin is tossed repeatedly and independently, the probability of observing a head on any toss being p. Compute the probability that a string of r consecutive heads is observed before a string of m consecutive tails.

3.12 In Computational Exercise 3.10 compute the probability that eventually all the genes become dominant.

3.13 Let $\{Z_n, n \geq 1\}$ be a sequence of iid random variables with common pmf $p_Z(1) = p_Z(2) = p_Z(3) = p_Z(4) = .25$. Let $X_0 = 0$ and $X_n = Z_1 + Z_2 + \cdots + Z_n$, $n \geq 1$. Define $T = \min\{n > 0 : X_n \text{ is divisible by } 7\}$. Note the strict inequality in $n > 0$. Compute $E(T|X_0 = 0)$.

3.14 Consider the binomial model of the stock fluctuation as described on page 28. Suppose $X_0 = 1$, and an individual owns one share of a stock at time 0. Suppose the individual has decided to sell the stock as soon as it reaches a value of 2 or more. Compute the expected time when such a sale would occur, assuming that $p = .5, u = .2$ and $d = .1$. Hint: It might be easier to deal with log of the stock price.

3.15 In the coin tossing experiment of Computational Exercise 3.11 compute the expected number of tosses needed to observe the sequence $HHTT$ for the first time.

3.16 Solve Computational Exercise 3.14 if the individual decides to sell the stock when it goes above 2 or below .7.

3.17 Let $\{X_n, n \geq 0\}$ be a simple random walk on $\{0, 1, \cdots, N\}$ with absorbing barriers at 0 and N. That is, $p_{0,0} = p_{N,N} = 1$, and $p_{i,i+1} = p, p_{i,i-1} = q$ for $1 \leq i \leq N - 1$. Let $T = \min\{n \geq 0 : X_n = 0 \text{ or } N\}$. Show that

$$E(T|X_0 = i) = \frac{i}{q-p} - \frac{N}{q-p} \cdot \frac{1 - (q/p)^i}{1 - (q/p)^N}$$

if $q \neq p$. What is the corresponding formula if $p = q$?

3.18 Suppose a DNA sequence is appropriately modeled by a 4-state DTMC with transition probability matrix $P(1)$ given on page 22. Suppose the first base is A. Compute the expected length of the sequence until we see the triplet ACT.

3.19 Let $\{X_n, n \geq 0\}$ be a simple random walk on $\{0, 1, 2, \cdots\}$ with $p_{0,0} = 1$ and $p_{i,i+1} = p, p_{i,i-1} = q$ for $i \geq 1$. Let $T = \min\{n \geq 0 : X_n = 0\}$. Show that, for $|z| \leq 1$,

$$E(z^T | X_0 = i) = \phi(z)^i$$

where

$$\phi(z) = (1 - \sqrt{1 - 4pqz^2})/2pz.$$

Give a probabilistic interpretation of $\phi(z)^i$ in terms of the generating functions of the convolutions of iid random variables.

3.20 Compute the expected number of bases between two consecutive appearances of the triplet CAG in the DNA sequence in Computational Exercise 3.18.

3.21 Let $\{X_n, n \geq 0\}$ be the DTMC of Example 2.16. Let $T = \min\{n \geq 0 : X_n = 0\}$. Show that

$$E(z^T | X_0 = i) = \phi(z)^i, \quad i \geq 0$$

where $\phi(z)$ is given by the smallest solution to

$$\phi(z) = z \sum_{i=0}^{\infty} \alpha_i \phi(z)^i.$$

3.22 In Computational Exercise 3.18 compute the probability that we see the triplet ACT before the triplet GCT.

3.23 In the Gambler's Ruin model of Example 2.11 on page 17 compute the expected number of bets won by the player A until the game terminates, assuming that the player A starts the game with i dollars, $1 \leq i \leq N - 1$.

3.24 Consider the discrete time queue with Bernoulli arrivals and departures as described in Example 2.12 on page 17. Suppose the queue has i customers in it initially. Compute the probability that the system will eventually become empty.

3.25 Consider the clinical trials of Example 2.5 on page 14 using the play the winner rule. Suppose the trial stops as soon as either drug produces r successes, with the drug producing the r successes first being declared the superior drug. Compute the probability that the better of the two drugs gets declared as the best.

3.26 Derive a recursive method to compute the expected number of patients needed to successfully conclude the clinical trial of the Computational Exercise 3.25.

3.27 Compute numerically the extinction probability of a branching process $\{X_n, n \geq 0\}$ with $X_0 = 1$ if each individual produces 20 offspring with probability 0.8 and 0 offspring with probability 0.2.

3.28 Consider a branching process $\{X_n, n \geq 0\}$ with $X_0 = 1$ and $E(X_1 | X_0 = 1) = \mu \leq 1$. Let N be the total number of individuals that ever live in this colony until the colony becomes extinct (which it does with probability 1, see Example 3.11

on page 67). N is called the total progeny and is given by $N = \sum_{n=0}^{\infty} X_n$. Let $\psi(z) = \mathsf{E}(z^{X_1}|X_0 = 1)$ and $\phi(z) = \mathsf{E}(z^N|X_0 = 1)$. Show that

$$\phi(z) = z\psi(\phi(z)).$$

Hence show that

$$\mathsf{E}(N|X_0 = 1) = \frac{1}{1 - \mu}.$$

Thus when $\mu = 1$, the process becomes extinct with probability one, but the expected total progeny is infinite!

3.29 Consider the DTMC of Modeling Exercise 2.17 with

$$\alpha_k = \frac{1}{M + 1}, \quad 0 \le k \le M.$$

Compute the expected time until the DTMC reaches the absorbing state M.

3.30 A clinical trial is designed to identify the better of two experimental treatments. The trial consists of several stages. At each stage two new patients are selected randomly from a common pool of patients and one is given treatment 1, and the other is given treatment 2. The stage is complete when the result of each treatment is known as a success or a failure. At the end of the nth stage we record X_n = the number of successes under treatment 1 minus those on treatment 2 observed on all the stages so far. The trial stops as soon as X_n reaches $+k$ or $-k$, where k is a given positive integer. If the trial stops with $X_n = k$, treatment 1 is declared to be better than 2, else treatment 2 is declared to be better than 1. Now suppose the probability of success of the ith treatment is p_i, $i = 1, 2$. Suppose $p_1 > p_2$. Assume that the results of the successive stages are independent. Compute the probability that the clinical trial reaches correct decision, as a function of p_1, p_2, and k.

3.31 In Computational Problem 3.30 compute the expected number of patients subjected to treatments during the entire clinical trial.

3.32 Consider the manufacturing set up described in Modeling Exercise 2.32. Suppose initially both bins are empty. Compute the expected time (in hours) until one of the machines is turned off.

3.33 Consider the two bar town of Modeling Exercise 2.11. Assume that the two transition probability matrices are identical. Compute the expected time when the boy meets the girl.

3.34 Solve Equation 3.19 for the special case when $\alpha_i = 0$ for $i \ge 4$.

3.35 Let $N \ge 2$ be a fixed positive integer. Suppose $2N$ nodes, labeled 1,2,...,$2N$, are arranged clockwise on a circle. A cat starts on node 1 and at each time period moves one step clockwise with probability p or one step counterclockwise with probability $q = 1 - p$. At the same time a mouse starts on node 3 and performs a similar random walk, independent of the cat. When the cat and the mouse land on the same node, the cat gets a nice meal. Compute the expected time when the cat catches the mouse.

3.36 Suppose a politically active student sends an email to five of her friends asking them to go to a web site and sign a petition. She tells them that she herself has done so. The email also exhorts the recipient to send the email to their friends and ask them to do the same. Suppose each recipient belongs to four categories:

1. ignores the email altogether (probability $a > 0$),
2. signs the petition but does not forward the email (probability $b > 0$),
3. does not sign the petition but sends the email out to a random number of friends asking them to do the same (probability $c > 0$),
4. signs the petition and sends the email out to a random number of friends asking them to do the same (probability $1 - a - b - c > 0$).

We call the student initiating this chain as the stage 0 person, and the first five students she contacted as stage one recipients. Assume that the stage $n + 1$ recipients contacted by a given stage n recipient ($n \geq 1$) who sends out emails are iid with mean μ, and disjoint from each other. All recipients behave identically and independently. This chain-email scheme can continue forever, or it can die out if all recipients at a given stage belong to category 1 or 2.

1. Compute the probability that the chain mail eventually dies out.
2. Given that the chain mail dies out with probability one, compute the expected number of signatures collected by the web site in the limit.

3.7 Conceptual Exercises

3.1 Let $\{X_n, n \geq 0\}$ be a DTMC on state-space $S = \{0, 1, 2, \cdots\}$ and transition probability matrix P. Let $A \subset S$ be a non-empty and define
$$T(A) = \min\{n \geq 0 : X_n \in A\}.$$
Show that
$$v_i(A) = P(T(A) = \infty | X_0 = i)$$
are given by the largest solution bounded above by 1 to the following equations:
$$v_i(A) = \sum_{j \notin A} p_{i,j} v_j(A), \quad i \notin A.$$
Hint: Use first-step analysis and follow the proof of Theorem 3.2.

3.2 For Conceptual Exercise 3.1 derive an analog of Theorem 3.3 for
$$m_i(A) = E(T(A) | X_0 = i).$$

3.3 Let $\{X_n, n \geq 0\}$ be a DTMC on state-space $S = \{0, 1, 2, \cdots\}$ and transition probability matrix P. Let

$$\tilde{T} = \min\{n > 0 : X_n = 0\}.$$

Show how to compute

$$\tilde{v}_i = \mathsf{P}(\tilde{T} = \infty | X_0 = i), \quad i \geq 0,$$

in terms of v_i, $i \geq 1$, as defined in Section 3.3.

3.4 For Conceptual Exercise 3.3 show how to compute

$$\tilde{m}_i = \mathsf{E}(\tilde{T} | X_0 = i), \quad i \geq 0,$$

in terms of m_i, $i \geq 1$, as defined in Section 3.4.

3.5 Let T be as in Equation 3.1, and define

$$m_{i,n} = \sum_{k=0}^{n} k \mathsf{P}(T = k | X_0 = i)$$

and $u_{i,n} = \mathsf{P}(T \leq n | X_0 = i)$. Using the first-step analysis show that

$$u_{i,n+1} = p_{i,0} + \sum_{j=1}^{\infty} p_{i,j} u_{j,n}, \quad i > 0, \ n \geq 0,$$

and

$$m_{i,n+1} = u_{i,n+1} + \sum_{j=1}^{\infty} p_{i,j} m_{j,n}, \quad i > 0, \ n \geq 0.$$

3.6 Let $\{X_n, n \geq 0\}$ be a DTMC on state-space $S = \{0, 1, 2, ..., N\}$ with transition probability matrix P. Let A be a non-empty subset of the state-space, and $T(A)$ be the time it takes the DTMC to visit all the states in A at least once. Thus when $A = \{k\}$, $T(A)$ is the usual first passage time to the state k.

1. Develop simultaneous equations for

$$\mu_j(k) = E(T(\{k\}) | X_0 = j), \quad j, k \in S.$$

2. Using the quantities $\mu_0(N)$ and $\mu_N(0)$ as known, develop simultaneous equations for

$$\mu_i(0, N) = E(T(\{0, N\}) | X_0 = i), \quad i \in S.$$

3.7 Let A be a fixed non-empty subset of the state-space $S = \{0, 1, 2, \cdots\}$. Derive a set of equations to compute the probability that the DTMC visits every state in A before visiting state 0, assuming $0 \notin A$.

3.8 Let $\{X_n, n \geq 0\}$ be a DTMC with state-space $\{0, 1, \cdots\}$, and transition probability matrix P. Let $w_{i,j}$ $(i, j \geq 1)$ be the expected number of visits to state j starting from state i before hitting state 0, counting the visit at time 0 if $i = j$. Using the first-step analysis show that

$$w_{i,j} = \delta_{i,j} + \sum_{k=1}^{\infty} p_{i,k} w_{k,j}, \quad i \geq 1$$

where $\delta_{i,j} = 1$ if $i = j$, and zero otherwise.

3.9 Let $\{X_n, n \geq 0\}$ be a DTMC with state-space $\{0, 1, \cdots\}$, and transition probability matrix P. Let $T = \min\{n \geq 0 : X_n = 0\}$. A state j is called a gateway state to state 0 if $X_{T-1} = j$. Derive equations to compute the probability $w_{i,j}$ that state j is a gateway state to state 0 if the DTMC starts in state i, $(i, j \geq 1)$.

3.10 Let $\{X_n, n \geq 0\}$ be a DTMC with state-space $\{0, 1, \cdots\}$, and transition probability matrix P. Let $(i_0, i_1, \cdots, i_k)$ be a sequence of states in the state-space. Define $T = \min\{n \geq 0 : X_n = i_0, X_{n+1} = i_1, \cdots, X_{n+k} = i_k\}$. Derive a method to compute $\mathsf{E}(T|X_0 = i)$.

3.11 Let $\{X_n, n \geq 0\}$ be a DTMC with state-space $\{0, 1, \cdots\}$, and transition probability matrix P. Let $T = \min\{n \geq 0 : X_n = 0\}$. Let M be the largest state visited by the DTMC until hits state 0. Derive a method to compute the distribution of M.

3.12 Show by induction that the sequence $\{v_i, i \geq 0\}$ satisfying Equation 3.21 is an increasing sequence.

CHAPTER 4

Discrete-Time Markov Chains: Limiting Behavior

"God does not play dice with the universe."
– Albert Einstein

"God not only plays dice, He also sometimes throws the dice where they cannot be seen."
– Stephen Hawking

4.1 Exploring the Limiting Behavior by Examples

Let $\{X_n, n \geq 0\}$ be a DTMC on $S = \{0, 1, 2, \cdots\}$ with transition probability matrix P. In Chapter 2 we studied two main aspects of the transient behavior of the DTMC: the n-step transition probability matrix $P^{(n)}$ and the occupancy matrix $M^{(n)}$. Theorem 2.4 showed that

$$P^{(n)} = P^n, \ n \geq 0,$$

and Theorem 2.6 showed that

$$M^{(n)} = \sum_{r=0}^{n} P^r, \ n \geq 0. \tag{4.1}$$

In this chapter we study the limiting behavior of $P^{(n)}$ as $n \to \infty$. Since the row sums of $M^{(n)}$ are $n+1$, we study the limiting behavior of $M^{(n)}/(n+1)$ as $n \to \infty$. Note that $[M^{(n)}]_{i,j}/(n+1)$ can be interpreted as the fraction of the time spent by the DTMC in state j starting from state i during $\{0, 1, \cdots, n\}$. Hence studying this limit makes practical sense. We begin with some examples illustrating the types of limiting behavior that can arise.

Example 4.1 Two-State Example. Let

$$P = \left[\begin{array}{cc} \alpha & 1 - \alpha \\ 1 - \beta & \beta \end{array} \right]$$

be the transition probability matrix of the two-state DTMC of Example 2.3 on page 13 with $\alpha + \beta < 2$. From Example 2.18 on page 34 we get

$$P^n = \frac{1}{2 - \alpha - \beta} \begin{bmatrix} 1 - \beta & 1 - \alpha \\ 1 - \beta & 1 - \alpha \end{bmatrix} + \frac{(\alpha + \beta - 1)^n}{2 - \alpha - \beta} \begin{bmatrix} 1 - \alpha & \alpha - 1 \\ \beta - 1 & 1 - \beta \end{bmatrix}, \quad n \geq 0.$$

Hence we get

$$\lim_{n \to \infty} P^n = \frac{1}{2 - \alpha - \beta} \begin{bmatrix} 1 - \beta & 1 - \alpha \\ 1 - \beta & 1 - \alpha \end{bmatrix}.$$

Thus the limit of $P^{(n)}$ exists and its row sums are one. It has an interesting feature in that both the rows of the limiting matrix are the same. This implies that the limiting distribution of X_n does not depend upon the initial distribution of the DTMC. We shall see that a large class of DTMCs share this feature. From Example 2.24 on page 39 we get, for $n \geq 0$,

$$M^{(n)} = \frac{n + 1}{2 - \alpha - \beta} \begin{bmatrix} 1 - \beta & 1 - \alpha \\ 1 - \beta & 1 - \alpha \end{bmatrix} + \frac{1 - (\alpha + \beta - 1)^{(n+1)}}{(2 - \alpha - \beta)^2} \begin{bmatrix} 1 - \alpha & \alpha - 1 \\ \beta - 1 & 1 - \beta \end{bmatrix}.$$

Hence, we get

$$\lim_{n \to \infty} \frac{M^{(n)}}{n + 1} = \frac{1}{2 - \alpha - \beta} \begin{bmatrix} 1 - \beta & 1 - \alpha \\ 1 - \beta & 1 - \alpha \end{bmatrix}.$$

Thus, curiously, the limit of $M^{(n)}/(n + 1)$ in this example is the same as that of $P^{(n)}$. This feature is also shared by a large class of DTMCs. We will identify this class in this chapter. ∎

Example 4.2   Three-State DTMC. Let

$$P = \begin{bmatrix} 0 & 1 & 0 \\ q & 0 & p \\ 0 & 1 & 0 \end{bmatrix},$$

where $0 < p < 1$ and $q = 1 - p$. We saw in Example 2.26 on page 42 that

$$P^{(2n)} = \begin{bmatrix} q & 0 & p \\ 0 & 1 & 0 \\ q & 0 & p \end{bmatrix}, \quad n \geq 1, \tag{4.2}$$

and

$$P^{(2n+1)} = \begin{bmatrix} 0 & 1 & 0 \\ q & 0 & p \\ 0 & 1 & 0 \end{bmatrix}, \quad n \geq 0. \tag{4.3}$$

Since the sequence $\{P^{(n)}, n \geq 1\}$ shows an oscillatory behavior, it is clear that the limit of $P^{(n)}$ does not exist. Now, direct calculations show that

$$M^{(2n)} = \begin{bmatrix} 1 + nq & n & np \\ nq & 1 + n & np \\ nq & n & 1 + np \end{bmatrix}, \quad n \geq 0, \tag{4.4}$$

and

$$M^{(2n+1)} = \begin{bmatrix} 1 + nq & 1 + n & np \\ (n+1)q & 1 + n & (n+1)p \\ nq & 1 + n & 1 + np \end{bmatrix}, \quad n \geq 0. \tag{4.5}$$

Hence

$$\lim_{n \to \infty} \frac{M^{(n)}}{n+1} = \begin{bmatrix} q/2 & 1/2 & p/2 \\ q/2 & 1/2 & p/2 \\ q/2 & 1/2 & p/2 \end{bmatrix}. \tag{4.6}$$

Thus, even if $M^{(n)}$ shows periodic behavior, $M^{(n)}/(n+1)$ has a limit, and all its rows are identical and add up to one! ∎

Example 4.3 Genotype Evolution Model. Consider the six-state DTMC of the genotype evolution model with the transition probability matrix as given in Equation 2.14 on page 24. Direct numerical calculations show that

$$\lim_{n \to \infty} P^{(n)} = \lim_{n \to \infty} \frac{M^{(n)}}{n+1} = \begin{bmatrix} 1 & 0 & 0 & 0 & 0 & 0 \\ 3/4 & 0 & 0 & 0 & 0 & 1/4 \\ 1/2 & 0 & 0 & 0 & 0 & 1/2 \\ 1/2 & 0 & 0 & 0 & 0 & 1/2 \\ 1/4 & 0 & 0 & 0 & 0 & 3/4 \\ 0 & 0 & 0 & 0 & 0 & 1 \end{bmatrix}. \tag{4.7}$$

Thus this example also shows that the limit of $P^{(n)}$ exists and is the same as the limit of $M^{(n)}/(n+1)$, and the row sums of the limiting matrix are one. However, in this example all the rows of the limiting matrix are *not* identical. Thus the limiting distribution of X_n exists, but depends upon the initial distribution of the DTMC. We will identify the class of DTMCs with this feature later on in this chapter. ∎

Example 4.4 Simple Random Walk. A simple random walk on all integers has the following transition probabilities

$$p_{i,i+1} = p, \quad p_{i,i-1} = q = 1 - p, \quad -\infty < i < \infty,$$

where $0 < p < 1$. We have seen in Example 2.19 on page 34 that

$$p_{0,0}^{(2n)} = \frac{(2n)!}{n!n!} p^n q^n, \quad n = 0, 1, 2, \cdots. \tag{4.8}$$

We show that the above quantity converges to zero as $n \to \infty$. We need to use the following asymptotic expression called Stirling's formula:

$$n! \sim \sqrt{2\pi} n^{n+1/2} e^{-n}$$

where $\sim$ indicates that the ratio of the two sides goes to one as $n \to \infty$. We have

$$\begin{aligned} p_{0,0}^{(2n)} &= \frac{(2n)!}{n!n!} p^n q^n \\ &\sim \frac{\sqrt{2\pi} e^{-2n} (2n)^{2n+1/2}}{(\sqrt{2\pi} e^{-n} n^{n+1/2})^2} p^n q^n \end{aligned}$$

$$= \frac{(4pq)^n}{\sqrt{\pi n}}$$

$$\leq \frac{1}{\sqrt{\pi n}} \tag{4.9}$$

where the last inequality follows because $4pq = 4p(1-p) \leq 1$ for $0 \leq p \leq 1$. Thus $p_{0,0}^{(2n)}$ approaches zero asymptotically. We also know that $p_{0,0}^{2n+1}$ is always zero. Thus we see that

$$\lim_{n \to \infty} p_{0,0}^{(n)} = 0.$$

Similar calculation shows that

$$\lim_{n \to \infty} [P^{(n)}]_{i,j} = 0$$

for all i and j. It is even more tedious, but possible, to show that

$$\lim_{n \to \infty} \frac{[M^{(n)}]_{i,j}}{n+1} = 0.$$

Thus in this example the two limits again coincide, but the row sums of the limiting matrix are not one, but zero! Since $P^{(n)}$ is a stochastic matrix, we have

$$\sum_{j=-\infty}^{\infty} p_{i,j}^{(n)} = 1, \quad n \geq 0.$$

Hence we must have

$$\lim_{n \to \infty} \sum_{j=-\infty}^{\infty} p_{i,j}^{(n)} = 1.$$

However, the above calculations show that

$$\sum_{j=-\infty}^{\infty} \lim_{n \to \infty} p_{i,j}^{(n)} = 0.$$

This implies that the interchange of limit and the sum in this infinite state-space DTMC is not allowed. We shall identify the class of DTMCs with this feature later in this chapter. ∎

Thus we have identified four cases:

Case 1: Limit of $P^{(n)}$ exists, has identical rows, and each row sums to one.

Case 2: Limit of $P^{(n)}$ exists, does not have identical rows, each row sums to one.

Case 3: Limit of $P^{(n)}$ exists, but the rows may not sum to one.

Case 4: Limit of $P^{(n)}$ does not exist.

We have also observed that limit of $M^{(n)}/(n+1)$ always exists, and it equals the limit of $P^{(n)}$ when it exists. We develop the necessary theory to help us classify the DTMCs so we can understand their limiting behavior better.

4.2 Classification of States

In this section we introduce the concepts of *communicating class, irreducibility, transience, recurrence, null recurrence, positive recurrence* and *periodicity*. These concepts will play an important role in the study of the limiting behavior of a DTMC. We will use these concepts to classify the states of a DTMC, that is,

1. identify the communicating classes in the state-space,
2. for each class specify whether it is transient, null recurrent, or positive recurrent, and
3. for the recurrent classes find the period and state if the class is periodic or aperiodic.

The meaning of the above statements will become clear by the end of this section.

4.2.1 Irreducibility

We begin with a series of definitions.

Definition 4.1   Accessibility. *A state j is said to be accessible from a state i if there is an $n \geq 0$ such that $p_{i,j}^{(n)} > 0$.*

Note that j is accessible from i if and only if there is a directed path from node i to node j in the transition diagram of the DTMC. This is so, since $p_{i,j}^{(n)} > 0$ implies that there is a sequence of states $i = i_0, i_1, \cdots, i_n = j$ such that $p_{i_k, i_{k+1}} > 0$ for $k = 0, 1, \cdots, n - 1$. This in turn is equivalent to the existence of a directed path $i = i_0, i_1, \cdots, i_n = j$ in the transition diagram.

We write $i \to j$ if state j is accessible from i. Since $p_{i,i}^{(0)} = 1$, it trivially follows that $i \to i$.

Definition 4.2   Communication. *States i and j are said to communicate if $i \to j$ and $j \to i$.*

If i and j communicate, we write $i \leftrightarrow j$. The following theorem states some important properties of the relation "communication."

Theorem 4.1   Properties of Communication.

(i) $i \leftrightarrow i$, (reflexivity).

(ii) $i \leftrightarrow j \Leftrightarrow j \leftrightarrow i$, (symmetry).

(iii) $i \leftrightarrow j, j \leftrightarrow k \Rightarrow i \leftrightarrow k$, (transitivity).

Proof: (*i*) and (*ii*) are obvious from the definition. To prove (*iii*) note that $i \to j$ and $j \to k$ implies that there are integers $n \geq 0$ and $m \geq 0$ such that

$$p_{i,j}^{(n)} > 0, \quad p_{j,k}^{(m)} > 0.$$

Hence

$$p_{i,k}^{(n+m)} \quad = \quad \sum_{r \in S} p_{i,r}^{(n)} p_{r,k}^{(m)} \quad \text{(Theorem 2.3)}$$

$$\geq \quad p_{i,j}^{(n)} p_{j,k}^{(m)} > 0.$$

Hence $i \to k$. Similarly $k \to j$ and $j \to i$ implies that $k \to i$. Thus

$$i \leftrightarrow j, \; j \leftrightarrow k \Rightarrow i \to j, \; j \to k \text{ and } k \to j, \; j \to i \Rightarrow i \to k, \; k \to i \Rightarrow i \leftrightarrow k. \quad \blacksquare$$

It is possible to write a computer program that can check the accessibility of one state from another in $O(q)$ time where q is the number of non-zero entries in P. The same algorithm can be used to check if two states communicate. See Conceptual Exercise 4.3.

From the above theorem it is clear that "communication" is a reflexive, symmetric, and transitive binary relation. (See Conceptual Exercise 4.1 for more examples of binary relations.) Hence we can use it to partition the state-space S into subsets known as communicating classes.

Definition 4.3 Communicating Class. *A set $C \subset S$ is said to be a communicating class if*

(*i*) $i \in C, \; j \in C \Rightarrow i \leftrightarrow j$,

(*ii*) $i \in C, \; i \leftrightarrow j \Rightarrow j \in C$.

Property (*i*) assures that any two states in a communicating class communicate with each other (hence the name). Property (*ii*) forces C to be a maximal set, that is, no strict superset of C can be a communicating class. Note that it is possible to have a state j outside C that is accessible from a state i inside C, but in this case, i cannot be accessible from j. Similarly, it is possible to have a state i inside C that is accessible from a state j outside C, but in this case, j cannot be accessible from i. This motivates the following definition.

Definition 4.4 Closed Communicating Class. *A communicating class is said to be closed if $i \in C$ and $j \notin C$ implies that j is not accessible from i.*

Note that once a DTMC visits a state in a closed communicating class C, it cannot leave it, that is,

$$X_n \in C \Rightarrow X_m \in C \text{ for all } m \geq n.$$

Two distinct communicating classes must be disjoint, see Conceptual Exercise 4.2. Thus we can uniquely partition the state-space S as follows:

$$S = C_1 \cup C_2 \cup \cdots \cup C_k \cup T, \tag{4.10}$$

where $C_1, C_2, \cdots, C_k$ are k disjoint closed communicating classes, and T is the union of all the other communicating classes. We do not distinguish between the communicating classes if they are not closed, and simply lump them together in T. Although it is possible to have an infinite number of closed communicating classes in a DTMC with a discrete state-space, in practice we shall always encounter DTMCs with a finite number of closed communicating classes.

It is possible to develop an algorithm to derive the above partition in time $O(q)$; see Conceptual Exercise 4.4. Observe that the classification depends only on which elements of P are positive and which are zero. It does not depend upon the actual values of the positive elements.

Definition 4.5   Irreducibility. *A DTMC with state-space S is said to be irreducible if S is a closed communicating class, else it is called reducible.*

It is clear that all states in an irreducible DTMC communicate with each other. We shall illustrate the above concepts by means of several examples below. In most cases we show the transition diagram of the DTMC. In our experience the transition diagram is the best visual tool to identify the partition of Equation 4.10.

Example 4.5   Two-State DTMC. Consider the two-state DTMC of Example 2.3 with $0 < \alpha, \beta < 1$. The transition diagram is shown in Figure 4.1. In this DTMC $1 \leftrightarrow 2$, and hence $\{1,2\}$ is a closed communicating class, and the DTMC is irreducible. Next suppose $\alpha = 1$ and $0 < \beta < 1$. The transition diagram is shown in Figure 4.2.

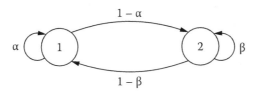

Figure 4.1 *The two-state DTMC: irreducible case.*

Here we have $1 \leftrightarrow 1$, $2 \leftrightarrow 2$, $2 \rightarrow 1$, but 2 is not accessible from 1. Thus $C_1 = \{1\}$

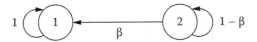

Figure 4.2 *The two-state DTMC: reducible case.*

is a closed communicating class, and $T = \{2\}$ is a communicating class that is not closed. Thus this DTMC is reducible. Finally, suppose $\alpha = \beta = 1$. In this case $C_1 = \{1\}$ and $C_2 = \{2\}$ are two closed communicating classes and $T = \phi$. The DTMC is reducible. ∎

Example 4.6 Genotype Evolution. Consider the six-state DTMC arising in the genotype evolution model of Example 4.3. Since $p_{1,1} = p_{6,6} = 1$, it is clear that $C_1 = \{1\}$ and $C_2 = \{6\}$ are two closed communicating classes. Also, $T = \{2, 3, 4, 5\}$ is a communicating class that is not closed, since states 1 and 6 are accessible from all the four states in T. The DTMC is reducible. ∎

Example 4.7 Simple Random Walk. Consider the simple random walk on $\{0, 1, \cdots, N\}$ with $p_{0,0} = p_{N,N} = 1$, $p_{i,i+1} = p$, $p_{i,i-1} = q$, $1 \leq i \leq N - 1$. Assume $p > 0$, $q > 0$, $p + q = 1$. The transition diagram is shown in Figure 4.3. In

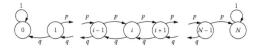

Figure 4.3 *Transition diagram of a random walk with absorbing barriers.*

this DTMC $C_1 = \{0\}$ and $C_2 = \{N\}$ are two closed communicating classes. Also, $T = \{1, 2, \cdots, N - 1\}$ is a communicating class that is not closed, since states 0 and N are accessible from all the states in T. The DTMC is reducible. What happens if p or q is 1? ∎

4.2.2 Recurrence and Transience

Next we introduce the concepts of recurrence and transience of states. They play an important role in the study of the limiting behavior of DTMCs. Let

$$\tilde{T}_i = \min\{n > 0 : X_n = i\}, \quad i \in S. \tag{4.11}$$

Define

$$\tilde{u}_i = \mathsf{P}(\tilde{T}_i < \infty | X_0 = i) \tag{4.12}$$

and

$$\tilde{m}_i = \mathsf{E}(\tilde{T}_i | X_0 = i). \tag{4.13}$$

When $\tilde{u}_i < 1$, $\tilde{m}_i = \infty$. However, as the following example suggests, $\tilde{m}_i$ can be infinite even if $\tilde{u}_i = 1$.

Example 4.8 Success Runs. Consider the success runs Markov chain on $\{0, 1, 2, \cdots\}$ with

$$p_{i,i+1} = \frac{i+1}{i+2}, \quad p_{i,0} = \frac{1}{i+2}, \quad i \geq 0.$$

Now, direct calculations as in Example 3.3 on page 60 show that

$$
\begin{aligned}
\mathsf{P}(\tilde{T}_0 = n | X_0 = 0) &= \mathsf{P}(X_i = i, \; i = 1, 2, \cdots, n-1, \; X_n = 0 | X_0 = 0) \\
&= \frac{1}{n(n+1)}, \quad n \geq 1.
\end{aligned}
$$

Hence,

$$\tilde{u}_0 = P(\tilde{T}_0 < \infty | X_0 = 0)$$

$$= \sum_{n=1}^{\infty} P(\tilde{T}_0 = n | X_0 = 0)$$

$$= \sum_{n=1}^{\infty} \frac{1}{n(n+1)} = 1.$$

The last equality above is seen by writing

$$\frac{1}{n(n+1)} = \frac{1}{n} - \frac{1}{n+1}$$

and using telescoping sums. However, we get

$$\tilde{m}_0 = \sum_{n=1}^{\infty} nP(\tilde{T}_0 = n | X_0 = 0)$$

$$= \sum_{n=1}^{\infty} \frac{n}{n(n+1)}$$

$$= \sum_{n=1}^{\infty} \frac{1}{n+1} = \infty.$$

Here, the last equality follows because the last sum is a harmonic series, which is known to diverge. ∎

With this discussion we make the following definitions:

Definition 4.6   Recurrence and Transience. *A state i is said to be*

(i) recurrent if $\tilde{u}_i = 1$,
(ii) transient if $\tilde{u}_i < 1$.

Definition 4.7   Positive and Null Recurrence. *A recurrent state i is said to be*

(i) positive recurrent if $\tilde{m}_i < \infty$,
(ii) null recurrent if $\tilde{m}_i = \infty$.

As a simple example we see that all absorbing states are positive recurrent. This follows since if i is an absorbing state then $P(X_1 = i | X_0 = i) = 1$. This implies that $\tilde{u}_i = 1$ and $\tilde{m}_i = 1$. Hence i is a positive recurrent state.

In Section 2.5 we studied $V_i^{(n)}$, the number of visits to state i by a DTMC over the finite time period $\{0, 1, 2, \cdots, n\}$. Here we study V_i, the number of visits by the DTMC to state i over the infinite time period $\{0, 1, 2, \cdots\}$. The next theorem gives the main result.

Theorem 4.2   Number of Visits. *If state i is recurrent*

$$P(V_i = \infty | X_0 = i) = 1.$$

If state i is transient

$$P(V_i = k | X_0 = i) = \tilde{u}_i^{k-1}(1 - \tilde{u}_i), \quad k \geq 1.$$

Proof: Follows from Markov property and time homogeneity. See Conceptual Exercise 4.6. ∎

The next theorem yields the necessary and sufficient condition for recurrence and transience.

Theorem 4.3   Criterion for Recurrence and Transience.

(i) State i is recurrent if and only if

$$\sum_{n=0}^{\infty} p_{i,i}^{(n)} = \infty.$$

(ii) State i is transient if and only if

$$\sum_{n=0}^{\infty} p_{i,i}^{(n)} < \infty.$$

Proof: From Theorem 4.2 we get

$$E(V_i | X_0 = i) = \begin{cases} \infty & \text{if } \tilde{u}_i = 1, \\ \frac{1}{1-\tilde{u}_i} < \infty & \text{if } \tilde{u}_i < 1. \end{cases}$$

Now, from the results in Section 2.5 we have

$$E(V_i^{(n)} | X_0 = i) = M_{i,i}^{(n)} = \sum_{m=0}^{n} p_{i,i}^{(m)}.$$

Hence

$$E(V_i | X_0 = i) = \sum_{n=0}^{\infty} p_{i,i}^{(n)}.$$

The theorem follows from this. ∎

Next we give a necessary and sufficient condition for the positive and null recurrence. Notice that $\lim_{n \to \infty} M_{i,i}^{(n)}/(n + 1)$ can be thought of as the long run number of visits to state i per unit time. Since the expected time between two consecutive visits to state i is $\tilde{m}_i$, it is intuitively clear that the long run number of visits to state i per unit time should be $1/\tilde{m}_i$. Thus if the limit is positive, that would imply a finite $\tilde{m}_i$, and if the limit is 0, that would imply an infinite $\tilde{m}_i$. This provides the intuition behind the next theorem.

Theorem 4.4   Criterion for Null and Positive Recurrence.

(i) *A recurrent state i is positive recurrent if and only if*

$$\lim_{n \to \infty} \frac{1}{n+1} \sum_{m=0}^{n} p_{i,i}^{(m)} > 0.$$

(ii) *A recurrent state i is null recurrent if and only if*

$$\lim_{n \to \infty} \frac{1}{n+1} \sum_{m=0}^{n} p_{i,i}^{(m)} = 0.$$

Proof: This theorem is a special case of a general theorem called the elementary renewal theorem, which will be proved in Chapter 8. Hence we do not include a formal proof here. ∎

Note that transience and recurrence are dependent upon the actual magnitudes of the transition probabilities $p_{i,j}$. The next two theorems give very useful properties of transience, null recurrence, and positive recurrence.

Theorem 4.5 Recurrence and Transience as Class Properties.

(i) $i \leftrightarrow j$, i *is recurrent* $\Rightarrow j$ *is recurrent.*

(ii) $i \leftrightarrow j$, i *is transient* $\Rightarrow j$ *is transient.*

Proof: (i) Suppose $i \leftrightarrow j$. Then there are integers n and m such that $p_{i,j}^{(n)} > 0$ and $p_{j,i}^{(m)} > 0$. Now,

$$\sum_{r=0}^{\infty} p_{j,j}^{(r)} \geq \sum_{r=0}^{\infty} p_{j,j}^{(r+n+m)}$$

$$= \sum_{r=0}^{\infty} \sum_{k \in S} p_{j,k}^{(m)} p_{k,k}^{(r)} p_{k,j}^{(n)}$$

$$\geq \sum_{r=0}^{\infty} p_{j,i}^{(m)} p_{i,i}^{(r)} p_{i,j}^{(n)}$$

$$= p_{j,i}^{(m)} p_{i,j}^{(n)} \sum_{r=0}^{\infty} p_{i,i}^{(r)} = \infty$$

since i is recurrent. Hence, from Theorem 4.3, j is recurrent.

(ii) Suppose i is transient and $i \leftrightarrow j$, but j is recurrent. By (i) above, j is recurrent, $i \leftrightarrow j$, implies that i is recurrent, a contradiction. Hence j must be transient. This proves the theorem. ∎

Theorem 4.6 Null and Positive Recurrence as Class Properties.

(i) $i \leftrightarrow j$, i *is positive recurrent* $\Rightarrow j$ *is positive recurrent.*

(ii) $i \leftrightarrow j$, i *is null recurrent* $\Rightarrow j$ *is null recurrent.*

Proof: (i) Let n and m be as in the proof of Theorem 4.5. We have

$$p_{j,j}^{(r+n+m)} = \sum_{k \in S} p_{j,k}^{(m)} p_{k,k}^{(r)} p_{k,j}^{(n)} \geq p_{j,i}^{(m)} p_{i,i}^{(r)} p_{i,j}^{(n)}.$$

Now,

$$\lim_{k \to \infty} \frac{1}{k+1} \sum_{r=0}^{k} p_{j,j}^{(r)} = \lim_{k \to \infty} \frac{1}{k+1} \sum_{r=0}^{k} p_{j,j}^{(r+n+m)}$$

$$\geq p_{j,i}^{(m)} \left(\lim_{k \to \infty} \frac{1}{k+1} \sum_{r=0}^{k} p_{i,i}^{(r)} \right) p_{i,j}^{(n)} > 0.$$

The last inequality follows because the positive recurrence of i implies that the last limit is positive (from Theorem 4.4), and we also have $p_{i,j}^{(n)} > 0$ and $p_{j,i}^{(m)} > 0$. Hence state j is positive recurrent.

(ii) Follows along the same lines as the proof of part (ii) in Theorem 4.5. ■

Theorems 4.5 and 4.6 greatly simplify the task of identifying the transient and recurrent states. If state i is recurrent then all states that belong to the same communicating class as i must be recurrent. The same conclusion holds for the transient or positive or null recurrent states. We say that transience, null recurrence, and positive recurrence are class properties. This enable us to make the following definitions.

Definition 4.8   Recurrent Class. *A communicating class is called recurrent (transient, positive recurrent, null recurrent) if all the states in it are recurrent (transient, positive recurrent, null recurrent).*

Definition 4.9   Recurrent DTMC. *An irreducible DTMC is called recurrent (transient, positive recurrent, null recurrent) if all its states are recurrent (transient, positive recurrent, null recurrent).*

Example 4.9   Simple Random Walk on $\{0, 1, \cdots, N\}$**.** Consider the simple random walk of Example 4.7. We saw there that it has two closed communicating classes $C_1 = \{0\}$ and $C_2 = \{N\}$, and a non-closed class $T = \{1, 2, \cdots, N-1\}$. This is a reducible DTMC. States 0 and N are absorbing and hence positive recurrent. Now consider state 1. If the DTMC starts in state 1 and visits state 0 at time 1 (which happens with probability $q > 0$), then it gets absorbed in state 0 and will never return to state 1. Hence we must have $\tilde{u}_1 \leq 1 - q < 1$. Hence state 1 is transient. Hence from Theorem 4.5 all states in T are transient. ■

Example 4.10   Simple Random Walk on $(-\infty, \infty)$**.** Consider the simple random walk of Example 4.4 with $p > 0, q > 0, p + q = 1$. We see that it is irreducible, since it is possible to go from any state to any other state in a finite number of steps. Here we establish the transience or recurrence of state 0. From Equation 4.9 we have the following asymptotic expression

$$p_{0,0}^{(2n)} \sim (4pq)^n / \sqrt{\pi n}.$$

Hence the tail of $\sum_{n=0}^{\infty} p_{0,0}^{(n)}$ has the same behavior as the tail of $\sum_{n=0}^{\infty} (4pq)^n / \sqrt{\pi n}$. Now, if $p \neq q$, we have $4pq < 1$ and hence the last sum converges, while, if $p = q = .5$, we have $4pq = 1$ and the last sum diverges. Hence, we have

$$\sum_{n=0}^{\infty} p_{0,0}^{(n)} \begin{cases} = \infty & \text{if } p = q \\ < \infty & \text{if } p \neq q. \end{cases}$$

Hence, from Theorem 4.3, we see that state 0 (and hence the DTMC) is recurrent if $p = q$, otherwise it is transient. This makes intuitive sense, since if $p > q$, the DTMC will drift to $+\infty$ and if $p < q$, it will drift to $-\infty$, thus making all states transient. However, when $p = q$, it has no drift, and is likely to return to the starting state with probability 1.

Next we focus on the recurrent case $p = q$. Let m_i be the expected time to reach state zero from state $i \neq 0$. Then the symmetry of the DTMC implies that $m_1 = m_{-1} = m$ (say). A first-step analysis shows that

$$\tilde{m}_0 = 1 + .5m_{-1} + .5m_1 = 1 + m.$$

Similarly,

$$m_1 = 1 + .5m_2.$$

Again, the space homogeneity of the DTMC suggests that $m_2 = 2m_1$, since the expected time to reach zero from state 2 is the sum of x, expected time to reach 1 from state 2, and y, the expected time to reach 0 from state 1. However, we must have $x = y = m$. Thus $m_2 = 2m$. Substituting in the above equation, we have

$$m = 1 + m,$$

which implies that $m = \infty$. Hence, $\tilde{m}_0 = \infty$, and by definition state 0 (and hence the entire DTMC) is null recurrent. ∎

In general, in spite of Theorems 4.5 and 4.6, it is difficult to establish recurrence properties of a state based solely on the definition or by the use of the criteria mentioned in Theorems 4.3 and 4.4. We shall develop better methods of doing so in the next two sections.

4.2.3 Periodicity

In this subsection we introduce another useful concept called *periodicity* for recurrent states. Although the concept can be defined for transient states as well, we will not need it for such states. We begin with a formal definition.

Definition 4.10 Periodicity. *Let i be a recurrent state and d be the largest positive integer such that*

$$\sum_{k=1}^{\infty} \mathsf{P}(\tilde{T}_i = kd) = 1,$$

where $\tilde{T}_i$ is as in Equation 4.11. If $d = 1$, the state i is said to be aperiodic. If $d > 1$, the state i is said to be periodic with period d.

The term periodic is self explanatory: the DTMC, starting in a state i with period d, can return to state i only at times that are integer multiples of d. It is clear that absorbing states are aperiodic. We leave it to the reader (see Conceptual Exercise 4.5) to prove the following equivalent characterization of periodicity.

Theorem 4.7 Let i be a recurrent state with period d. Then $p_{i,i}^{(n)} = 0$ for all n that are not positive-integer multiples of d.

We will use the above theorem to show that, like recurrence and transience, periodicity is a class property.

Theorem 4.8 Periodicity is a Class Property. *If $i \leftrightarrow j$, then i and j have the same period.*

Proof: Let d be the period of state i. Now, $i \leftrightarrow j$ implies that there are two integers n and m such that

$$p_{i,j}^{(n)} > 0, \quad \text{and} \quad p_{j,i}^{(m)} > 0.$$

We have

$$p_{i,i}^{(n+r+m)} = \sum_{k,l} p_{i,k}^{(n)} p_{k,l}^{(r)} p_{l,i}^{(m)} \geq p_{i,j}^{(n)} p_{j,j}^{(r)} p_{j,i}^{(m)}.$$

For $r = 0$, we get

$$p_{i,i}^{(n+m)} \geq p_{i,j}^{(n)} p_{j,i}^{(m)} > 0.$$

Hence $n + m$ must be an integer multiple of d. Now suppose r is not divisible by d. Then $n + m + r$ is not divisible by d. Hence $p_{i,i}^{(n+r+m)} = 0$ since i has period d. Thus $p_{i,j}^{(n)} p_{j,j}^{(r)} p_{j,i}^{(m)} = 0$. However, $p_{i,j}^{(n)}$ and $p_{j,i}^{(m)}$ are not zero. Hence we must have $p_{j,j}^{(r)} = 0$ if r is not divisible by d. Thus the period of state j must be an integer multiple of d. Suppose it is kd for some integer $k \geq 1$. However, by the same argument, if the period of j is kd, that of i must be an integer multiple of kd. However, we have assumed that the period of i is d. Hence we must have $k = 1$. This proves the theorem. ∎

The above theorem implies that all states in a communicating class have the same period. Thus periodicity is a class property. This enables us to talk about the period of a communicating class or an irreducible DTMC. A communicating class or an irreducible DTMC is said to be periodic with period d if any (and hence every) state in it has period $d > 1$. If $d = 1$, it is called aperiodic. It should be noted that periodicity, like communication, depends only upon which elements of P are positive, and which are zero; and not on the actual magnitudes of the positive elements.

Example 4.11 Two-State DTMC. Consider the two-state DTMC of Example 4.1. In each case considered there $p_{i,i} > 0$ for $i = 1, 2$, hence both the states are aperiodic. Now consider the case with $\alpha = \beta = 0$. The transition probability matrix in this case is

$$P = \begin{bmatrix} 0 & 1 \\ 1 & 0 \end{bmatrix}.$$

Thus we see that if $X_0 = 1$ we have

$$X_{2n} = 1, \quad X_{2n-1} = 2, \quad n \geq 1.$$

Hence the DTMC returns to state 1 only at even times, and hence it has period 2. Since this is an irreducible DTMC, both states must have period 2. Thus state 2 has period 2. This is a periodic DTMC. ∎

Example 4.12   Simple Random Walk. Consider the simple random walk of Example 4.4. Now suppose the DTMC starts in state 0. Clearly, DTMC cannot return to state 0 at odd times, and can return at every even time. Hence state 0 has period 2. Since this is an irreducible DTMC, all the states must have period 2. We say the simple random walk is periodic with period 2. ∎

It is straightforward to develop an $O(N^3)$ algorithm to find the periodicity of the states in a DTMC with N states. By using more sophisticated data structures one can develop a more efficient algorithm to do the same in $O(q)$ steps, where q is the number of non-zero entries in its transition probability matrix.

It is relatively easy to find the communicating classes and their periods (in case they are recurrent). However, the task of determining the transience and recurrence is not so. We spend the next two sections developing methods of doing so.

4.3 Determining Recurrence and Transience: Finite DTMCs

In this section we assume that the state-space of the DTMC is finite. Hence all communicating classes of the DTMC are finite. Determining the transience or recurrence of a finite communicating class is particularly easy, as seen from the next theorem.

Theorem 4.9   Finite Closed Classes. *All states in a finite closed communicating class are positive recurrent.*

Proof: Let C be a finite closed communicating class. Then for all $i \in C$, we have

$$1 = P(X_k \in C | X_0 = i) = \sum_{j \in C} p_{i,j}^{(k)}, \quad k \geq 0.$$

Hence

$$\frac{1}{n+1} \sum_{k=0}^{n} \sum_{j \in C} p_{i,j}^{(k)} = 1.$$

Taking limits, we get

$$\lim_{n \to \infty} \frac{1}{n+1} \sum_{k=0}^{n} \sum_{j \in C} p_{i,j}^{(k)} = \sum_{j \in C} \lim_{n \to \infty} \frac{1}{n+1} \sum_{k=0}^{n} p_{i,j}^{(k)} = 1,$$

where the interchange of the limit and the sum is allowed since C is finite. Thus there must be at least one j for which

$$\lim_{n\to\infty} \frac{1}{n+1} \sum_{k=0}^{n} p_{i,j}^{(k)} > 0. \tag{4.14}$$

Since C is a communicating class, there is an r such that $p_{j,i}^{(r)} > 0$, (this r may depend on the pair i, j). Now,

$$p_{j,j}^{(k+r)} \geq p_{j,i}^{(r)} p_{i,j}^{(k)}. \tag{4.15}$$

We have

$$\lim_{n\to\infty} \frac{1}{n+1} \sum_{k=0}^{n} p_{j,j}^{(k)} = \lim_{n\to\infty} \frac{1}{n+1} \sum_{k=0}^{n} p_{j,j}^{(k+r)}$$

$$\geq p_{j,i}^{(r)} \lim_{n\to\infty} \frac{1}{n+1} \sum_{k=0}^{n} p_{i,j}^{(k)} > 0,$$

where the last inequality follows from Equation 4.14. Hence, from Theorem 4.4 it follows that state j is positive recurrent. Hence all states in C must be positive recurrent. ∎

Theorem 4.10 Non-closed Classes. *All states in a non-closed communicating class are transient.*

Proof: Let C be a non-closed communicating class. Then there exists an $i \in C$ and a $j \notin C$ such that i is not accessible from j and $p_{i,j} > 0$. Thus, if the DTMC visits state j starting from state i, it will never return to state i. Hence

$$1 - \tilde{u}_i = \mathsf{P}(\tilde{T}_i = \infty | X_0 = i) \geq p_{i,j} > 0.$$

Thus $\tilde{u}_i < 1$, and hence state i is transient. Since C is a communicating class, all the states in it must be transient. ∎

Since all communicating classes in a finite state DTMC must be finite, and they must be all closed or not, Theorems 4.9 and 4.10 are sufficient to determine the transience and recurrence of any state in such a DTMC. We leave it to the reader to show that there are no null recurrent states in a finite state DTMC (see Conceptual Exercise 4.9) and that not all states in a finite state DTMC can be transient (see Conceptual Exercise 4.11).

We illustrate with two examples.

Example 4.13 Genotype Evolution. Consider the six-state DTMC of Example 4.3. We saw in Example 4.6 that it has two closed communicating classes $C_1 = \{1\}$ and $C_2 = \{6\}$, and a non-closed class $T = \{2, 3, 4, 5\}$. Thus states 1 and 6 are positive recurrent and states 2,3,4,5 are transient. ∎

Example 4.14   Simple Random Walk. Consider the simple random walk of Example 4.7. We saw there that it has two closed communicating classes $C_1 = \{0\}$ and $C_2 = \{N\}$, and a non-closed class $T = \{1, 2, \cdots, N - 1\}$. Hence C_1 and C_2 are positive recurrent and T is transient. This is consistent with the conclusions we reached in Example 4.9. ∎

Determining the transience or recurrence of states in a communicating class with an infinite number of states is a much harder problem. We discuss several methods of doing so in the next section.

4.4 Determining Recurrence and Transience: Infinite DTMCs

Without loss of generality we consider an irreducible DTMC on state-space $S = \{0, 1, 2, \cdots\}$ with transition probability matrix P. The states in such a DTMC may be all positive recurrent, or all null recurrent, or all transient. There is no easy way of determining this. We shall illustrate with several examples.

Example 4.15   General Success Runs. Consider the general success runs DTMC of Example 3.3 on page 60. This DTMC has state-space $S = \{0, 1, 2, \cdots\}$ and transition probabilities

$$p_{i,0} = q_i, \quad p_{i,i+1} = p_i, \quad i \in S.$$

Assume that $p_i > 0$, $q_i > 0$ for all $i \geq 0$. Then the DTMC is irreducible. Thus if we determine the transience or recurrence of state 0, that will automatically determine the transience and recurrence of all the states in S. From the first-step analysis we get

$$\tilde{u}_0 = 1 - p_0 v_1,$$

where v_1 is the probability that the DTMC never visits state 0 starting from state 1. From the analysis in Example 3.8 on page 64 we have

$$v_1 = 0 \Leftrightarrow \sum_{i=1}^{\infty} q_i = \infty,$$

and

$$v_1 > 0 \Leftrightarrow \sum_{i=1}^{\infty} q_i < \infty.$$

It follows that the general success runs DTMC is

(i) recurrent if and only if $\sum_{i=1}^{\infty} q_i = \infty$, and
(ii) transient if and only if $\sum_{i=1}^{\infty} q_i < \infty$.

Now suppose the DTMC is recurrent. Then we have

$$\tilde{m}_0 - 1 + p_0 m_1,$$

where m_1 is the expected first passage time T to state 0 starting from state 1. Using the results of Example 3.3 we get

$$
\begin{aligned}
m_1 &= \sum_{n=1}^{\infty} n P(T = n | X_0 = 1) \\
&= \sum_{n=1}^{\infty} P(T \geq n | X_0 = 1) \\
&= 1 + \sum_{n=2}^{\infty} \prod_{i=1}^{n-1} p_i.
\end{aligned}
$$

Thus the recurrent DTMC is positive recurrent if the last sum converges, otherwise it is null recurrent. ∎

Example 4.16   General Simple Random Walk. Consider the random walk on $\{0, 1, 2, \cdots\}$ with the following transition probabilities:

$$
\begin{aligned}
p_{i,i+1} &= p_i \text{ for } i \geq 0, \\
p_{0,0} &= q_0 = 1 - p_0, \\
p_{i,i-1} &= q_i = 1 - p_i \text{ for } i \geq 1.
\end{aligned}
$$

We assume that $p_i > 0$, $i \geq 0$ and $q_i > 0$, $i \geq 1$ so that the DTMC is irreducible. We have

$$
\tilde{u}_0 = 1 - p_0 v_1,
$$

where v_1 is the probability that the DTMC never visits state 0 starting from state 1. Using the results from Example 3.10 we get

$$
\begin{aligned}
v_1 > 0 &\quad \text{if } \sum_{i=0}^{\infty} \alpha_i < \infty, \\
v_1 = 0 &\quad \text{if } \sum_{i=0}^{\infty} \alpha_i = \infty,
\end{aligned}
$$

where $\alpha_0 = 1$

$$
\alpha_i = \frac{q_1 q_2 \cdots q_i}{p_1 p_2 \cdots p_i}, \quad i \geq 1.
$$

Thus state 0 (and hence the entire DTMC) is recurrent if and only if $\sum_{i=0}^{\infty} \alpha_i = \infty$, and transient if and only if $\sum_{i=0}^{\infty} \alpha_i < \infty$. Next, assuming the DTMC is recurrent, we can use the first-step analysis to obtain

$$
\tilde{m}_0 = 1 + p_0 m_1,
$$

where m_1 is the expected first passage time into state 0 starting from state 1. From Example 3.16 we have

$$
m_1 = \sum_{j=1}^{\infty} \frac{1}{p_j \alpha_j}.
$$

Now let $\rho_0 = 1$ and

$$
\rho_i = \frac{p_0 p_1 \cdots p_{i-1}}{q_1 q_2 \cdots q_i}, \quad i \geq 1.
$$

Thus

$$\tilde{m}_0 = \sum_{i=0}^{\infty} \rho_i.$$

Thus state 0 (and hence the whole DTMC) is positive recurrent if the series $\sum \rho_i$ converges, and null recurrent if it diverges. Combining all these results, we get the following complete classification: the state 0 (and hence the entire DTMC) is

(i) positive recurrent if and only if $\sum_{i=0}^{\infty} \rho_i < \infty$,

(ii) null recurrent if and only if $\sum_{i=0}^{\infty} \rho_i = \infty$, and $\sum_{i=0}^{\infty} \alpha_i = \infty$,

(iii) transient if and only if $\sum_{i=0}^{\infty} \alpha_i < \infty$. ∎

Example 4.17   Production-Inventory System: Batch Production. Consider the DTMC of Example 2.16 on page 20. It has state-space $\{0, 1, 2, \cdots\}$ and transition probability matrix as given in Equation 2.11. We assume that $\alpha_0 > 0$ and $\alpha_0 + \alpha_1 < 1$ so that the DTMC is irreducible. We study the recurrence properties of state 0. From the first-step analysis we get

$$\tilde{u}_0 = \sum_{j=0}^{\infty} \alpha_j u_j,$$

where u_j is the probability that the DTMC eventually visits state 0 starting from state j. Using the results from Example 3.12 we see that $u_j = 1$ for all $j \geq 0$ (and hence $\tilde{u}_0 = 1$) if the mean production batch size $\mu = \sum_{k=0}^{\infty} k\alpha_k \leq 1$, otherwise $u_j < 1$ for all $j \geq 1$ (and hence $\tilde{u}_0 < 1$). This implies that state 0 (and hence the DTMC) is recurrent if $\mu \leq 1$ and transient if $\mu > 1$. Next, assuming the DTMC is recurrent, the first-step analysis yields

$$\tilde{m}_0 = 1 + \sum_{j=0}^{\infty} \alpha_j m_j,$$

where m_j is the expected time it takes the DTMC to reach state zero starting from state j. Using the results of Example 3.17, we see that $\tilde{m}_0 < \infty$ if $\mu < 1$. Hence the DTMC is

(i) positive recurrent if and only if $\mu < 1$,

(ii) null recurrent if and only if $\mu = 1$,

(iii) transient if and only if $\mu > 1$.

This completes the classification of the DTMC. This makes intuitive sense since this DTMC models the situation where exactly one item is removed from the inventory every time period, while μ items are added to the inventory per period on the average. ∎

Example 4.18   Production-Inventory System: Batch Demands. Consider the DTMC of Example 2.17 on page 21. It has state-space $\{0, 1, 2, \cdots\}$ and transition probability matrix as given in Equation 2.13. We assume that $\alpha_0 > 0$ and $\alpha_0 + \alpha_1 < 1$

so that the DTMC is irreducible. We study the recurrence properties of state 0. From the first-step analysis we get

$$\tilde{u}_0 = \beta_0 + \alpha_0 u_1,$$

where u_1 is the probability that the DTMC eventually visits state 0 starting from state 1. Using the results from Example 3.13 we see that $u_1 = 1$ (and hence $\tilde{u}_0 = 1$) if the mean production batch size $\mu = \sum_{k=0}^{\infty} k\alpha_k \geq 1$, otherwise $u_1 < 1$ (and hence $\tilde{u}_0 < 1$). This implies that state 0 (and hence the DTMC) is recurrent if $\mu \geq 1$ transient if $\mu < 1$. Next, assuming the DTMC is recurrent, the first-step analysis yields

$$\tilde{m}_0 = 1 + \alpha_0 m_1,$$

where m_1 is the expected time it takes the DTMC to reach state zero starting from state 1. Using the results of Example 3.18, we see that

$$\tilde{m}_0 = \frac{1}{1 - \alpha},$$

where α is the unique solution in $(0, 1)$ if $\mu > 1$. We can also conclude that $\tilde{m}_0 = \infty$, if $\mu \leq 1$. Hence the DTMC is

   (i) positive recurrent if and only if $\mu > 1$,
  (ii) null recurrent if and only if $\mu = 1$,
 (iii) transient if and only if $\mu < 1$.

This completes the classification of the DTMC. ∎

4.4.1 Foster's Criterion

In this section we shall derive a sufficient condition for the positive recurrence of an irreducible DTMC $\{X_n, n \geq 0\}$ with state-space S and transition probability matrix P. Let $\nu : S \to [0, \infty)$ and, for $i \in S$, define

$$\begin{aligned} d(i) &= \mathsf{E}(\nu(X_{n+1}) - \nu(X_n)|X_n = i) \\ &= \sum_{j \in S} p_{i,j}\nu(j) - \nu(i). \end{aligned}$$

The function ν is called a potential function, and the quantity $d(i)$ is called the generalized drift in state i. The main result, called Foster's criterion, is given below.

Theorem 4.11 Foster's Criterion. *Let* $\{X_n, n \geq 0\}$ *be an irreducible DTMC on a countable state-space* S. *If there exists a potential function* $\nu : S \to [0, \infty)$, *an* $\epsilon > 0$, *and a finite set* $H \subset S$ *such that*

$$|d(i)| < \infty \quad \text{for } i \in H \tag{4.16}$$
$$d(i) < -\epsilon \quad \text{for } i \notin H \tag{4.17}$$

then the DTMC is positive recurrent.

Before we prove the above result, we give the intuition behind it. Suppose the hypothesis of the above theorem holds. Since ν is a non-negative function, $E(\nu(X_n)) \geq 0$ for all $n \geq 0$. However, whenever $X_n \notin H$, we have $E(\nu(X_{n+1})) < E(\nu(X_n)) - \epsilon$. Hence the DTMC cannot stay outside of H for too long, else $E(\nu(X_n))$ will become negative. Hence the DTMC must enter the finite set H often enough. The theorem says that the visits to H are sufficiently frequent to make the states in H (and hence, due to irreducibility, the whole chain) positive recurrent. A formal proof follows.

Proof: Define

$$w(k) = \sum_{j \in S} p_{k,j} \nu(j), \quad k \in H.$$

Then, from Equation 4.16, $|w(k) - \nu(k)| < \infty$ for all $k \in H$. Hence $w(k) < \infty$ for all $k \in H$. Now, for $i \in S$, let

$$y_i^{(0)} = \nu(i),$$

and

$$y_i^{(n)} = \sum_{j \in S} p_{i,j}^{(n)} \nu(j).$$

Then

$$
\begin{aligned}
y_i^{(r+1)} &= \sum_{j \in S} p_{i,j}^{(r+1)} \nu(j) \\
&= \sum_{j \in S} \sum_{k \in S} p_{i,k}^{(r)} p_{k,j} \nu(j) \\
&= \sum_{k \in H} p_{i,k}^{(r)} \sum_{j \in S} p_{k,j} \nu(j) + \sum_{k \notin H} p_{i,k}^{(r)} \sum_{j \in S} p_{k,j} \nu(j) \\
&\leq \sum_{k \in H} p_{i,k}^{(r)} w(k) + \sum_{k \notin H} p_{i,k}^{(r)} (\nu(k) - \epsilon) \\
&\leq \sum_{k \in H} p_{i,k}^{(r)} (w(k) + \epsilon) + \sum_{k \in S} p_{i,k}^{(r)} \nu(k) - \epsilon \text{ (since } \nu(k) \geq 0 \text{ for all } k) \\
&= \sum_{k \in H} p_{i,k}^{(r)} (w(k) + \epsilon) + y_i^{(r)} - \epsilon.
\end{aligned}
$$

Adding the above inequalities for $r = 0, 1, 2, \cdots, n$ yields

$$y_i^{(n+1)} \leq \sum_{k \in H} \sum_{r=0}^{n} p_{i,k}^{(r)} (w(k) + \epsilon) + y_i^{(0)} - (n+1)\epsilon.$$

However, $y_i^{(n+1)} \geq 0$ and $y_i^{(0)}$ is finite for each i. Hence, dividing by $n+1$ and letting $n \to \infty$ yields

$$\sum_{k \in H} \left(\lim_{n \to \infty} \frac{1}{n+1} \sum_{r=0}^{n} p_{i,k}^{(r)} \right) (w(k) + \epsilon) \geq \epsilon.$$

Since $0 \leq w(k) < \infty$ for all $k \in H$, the above inequality implies that there exists a $k \in H$ for which

$$\lim_{n \to \infty} \frac{1}{n+1} \sum_{r=0}^{n} p_{i,k}^{(r)} > 0. \tag{4.18}$$

Using the fact that $k \to i$, and arguments similar to those in Theorem 4.6, we can show that

$$\lim_{n \to \infty} \frac{1}{n+1} \sum_{r=0}^{n} p_{k,k}^{(r)} > 0. \tag{4.19}$$

(See Conceptual Exercise 4.12.) Hence state k is positive recurrent, as a consequence of part (i) of Theorem 4.4. Since the DTMC is assumed irreducible, all the states are positive recurrent. ■

Foster's criterion is especially easy to apply since checking Equations 4.16 and 4.17 is a simple task. In many applications $S = \{0, 1, 2, \cdots\}$ and usually $\nu(i) = i$ suffices to derive useful sufficient conditions for positive recurrence. In this case $d(i)$ is called the drift in state i, and we have the following result, called Pakes' lemma, whose proof we leave to the reader; see Conceptual Exercise 4.13.

Theorem 4.12 Pakes' Lemma. *Let* $\{X_n, n \geq 0\}$ *be an irreducible DTMC on* $S = \{0, 1, 2, \cdots\}$ *and let*

$$d(i) = \mathsf{E}(X_{n+1} - X_n | X_n = i), \quad i \in S.$$

The DTMC is positive recurrent if

(i) $d(i) < \infty$ *for all* $i \in S$, *and*

(ii) $\limsup_{i \in S} d(i) < 0$.

Pakes' lemma is useful in many DTMCs arising out of inventory or queueing context. We show its usefulness in the following example.

Example 4.19 Production-Inventory System: Batch Production. Consider the DTMC of Example 4.17. Assume that μ, the mean production batch size is finite. Then,

$$d(0) = \mathsf{E}(X_{n+1} - X_n | X_n = 0) = \mu,$$

and

$$d(i) = \mathsf{E}(X_{n+1} - X_n | X_n = i) = \mu - 1, \quad i \geq 1.$$

Hence, by Pakes' lemma, the DTMC is positive recurrent if $\mu < 1$. Note that Foster's criterion or Pakes' lemma does not say whether the DTMC is null recurrent or transient if $\mu \geq 1$. ■

Theorem 4.12 is so intuitive that it is tempting to assume that we will get a sufficient condition for transience if we reverse the inequality in condition (ii) of that theorem. Unfortunately, this turns out not to be true. However, we do get the following more restricted result, which we state without proof:

Theorem 4.13 *Let $\{X_n, n \geq 0\}$ be an irreducible DTMC on $S = \{0, 1, 2, \cdots\}$. The DTMC is transient if there exists a $k > 0$ such that*

(i) $-k < d(i) < \infty$ *for all* $i \in S$, *and*

(ii) $\limsup_{i \in S} d(i) > 0.$

We end this section with the remark that the recurrence and transience properties of an infinite state DTMC are dependent on the actual magnitudes on the elements of P, and not just whether they are zero or positive. Using the concepts developed in this section we shall study the limiting behavior of DTMCs in the next section.

4.5 Limiting Behavior of Irreducible DTMCs

In this section we derive the main results regarding the limiting distribution of an irreducible DTMC $\{X_n, n \geq 0\}$ on state-space $S = \{0, 1, 2, \cdots\}$ and transition probability matrix P. We treat the four cases separately: the DTMC is transient, null recurrent, aperiodic positive recurrent, and periodic positive recurrent.

4.5.1 The Transient Case

We begin with the main result in the following theorem.

Theorem 4.14 **Transient DTMC.** *Let $\{X_n, n \geq 0\}$ be an irreducible transient DTMC. Then*

$$\lim_{n \to \infty} p_{i,j}^{(n)} = 0, \quad i, j \in S. \tag{4.20}$$

Proof: Since the DTMC is transient, we have from Theorem 4.3,

$$\sum_{n=0}^{\infty} p_{j,j}^{(n)} < \infty, \quad j \in S.$$

Since $p_{j,j}^{(n)} \geq 0$, this implies that

$$\lim_{n \to \infty} p_{j,j}^{(n)} = 0, \quad j \in S.$$

Now, consider a state $i \in S$, $i \neq j$. Since the DTMC is irreducible, $j \to i$, and hence there exists an $m > 0$ such that $p_{j,i}^{(m)} > 0$. Hence, for $n \geq m$,

$$
\begin{aligned}
p_{j,j}^{(n)} &= \sum_{r \in S} p_{j,r}^{(m)} p_{r,j}^{(n-m)} \quad \text{(Theorem 2.3)} \\
&\geq p_{j,i}^{(m)} p_{i,j}^{(n-m)}.
\end{aligned}
$$

Since $p_{j,j}^{(n)}$ converges to zero, and $p_{j,i}^{(m)} > 0$, the above inequality implies Equation 4.20. ∎

By using dominated convergence theorem, we see that the above theorem implies that

$$\lim_{n \to \infty} P(X_n = j) = 0, \quad j \in S,$$

or more generally, for any finite set $A \subset S$,

$$\lim_{n \to \infty} P(X_n \in A) = 0.$$

Even more importantly, we can show that

$$\sum_{n=0}^{\infty} p_{i,j}^{(n)} < \infty, \tag{4.21}$$

which implies that

$$\sum_{n=0}^{\infty} P(X_n \in A) < \infty,$$

which implies that

$$P(X_n \in A \text{ infinitely often}) = 0. \tag{4.22}$$

See Conceptual Exercise 4.14. Thus a transient DTMC will eventually permanently exit any finite set with probability 1.

4.5.2 The Discrete Renewal Theorem

We begin the study of the convergence results for recurrent DTMCs with the discrete renewal theorem. This is the discrete analog of a general theorem called the key renewal theorem, which is presented in Chapter 8. We first give the main result.

Theorem 4.15 Discrete Renewal Theorem. *Let $\{u_n, n \geq 1\}$ be a sequence of real numbers with*

$$u_n \geq 0, \quad \sum_{n=1}^{\infty} u_n = 1,$$

and let $\mu = \sum_{n=1}^{\infty} n u_n$. Let d, called the period, be the largest integer such that

$$\sum_{n=1}^{\infty} u_{nd} = 1.$$

Let $\{v_n, n \geq 0\}$ be another given sequence with

$$\sum_{n=0}^{\infty} |v_n| < \infty. \tag{4.23}$$

Suppose the sequence $\{g_n, n \geq 0\}$ satisfies

$$g_n = v_n + \sum_{m=1}^{n} u_m g_{n-m}, \quad n \geq 0. \tag{4.24}$$

(i) If $d = 1$, the sequence $\{g_n, n \geq 0\}$ converges and

$$\lim_{n \to \infty} g_n = \frac{1}{\mu} \sum_{n=0}^{\infty} \nu_n. \tag{4.25}$$

(ii) If $d > 1$, the sequence $\{g_n, n \geq 0\}$ has d convergent subsequences $\{g_{nd+k}, n \geq 0\}$ $(0 \leq k \leq d - 1)$ and

$$\lim_{n \to \infty} g_{nd+k} = \frac{d}{\mu} \sum_{n=0}^{\infty} \nu_{nd+k}. \tag{4.26}$$

If $\mu = \infty$, the limits are to be interpreted as 0.

Proof: (i). The hard part is to prove that the limit exists. We refer the reader to Karlin and Taylor (1975) or Kohlas (1982) for the details. Here we assume that the limit exists and show that it is given as stated in Equation 4.25.

First, it is possible to show by induction that

$$|g_n| \leq \sum_{k=0}^{n} |\nu_k|, \quad n \geq 0.$$

From Equation 4.23, we conclude that $\{g_n, n \geq 0\}$ is a bounded sequence. Summing Equation 4.24 from $n = 0$ to k we get

$$\sum_{n=0}^{k} g_n = \sum_{n=0}^{k} \nu_n + \sum_{n=0}^{k} \sum_{m=1}^{n} u_m g_{n-m}, \quad k \geq 0.$$

Rearranging the terms and using $f_n = \sum_{k=n+1}^{\infty} u_k$, we get

$$\sum_{r=0}^{k} g_{k-r} f_r = \sum_{r=0}^{k} \nu_r.$$

Assuming $\lim_{k \to \infty} g_k = g$ exists, we use the dominated convergence theorem to get

$$\begin{aligned}
\sum_{r=0}^{\infty} \nu_r &= \lim_{k \to \infty} \sum_{r=0}^{k} \nu_r \\
&= \lim_{k \to \infty} \sum_{r=0}^{k} g_{k-r} f_r \\
&= g \sum_{r=0}^{\infty} f_r \\
&= g \sum_{r=1}^{\infty} r u_r \\
&= g\mu,
\end{aligned}$$

which gives the desired result.

(ii). From the definition of d it follows that $u_r = 0$ if r is not an integer multiple of d, and $\{u'_r = u_{rd}, r \geq 1\}$ has period 1. We also have

$$\mu' = \sum_{r=1}^{\infty} r u'_r = \sum_{r=1}^{\infty} r u_{rd} = \frac{\mu}{d}.$$

Now, fix a $0 \leq k < d$ and define

$$\nu'_n = \nu_{nd+k}, \quad g'_n = g_{nd+k}, \quad n \geq 0.$$

Then, Equation 4.24 reduces to

$$g'_n = \nu'_n + \sum_{r=0}^{n} u'_r g'_{n-r}.$$

Since $\{u'_r, r \geq 1\}$ has period one, we can use Equation 4.25 to get

$$\lim_{n \to \infty} g'_n = \frac{1}{\mu'} \sum_{n=0}^{\infty} \nu'_n.$$

This yields Equation 4.26 as desired. ■

The case $d = 1$ is called the aperiodic case, and the case $d > 1$ is called the periodic case. Equation 4.24 is called the discrete renewal equation.

4.5.3 The Recurrent Case

Let $\{X_n, n \geq 0\}$ be an irreducible recurrent DTMC. The next theorem shows that $\{p_{j,j}^{(n)}, n \geq 0\}$ satisfy a discrete renewal equation.

Theorem 4.16   Discrete Renewal Equation for $p_{j,j}^{(n)}$. *Fix a $j \in S$. Let*

$$
\begin{aligned}
\tilde{T}_j &= \min\{n > 0 : X_n = j\}, \quad j \in S, &\text{(4.27)}\\
g_n &= p_{j,j}^{(n)}, \quad n \geq 0,\\
\nu_0 &= 1, \quad \nu_n = 0, \quad n \geq 1,\\
u_n &= \mathsf{P}(\tilde{T}_j = n | X_0 = j), \quad n \geq 1.
\end{aligned}
$$

Then Equation 4.24 is satisfied.

Proof: We have

$$g_0 = p_{j,j}^{(0)} = 1 = \nu_0.$$

For $n \geq 1$, conditioning on $\tilde{T}_j$,

$$
\begin{aligned}
g_n = p_{j,j}^{(n)} &= \mathsf{P}(X_n = j | X_0 = j)\\
&= \sum_{m=1}^{n} \mathsf{P}(X_n = j | X_0 = j, \tilde{T}_j = m) \mathsf{P}(\tilde{T}_j = m | X_0 = j)
\end{aligned}
$$

$$= \sum_{m=1}^{n} \mathsf{P}(X_n = j | X_0 = j, X_r \neq j, 1 \leq r \leq m-1, X_m = j) u_m$$

$$= \sum_{m=1}^{n} u_m \mathsf{P}(X_n = j | X_m = j)$$

$$= \sum_{m=1}^{n} u_m g_{n-m},$$

where we have used time homogeneity to get the last equality and Markov property to get the one before that. This proves the theorem. ∎

Using the above theorem we get the next important result.

Theorem 4.17 Limiting Behavior of Recurrent States. *Let j be recurrent, and $\tilde{T}_j$ be as defined in Equation 4.27 and*

$$\tilde{m}_j = \mathsf{E}(\tilde{T}_j | X_0 = j).$$

(i) If state j is aperiodic

$$\lim_{n \to \infty} p_{j,j}^{(n)} = 1/\tilde{m}_j.$$

(ii) If state j is periodic with period $d > 1$

$$\lim_{n \to \infty} p_{j,j}^{(nd)} = d/\tilde{m}_j.$$

If $\tilde{m}_j = \infty$ the limits are to be interpreted as 0.

Proof: We see from Theorem 4.16 that $\{p_{j,j}^{(n)}, n \geq 0\}$ satisfy the discrete renewal equation. Since j is recurrent, we have

$$\sum_{n=1}^{\infty} \mathsf{P}(\tilde{T}_j = n | X_0 = j) = 1,$$

and

$$\sum_{n=0}^{\infty} |\nu_n| = \nu_0 = 1.$$

We also have

$$\mu = \sum_{n=1}^{\infty} n u_n = \mathsf{E}(\tilde{T}_j | X_0 = j) = \tilde{m}_j.$$

Now suppose state j is aperiodic. Hence $d = 1$, and we can apply Theorem 4.15 part (i). We get

$$\lim_{n \to \infty} p_{j,j}^{(n)} = \frac{1}{\mu} \sum_{n=0}^{\infty} \nu_n = 1/\tilde{m}_j.$$

This yields part (i). Part (ii) follows similarly from part (ii) of Theorem 4.15. ∎

4.5.4 The Null Recurrent Case

In this subsection we consider the limiting behavior of an irreducible null recurrent DTMC. Such a DTMC necessarily has infinite state-space. The main result is given in the next theorem.

Theorem 4.18   The Null Recurrent DTMC. *For an irreducible null recurrent DTMC*

$$\lim_{n\to\infty} p_{i,j}^{(n)} = 0, \quad i,j \in S.$$

Proof: Since the DTMC is null recurrent, we know that

$$\tilde{m}_j = \infty, \quad j \in S.$$

If state j is aperiodic, the theorem follows from part (i) of Theorem 4.17. If it is periodic with period $d > 1$, part (ii) of Theorem 4.17 yields

$$\lim_{n\to\infty} p_{j,j}^{(nd)} = 0.$$

In this case, periodicity implies

$$p_{j,j}^{(nd+k)} = 0, \quad 1 \le k \le d-1, \ n \ge 0.$$

Following the proof of Theorem 4.14 and using the assumption of irreducibility we can show that

$$\lim_{n\to\infty} p_{i,j}^{(n)} = 0, \quad i,j \in S.$$

This proves the theorem. ∎

    By using dominated convergence theorem we see that the above theorem implies that

$$\lim_{n\to\infty} P(X_n = j) = 0, \quad j \in S,$$

or more generally, for any finite set $A \subset S$,

$$\lim_{n\to\infty} P(X_n \in A) = 0.$$

Unlike in the transient case, Theorem 4.3 implies that

$$\sum_{n=0}^{\infty} p_{i,j}^{(n)} = \infty.$$

Since the DTMC is recurrent each state is visited infinitely often. Hence, in contrast to the transient case,

$$P(X_n \in A \text{ infinitely often}) = 1.$$

Thus a null recurrent DTMC will visit every finite set infinitely often over the infinite horizon, even though the limiting probability that the DTMC is in the set A is zero. This non-intuitive behavior is the result of the fact that although each state is visited infinitely often, the expected time between two consecutive visits to the state is infinite.

4.5.5 The Positive Recurrent Aperiodic Case

In this subsection we assume that $\{X_n, n \geq 0\}$ is an irreducible positive recurrent aperiodic DTMC. Such DTMCs are also called ergodic. Now, for a positive recurrent DTMC, $\tilde{m}_j < \infty$ for all $j \in S$. Hence, from part (i) of Theorem 4.17 we get

$$\lim_{n \to \infty} p_{j,j}^{(n)} = 1/\tilde{m}_j > 0, \quad j \in S.$$

The next theorem yields the limiting behavior of $p_{i,j}^{(n)}$ as $n \to \infty$.

Theorem 4.19 The Positive Recurrent Aperiodic DTMC. *For an irreducible positive recurrent DTMC, there exist $\{\pi_j > 0, \ j \in S\}$ such that*

$$\lim_{n \to \infty} p_{i,j}^{(n)} = \pi_j, \quad i, j \in S. \tag{4.28}$$

The $\{\pi_j, j \in S\}$ are the unique solution to

$$\pi_j = \sum_{i \in S} \pi_i p_{i,j}, \quad j \in S, \tag{4.29}$$

$$\sum_{j \in S} \pi_j = 1. \tag{4.30}$$

Proof: We have already seen that Equation 4.28 holds when $i = j$ with $\pi_j = 1/\tilde{m}_j > 0$. Hence assume $i \neq j$. Following the proof of Theorem 4.16 we get

$$p_{i,j}^{(n)} = \sum_{m=1}^{n} u_m p_{j,j}^{(n-m)}, \quad n \geq 0,$$

where

$$u_m = \mathsf{P}(\tilde{T}_j = m | X_0 = i).$$

Since $i \leftrightarrow j$, it follows that

$$\sum_{m=1}^{\infty} u_m = \mathsf{P}(X_m = j \text{ for some } m \geq 1 \,|X_0 = i) = 1.$$

Now let $0 < \epsilon < 1$ be given. Thus it is possible to pick an N such that

$$\sum_{m=N+1}^{\infty} u_m \leq \epsilon/4, \tag{4.31}$$

and

$$|p_{j,j}^{(n)} - \pi_j| \leq \epsilon/2, \text{ for all } n \geq N. \tag{4.32}$$

Then, for $n \geq 2N$, we get

$$|p_{i,j}^{(n)} - \pi_j|$$

$$= \left| \sum_{m=1}^{n} u_m p_{j,j}^{(n-m)} - \pi_j \right|$$

$$= \left| \sum_{m=1}^{n-N} u_m(p_{j,j}^{(n-m)} - \pi_j) + \sum_{m=n-N+1}^{n} u_m p_{j,j}^{(n-m)} - \sum_{m=n-N+1}^{\infty} u_m \pi_j \right|$$

$$\leq \sum_{m=1}^{n-N} u_m |p_{j,j}^{(n-m)} - \pi_j| + \sum_{m=n-N+1}^{n} u_m p_{j,j}^{(n-m)} + \sum_{m=n-N+1}^{\infty} u_m \pi_j$$

(follows from the triangle inequality,)

$$\leq \sum_{m=1}^{n-N} u_m \epsilon/2 + 2 \sum_{m=N+1}^{\infty} u_m$$

(follows using Equation 4.32, $u_m \geq 0$, $p_{j,j}^{(n)} \leq 1$, $\pi_j \leq 1$,)

$$\leq \epsilon/2 + \epsilon/2$$

(follows from Equation 4.31 and $\sum u_m = 1$,)

$$\leq \epsilon.$$

This proves Equation 4.28. Next we derive Equations 4.29 and 4.30. Now let $a_j^{(n)} = P(X_n = j)$. Then Equation 4.28 implies

$$\lim_{n \to \infty} a_j^{(n)} = \pi_j, \quad j \in S.$$

Now, the Chapman–Kolmogorov equations 2.22 yield

$$a_j^{(n+m)} = \sum_{i \in S} a_i^{(m)} p_{i,j}^{(n)}, \quad n, m \geq 0.$$

Let $m \to \infty$ on both sides. The interchange of the limit and the sum on the right-hand side is justified due to bounded convergence theorem. Hence we get

$$\pi_j = \sum_{i \in S} \pi_i p_{i,j}^{(n)}.$$

Equation 4.29 results from the above by setting $n = 1$. Now let $n \to \infty$. Again, bounded convergence theorem can be used to interchange the sum and the limit on the right-hand side to get

$$\pi_j = \left(\sum_{i \in S} \pi_i \right) \pi_j.$$

Since $\pi_j > 0$, we must have $\sum \pi_i = 1$, yielding Equation 4.30.

Now suppose $\{\pi_i', i \in S\}$ is another solution to Equations 4.29 and 4.30. Using the same steps as before we get

$$\pi_j' = \sum_{i \in S} \pi_i' p_{i,j}^{(n)}, \quad n \geq 0.$$

Letting $n \to \infty$ we get (by using the bounded convergence theorem)

$$\pi_j' = \left(\sum_{i \in S} \pi_i' \right) \pi_j = \pi_j.$$

Thus Equations 4.29 and 4.30 have a unique solution. ∎

Equation 4.29 is called the balance equation, since it balances the probability of entering a state with the probability of exiting a state. Equation 4.30 is called the normalizing equation, for obvious reasons. The solution $\{\pi_j, j \in S\}$ satisfying balance and the normalizing equations is called the *limiting distribution*, since it is the limit of the distribution of X_n as $n \to \infty$. It is also called the *steady state distribution*. It should be noted that it is the state-distribution that is steady, not the state itself.

Now suppose the DTMC starts with initial distribution

$$P(X_0 = j) = \pi_j, \quad j \in S.$$

Then it can be shown that (see Conceptual Exercise 4.15)

$$P(X_n = j) = \pi_j, \quad j \in S, \text{ for all } n \geq 1.$$

Thus the distribution of X_n is independent of n if the DTMC starts with the initial distribution $\{\pi_j, j \in S\}$. Hence $\{\pi_j, j \in S\}$ is also called the *stationary distribution* of the DTMC.

As a consequence of Equation 4.28 (see Abel's theorem in Marsden (1974)) we have

$$\lim_{n \to \infty} \frac{1}{n+1} \sum_{r=0}^{n} p_{i,j}^{(r)} = \pi_j.$$

Thus in a positive recurrent aperiodic DTMC, the limiting fraction of the time spent in state j (called the limiting *occupancy distribution*) is also given by π_j. Thus for a DTMC the limiting distribution, the stationary distribution, and the limiting occupancy distribution all coincide.

4.5.6 The Positive Recurrent Periodic Case

Let $\{X_n, n \geq 0\}$ be an irreducible positive recurrent DTMC with period $d > 1$. From part (ii) of Theorem 4.17 we get

$$\lim_{n \to \infty} p_{j,j}^{(nd)} = d/\tilde{m}_j = d\pi_j > 0, \quad j \in S. \tag{4.33}$$

Now let

$$
\begin{aligned}
\alpha_{i,j}(r) &= P(\tilde{T}_j \equiv r(\bmod d)|X_0 = i), \ 0 \leq r \leq d-1 \\
&= \sum_{n=0}^{\infty} P(\tilde{T}_j = nd + r|X_0 = i).
\end{aligned}
$$

The next theorem shows that $p_{i,j}^{(n)}$ does not have a limit, but does have d convergent subsequences.

Theorem 4.20 *Let $\{X_n, n \geq 0\}$ be an irreducible positive recurrent DTMC with period $d > 1$. Let $\{\pi_j, j \geq 0\}$ be the solution to Equations 4.29 and 4.30. Then*

$$\lim_{n \to \infty} p_{i,j}^{(nd+r)} = d\pi_j \alpha_{i,j}(r) \tag{4.34}$$

for $i, j \in S$ and $0 \leq r \leq d - 1$.

Proof: Follows along the same lines as that of Theorem 4.19 by writing

$$p_{i,j}^{(nd+r)} = \sum_{k=0}^{n} \mathsf{P}(\tilde{T}_j = kd + r | X_0 = i) p_{j,j}^{(n-k)d}$$

and using Equation 4.33 and the definition of $\alpha_{i,j}(r)$. ∎

The next theorem gives the result about the limiting occupancy distribution.

Theorem 4.21   Limiting Occupancy Distribution (for periodic and aperiodic DTMCs).

(i)

$$\lim_{n \to \infty} \frac{M_{i,j}^{(n)}}{n+1} = \pi_j, \quad i, j \in S. \tag{4.35}$$

(ii) $\{\pi_j, j \in S\}$ *are given by the unique solution to*

$$\pi_j = \sum_{i \in S} \pi_i p_{i,j}, \quad j \in S, \tag{4.36}$$

$$\sum_{j \in S} \pi_j = 1. \tag{4.37}$$

Proof: (i) As an easy consequence of Equation 4.34 we get

$$\lim_{n \to \infty} \frac{1}{n+1} \sum_{k=0}^{n} p_{i,j}^{(kd+r)} = d\pi_j \alpha_{i,j}(r), \quad i, j \in S.$$

(This just says that the Cesaro limit agrees with the usual limit when the latter exists. See Marsden (1974).) Using this we get, for $0 \leq m \leq d - 1$,

$$\lim_{n \to \infty} \frac{1}{nd + m + 1} \sum_{k=0}^{nd+m} p_{i,j}^{(k)} = \frac{1}{d} \sum_{r=0}^{d-1} \lim_{n \to \infty} \frac{d}{nd + m + 1} \sum_{k=0}^{n-1} p_{i,j}^{(kd+r)}$$

$$+ \lim_{n \to \infty} \frac{1}{nd + m + 1} \sum_{r=0}^{m} p_{i,j}^{(nd+r)}$$

$$= \frac{1}{d} \sum_{r=0}^{d-1} d\pi_j \alpha_{i,j}(r) = \pi_j.$$

The Equation 4.35 follows from this.

(ii) We have

$$\sum_{j \in S} \frac{M_{i,j}^{(n)}}{n+1} p_{j,m} = \sum_{j \in S} \frac{1}{n+1} \sum_{k=0}^{n} p_{i,j}^{(k)} p_{j,m}$$

$$= \frac{1}{n+1} \sum_{k=0}^{n} p_{i,m}^{(k+1)}$$

$$= \frac{n+2}{n+1} \frac{1}{n+2} \sum_{k=0}^{n+1} p_{i,m}^{(k)} - \frac{1}{n+1} p_{i,m}^{(0)}$$

$$= \frac{n+2}{n+1} \frac{M_{i,m}^{(n+1)}}{n+2} - \frac{1}{n+1} p_{i,m}^{(0)}.$$

Letting $n \to \infty$ on both sides of the above equation we get

$$\sum_{j \in S} \pi_j p_{j,m} = \pi_m,$$

which is Equation 4.36. By repeating the above step k times we get

$$\sum_{j \in S} \pi_j p_{j,m}^{(k)} = \pi_m, \quad k \geq 1.$$

By summing the above equation over $k = 0$ to n and dividing by $n+1$ we get

$$\sum_{j \in S} \pi_j \frac{M_{j,m}^{(n)}}{n+1} = \pi_m.$$

Now let $n \to \infty$. Using Equation 4.35 we get

$$\left(\sum_{j \in S} \pi_j \right) \pi_m = \pi_m.$$

Since $\pi_m > 0$ this implies Equation 4.37. Uniqueness can be proved in a manner similar to the proof of Theorem 4.19. ∎

4.5.7 Necessary and Sufficient Condition for Positive Recurrence

The next theorem provides a necessary and sufficient condition for positive recurrence. It also provides a sort of converse to Theorems 4.19 and 4.21.

Theorem 4.22 Positive Recurrence. *Let* $\{X_n, n \geq 0\}$ *be an irreducible DTMC. It is positive recurrent if and only if there is a non-negative solution to*

$$\pi_j = \sum_{i \in S} \pi_i p_{i,j}, \quad j \in S, \tag{4.38}$$

$$\sum_{j \in S} \pi_j = 1. \tag{4.39}$$

If there is a solution to the above equations, it is unique.

Proof: Theorems 4.19 and 4.21 give the "if" part. Here we prove the "only if" part. Suppose $\{X_n, n \geq 0\}$ is an irreducible DTMC that is either null recurrent or transient. Suppose there is a non-negative solution to Equation 4.38. Then following the same argument as before, we have

$$\pi_j = \sum_{i \in S} \pi_i p_{i,j}^{(n)}, \quad n \geq 1.$$

Letting $n \to \infty$ and using Theorems 4.14 and 4.18 we get

$$\pi_j = \lim_{n \to \infty} \sum_{i \in S} \pi_i p_{i,j}^{(n)} = \sum_{i \in S} \pi_i \lim_{n \to \infty} p_{i,j}^{(n)} = 0.$$

Here the interchange of the limits and the sum is justified by dominated convergence theorem. Thus the solution cannot satisfy Equation 4.39. This proves the theorem. Uniqueness follows from the uniqueness of the limiting distribution of the positive recurrent DTMCs. ∎

The above theorem is very useful. For irreducible DTMCs it allows us to directly solve Equations 4.38 and 4.39 without first checking for positive recurrence. If we can solve these equations, the DTMC is positive recurrent. Note that we do not insist on aperiodicity in this result. However, the interpretation of the solution $\{\pi_j, j \in S\}$ depends upon whether the DTMC is aperiodic or periodic. In an aperiodic DTMC there are three possible interpretations:

 (i) It is the limiting distribution of the DTMC,
 (ii) It is the stationary distribution of the DTMC,
(iii) It is the limiting occupancy distribution for the DTMC.

When the DTMC is periodic, the second and the third interpretations continue to hold, whereas the first one fails, since a periodic DTMC does not have a limiting distribution.

4.5.8 Examples

We end this section with three examples.

Example 4.20 Two-State DTMC. Consider the two-state DTMC of Example 4.1 on state-space $\{1, 2\}$. Assume that $0 < \alpha + \beta < 2$, so that the DTMC is positive recurrent and aperiodic. Equation 4.38 yields

$$\begin{aligned}
\pi_1 &= \alpha \pi_1 + (1 - \beta)\pi_2, \\
\pi_2 &= (1 - \alpha)\pi_1 + \beta \pi_2.
\end{aligned}$$

Note that these two equations are identical. Using the normalizing equation

$$\pi_1 + \pi_2 = 1$$

we get the following unique solution:

$$\pi_1 = \frac{1 - \beta}{2 - \alpha - \beta}, \quad \pi_2 = \frac{1 - \alpha}{2 - \alpha - \beta}.$$

We can verify Theorem 4.19 directly from the results of Example 4.1. ∎

Example 4.21 Three-State DTMC. Consider the three-state DTMC of Example 4.2. The DTMC is periodic with period 2. Equations 4.38 and 4.39 yield:

$$
\begin{aligned}
\pi_1 &= q\pi_2 \\
\pi_2 &= \pi_1 + \pi_3 \\
\pi_3 &= p\pi_2 \\
\pi_1 + \pi_2 + \pi_3 &= 1.
\end{aligned}
$$

These have a unique solution given by

$$\pi_1 = q/2, \quad \pi_2 = 1/2, \quad \pi_3 = p/2.$$

The above represents the stationary distribution, and the limiting occupancy distribution of the DTMC. Since the DTMC is periodic, it has no limiting distribution. These results are consistent with the results of Example 4.2. Using the matrix notation $\alpha(r) = [\alpha_{i,j}(r)]$, we see that

$$
\alpha(0) = \begin{bmatrix} 1 & 0 & 1 \\ 0 & 1 & 0 \\ 1 & 0 & 1 \end{bmatrix}, \quad \alpha(1) = \begin{bmatrix} 0 & 1 & 0 \\ 1 & 0 & 1 \\ 0 & 1 & 0 \end{bmatrix}.
$$

Using this we can verify that Theorem 4.20 produces the results that are consistent with Equations 4.2 and 4.3. ∎

Example 4.22 Genotype Evolution. Consider the six-state DTMC of Example 4.3. This is a reducible DTMC, and one can see that Equations 4.38 and 4.39 have an infinite number of solutions. We shall deal with the reducible DTMCs in Section 4.7. ∎

It should be clear by now that the study of the limiting behavior of irreducible positive recurrent DTMCs involves solving the balance equations and the normalizing equation. If the state-space is finite and the transition probabilities are given numerically, one can use standard numerical procedures to solve them. When the DTMC has infinite state-space or the transition probabilities are given algebraically, numerical methods cannot be used. In such cases one obtains the solution by analytical methods. The analytical methods work only if the transition probabilities have a special structure. We shall describe various examples of analytical solutions in the next section.

4.6 Examples: Limiting Behavior of Infinite State-Space Irreducible DTMCs

In this section we consider several examples of infinite state DTMCs with special structure, and study their limiting behavior. The methods of solving the balance and normalizing equations can generally be classified into two groups: the recursive methods, and the generating function methods. In the recursive methods one uses the balance equations to express each π_i in terms of π_0, and then computes π_0 using the normalizing equation. In the generating function methods one obtains the generating function of the π_i's, which in theory can be used to compute π_i's. We have encountered these two methods in Chapter 3.

Example 4.23   Success Runs. Consider the success runs DTMC of Example 2.15 on page 19 with transition probabilities

$$p_{i,0} = q_i, \quad p_{i,i+1} = p_i, \quad i = 0, 1, 2, \cdots.$$

We assume that $p_i > 0$ and $q_i > 0$ for all $i \geq 0$, making the DTMC irreducible and aperiodic. The balance equations yield

$$\pi_{i+1} = p_i \pi_i, \quad i \geq 0.$$

Solving the above equation recursively yields

$$\pi_i = \rho_i \pi_0, \quad i \geq 0,$$

where

$$\rho_0 = 1, \quad \rho_i = p_0 p_1 \cdots p_{i-1}, \quad i \geq 1.$$

The normalizing equation yields

$$1 = \sum_{i=0}^{\infty} \pi_i = \pi_0 \left(\sum_{i=0}^{\infty} \rho_i \right).$$

Thus, if $\sum \rho_i$ converges, we have

$$\pi_0 = \left(\sum_{i=0}^{\infty} \rho_i \right)^{-1}.$$

Thus, from Theorem 4.21, we see that the success runs DTMC is positive recurrent if and only if $\sum \rho_i$ converges. Combining with the results of Example 4.15 we get the complete classification as follows: The success runs DTMC is

(i)   positive recurrent if $\sum \rho_i < \infty$,

(ii) null recurrent if $\sum \rho_i = \infty$ and $\sum q_i = \infty$,

(iii) transient if $\sum q_i < \infty$.

When the DTMC is positive recurrent, its limiting distribution is given by

$$\pi_i = \frac{\rho_i}{\sum_{j=0}^{\infty} \rho_j}, \quad i \geq 0.$$

This is also the stationary distribution and the limiting occupancy distribution of the DTMC.

Special Case. Suppose $p_i = p$, for all $i \geq 0$, and $0 < p < 1$. Then

$$\rho_i = p^i, \quad i \geq 0,$$

and hence $\sum \rho_i = 1/(1-p) < \infty$. Hence the DTMC is positive recurrent and its limiting distribution is given by

$$\pi_i = p^i(1-p), \quad i \geq 0.$$

Thus, in the limit, X_n is a modified geometric distribution with parameter $1 - p$. ∎

Example 4.24 General Simple Random Walk. Now consider the general simple random walk with the following transition probabilities:

$$\begin{aligned}
p_{i,i+1} &= p_i \text{ for } i \geq 0, \\
p_{i,i-1} &= q_i = 1 - p_i \text{ for } i \geq 1, \\
p_{0,0} &= 1 - p_0.
\end{aligned}$$

Assume that $0 < p_i < 1$ for all $i \geq 0$, so that the DTMC is irreducible and aperiodic. The balance equations for this DTMC are:

$$\begin{aligned}
\pi_0 &= (1 - p_0)\pi_0 + q_1\pi_1, \\
\pi_j &= p_{j-1}\pi_{j-1} + q_{j+1}\pi_{j+1}, \quad j \geq 1.
\end{aligned}$$

It is relatively straightforward to prove by induction that the solution is given by (see Conceptual Exercise 4.16).

$$\pi_i = \rho_i \pi_0, \quad i \geq 0, \tag{4.40}$$

where

$$\rho_0 = 1, \quad \rho_i = \frac{p_0 p_1 \cdots p_{i-1}}{q_1 q_2 \cdots q_i}, \quad i \geq 1. \tag{4.41}$$

The normalizing equation yields

$$1 = \sum_{i=0}^{\infty} \pi_i = \pi_0 \left(\sum_{i=0}^{\infty} \rho_i \right).$$

Thus, if $\sum \rho_i$ converges, we have

$$\pi_0 = \left(\sum_{i=0}^{\infty} \rho_i \right)^{-1}.$$

Thus, from Theorem 4.21, we see that the general simple random walk is positive recurrent if and only if $\sum \rho_i$ converges. This is consistent with the results of Example 4.16. When the DTMC is positive recurrent, the limiting distribution is given by

$$\pi_i = \frac{\rho_i}{\sum_{j=0}^{\infty} \rho_j}, \quad i \geq 0. \tag{4.42}$$

This is also the stationary distribution and the limiting occupancy distribution of the DTMC. Note that the DTMC is periodic with period 2 if $p_0 = 1$. In this case the expressions for π_i remain valid, but now $\{\pi_i, i \geq 0\}$ is not a limiting distribution.

Special Case 1: Suppose $p_N = 0$, and $p_{N,N} = 1 - q_N$, for a given $N \geq 0$. In this case we can restrict our attention to the irreducible DTMC over $\{0, 1, 2, \cdots, N\}$. In this case $\rho_i = 0$ for $i > N$ and the above results reduce to

$$\pi_i = \frac{\rho_i}{\sum_{j=0}^{N} \rho_j}, \quad 0 \leq i \leq N.$$

Special Case 2. Suppose $p_i = p$ for all $i \geq 0$, and $0 < p < 1$, and let $q = 1 - p$. In this case the DTMC is irreducible and aperiodic, and we have

$$\rho_i = \rho^i, \quad i \geq 0,$$

where $\rho = p/q$. Hence $\sum \rho_i$ converges if $p < q$ and diverges if $p \geq q$. Combining this with results from Example 4.16, we see that this random walk is

(i)   positive recurrent if $p < q$,

(ii) null recurrent if $p = q$,

(iii) transient if $p > q$.

In case $p < q$, the limiting distribution is given by

$$\pi_i = \rho^i(1 - \rho), \quad i \geq 0.$$

Thus, in the limit, X_n is a modified geometric random variable with parameter $1 - \rho$. ■

Example 4.25   Production-Inventory System: Batch Production. Consider the DTMC of Example 2.16 on page 20. It has state-space $S = \{0, 1, 2, \cdots\}$ and transition probabilities:

$$p_{0,j} = \alpha_j, \quad j \geq 0,$$
$$p_{i,j} = \alpha_{j-i+1}, \quad j \geq i - 1 \geq 0,$$

where $\alpha_j \geq 0$ and $\sum \alpha_j = 1$. We assume that $\alpha_0 > 0, \alpha_0 + \alpha_1 < 1$ so that the DTMC is irreducible and aperiodic. The balance equations for this DTMC are given by

$$\pi_i = \sum_{j=0}^{i} \alpha_j \pi_{i-j+1} + \alpha_i \pi_0, \quad i \geq 0. \tag{4.43}$$

This equation looks quite similar to Equation 3.21 on page 70. Hence we follow a similar generating function approach to solve them. Thus we obtain

$$\phi(z) = \sum_{i=0}^{\infty} z^i \pi_i$$

in terms of the known generating function

$$\psi(z) = \sum_{i=0}^{\infty} z^i \alpha_i.$$

Following the steps as in Example 3.13 we get

$$\phi(z) = \pi_0 \frac{\psi(z)(1-z)}{\psi(z) - z}.$$

The normalizing equation yields

$$1 = \sum_{i=0}^{\infty} \pi_i = \lim_{z \to 1} \phi(z)$$

which can be used to compute π_0 as follows: We have

$$
\begin{aligned}
1 &= \pi_0 \lim_{z \to 1} \frac{\psi(z)(1-z)}{\psi(z) - z} \\
&= \pi_0 \frac{-1}{\psi'(1) - 1}, \quad \text{(by L'Hopital's rule)}
\end{aligned}
$$

where

$$\psi'(1) = \frac{d}{dz} \psi(z) \Big|_{z=1} = \sum_{k=0}^{\infty} k \alpha_k = \mu.$$

Thus, if $\mu < 1$, we have $\pi_0 = 1 - \mu > 0$. Hence from Theorem 4.22, we see that the DTMC is positive recurrent if and only if $\mu < 1$. This is consistent with the results of Example 4.17. When the DTMC is positive recurrent, the generating function of its limiting distribution is given by

$$\phi(z) = (1 - \mu) \frac{\psi(z)(1-z)}{\psi(z) - z}. \tag{4.44}$$

We also get

$$\pi_0 = 1 - \mu$$

as the long run probability that the system is empty. A numerical way of computing the limiting distribution is described in Computational Exercise 4.6. ∎

Example 4.26 Production-Inventory System: Batch Demands. Consider the DTMC of Example 4.18. It has state-space $S = \{0, 1, 2, \cdots\}$ and transition probabilities:

$$
\begin{aligned}
p_{i,j} &= \alpha_{i-j+1}, \quad 0 < j \leq i+1, \\
p_{i,0} &= \beta_i = \sum_{k=i+1}^{\infty} \alpha_k, \quad i \geq 0,
\end{aligned}
$$

where $\alpha_j \geq 0$ and $\sum \alpha_j = 1$. We assume that $\alpha_0 > 0, \alpha_0 + \alpha_1 < 1$ so that the DTMC is irreducible. The balance equations for this DTMC are given by

$$\pi_i = \sum_{j=0}^{\infty} \alpha_j \pi_{i+j-1}, \quad i \geq 1.$$

This equation looks the same as Equation 3.20 on page 70. Using the results of Example 3.12, we see that the solution to the balance equations is given by

$$\pi_i = c\rho^i, \quad i \geq 0,$$

where $c > 0$ is a constant, and ρ satisfies

$$\rho = \psi(\rho) = \sum_{k=0}^{\infty} \alpha_k \rho^k. \tag{4.45}$$

From Example 3.11, we know that there is a unique solution to the above equation in $(0,1)$ if and only if $\mu > 1$. This is then the necessary and sufficient condition for positive recurrence. This is consistent with the results of Example 4.18. When the DTMC is positive recurrent (i.e., there is a solution $\rho < 1$ satisfying the above equation), the normalizing equation yields

$$1 = \sum_{i=0}^{\infty} \pi_i = c \sum_{i=0}^{\infty} \rho^i = \frac{c}{1 - \rho}.$$

This yields the limiting distribution as

$$\pi_i = (1 - \rho)\rho^i, \quad i \geq 0. \tag{4.46}$$

Thus, in the limit, X_n is a modified geometric random variable with parameter $1 - \rho$. ∎

Example 4.27   Google PageRank Algorithm. As explained in Section 2.3.7, Google uses irreducible DTMCs in an ingenious way to rank the webpages in the order of their importance so that the most important ones will be ranked at the top when Google returns the search results. Here we explain a simplified version of the algorithm.

Suppose there are N webpages in total that we want to rank. Let $c_{i,j} = 1$ if page i has at least one link pointing to page j, and zero otherwise. Also let $c_i = \sum_{k=1}^{N} c_{i,k}$ be the total number of pages that page i links to. Using the user behavior of Model 2 in Section 2.3.7 we define

$$p_{i,j} = \begin{cases} c_{i,j}/c_i & \text{if } c_i > 0, \\ 1/N & \text{if } c_i = 0. \end{cases}$$

Now suppose a user visits these N webpages in a Markovian fashion as follows: if the user visits webpage i on the nth click, she will visit page j with probability $p_{i,j}$ on the $(n + 1)$st click, regardless of the click history so far. Thus, if the webpage connects to other pages, the user will click on one of them with equal probability. If the current webpage has no outgoing links, the user will jump to a random webpage from among the N webpages. Suppose that the matrix $P = [p_{i,j}]$ is irreducible. Let π_j be the long run occupancy distribution of page j. The theory developed here says that this is independent of the initial visit by the user, and is given by the unique solution to Equations 4.36 and 4.37. Google PageRank algorithm ranks the pages according to π_j's, with the highest occupancy distribution page at the top of the list.

For example, consider a set of four pages $\{1, 2, 3, 4\}$ with links matrix as follows:

$$C = \begin{bmatrix} 0 & 1 & 0 & 1 \\ 1 & 0 & 1 & 1 \\ 1 & 0 & 0 & 0 \\ 0 & 0 & 0 & 0 \end{bmatrix}.$$

Thus page 1 links to pages 2 and 4, page 2 links to 1, 3 and 4, page 3 links to page 1, and page 4 does not link to any other page. The transition probability matrix describing the user behavior is then given by

$$P = \begin{bmatrix} 0 & 1/2 & 0 & 1/2 \\ 1/3 & 0 & 1/3 & 1/3 \\ 1 & 0 & 0 & 0 \\ 1/4 & 1/4 & 1/4 & 1/4 \end{bmatrix}.$$

Note that once the user comes to page 4 she will jump to any of the four pages with equal probability. This is an irreducible DTMC and the limiting occupancy distribution is given by

$$[0.3077 \ 0.2308 \ 0.1538 \ 0.3077].$$

Thus Google will rank pages 1 and 4 as the top two pages, and then page 2 and lastly page 3. We shall see another variation of this algorithm in Example 4.28.

Example 4.28 Google's Damped PageRank Algorithm. We continue the above example with user behavior Model 3 developed in Section 2.3.7, involving a damping parameter d (usually assumed to be around .85). This variation removes the possibility that the P matrix may not be irreducible. The damping factor affects the user behavior as follows. Let the transition probability matrix P be as defined in Example 4.27. After visiting page i, the user flips a coin which turns up heads with probability d. If the coin turns up heads the user decides to visit page j with probability $p_{i,j}$. If the coin turns up tails (which happens with probability $1 - d$), the user decides to visit a random page with equal probability. Thus the user behavior is Markovian, but with a modified transition matrix $\tilde{P} = [\tilde{p}_{i,j}]$ given by

$$\tilde{p}_{i,j} = (1 - d)/N + dp_{i,j}, \quad 1 \le i, j \le N.$$

Now let $\tilde{\pi}_j$ be the limiting probability that the user is on page j in the long run, which we know exists, since the $\tilde{P}$ matrix is irreducible and aperiodic (assuming $d < 1$). Then Google damped PageRank algorithm uses $\tilde{\pi}_j$ as the rank of page j.

For the numerical values in Example 4.27, and $d = .85$, we see that

$$\tilde{P} = \begin{bmatrix} 0.0375 & 0.4625 & 0.0375 & 0.4625 \\ 0.3208 & 0.0375 & 0.3208 & 0.3208 \\ 0.8875 & 0.0375 & 0.0375 & 0.0375 \\ 0.2500 & 0.2500 & 0.2500 & 0.2500 \end{bmatrix}.$$

The limiting distribution $\tilde{\pi}$ is computed to be

$$[0.3069 \; 0.2309 \; 0.1659 \; 0.2963].$$

Thus now the pages ranked as $1, 4, 2, 3$. Other interpretations of this page rank are discussed in Computational Exercises 4.55 and 4.56. Note that if we set $d = 1$, we get back the PageRank algorithm of Example 4.27.

4.7 Limiting Behavior of Reducible DTMCs

In this section we consider a reducible DTMC $\{X_n, n \geq 0\}$ with state-space S. Assume that there are k closed communicating classes $C_i, 1 \leq i \leq k$, and T is the set of states that do not belong to any closed communicating class. Then, as in Equation 4.10, the state-space is partitioned as follows :

$$S = C_1 \cup C_2 \cup \cdots \cup C_k \cup T.$$

Now, if $i \in C_r$ and $j \in C_s$, with $r \neq s$, then i is not accessible from j and vice versa. Hence $p_{i,j} = p_{j,i} = 0$. Now, relabel the states in S by integers such that $i \in C_r$ and $j \in C_s$ with $r < s$ implies that $i < j$, and $i \in C_r$ and $j \in T$ implies that $i < j$. With this relabeling, the matrix P has the following canonical block structure:

$$P = \begin{bmatrix} P(1) & 0 & \cdots & 0 & 0 \\ 0 & P(2) & \cdots & 0 & 0 \\ \vdots & \vdots & \ddots & \vdots & \vdots \\ 0 & 0 & \cdots & P(k) & 0 \\ & D & & & Q \end{bmatrix}. \tag{4.47}$$

Here $P(i)$ is a $|C_i| \times |C_i|$ stochastic matrix $(1 \leq i \leq k)$, Q is a $|T| \times |T|$ sub-stochastic matrix (i.e., all row sums of Q being less than or equal to one, with at least one being strictly less than one), and D is a $|T| \times |S - T|$ matrix. Elementary matrix algebra shows that the nth power of P has the following structure:

$$P^n = \begin{bmatrix} P(1)^n & 0 & \cdots & 0 & 0 \\ 0 & P(2)^n & \cdots & 0 & 0 \\ \vdots & \vdots & \ddots & \vdots & \vdots \\ 0 & 0 & \cdots & P(k)^n & 0 \\ & D_n & & & Q^n \end{bmatrix}.$$

Since $P(r)$ $(1 \leq r \leq k)$ is a transition probability matrix of an irreducible DTMC with state-space C_r, we already know how $P(r)^n$ behaves as $n \to \infty$. Similarly, since all states in T are transient, we know that $Q^n \to 0$ as $n \to \infty$. Thus the study of the limiting behavior of P^n reduces to the study of the limiting behavior of D_n as $n \to \infty$. This is what we proceed to do.

Let $T(r)$ be the first passage time to the set C_r, i.e.,

$$T(r) = \min\{n \geq 0 : X_n \in C_r\}, \quad 1 \leq r \leq k.$$

Let

$$u_i(r) = P(T(r) < \infty | X_0 = i), \quad 1 \leq r \leq k, \ i \in T. \tag{4.48}$$

The next theorem gives a method of computing the above probabilities.

Theorem 4.23   Absorption Probabilities. *The quantities $\{u_i(r), i \in T, 1 \leq r \leq k\}$ are given by the smallest non-negative solution to*

$$u_i(r) = \sum_{j \in C_r} p_{i,j} + \sum_{j \in T} p_{i,j} u_j(r). \tag{4.49}$$

Proof: This proof is similar to the proof of Theorem 3.2 on page 62. See also Conceptual Exercise 3.1. ∎

Using the quantities $\{u_i(r), i \in T, 1 \leq r \leq k\}$ we can describe the limiting behavior of D_n as $n \to \infty$. This is done in the theorem below.

Theorem 4.24   Limit of D_n. *Let $\{u_i(r), i \in T, 1 \leq r \leq k\}$ be as defined in Equation 4.48. Let $i \in T$ and $j \in C_r$.*

(i) If C_r is transient or null recurrent,

$$\lim_{n \to \infty} D_n(i, j) = 0. \tag{4.50}$$

(ii) If C_r is positive recurrent and aperiodic,

$$\lim_{n \to \infty} D_n(i, j) = u_i(r) \pi_j,$$

where $\{\pi_j, j \in C_r\}$ is the unique solution to

$$\pi_j = \sum_{m \in C_r} \pi_m p_{m,j},$$

$$\sum_{m \in C_r} \pi_m = 1.$$

(iii) If C_r is positive recurrent and periodic, $D_n(i, j)$ does not have a limit. However,

$$\lim_{n \to \infty} \frac{1}{n+1} \sum_{m=0}^{n} D_m(i, j) = u_i(r) \pi_j,$$

where $\{\pi_j, j \in C_r\}$ is as in part (ii) above.

Proof: (i). If $u_i(r) = 0$, then $D_n(i, j) = 0$ for all $n \geq 1$, and hence Equation 4.50 follows. If $u_i(r) > 0$, then it is possible to go from i to any state $j \in C_r$, i.e., there is an $m \geq 1$ such that $p_{i,j}^{(m)} = D_m(i, j) > 0$. Since state j is null recurrent or transient, $p_{j,j}^{(n)} \to 0$ as $n \to \infty$. The proof follows along the same lines as that of Theorem 4.14.

(ii). Let $i \in T$ and $j \in C_r$ be fixed. Following Theorem 4.19 let

$$\tilde{T}_j = \min\{n > 0 : X_n = j\},$$

and

$$u_m = \mathsf{P}(\tilde{T}_j = m | X_0 = i).$$

Then, since C_r is a closed recurrent class, $\tilde{T}_j < \infty$ if and only if $T(r) < \infty$. Hence

$$\sum_{m=1}^{\infty} u_m = u_i(r).$$

Thus, given an $\epsilon > 0$, it is possible to find an $N > 0$ such that

$$u_i(r) - \sum_{m=N+1}^{\infty} u_m \leq \epsilon/2$$

and

$$|p_{j,j}^{(m)} - \pi_j| \leq \epsilon/2$$

for all $m \geq N$. The rest of the proof is similar to that of Theorem 4.19.

(iii). Similar to the proof of Theorem 4.21. ∎

We illustrate with an example.

Example 4.29 Genotype Evolution Model. Consider the six-state DTMC of the genotype evolution model with the transition probability matrix as given in Equation 2.14 on page 24. We do not relabel the states. From Example 4.6 we see that this is a reducible DTMC with two closed communicating classes $C_1 = \{1\}, C_2 = \{6\}$, and the transient class $T = \{2, 3, 4, 5\}$. We also have

$$P(1) = [1], \quad P(2) = [1],$$

$$Q = \begin{bmatrix} 1/2 & 0 & 1/4 & 0 \\ 0 & 0 & 1 & 0 \\ 1/4 & 1/8 & 1/4 & 1/4 \\ 0 & 0 & 1/4 & 1/2 \end{bmatrix},$$

and

$$D = \begin{bmatrix} 1/4 & 0 \\ 0 & 0 \\ 1/16 & 1/16 \\ 0 & 1/4 \end{bmatrix}.$$

Remember that the rows of D are indexed $\{2, 3, 4, 5\}$ and the columns are indexed $\{1, 6\}$.

We have

$$P(1)^n \to [1], \quad P(2)^n \to [1],$$

$$Q^n \to \begin{bmatrix} 0 & 0 & 0 & 0 \\ 0 & 0 & 0 & 0 \\ 0 & 0 & 0 & 0 \\ 0 & 0 & 0 & 0 \end{bmatrix}.$$

Equations for $\{u_i(1), i \in T\}$ are given by

$$\begin{aligned}
u_2(1) &= .25 + .5u_2(1) + .25u_4(1) \\
u_3(1) &= u_4(1) \\
u_4(1) &= .0625 + .25u_2(1) + .125u_3(1) + .25u_4(1) + .25u_5(1) \\
u_5(1) &= .25u_4(1) + .5u_5(1).
\end{aligned}$$

The solution is given by

$$[u_2(1), \ u_3(1), \ u_4(1), \ u_5(1)] = [.75, \ .5, \ .5, \ .25].$$

Similar calculations yield

$$[u_2(2), \ u_3(2), \ u_4(2), \ u_5(2)] = [.25, \ .5, \ .5, \ .75].$$

Hence we get

$$D_n \to \begin{bmatrix} 3/4 & 1/4 \\ 1/2 & 1/2 \\ 1/2 & 1/2 \\ 1/4 & 3/4 \end{bmatrix},$$

and

$$\lim_{n \to \infty} P^{(n)} = \begin{bmatrix} 1 & 0 & 0 & 0 & 0 & 0 \\ 3/4 & 0 & 0 & 0 & 0 & 1/4 \\ 1/2 & 0 & 0 & 0 & 0 & 1/2 \\ 1/2 & 0 & 0 & 0 & 0 & 1/2 \\ 1/4 & 0 & 0 & 0 & 0 & 3/4 \\ 0 & 0 & 0 & 0 & 0 & 1 \end{bmatrix}.$$

This matches the result in Example 4.6. ■

4.8 DTMCs with Costs and Rewards

Let X_n be the state of a system at time n. Suppose $\{X_n, n \geq 0\}$ is a DTMC with state-space S and transition probability matrix P. Furthermore, the system incurs an expected cost of $c(i)$ at time n if $X_n = i$. For other cost models, see Conceptual Exercises 4.17 and 4.18. Rewards can be thought of as negative costs. We consider costs incurred over infinite horizon. For the analysis of costs over finite horizon, see Conceptual Exercises 4.19 and 4.20.

4.8.1 Discounted Costs

Suppose the costs are discounted at rate α, where $0 \leq \alpha < 1$ is a fixed discount factor. Thus if the system incurs a cost c at time n, its present value at time 0 is $\alpha^n c$, i.e., it is equivalent to incurring a cost of $\alpha^n c$ at time zero. Let C be the total discounted cost over the infinite horizon, i.e.,

$$C = \sum_{n=0}^{\infty} \alpha^n c(X_n).$$

Let $\phi(i)$ be the expected total discounted cost (ETDC) incurred over the infinite horizon starting with $X_0 = i$. That is,

$$\phi(i) = \mathsf{E}(C|X_0 = i).$$

The next theorem gives the main result regarding the ETDC. We introduce the following column vectors

$$c = [c(i)]_{i \in S}, \quad \phi = [\phi(i)]_{i \in S}.$$

Theorem 4.25 ETDC. *Suppose $0 \leq \alpha < 1$. Then ϕ is given by*

$$\phi = (I - \alpha P)^{-1} c. \tag{4.51}$$

Proof: Let C_1 be the total discounted cost incurred over $\{1, 2, \cdots\}$ discounted back to time 0, i.e.,

$$C_1 = \sum_{n=1}^{\infty} \alpha^n c(X_n).$$

From time-homogeneity it is clear that

$$\mathsf{E}(C_1|X_1 = j) = \alpha\phi(j), \quad j \in S.$$

Using the first-step analysis we get

$$\begin{aligned}
\mathsf{E}(C_1|X_0 = i) &= \sum_{j \in S} p_{i,j} \mathsf{E}(C_1|X_0 = i, X_1 = j) \\
&= \alpha \sum_{j \in S} p_{i,j} \phi(j).
\end{aligned}$$

Hence,

$$\begin{aligned}
\phi(i) &= \mathsf{E}(C|X_0 = i) \\
&= \mathsf{E}(c(X_0) + C_1|X_0 = i) \\
&= c(i) + \alpha \sum_{j \in S} p_{i,j} \phi(j).
\end{aligned}$$

In matrix form the above equation becomes

$$\phi = c + \alpha P \phi. \tag{4.52}$$

The matrix $I - \alpha P$ is invertible for $0 \leq \alpha < 1$. Hence we get Equation 4.51. ∎

Note that there is no assumption of transience or recurrence or periodicity or irreducibility behind the above theorem. Equation 4.51 is valid for any transition probability matrix P.

Example 4.30 Two-State Machine. Consider the two-state machine of Example 2.6 on page 15. It was modeled by a DTMC $\{X_n, n \geq 0\}$ with state-space $\{0, 1\}$ (0 being down, and 1 being up), and transition probability matrix

$$P = \begin{bmatrix} p_d & 1 - p_d \\ 1 - p_u & p_u \end{bmatrix},$$

where $0 \leq p_u, p_d \leq 1$. Now suppose the machine produces a revenue of \$$r$ per day when it is up, and it costs \$$d$ in repair costs per day when the machine is down. Suppose a new machine in working order costs \$$m$. Is it profitable to purchase it if the discount factor is $0 \leq \alpha < 1$?

Let $c(i)$ be the expected cost of visiting state i. We have

$$c = [c(0) \ \ c(1)]^\top = [d \ \ -r]^\top.$$

Then, using Theorem 4.25 we get

$$\phi = [\phi(0) \ \ \phi(1)]^\top = (I - \alpha P)^{-1} c.$$

Direct calculations yield

$$\phi = \frac{1}{(1 - \alpha p_u)(1 - \alpha p_d) - \alpha^2(1 - p_u)(1 - p_d)} \begin{bmatrix} d(1 - \alpha p_u) - r\alpha(1 - p_d) \\ d\alpha(1 - p_u) - r(1 - \alpha p_d) \end{bmatrix}.$$

Thus it is profitable to buy a new machine if the expected total discounted net revenue from a new machine over the infinite horizon is greater than the initial purchase price of m, i.e., if

$$m \leq \frac{r(1 - \alpha p_d) - d\alpha(1 - p_u)}{(1 - \alpha p_u)(1 - \alpha p_d) - \alpha^2(1 - p_u)(1 - p_d)}.$$

Note that the above inequality reduces to $m \leq r$ if $\alpha = 0$, as expected. How much should you be willing to pay for a machine in down state? ∎

4.8.2 Average Costs

The discounted costs have the disadvantage that they depend upon the discount factor and the initial state, thus making decisions more complicated. These issues are addressed by considering the long run cost per period, called the average cost. The expected total cost up to time N, starting from state i, is given by $\mathsf{E}(\sum_{n=0}^{N} c(X_n)|X_0 = i)$. Dividing it by $N + 1$ gives the cost per period. Hence the long run expected cost per period is given by:

$$g(i) = \lim_{N \to \infty} \frac{1}{N + 1} \mathsf{E}\left(\sum_{n=0}^{N} c(X_n) \mid X_0 = i\right),$$

assuming that the above limit exists. To keep the analysis simple, we will assume that the DTMC is irreducible and positive recurrent with limiting occupancy distribution given by $\{\pi_j, j \in S\}$. Intuitively, it makes sense that the long run cost per period should be given by $\sum \pi_j c(j)$, independent of the initial state i. This intuition is formally proved in the next theorem:

Theorem 4.26 Average Cost. Suppose $\{X_n, n \geq 0\}$ is an irreducible positive recurrent DTMC with limiting occupancy distribution $\{\pi_j, j \in S\}$. Suppose

$$\sum_{j \in S} \pi_j |c(j)| < \infty.$$

Then

$$g(i) = g = \sum_{j \in S} \pi_j c(j).$$

Proof: Let $M_{i,j}^{(N)}$ be the expected number of visits to state j over $\{0, 1, 2, \cdots, N\}$ starting from state i. See Section 2.5. Then, we see that

$$
\begin{aligned}
g(i) &= \lim_{N \to \infty} \frac{1}{N+1} \sum_{j \in S} M_{i,j}^{(N)} c(j) \\
&= \lim_{N \to \infty} \sum_{j \in S} \frac{M_{i,j}^{(N)}}{N+1} c(j) \\
&= \sum_{j \in S} \lim_{N \to \infty} \frac{M_{i,j}^{(N)}}{N+1} c(j) \\
&= \sum_{j \in S} \pi_j c(j).
\end{aligned}
$$

Here the last interchange of sum and the limit is allowed because the DTMC is positive recurrent. The last equality follows from Theorem 4.21. ∎

We illustrate with an example.

Example 4.31 Brand Switching. Consider the model of brand switching as described in Example 2.14, where a customer chooses among three brands of beer, say A, B, and C, every week when he buys a six-pack. Let X_n be the brand he purchases in week n. We assume that $\{X_n, n \geq 0\}$ is a DTMC with state-space $S = \{A, B, C\}$ and transition probability matrix given below:

$$P = \begin{bmatrix} 0.1 & 0.2 & 0.7 \\ 0.2 & 0.4 & 0.4 \\ 0.1 & 0.3 & 0.6 \end{bmatrix}.$$

Now suppose a six-pack costs \$6.00 for brand A, \$5.00 for brand B, and \$4.00 for brand C. What is the weekly expenditure on beer by the customer in the long run?

We have
$$c(A) = 6, \quad c(B) = 5, \quad c(C) = 4.$$

Also solving the balance and normalizing equations we get
$$\pi_A = 0.132, \quad \pi_B = 0.319, \quad \pi_C = 0.549.$$

Hence the long run cost per week is
$$g = 6\pi_A + 5\pi_B + 4\pi_C = 4.583.$$

Thus the customer spends $4.58 per week on beer. ∎

It is possible to use the results in Section 4.7 to extend this analysis to reducible DTMCs. However, the long run cost rate may depend upon the initial state in that case.

4.9 Reversibility

In this section we study a special class of DTMCs called the reversible DTMCs. Intuitively, if we watch a movie of a reversible DTMC we will not be able to tell whether the time is running forward or backward. Thus, the probability of traversing a cycle of $r + 1$ states $i_0 \to i_1 \to i_2 \to \cdots \to i_{r-1} \to i_r \to i_0$ is the same as traversing it in reverse order $i_0 \to i_r \to i_{r-1} \to \cdots \to i_2 \to i_1 \to i_0$. We make this more precise in the definition below.

Definition 4.11   Reversibility. *A DTMC with state-space S and transition probability matrix P, is called reversible if for every $r \geq 1$, and $i_0, i_1, \cdots, i_r$,*

$$p_{i_0,i_1} p_{i_1,i_2} \cdots p_{i_{r-1},i_r} p_{i_r,i_0} = p_{i_0,i_r} p_{i_r,i_{r-1}} p_{i_{r-1},i_{r-2}} \cdots p_{i_1,i_0}. \tag{4.53}$$

The next theorem gives a simple necessary and sufficient condition for reversibility.

Theorem 4.27   *An irreducible, positive recurrent DTMC with state-space S, transition probability matrix P and stationary distribution $\{\pi_i, i \in S\}$ is reversible if and only if*

$$\pi_i p_{i,j} = \pi_j p_{j,i}, \quad i, j \in S. \tag{4.54}$$

Proof: Suppose the DTMC is irreducible and positive recurrent, and Equation 4.54 holds. Consider a cycle of states $i_0, i_1, \cdots, i_r, i_0$. Then we get

$$
\begin{aligned}
\pi_{i_0} p_{i_0,i_1} p_{i_1,i_2} \cdots p_{i_{r-1},i_r} p_{i_r,i_0} &= p_{i_1,i_0} \pi_{i_1} p_{i_1,i_2} \cdots p_{i_{r-1},i_r} p_{i_r,i_0} \\
&\quad\vdots \\
&= p_{i_1,i_0} p_{i_2,i_1} \cdots p_{i_r,i_{r-1}} p_{i_0,i_r} \pi_{i_0}.
\end{aligned}
$$

Since the DTMC is positive recurrent, $\pi_{i_0} > 0$. Hence the above equation implies Equation 4.53.

To prove necessity, suppose Equation 4.53 holds. Summing over paths of length r from i_1 to i_0 we get

$$p_{i_0,i_1} p_{i_1,i_0}^{(r)} = p_{i_0,i_1}^{(r)} p_{i_1,i_0}.$$

Now sum over $r = 0$ to n, and divide by $n+1$ to get

$$p_{i_0,i_1} \left(\frac{1}{n+1} \sum_{r=0}^{n} p_{i_1,i_0}^{(r)} \right) = \left(\frac{1}{n+1} \sum_{r=0}^{n} p_{i_0,i_1}^{(r)} \right) p_{i_1,i_0},$$

or

$$p_{i_0,i_1} \frac{M_{i_1,i_0}^{(n)}}{n+1} = \frac{M_{i_0,i_1}^{(n)}}{n+1} p_{i_1,i_0}.$$

Now let $n \to \infty$. Since the DTMC is irreducible and positive recurrent, we can use Theorem 4.21 to get

$$p_{i_0,i_1} \pi_{i_0} = p_{i_1,i_0} \pi_{i_1},$$

which is Equation 4.54. ∎

The Equations 4.54 are called the *local balance* or *detailed balance* equations, as opposed to Equations 4.38, which are called *global balance* equations. Intuitively, the local balance equations say that, in steady state, the expected number of transitions from state i to j per period is the same as the expected number of transitions per period from j to i. This is in contrast to stationary DTMCs that are not reversible: for such DTMCs the global balance equations imply that, in steady state, the expected number of transitions out of a state per period is the same as the expected number of transitions into that state per period. It can be shown that the local balance equations imply global balance equations, but not the other way. See Conceptual Exercise 4.21.

Example 4.32 Symmetric DTMCs. Consider an irreducible DTMC on a finite state-space $\{1, 2, \cdots, N\}$ with symmetric transition probability matrix P. Show that it is reversible.

Since P is symmetric, it must be doubly stochastic. Since the DTMC is assumed to be irreducible, it has a unique stationary distribution given by

$$\pi_j = 1/N, \quad j \in S.$$

See Conceptual Exercise 4.22. Thus

$$\pi_i p_{i,j} = (1/N)p_{i,j} = (1/N)p_{j,i} = \pi_j p_{j,i}, \quad i, j \in S.$$

Here the second equality is a consequence of symmetry. Thus the DTMC satisfies local balance equations, and hence is reversible. ∎

Example 4.33 General Simple Random Walk. Consider a positive recurrent general simple random walk as described in Example 4.24. Show that it is a reversible DTMC.

From Example 4.24 we see that the limiting occupancy distribution is given by

$$\pi_i = \rho_i \pi_0, \quad i \geq 0,$$

where

$$\rho_0 = 1, \quad \rho_i = \frac{p_0 p_1 \cdots p_{i-1}}{q_1 q_2 \cdots q_i}, \quad i \geq 1.$$

Hence we have

$$\pi_i p_{i,i+1} = \pi_0 \frac{p_0 p_1 \cdots p_{i-1}}{q_1 q_2 \cdots q_i} p_i = \pi_0 \frac{p_0 p_1 \cdots p_{i-1} p_i}{q_1 q_2 \cdots q_{i+1}} q_{i+1} = \pi_{i+1} p_{i+1,i}, \ i \geq 0.$$

Since the only transitions in a simple random walk are from i to $i+1$, and i to $\max(i-1, 0)$, we see that Equations 4.54 are satisfied. Hence the DTMC is reversible. ∎

Reversible DTMCs are nice because it is particularly easy to compute their stationary distribution. A large class of reversible DTMCs are the tree DTMCs, of which the general simple random walk is a special case. See Conceptual Exercise 4.23 for details.

We end this section with an interesting result about the eigenvalues of a reversible DTMC with finite state-space.

Theorem 4.28 Eigenvalues and Reversibility. *Let $\{X_n, n \geq 0\}$ be an irreducible, positive recurrent, reversible DTMC with finite state-space $S = \{1, 2, \cdots, N\}$ and transition probability matrix P. Then all the eigenvalues of P are real.*

Proof: Since $\{X_n, n \geq 0\}$ is irreducible, positive recurrent , and reversible, we have

$$\pi_i p_{i,j} = \pi_j p_{j,i},$$

where $\{\pi_i, i \in S\}$ is the stationary distribution. Now let $\Pi = \text{diag}[\pi_1, \pi_2, \cdots, \pi_N]$. Then the above equation can be written in matrix form as

$$\Pi P = P^\top \Pi,$$

where $P^\top$ is the transpose of P. This can be rewritten as

$$\Pi^{1/2} P \Pi^{-1/2} = \Pi^{-1/2} P^\top \Pi^{1/2}, \tag{4.55}$$

where $\Pi^m = \text{diag}[\pi_1^m, \pi_2^m, \cdots, \pi_N^m]$ for $m = \pm 1/2$. Now let λ be an eigenvalue of $\Pi^{1/2} P \Pi^{-1/2}$ and x be the corresponding right eigenvector, i.e.,

$$\Pi^{1/2} P \Pi^{-1/2} x = \lambda x.$$

The above equation can be written as

$$P \Pi^{-1/2} x = \lambda \Pi^{-1/2} x.$$

Thus λ is an eigenvalue of P. Similarly, any eigenvalue of P is an eigenvalue of $\Pi^{1/2} P \Pi^{-1/2}$. However, Equation 4.55 implies that $\Pi^{1/2} P \Pi^{-1/2}$ is symmetric, and hence all its eigenvalues are real. Thus all eigenvalues of P must be real. ∎

4.10 Computational Exercises

4.1 Numerically study the limiting behavior of the marginal distribution and the occupancy distribution in the brand switching example of Example 2.14 on page 19, by studying P^n and $M^{(n)}/(n+1)$ for $n = 1$ to 20.

4.2 Study the limiting behavior of the two-state DTMC of Example 2.3 if

1. $\alpha + \beta = 0$,
2. $\alpha + \beta = 2$.

4.3 Numerically study the limiting behavior of the random walk on $\{0, 1, \cdots, N\}$ with

$$p_{0,1} = p_{0,0} = p_{N,N} = p_{N,N-1} = 1/2, \quad p_{i,i+1} = p_{i,i-1} = 1/2, \ 1 \leq i \leq N - 1,$$

by studying P^n and $M^{(n)}/(n+1)$ for $n = 1$ to **20**, and $N = 1$ to **5**. What is your guess for the limiting values for a general N?

4.4 Establish the conditions of transience, null recurrence, and positive recurrence for the space-nonhomogeneous simple random walk on $\{0, 1, 2, \cdots\}$ with the following transition probabilities:

$$
\begin{aligned}
p_{i,i+1} &= p_i \ \text{for } i \geq 0, \\
p_{i,i-1} &= q_i \ \text{for } i \geq 1, \\
p_{i,i} &= r_i \ \text{for } i \geq 0.
\end{aligned}
$$

Assume that $0 < p_i < 1 - r_i$ for all $i \geq 0$.

4.5 Establish the conditions of transience, null recurrence, and positive recurrence for the simple random walk on $\{0, 1, 2, \cdots\}$ with reflecting boundary at zero, i.e.,

$$
\begin{aligned}
p_{i,i+1} &= p \ \text{for } i \geq 1, \\
p_{i,i-1} &= q \ \text{for } i \geq 1, \\
p_{0,1} &= 1.
\end{aligned}
$$

Assume that $0 < p < 1$.

4.6 Consider the DTMC of Example 4.17. Suppose it is positive recurrent. Show that the limiting distribution $\{\pi_j, j \geq 0\}$ can be recursively calculated by

$$
\begin{aligned}
\pi_0 &= 1 - \mu, \\
\pi_1 &= \frac{1 - \alpha_0}{\alpha_0} \pi_0, \\
\pi_2 &= \frac{1 - \alpha_0 - \alpha_1}{\alpha_0} (\pi_0 + \pi_1), \\
\pi_{j+1} &= \frac{1 - \sum_{i=0}^{j} \alpha_i}{\alpha_0} \left(\sum_{i=0}^{j} \pi_i \right) \\
&\quad + \sum_{i=2}^{j} \pi_i \sum_{k=j-i+2}^{j} \frac{\alpha_k}{\alpha_0}, \ j \geq 2.
\end{aligned}
$$

This is a stable recursion since it involves adding positive terms.

4.7 Study the limiting behavior of the random walk in Computational Exercise 4.5, assuming it is positive recurrent.

4.8 Study the limiting behavior of the random walk in Computational Exercise 4.4, assuming it is positive recurrent.

4.9 Classify the states of the brand switching model of Example 4.31. That is, identify the communicating classes, and state whether they are transient, null recurrent, or positive recurrent, and whether they are periodic or aperiodic.

4.10 Study the limiting behavior of P^n as $n \to \infty$ for the following transition probability matrices by using the method of diagonalization to compute P^n and then letting $n \to \infty$.

$$(a). \begin{bmatrix} .5 & .5 & 0 & 0 \\ .5 & 0 & .5 & 0 \\ 0 & .5 & 0 & .5 \\ 0 & 0 & .5 & .5 \end{bmatrix}, \quad (b). \begin{bmatrix} 0 & 1 & 0 & 0 \\ .5 & 0 & .5 & 0 \\ 0 & .5 & 0 & .5 \\ 0 & 0 & 1 & 0 \end{bmatrix}.$$

4.11 Compute the limit of $M^{(n)}/(n+1)$ as $n \to \infty$ for the P matrices in Computational Exercise 4.10, using the method of diagonalization to compute the P^n.

4.12 Do a complete classification of the states of the DTMC in

1. Example 2.13 on page 18,
2. The Wright–Fisher model on page 25,
3. The Moran model on page 25.

4.13 Do a complete classification of the states of the DTMCs with the following transition probability matrices:

$$(a). \begin{bmatrix} .5 & .5 & 0 & 0 \\ 0 & .5 & .5 & 0 \\ 0 & 0 & .5 & .5 \\ 0 & 0 & 0 & 1 \end{bmatrix}, \quad (b). \begin{bmatrix} .2 & .8 & 0 & 0 \\ .6 & .4 & 0 & 0 \\ .2 & .3 & .5 & 0 \\ 0 & 0 & 0 & 1 \end{bmatrix}.$$

4.14 Consider a DTMC on all integers with following transition probabilities

$$p_{i,i+1} = \alpha_i, p_{i,0} = 1 - \alpha_i, \quad i > 0,$$
$$p_{i,i-1} = \beta_{-i}, p_{i,0} = 1 - \beta_{-i}, \quad i < 0,$$
$$p_{0,1} = p, p_{0,-1} = 1 - p.$$

Find the condition for transience, null recurrence, and positive recurrence.

4.15 Classify the following success runs DTMCs as transient, null recurrent, or positive recurrent

1. $p_i = 1/(1+i), \quad i \geq 0,$

2. $q_0 = 0$, $q_i = 1/(1+i)$, $i \geq 1$,

3. $p_i = \begin{cases} 1/(1+i) & \text{if } i \text{ even,} \\ i/(1+i) & \text{if } i \text{ odd.} \end{cases}$

4.16 State whether the following simple random walks on non-negative integers are transient, null recurrent, or positive recurrent.

1. $p_{0,1} = 1$, $p_{i,i+1} = p$, $p_{i,i-1} = q = 1 - p$, $i \geq 1$,
2. $p_{0,1} = 1$, $p_{i,i+1} = 1/(1+i)$, $p_{i,i-1} = i/(1+i)$, $i \geq 1$,
3. $p_{0,1} = 1$, $p_{i,i+1} = i/(1+i)$, $p_{i,i-1} = 1/(1+i)$, $i \geq 1$.

4.17 Consider a DTMC on non-negative integers with following transition probabilities

$$p_{0,i} = p_i, \qquad i \geq 0,$$
$$p_{i,i-1} = 1, \qquad i \geq 1.$$

Assume that $\sum p_i = 1$. Establish the condition for transience, null recurrence, and positive recurrence.

4.18 Use Pakes' lemma to derive a sufficient condition for the positive recurrence of the DTMC in Example 2.17 on page 21.

4.19 Show that the discrete time queue of Example 2.12 is positive recurrent if $p < q$, null recurrent if $p = q$, and transient if $p > q$.

4.20 Compute the limiting distribution of the DTMC of Computational Exercise 2.9.

4.21 Compute the limiting distributions and the limiting occupancy distributions of the DTMCs in Computational Exercise 4.10.

4.22 Is the DTMC in Computational Exercise 3.9 positive recurrent? If so, compute its limiting distribution.

4.23 Numerically compute the limiting distribution of the number of passengers on the bus in the Modeling Exercise 2.21. Assume that the bus capacity is 20, the number of passengers waiting for the bus at any stop is a Poisson random variable with mean 10, and that a passenger on the bus alights at any stop with probability .4.

4.24 Consider the following modification of the transition probabilities of the DTMC of Example 2.16:

$$p_{0,j} = \beta_j, \quad j \geq 0$$
$$p_{i,j} = \alpha_{j-i+1}, \quad j \geq i - 1 \geq 0,$$

with $\alpha_j, \beta_j \geq 0$, $\sum_{j=0}^{\infty} \alpha_j = \sum_{j=0}^{\infty} \beta_j = 1$, $\mu = \sum_{j=0}^{\infty} j\alpha_j$, $\nu = \sum_{j=0}^{\infty} j\beta_j$.

Assume the chain is irreducible. Show that this DTMC is positive recurrent if and only if $\mu < 1$. Let $\{\pi_j, j \in S\}$ be the limiting distribution, assuming it is positive recurrent. Show that

$$\phi(z) = \sum_{j=0}^{\infty} z^j \pi_j = \pi_0 \frac{A(z) - zB(z)}{A(z) - z},$$

where

$$A(z) = \sum_{j=0}^{\infty} z^j \alpha_j, \quad B(z) = \sum_{j=0}^{\infty} z^j \beta_j,$$

and

$$\pi_0 = \frac{1 - \mu}{1 - \mu + \nu}.$$

4.25 Compute the long run fraction of patients who receive drug i if we follow the play the winner policy of Modeling Exercise 2.13.

4.26 Let S be a set of all permutations of the integers $1, 2, \cdots, N$. Let $X_0 = (1, 2, \cdots, N)$ be the initial permutation. Construct X_{n+1} from X_n by interchanging the I-th and J-th component of X_n, where I and J are independent and uniformly distributed over $\{1, 2, \cdots, N\}$. (If $I = J$, then $X_{n+1} = X_n$.) Show that $\{X_n, n \geq 0\}$ is a DTMC and compute its limiting distribution. (Conceptual Exercise 4.22 may be useful here.) This gives one method of generating a permutation uniformly from the set of all permutations.

4.27 Compute the limiting distribution of the DTMC in Example 2.17 in the following cases

1. $\alpha_i = \alpha^i (1 - \alpha), \quad i \geq 0$, for a fixed constant $\alpha > 1/2$,
2. $\alpha_i = \frac{1}{m+1}, \quad 0 \leq i \leq m$, for a constant $m > 2$.

You may need to solve Equation 4.45 numerically.

4.28 Let X_n be the sum of the first n outcomes of tossing a six-sided fair die repeatedly and independently. Compute

$$\lim_{n \to \infty} \mathsf{P}(X_n \text{ is dividible by } 7).$$

4.29 (This is a generalization of Computational Exercise 4.28.) Let $\{Y_n, n \geq 1\}$ be a sequence of iid random variables with common pmf $\mathsf{P}(Y_n = k) = \alpha_k, \quad k = 1, 2, 3, \cdots$. Let $X_n = Y_1 + \cdots + Y_n$. Compute

$$\lim_{n \to \infty} \mathsf{P}(X_n \text{ is dividible by } r),$$

where r is fixed positive integer.

4.30 Consider the discrete time queue of Example 2.12 on page 17. Assume that the queue is positive recurrent and compute its limiting distribution.

4.31 Compute the limiting distribution of the number of white balls in urn A in the urn model described in Example 2.13 on page 18.

4.32 Consider Modeling Exercise 2.18 on page 47. Assume that u, the expected up time and d, the expected down time are finite.

1. Is this DTMC irreducible? Positive recurrent? Aperiodic?
2. Show that the long run probability that the machine is up is given by $u/(u + d)$.

4.33 Consider Modeling Exercise 2.1 on page 44. Assume that τ, the expected lifetime of each light bulb, is finite. Show that the DTMC is positive recurrent and compute its limiting distribution.

4.34 Consider Modeling Exercise 2.19 on page 47. Let $\mu = E(Y_n)$, and $v = \text{Var}(Y_n)$. Suppose $X_0 = 0$ with probability 1. Compute $\mu_n = E(X_n)$ as a function of n.

4.35 Show that the DTMC in Modeling Exercise 2.19 on page 47 is positive recurrent if $\mu < \infty$. Let $\psi(z)$ be the generating function of Y_n and $\phi_n(z)$ be the generating function of X_n. Compute $\phi(z)$, the generating function of the limiting distribution of X_n.

4.36 Consider Modeling Exercise 2.22 on page 48.

1. Is this DTMC irreducible? Aperiodic?
2. Compute the long run average fraction of items inspected.
3. Suppose items found to be defective are instantaneously repaired before being shipped. Compute the long run fraction of defective items in the shipped items.

4.37 Consider Modeling Exercise 2.20 on page 47. Compute the long run fraction of the time the audio quality is impaired.

4.38 Consider Modeling Exercise 2.32 on page 50. Compute

1. the long run fraction of the time both machines are working,
2. the average number of assemblies shipped per period,
3. the fraction of the time the i-th machine is off.

4.39 Consider the DTMC in Modeling Exercise 2.24 on page 48. Let τ_i be the mean lifetime of the machine from vendor i, $i = 1, 2$. Assume $\tau_i < \infty$. Show that the DTMC is positive recurrent and compute the long run fraction of the time the machine from vendor 1 is in use.

4.40 Compute the limiting distribution of the DTMC in Modeling Exercise 2.27 on page 49.

4.41 Compute the limiting distribution of the DTMC in Modeling Exercise 2.28 on page 49.

4.42 Compute the limiting distribution of the DTMC in Modeling Exercise 2.6 on page 44.

4.43 Compute $\lim_{n\to\infty} P^n$ for the following probability transition matrices:

$$
(a).\begin{bmatrix} .3 & .7 & 0 & 0 \\ .4 & .6 & 0 & 0 \\ .2 & .3 & .5 & 0 \\ .5 & 0 & .4 & .1 \end{bmatrix}, \quad (b).\begin{bmatrix} .3 & .3 & .4 & 0 & 0 & 0 \\ .5 & 0 & .5 & 0 & 0 & 0 \\ .9 & 0 & .1 & 0 & 0 & 0 \\ 0 & 0 & 0 & 1 & 0 & 0 \\ 0 & 0 & .4 & .2 & .2 & .2 \\ 0 & .1 & 0 & .4 & .1 & .4 \end{bmatrix}.
$$

4.44 Let P be the transition probability matrix of the gambler's ruin DTMC in the Example 2.11 on page 17. Compute $\lim_{n\to\infty} P^n$.

4.45 Let $\{X_n, n \geq 0\}$ be the DTMC of Modeling Exercise 2.33. Suppose it costs C_1 dollars to replace a failed light bulb, and C_2 dollars to replace a working light bulb. Compute the long run cost of this replacement policy.

4.46 A machine produces two items per day. Each item is non-defective with probability p, the quality of the successive items being independent. Defective items are thrown away immediately, and the non-defective items are stored to satisfy the demand of one item per day. Any demand that cannot be satisfied immediately is lost. Let X_n be the number of items in storage at the beginning of each day (before the production and demand for that day is taken into account). Show that $\{X_n, n \geq 0\}$ is a DTMC. When is it positive recurrent? Compute its limiting distribution when it exists.

4.47 Suppose it costs $\$c$ to store an item for one day and $\$d$ for every demand that is lost. Compute the long run cost per day of operating the facility in Computational Exercise 4.46 assuming it is stable.

4.48 Consider the production facility of Computational Exercise 4.46. Suppose the following operating procedure is followed. The machine is turned off as soon the inventory level (that is, the number of items in storage) reaches K, a fixed positive integer. It then remains off until the inventory reduces to k, another fixed positive integer less than K, at which point it is turned on again. Model this by an appropriate DTMC and compute

1. the steady state probability that the machine is off,
2. the steady state probability that there are i items in storage, $0 \leq i \leq K$.

4.49 Consider the production facility of Computational Exercise 4.48. Suppose shutting the machine down costs $\$A$ and turning it on costs $\$B$. Compute the long run cost per day of operating this system.

4.50 Consider the DTMC developed in the Modeling Exercise 2.14.

1. Is the DTMC irreducible and aperiodic?

2. What is the long run probability that the machine is up and running?

3. Suppose repairing the ith component costs c_i dollars per day ($i = 1, 2$). The daily output of a working machine generates a revenue of R dollars. What is the long run net revenue per day of this machine? Find the minimum value of R for which it is profitable to operate this machine.

4.51 Consider the beer-guzzling Mr. Al Anon of Modeling Exercise 2.16 on page 46. Assume the following data is given:

$$\alpha = .6, \ \beta = .2, \ H = \$4.00, \ L = \$3.00.$$

Compute the long run daily beer expense of Mr. Al Anon.

4.52 Price of an item fluctuates between two levels: high ($\$H$) and low ($\$L$). Let Y_n be the price of the item at the beginning of day n. Suppose $\{Y_n, n \geq 0\}$ is a DTMC with state-space $\{0, 1\}$, where state 0 is low, and state 1 is high. The transition probability matrix is

$$\begin{bmatrix} 1 - \alpha & \alpha \\ \beta & 1 - \beta \end{bmatrix}.$$

A production facility consumes one such item at the beginning of each day and has a storage capacity of K. The production manager uses the following procurement policy: If the price of the item at the beginning of day n is low, then procure enough items instantly to bring the number of items in stock to K. If the price level is high, and there is at least one item in stock, do nothing; else procure one item instantly. Let X_n be the number of items in stock at the beginning of day n. Suppose the inventory costs are charged at the rate of $\$h$ per item in stock at the beginning of a day per day. (Assume the following sequence of events: at the beginning of the nth day, we first consume an item, then observe the new price level Y_n, then procure the quantity determined by the above rule, then observe the new stock level X_n.)

1. Compute the long run procurement plus holding cost per day of following this policy.

2. Suppose the price is known to be $25 on 20% of the days, and $15 for 80% of the days on the average. The price cycle lasts on the average 25 days. The holding cost is $.50 per item per day. Plot or tabulate the long run average cost per day as a function of K. What is the value of K that minimizes the cost rate?

4.53 Consider the DTMC model of a buffer of size B as described in Modeling Exercise 2.20.

1. At the beginning, when the play program is initiated, the system waits until the buffer has a fixed number b of bytes, and then starts playing. Let T be the first time either the buffer becomes empty, or some bytes are lost. A song of K bytes plays flawlessly if $T > K$. Show how to compute the probability that the song of K bytes plays flawlessly.

2. Consider the following parameters:

$$B = 100, \quad \alpha_0 = .2, \quad \alpha_1 = .5, \quad K = 512.$$

What value of b should be used to maximize the probability that this song is played flawlessly? What is this maximum probability?

4.54 Consider the DTMC developed in Modeling Exercise 2.12 using the function f given there.

1. Compute its steady state distribution.

2. Compute the long run gasoline cost per day to the student. Compare this with the long run gasoline cost per day if the student filled up the tank (capacity = 11 gallons) whenever it became empty, regardless of the price of gasoline.

4.55 Consider the damped page rank algorithm of Example 4.28. Consider the following model of user behavior. She starts at a random page, picked uniformly from the N pages. Then she clicks according to a DTMC determined by the transition matrix P of Example 4.27 for a random number T of times with $P(T = k) = d^k(1-d)$, for $k \geq 0$. Define the page rank $\hat{\pi}_j$ as the probability that at time T (when she stops clicking) she ends up on page j $(1 \leq j \leq N)$. Show that $\hat{\pi}_j = \tilde{\pi}_j$ as defined in Example 4.28.

4.56 Consider Computational Exercise 4.55. Define the page rank $\bar{\pi}_j$ as the expected number of times the user visits page j until she stops clicking. Show that $\bar{\pi}_j$ is proportional to $\tilde{\pi}_j$ as defined in Example 4.28. Find the proportionality constant.

4.57 Consider the citation model of Modeling Exercise 2.34. Suppose there are 5 papers indexed 1, 2, 3, 4, and 5. Paper 2 cites paper 1, papers 3 and 4 cite paper 2, and paper 5 cites papers 2 and 3. Using Model 2 (see Section 2.3.7) compute the appropriate ranking of the five papers. Do the same analysis using Model 3 (see Section 2.3.7), with damping factor $d = .6$.

4.58 A company manufacturing digital cameras introduces a new model every year (and discontinues all the previous models). Its marketing department has developed the following model of consumer behavior. Consumers make camera buying decisions once a year. A consumer who has a k year old camera from this company will, independently of the past, buy a new camera from this company with probability α_k, or keep the current camera with probability β_k, or switch to another company with probability $1 - \alpha_k - \beta_k$, and be lost forever. We assume that $\beta_K = 0$ for a given $K > 0$. That is, a customer with K year old camera will either upgrade to a new one from this company, or will be lost to the competition. Let B_n be the number of first time buyers of the camera from this company in year n. (Buyers of a new camera are said to have a zero-year-old camera.) Suppose $\{B_n, n \geq 0\}$ is a sequence of iid non-negative integer-valued random variables with common mean b. Let $X_n(k)$ be the number of customers who have a k year old camera from this company in year n $(0 \leq k \leq K, \quad n \geq 0)$. Let $X_n = [X_n(0), X_n(1), \cdots, X_n(K)]$.

1. Show that $\{X_n, n \geq 0\}$ is a DTMC. What is its state-space? Is it irreducible and aperiodic? Assume that $\alpha_k > 0$ for $0 \leq k \leq K$ and $\beta_k > 0$ for $0 \leq k < K$, with $\beta_K = 0$.

2. Suppose the company enters the market for the first time in year 0. Let $x_n(k) = E(X_n(k))$ and $x_n = [x_n(0), x_n(1), \cdots, x_n(K)]$. Derive a recursive equation for x_n. Compute

$$x = \lim_{n \to \infty} x_n.$$

You may find the following matrix notation useful:

$$M = \begin{bmatrix} \alpha_0 & \beta_0 & 0 & \cdots & 0 & 0 \\ \alpha_1 & 0 & \beta_1 & \cdots & 0 & 0 \\ \alpha_2 & 0 & 0 & \cdots & 0 & 0 \\ \vdots & \vdots & \vdots & \ddots & \vdots & \vdots \\ \alpha_{K-1} & 0 & 0 & \cdots & 0 & \beta_{K-1} \\ \alpha_K & 0 & 0 & \cdots & 0 & 0 \end{bmatrix}.$$

3. Let s_n be the expected number of cameras the company sells in year n. Compute s_n in terms of x_n, and its limit as $n \to \infty$.

4. Suppose the conditions of part 1 hold. Use Foster's criterion to show that this DTMC is always positive recurrent. (You may try $\nu(x) = \sum_{k=0}^{K} x(k)$ and use the result derived in part 2 above.)

4.59 A machine produces items at a rate of one per day which are stored in a warehouse. The items are perishable with a fixed lifetime of d days, that is, if an item does not get used within d days after production, it has to be discarded. Let D_n be the demand on day n, and assume that $\{D_n, n \geq 0\}$ is a sequence of iid random variables with

$$a_i = P(D_n = i), \quad i = 0, 1, 2, \cdots.$$

Assume that the production occurs at the beginning of a day, and the demand occurs at the end of a day, and the manager uses the oldest items first to satisfy the demand. Any unsatisfied demand is lost.

1. Model the system described above as a DTMC. What is its state-space? What is the transition probability matrix?

2. Does the limiting distribution exist? Why or why not? Write the balance equations and the normalizing equation. Derive a recursive method of computing the limiting distribution that avoids solving simultaneous equations.
   Now consider a special case where $a_i = 0$ for $i > 2$ for the rest of the exercise.

3. Write an explicit expression for the limiting distribution.

4. Compute the average age of the items in the warehouse in steady state.

5. Compute the average age of the items that are issued to satisfy demand on a day on which there is a positive demand (in steady state).

6. Compute the fraction of items discarded due to old age.

4.60 Suppose you are in charge of managing the inventory of an item that has a deterministic demand of one per day. You order N items from a supplier per day, where $N \geq 2$ is a fixed integer. However, the supplier is unreliable, and supplies the entire order right away with probability p $(0 < p < 1/2)$ or supplies nothing with probability $1 - p$. (This happens independently on each day.) Assume that the demand on each day occurs before the supply arrives for that day, and any unmet demand is lost. Let X_n be the number of items in the warehouse on day n (after the supply and demand for the day are accounted for).

1. Show that $\{X_n, n \geq 0\}$ is a DTMC. What is the state-space and the transition probability matrix? Is it irreducible and aperiodic? Why or why not?

2. For a fixed p, what are the possible values of N that will ensure that $\{X_n, n \geq 0\}$ is a positive recurrent DTMC?

3. Suppose the DTMC is positive recurrent. What is the expected number of items in the warehouse in steady state?

4.61 Consider a system with two servers. The arrivals occur one at a time and the inter-arrival times are iid geometric random variables with parameter a. An incoming arrival joins the queue in front of server i with probability $p_i > 0$ $(i = 1, 2)$, and stays there until its service is complete. Clearly $p_1 + p_2 = 1$. The service times at server i are iid geometric random variables with parameter d_i, $(0 < d_1, d_2 < 1)$. Let X_n^i be the number of customers in the queue in front of server i at time n (after accounting for all the arrivals and departures at time n).

1. Show that, for a given $i = 1, 2$, $\{X_n^i, n \geq 0\}$ is a DTMC. Derive its state-space and transition probability matrix. Is it irreducible and aperiodic? Why or why not? When is it positive recurrent?

2. Are the two queues independent of each other?

3. Find the optimal routing probabilities p_1 and p_2 that will minimize the sum of the expected number of customers in the two queues in steady state.

4.62 Consider a service system with an infinite number of servers where customers arrive in batches at discrete times $n = 0, 1, 2, \cdots$. Successive batch sizes are iid non-negative integer valued random variables with mean τ. The sojourn times of the customers in the system are iid geometric random variables with parameter p. Each customer leaves at the end of his/her sojourn time. Suppose each customer pays a fee of f per period to the system operator while in the system. Let X_n be the number of customers in the system at time n after all arrivals and departures at time n are taken into account.

1. Show that $\{X_n, n \geq 0\}$ is a DTMC. Is this DTMC irreducible? Aperiodic? Positive recurrent, transient or null recurrent?

2. Compute $E(X_n)$ as a function of n, assuming that initially the system is empty.

3. Let $0 < \alpha < 1$ be the discrete discount factor. Compute ϕ, the expected total discounted fees collected by the system operator over infinite horizon.

4. Now suppose the system operator wants to maximize ϕ by choosing an optimal fee f^*. He knows that changing the fee will change the demand, but not the sojourn time of a customer in the system. Suppose the dependence of the demand on the fee is captured by the relation $\tau = Ae^{-\theta f}$ for some known parameters $\theta > 0$ and $A > 0$. Compute f^*.

5. Suppose the successive batch sizes are Poisson random variables. Compute the limiting distribution of X_n as $n \to \infty$.

4.63 A housing developer builds K homes per period (say per quarter). Each home that is ready at the beginning of a period sells by the end of that period with probability p. Let X_n be the number of unsold homes at the beginning of the nth period.

1. Show that $\{X_n, n \geq 0\}$ is a DTMC.

2. Compute $m_n = E(X_n | X_0 = 0)$.

3. Suppose each home costs $\$c$ in material and labor and sells for $\$r$, and thus produces a $\$(r - c)$ revenue at the end of the period. Every home that stays unsold for a quarter costs h dollars in interest + insurance costs, and this cost is incurred at the beginning of the period. Let α be the discount factor per period. Compute the expected total discounted net revenue over the infinite horizon.

4. Compute the expected net revenue per period in steady state.

5. The developer can influence the sales by reducing the selling price. In particular, she has developed the following linear model for price-demand dependence:

$$p = (2c - r)/c, \quad c \leq r \leq 2c.$$

Thus if the developer sells the houses at cost, $(r = c)$, it will sell within one period with probability 1, and if she tries a 100% markup $(r = 2c)$, it won't sell at all. Compute the optimal markup $r - c$ to maximize the expected net revenue per period in steady state.

4.64 Consider a discrete time production system which produces one item at times $n = 0, 1, 2, \cdots$ and holds them in a warehouse. Let A_n be the demand at time n and assume that $\{A_n, n \geq 0\}$ is a sequence of iid random variables with

$$P(A_n = 0) = 1 - p, \quad P(A_n = K) = p,$$

where K is a fixed positive integer and $p \in (0, 1)$ is a fixed real number. Assume that the demand at time n occurs after the production at time n. If the warehouse has less than K items and a demand for K items occurs, that demand is lost. If there are K or more items in the warehouse when a demand occurs, K items are removed from the warehouse instantaneously. Let X_n be the number of items in the warehouse after the production and demand are accounted for at time n.

1. Show that $\{X_n, n \geq 0\}$ is a DTMC.

2. What is its state-space? What is the transition probability matrix?

3. Is it irreducible? Aperiodic?

4. Derive a condition for positive recurrence of the DTMC and compute the stationary distribution in terms of the root in (0,1) to the equation $\rho = 1 - p + p\rho^K$.

5. Compute the expected number of items in the warehouse in steady state.

4.11 Conceptual Exercises

4.1 Let S be a finite set. A (binary) relation on S is a function $R : S \times S \to \{0,1\}$. If x and y in S satisfy the relation R, define $R(x,y) = 1$, and 0 otherwise. A relation R is said to be

(i) reflexive if $R(x,x) = 1$ for all $x \in S$,

(ii) symmetric if $R(x,y) = R(y,x)$ for all $x, y \in S$,

(iii) transitive if $R(x,y) = 1$, $R(y,z) = 1 \Rightarrow R(x,z) = 1$, for all $x, y, z \in S$.

Investigate the properties of the following relations:

1. S = set of all humans, $R(x,y) = 1$ if and only if x and y are blood relatives,

2. S = set of all humans, $R(x,y) = 1$ if and only if x is a brother of y,

3. S = set of all students at a university, $R(x,y) = 1$ if x and y are attending at least one class together,

4. S = set of all polynomials, $R(x,y) = 1$ if the degree of the product of x and y is 10.

4.2 Let C_1 and C_2 be two communicating classes of a DTMC. Show that either $C_1 = C_2$ or $C_1 \cap C_2 = \phi$.

4.3 Let $\{X_n, n \geq 0\}$ be a DTMC with state-space $\{1, 2, \cdots N\}$ and transition probability matrix P with q non-zero entries. Develop an algorithm to check whether or not state i is accessible from state j in $O(q)$ steps.

4.4 For the DTMC in Conceptual Exercise 4.3 develop an algorithm to identify the partition in Equation 4.10 in $O(q)$ steps.

4.5 Prove or disprove the following:

1. State i is aperiodic if $p_{i,i} > 0$.

2. State i is aperiodic if and only if $p_{i,i} > 0$.

3. State i has period d implies that $p_{i,i}^{(n)} = 0$ for all n that are not positive-integer multiples of d.

4. State i has period d implies that $p_{i,i}^{(n)} > 0$ for all n that are positive-integer multiples of d.

4.6 Complete the proof of Theorem 4.2.

4.7 Show that in an irreducible DTMC with N states, it is possible to go from any state to any other state in N steps or less.

4.8 Show that the period of an irreducible DTMC with N states is N or less.

4.9 Show that there are no null recurrent states in a finite state DTMC.

4.10 Show by example that it is possible for an irreducible DTMC with N states to have any period $d \in \{1, 2, \cdots, N\}$.

4.11 Show that not all states in a finite state DTMC can be transient.

4.12 Deduce Equation 4.19 from Equation 4.18 by following an argument similar to the one in the proof of Theorem 4.6.

4.13 Prove Theorem 4.12 (Pakes' lemma) using Theorem 4.11 (Foster's criterion).

4.14 Show that a transient DTMC eventually permanently exits any finite set with probability 1. (Use Borel–Cantelli lemma of Appendix A.)

4.15 Let $\{\pi_j, j \in S\}$ be a solution to the balance and normalizing equations. Now suppose the DTMC starts with initial distribution

$$P(X_0 = j) = \pi_j, \quad j \in S.$$

Show that

$$P(X_n = j) = \pi_j, \quad j \in S, \text{ for all } n \geq 1.$$

4.16 Establish Equation 4.40 by induction.

4.17 Let $\{X_n, n \geq 0\}$ be a DTMC with state-space S and transition probability matrix P. Suppose the system incurs an expected cost of $c(i, j)$ at time n if $X_n = i$ and $X_{n+1} = j$. Let

$$c(i) = \sum_{j \in S} c(i, j) p_{i,j}, \quad c = [c(i)]_{i \in S}.$$

Let

$$\phi(i) = E(\sum_{n=0}^{\infty} \alpha^n c(X_n, X_{n+1}) | X_0 = i), \quad \phi = [\phi(i)]_{i \in S}.$$

Show that ϕ satisfies Equation 4.52.

4.18 Consider the cost model of Conceptual Exercise 4.17, and assume that the DTMC is irreducible and positive recurrent with limiting occupancy distribution $\{\pi_i, i \in S\}$. Show that g, the long run expected cost per unit time, is given by

$$g = \sum_{i \in S} \sum_{j \in S} \pi_i c(i, j) p_{i,j}.$$

4.19 Consider the cost set up of Section 4.8. Let $\phi(N, i)$ be the expected discounted cost incurred over time $\{0, 1, 2, \cdots, N\}$ starting from state i, i.e.,

$$\phi(N, i) = \mathsf{E}\left(\sum_{n=0}^{N} \alpha^n c(X_n) \,\Big|\, X_0 = i\right).$$

Show that $\phi(N, i)$ can be computed recursively as follows:

$$\phi(0, i) = c(i),$$
$$\phi(N, i) = c(i) + \alpha \sum_{j \in S} p_{i,j} \phi(N - 1, j), \quad N \geq 1.$$

4.20 Consider the cost set up of Section 4.8. Let $g(N, i)$ be the expected cost per period incurred over time $\{0, 1, 2, \cdots, N\}$ starting from state i, i.e.,

$$g(N, i) = \frac{1}{N+1} \mathsf{E}\left(\sum_{n=0}^{N} c(X_n) \,\Big|\, X_0 = i\right).$$

Show that $g(N, i)$ can be computed recursively as follows:

$$g(0, i) = c(i),$$
$$g(N, i) = \frac{1}{N+1}\left\{c(i) + N \sum_{j \in S} p_{i,j} g(N - 1, j)\right\}, \quad N \geq 1.$$

4.21 Show that if $\{\pi_j, j \in S\}$ satisfy the local balance Equations 4.54, they satisfy the global balance Equations 4.38. Show by a counter example that the reverse is not true.

4.22 A transition probability matrix is called doubly stochastic if

$$\sum_{i \in S} p_{i,j} = 1, \quad j \in S.$$

Let $\{X_n, n \geq 0\}$ be an irreducible DTMC on a finite state-space $\{1, 2, \cdots, N\}$ with a doubly stochastic transition probability matrix P. Show that its stationary distribution is given by

$$\pi_j = 1/N, \quad j \in S.$$

4.23 A DTMC is said to be tree if between any two distinct states i and j there is exactly one sequence of distinct states $i_1, i_2, \cdots i_r$ such that

$$p_{i,i_1} p_{i_1,i_2} \cdots p_{i_r,j} > 0.$$

Show that a positive recurrent tree DTMC is reversible.

4.24 Let P be a diagonalizable $N \times N$ transition probability matrix. Using the representation in Equation 2.34 study the limiting behavior of P^n and $M^{(n)}/(n+1)$ in terms of the eigenvectors. Consider the following cases

1. $\lambda = 1$ is an eigenvalue of multiplicity one, and it is the only eigenvalue with $|\lambda| = 1$.

2. $\lambda = 1$ has multiplicity k, and these are the only eigenvalues with $|\lambda| = 1$.

3. There are eigenvalues with $|\lambda| = 1$ other than $\lambda = 1$.

4.25 Let $\{X_n, n \geq 0\}$ be an irreducible and positive recurrent DTMC on state-space S. Let P be its transition probability matrix and π be its stationary distribution. Suppose the Markov chain earns a reward of r_i every time it visits state i. Let $R_i(n), n \geq 1$, be the total expected reward earned by the DTMC over time periods $\{0, 1, ..., n - 1\}$ starting from state i. Let $R(n) = [R_i(n), i \in S]$ be the column vector of these expected rewards.

1. Derive a set of recursive equations satisfied by $R_i(n), i \in S$. Write them in matrix form using P and $r = [r_i, i \in S]$.

2. It is known that (you do not need to prove this) there exists a constant g and a column vector $h = [h_i, i \in S]$ such that

$$R(n) = h + nge + o(n), \tag{4.56}$$

where e is a column vector of ones, and $o(n)$ is a vector with the following property:

$$\lim_{n \to \infty} o(n) = 0.$$

Compute the value of g. **Hint:** Substitute Equation 4.56 in the result obtained in part 1 above. Pre-multiply by π.

3. Derive the equations that uniquely determine h.

4.26 Suppose j is recurrent, and $i \to j$. Show that

$$\sum_{n=0}^{\infty} p_{i,j}^{(n)} = \infty.$$

4.27 Suppose $i \to j$. Show that

1. $\lim_{n \to \infty} \frac{1}{n+1} \sum_{r=0}^{n} p_{i,j}^{(r)} > 0$ if j is positive recurrent,

2. $\lim_{n \to \infty} \frac{1}{n+1} \sum_{r=0}^{n} p_{i,j}^{(r)} = 0$ if j is null recurrent.

4.28 Let $\{X_n, n \geq 0\}$ be an irreducible positive recurrent aperiodic DTMC with state-space S and limiting distribution $\{\pi_j, j \in S\}$. Show by a counter example that $P(X_n = j) = \pi_j$, for all $j \in S$, for a given $n > 0$ does not imply that $P(X_m = j) = \pi_j$ for all $j \in S$ and $m \geq 0$.

4.29 Let $\{X_n, n \geq 0\}$ be an irreducible positive recurrent DTMC with state-space S and stationary distribution $\{\pi_j, j \in S\}$. Show that the long run fraction of transitions that take the DTMC from state i to j is given by $\pi_i p_{i,j}$.

4.30 Let P be a transition probability matrix of an irreducible positive recurrent DTMC. Show that a DTMC with transition probability matrix $(P + P^\top)/2$ is reversible.

4.31 Let $\{X_n, n \geq 0\}$ be an irreducible recurrent DTMC with state-space $\{0, 1, \cdots\}$, and transition probability matrix P. Let $T = \min\{n \geq 0 : X_n = 0\}$. Now suppose the DTMC earns a reward r_i every time it visits state i. Let R be the total reward earned over times $\{0, 1, ...T - 1\}$. Using the first-step analysis derive a set of linear equations satisfied by

$$g(i) = E(R|X_0 = i)$$

for $i = 0, 1, 2, \cdots$. (Note that $g(0) = 0$.)

4.32 Suppose $\{X_n, n \geq 0\}$ is a positive recurrent DTMC with state-space S. Suppose also that $\{Y_i, i \geq 1\}$ are iid geometric random variables with parameter $\alpha \in (0, 1)$. Let $S_0 = 0$ and $S_n = \sum_{i=1}^{n} Y_i$, for $n \geq 1$. Define $Z_n = X_{S_n}, n \geq 0$.

1. Show that if $\{X_n, n \geq 0\}$ is aperiodic, then so is $\{Z_n, n \geq 0\}$.

2. Prove or disprove the converse to the statement in part 1.

3. Suppose $\{X_n, n \geq 0\}$ has steady state distribution $\{\pi_j, j \in S\}$. What is the steady state distribution of $\{Z_n, n \geq 0\}$?

4.33 Let $\{X_n, n \geq 0\}$ be an irreducible DTMC with state-space $S = \{0, 1, 2, \cdots\}$ and transition probability matrix P. Let

$$T = \min\{n \geq 0 : X_n = 0\}.$$

Let $s, t \in S - \{0\}$ be fixed states. The DTMC is said to undergo a transition from s to t at time n if $X_n = s$ and $X_{n+1} = t, n \geq 0$. Let $g(i)$ be the expected number of transitions from s to t starting with $X_0 = i$ over $0, 1, \cdots, T$. Use the first-step analysis to derive a set of linear equations satisfied by $\{g(i), i \in S\}$

CHAPTER 5

Poisson Processes

"Basic research is what I am doing when I don't know what I am doing."
— *Wernher von Braun*

5.1 Exponential Distributions

In this chapter we study an important special class of continuous-time stochastic processes called Poisson processes. These processes are defined in terms of random variables with exponential distribution, called "exponential random variables" for short. We shall see in this and the next few chapters that exponential random variables are used to build a large number of stochastic models. It is critical for a student to develop an ability to deal with the exponential distribution with ease, without having to think too much. With this in mind we have collected several relevant properties of the exponential distribution in this section. We begin with the definition.

Definition 5.1 Exponential Distribution. *A non-negative random variable X is said to be an exponential random variable with parameter λ, denoted as $X \sim \exp(\lambda)$, if*

$$F_X(x) = P(X \le x) = \begin{cases} 0 & \text{if } x < 0 \\ 1 - e^{-\lambda x} & \text{if } x \ge 0, \end{cases} \tag{5.1}$$

where $\lambda > 0$ is a fixed constant. The cumulative distribution function (cdf) F_X is called the exponential distribution.

The cdf F_X is plotted in Figure 5.1. The probability density function (pdf) f_X of an $\exp(\lambda)$ random variable is called the exponential density and is given by

$$f_X(x) = \frac{d}{dx} F_X(x) = \begin{cases} 0 & \text{if } x < 0 \\ \lambda e^{-\lambda x} & \text{if } x \ge 0. \end{cases}$$

The density function is plotted in Figure 5.2. The Laplace Stieltjes transform (LST) of $X \sim \exp(\lambda)$ is given by

$$\tilde{F}_X(s) = \mathsf{E}(e^{-sX}) = \int_0^\infty e^{-sx} f_X(x) dx = \frac{\lambda}{\lambda + s}, \quad \text{Re}(s) > -\lambda, \tag{5.2}$$

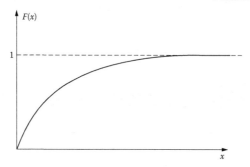

Figure 5.1 *The cdf of an exponential random variable.*

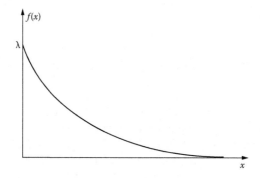

Figure 5.2 *The pdf of an exponential random variable.*

where the $Re(s)$ denotes the real part of the complex number s. Taking the derivatives of $\tilde{F}_X(s)$ we can compute the rth moments of X for all positive integer values of r as follows:

$$E(X^r) = (-1)^r \frac{d^r}{ds^r} \tilde{F}_X(s) \bigg|_{s=0} = \frac{r!}{\lambda^r}.$$

In particular we have

$$E(X) = \frac{1}{\lambda}, \quad \text{Var}(X) = \frac{1}{\lambda^2}.$$

Thus the coefficient of variation of X, $\text{Var}(X)/E(X)^2$, is 1. We now study many special and interesting properties of the exponential random variable.

5.1.1 Memoryless Property

We begin with the definition of the memoryless property.

Definition 5.2 *A non-negative random variable X is said to have the memoryless* *property if*

$$P(X > s + t | X > s) = P(X > t), \quad s, t \geq 0. \tag{5.3}$$

Thus if X represents the lifetime of a component (say a computer hard drive), the memoryless property says that the probability that an s year old hard drive will last an additional t years is the same as the probability that a new hard drive will last t years. It is as if the hard drive has no memory that it has already been functioning for s years! The next theorem gives an important characterization of an exponential random variable.

Theorem 5.1 Memoryless Property. *A continuous non-negative random variable has memoryless property if and only if it is an $exp(\lambda)$ random variable for some $\lambda > 0$.*

Proof: We first show the "if" part. So, suppose $X \sim exp(\lambda)$ for some $\lambda > 0$. Then, for $s, t \geq 0$,

$$
\begin{aligned}
P(X > s + t | X > s) &= \frac{P(X > s + t, X > s)}{P(X > s)} \\
&= \frac{P(X > s + t)}{P(X > s)} = \frac{e^{-\lambda(s+t)}}{e^{-\lambda s}} \\
&= e^{-\lambda t} = P(X > t).
\end{aligned}
$$

Hence, by definition, X has memoryless property. Next we show the "only if" part. So, let X be a non-negative random variable with complementary cdf

$$F^c(x) = P(X > x), \quad x \geq 0.$$

Then, from Equation 5.3, we must have

$$F^c(s + t) = F^c(s)F^c(t), \quad s, t \geq 0.$$

This implies

$$F^c(2) = F^c(1)F^c(1) = (F^c(1))^2,$$

and

$$F^c(1/2) = (F^c(1))^{1/2}.$$

In general, for all positive rational a we get

$$F^c(a) = (F^c(1))^a.$$

The only continuous function that will satisfy the above equation for all rational numbers is

$$F^c(x) = (F^c(1))^x = e^{(\ln F^c(1))x} = e^{-\lambda x}, \quad x \geq 0,$$

where $\lambda = -\ln(F^c(1))$. Since $F^c(x)$ is a probability we must have $\lambda > 0$, and hence $F_X(x) = 1 - F^c(x)$ satisfies Equation 5.1. Thus X is an $exp(\lambda)$ random variable. ∎

5.1.2 Hazard Rate

Let X be a non-negative continuous random variable with pdf f and complementary cdf $F^c(x) = P(X > x)$. Hazard (or failure) rate of X is defined as

$$r(x) = \frac{f(x)}{F^c(x)}, \quad x \geq 0.$$

Note that, for small h,

$$r(x)h + o(h) = \frac{f(x)h + o(h)}{F^c(x)} = P(X \in (x, x + h)|X > x).$$

Here $o(h)$ is a function such that $o(h)/h \rightarrow 0$ as $h \rightarrow 0$. Thus, in the limit as $h \rightarrow 0$, $r(x)h$ can be interpreted as the conditional probability that a machine with lifetime X fails in the interval $(x, x + h)$ given that it has not failed until x. Hence the name "failure rate" or "hazard rate." In insurance literature it is also known as mortality rate. Note that $r(x)$ is not a probability density function, since its integral from 0 to ∞ is infinity, as follows from Equation 5.4 below.

The next theorem characterizes the exponential distribution via its hazard rate.

Theorem 5.2 *A continuous non-negative random variable has a constant hazard rate $r(x) = \lambda > 0$, $x \geq 0$, if and only if it is an exp(λ) random variable.*

Proof: We first show the "if" part. So, suppose $X \sim \exp(\lambda)$ for some $\lambda > 0$. Then its hazard rate is given by

$$\begin{aligned} r(x) &= \frac{f(x)}{F^c(x)} \\ &= \frac{\lambda e^{-\lambda x}}{e^{-\lambda x}} = \lambda. \end{aligned}$$

To show the "only if" part, note that the hazard rate completely determines the complementary cdf by the following formula (see Conceptual Exercise 5.1)

$$F^c(x) = \exp\left(-\int_0^x r(u)du\right), \quad x \geq 0. \tag{5.4}$$

Hence, if the random variable X has hazard rate $r(u) = \lambda$ for all $u \geq 0$, we must have

$$F^c(x) = e^{-\lambda x}, \quad x \geq 0.$$

Thus $X \sim \exp(\lambda)$. ∎

5.1.3 Probability of First Failure

Let $X_i \sim \exp(\lambda_i), i = 1, 2$, be two independent random variables representing the lifetimes of two machines. Then the probability of machine 1 failing before machine

2 is given by

$$
\begin{aligned}
P(X_1 < X_2) &= \int_0^\infty P(X_1 < X_2 | X_1 = x)\lambda_1 e^{-\lambda_1 x} dx \\
&= \int_0^\infty P(X_2 > x)\lambda_1 e^{-\lambda_1 x} dx \\
&= \int_0^\infty e^{-\lambda_2 x}\lambda_1 e^{-\lambda_1 x} dx \\
&= \frac{\lambda_1}{\lambda_1 + \lambda_2}.
\end{aligned}
$$

Thus the probability of first failure is proportional to the failure rate of the exponential random variables. Note that the assumption of independence is critical. The above result can be easily extended to more than two exponential random variables; see Conceptual Exercise 5.3.

Example 5.1 A running track is 1 km long. Two runners start on it at the same time. The speed of runner i is X_i, $i = 1, 2$. Suppose $X_i \sim \exp(\lambda_i)$, $i = 1, 2$, and X_1 and X_2 are independent. The mean speed of runner 1 is 20 km per hour and that of runner 2 is 22 km per hour. What is the probability that runner 1 wins the race?

The required probability is given by

$$
\begin{aligned}
P\left(\frac{1}{X_1} < \frac{1}{X_2}\right) &= P(X_2 < X_1) = \frac{\lambda_2}{\lambda_1 + \lambda_2} \\
&= \frac{1/22}{1/22 + 1/20} = \frac{20}{20 + 22} = 0.476.
\end{aligned}
$$

5.1.4 Minimum of Exponentials

Let $\{X_i, 1 \leq i \leq n\}$ be n non-negative random variables. Define

$$
Z = \min\{X_1, X_2, \cdots, X_n\} \tag{5.5}
$$

and

$$
N = i \text{ if } X_i = Z, \quad 1 \leq i \leq n. \tag{5.6}
$$

Note that N is unambiguously defined if the random variables $\{X_i, 1 \leq i \leq n\}$ are continuous, since in that case exactly one among the n random variables will equal Z. Thus, if X_i represents the time when an event of type i occurs ($1 \leq i \leq n$), then Z represents the time when the first of these n events occurs, and N represents the type of the event that occurs first. The next theorem gives the joint distribution of N and Z when the X_i's are independent exponential random variables.

Theorem 5.3 Let $X_i \sim \exp(\lambda_i)$, $1 \leq i \leq n$, be independent and let N and Z be as defined by Equations 5.6 and 5.5. Then

$$
P(N = i, Z > z) = \frac{\lambda_i}{\lambda} e^{-\lambda z}, \quad z \geq 0, \ 1 \leq i \leq n,
$$

where

$$\lambda = \sum_{i=1}^{n} \lambda_i.$$

Proof: For a fixed $i \in \{1, 2, \cdots, n\}$ and $z \geq 0$, we have

$$
\begin{aligned}
P(N = i, Z > z) &= P(X_j > X_i > z, \; j \neq i, 1 \leq j \leq n) \\
&= \int_0^\infty P(X_j > X_i > z, \; j \neq i, 1 \leq j \leq n | X_i = x) \lambda_i e^{-\lambda_i x} dx \\
&= \int_z^\infty P(X_j > x, \; j \neq i, 1 \leq j \leq n) \lambda_i e^{-\lambda_i x} dx \\
&= \int_z^\infty \left[\prod_{j=1, j \neq i}^{n} e^{-\lambda_j x} \right] \lambda_i e^{-\lambda_i x} dx \\
&= \lambda_i \int_z^\infty e^{-\lambda x} dx \\
&= \frac{\lambda_i}{\lambda} e^{-\lambda z}
\end{aligned}
$$

as desired. ■

An important implication of the above theorem is that N and Z are *independent* random variables with the following marginal distributions:

$$P(N = i) \;=\; \frac{\lambda_i}{\lambda}, \quad 1 \leq i \leq n, \tag{5.7}$$

$$P(Z \leq z) \;=\; 1 - e^{-\lambda z}, \quad z \geq 0. \tag{5.8}$$

Thus the minimum of n *independent* exponential random variables is an exponential random variable whose parameter is the sum of the parameters of the original random variables. These properties will play a critical role in later development of continuous time Markov chains in Chapter 6.

5.1.5 Strong Memoryless Property

In this section we generalize the memoryless property of Equation 5.3 from a fixed t to a random T. We call this the strong memoryless property. The precise result is given in the next theorem.

Theorem 5.4 *Let $X \sim \exp(\lambda)$ and T be another non-negative random variable that is independent of X. Then*

$$P(X > s + T | X > T) = e^{-\lambda s}, \quad s \geq 0.$$

Proof: We have

$$P(X > s + T, X > T) \;=\; \int_0^\infty P(X > s + T, X > T | T = t) dF_T(t)$$

$$= \int_0^\infty \mathsf{P}(X > s + t, X > t) dF_T(t)$$

$$= \int_0^\infty e^{-\lambda(s+t)} dF_T(t)$$

$$= e^{-\lambda s} \int_0^\infty e^{-\lambda t} dF_T(t)$$

$$= e^{-\lambda s} \mathsf{P}(X > T).$$

Then

$$\mathsf{P}(X > s + T | X > T) = \frac{\mathsf{P}(X > s + T, X > T)}{\mathsf{P}(X > T)} = e^{-\lambda s},$$

as desired. ∎

Another way of interpreting the strong memoryless property is that, given $X > T$, $X - T$ is an $\exp(\lambda)$ random variable. Indeed it is possible to prove a multivariate extension of this property as follows. Let $X_i \sim \exp(\lambda_i)$, $1 \le i \le n$, be independent and let T be a non-negative random variable that is independent of them. Then

$$\mathsf{P}(X_i > s_i + T, \ 1 \le i \le n | X_i > T, \ 1 \le i \le n) = \prod_{i=1}^n e^{-\lambda_i s_i}, \quad s_i \ge 0. \quad (5.9)$$

A simple application of this property is given in the following example.

Example 5.2 Parallel System. Consider a system of n components in parallel, i.e., the system functions as long as at least one of the n component functions, and it fails as soon as all the components fail. Let X_i be the lifetime of component i and assume that $X_i \sim \exp(\lambda)$ are iid random variables. Then the lifetime of the system is given by

$$Z = \max\{X_1, X_2, \cdots, X_n\}.$$

Compute $\mathsf{E}(Z)$.

One can solve this problem by first computing the cdf of Z as follows:

$$\mathsf{P}(Z \le z) = \mathsf{P}(X_i \le z, \ 1 \le i \le n) = \left(1 - e^{-\lambda z}\right)^n,$$

and then computing its expected value. Here we show an alternate method using the properties of the exponential distribution. Let Z_k be the time of the k-th failure, $1 \le k \le n$. Thus

$$Z_1 = \min\{X_1, X_2, \cdots, X_n\} \sim \exp(n\lambda),$$

$$Z_n = \max\{X_1, X_2, \cdots, X_n\}.$$

Now, at time Z_1, one component fails, and the remaining $n - 1$ components survive. Due to Equation 5.9, the remaining lifetimes of the surviving components are iid $\exp(\lambda)$ random variables. Hence the time until the next failure, namely $Z_2 - Z_1$, equals the minimum of the $n - 1$ iid $\exp(\lambda)$ random variables. Hence

$$Z_2 - Z_1 \sim \exp((n - 1)\lambda).$$

Proceeding in this way we see that

$$Z_{k+1} - Z_k \sim \exp((n-k)\lambda), \quad 1 \le k \le n-1.$$

Thus, writing $Z_n = \sum_{k=0}^{n-1}(Z_{k+1} - Z_k)$, (with $Z_0 = 0$), we get

$$E(Z) = E(Z_n) = \sum_{k=0}^{n-1} E(Z_{k+1} - Z_k) = \sum_{k=0}^{n-1} \frac{1}{(n-k)\lambda} = \sum_{k=1}^{n} \frac{1}{k\lambda}.$$

The above equation is an example of the law of diminishing returns. A system of one component (with $\exp(\lambda)$ lifetime) has an expected lifetime of $1/\lambda$. A system of two independent $\exp(\lambda)$ components in parallel has an expected lifetime of $1/\lambda + 1/(2\lambda) = 1.5/\lambda$. Thus doubling the number of components increases the mean lifetime by only 50%. One reason behind this diminishing return is that all components are subject to failure, even though only one is needed for the system to function. We say that the system is operating with "warm standbys." Another way is to operate the system with only one component and $(n-1)$ spares. When the working item fails, replace it with a spare item. Since the spares do not fail while they are not in use, they are said to be "cold standbys." In this case the expected lifetime of the system is n/λ. Obviously cold standby strategy is a more efficient system than the warm standby strategy. However, there are cases (in life-critical systems) where we cannot use the cold standby strategy since the process of replacing the failed component by a new one will interrupt the system operation. ∎

Example 5.3 System A has two components in parallel, with iid $\exp(\lambda)$ lifetimes. System B has a single component with $\exp(\mu)$ lifetime, independent of system A. What is the probability that system A fails before system B?

Let Z_i be the time of the ith failure in system A. System A fails at time Z_2. From Example 5.2, we have $Z_1 \sim \exp(2\lambda)$ and $Z_2 - Z_1 \sim \exp(\lambda)$. Let $X \sim \exp(\mu)$ be the lifetime of the component in system B. Then

$$
\begin{aligned}
P(\text{System } A \text{ fails before System } B) &= \\
&= P(Z_1 < X)P(Z_2 - Z_1 < X - Z_1 | X > Z_1) \\
&= P(\exp(2\lambda) < \exp(\mu))P(\exp(\lambda) < \exp(\mu)) \\
&= \frac{2\lambda}{2\lambda + \mu} \cdot \frac{\lambda}{\lambda + \mu}. \quad \blacksquare
\end{aligned}
$$

5.1.6 Sum of iid Exponentials

Let $\{X_i, 1 \le i \le n\}$ be iid random variables, representing the lifetimes of n components. Suppose we start by putting component 1 in use, and when it fails, replace it with component 2, and so on until all components fail. The replacements are instantaneous. The lifetime of the system is thus given by

$$Z = X_1 + X_2 + \cdots + X_n.$$

The next theorem gives the pdf of Z if the lifetimes are iid $\exp(\lambda)$ random variables.

Theorem 5.5 *Suppose $\{X_i, 1 \le i \le n\}$ are iid $\exp(\lambda)$ random variables. Then Z is an Erlang (or Gamma) (n, λ) random variable (denoted as $\mathrm{Erl}(n, \lambda)$), with density*

$$f_Z(z) = \begin{cases} 0 & \text{if } z < 0 \\ \lambda e^{-\lambda z} \frac{(\lambda z)^{n-1}}{(n-1)!} & \text{if } z \ge 0, \end{cases}$$

and cdf

$$F_Z(z) = \begin{cases} 0 & \text{if } z < 0 \\ 1 - e^{-\lambda z} \sum_{r=0}^{n-1} \frac{(\lambda z)^r}{r!} & \text{if } z \ge 0. \end{cases}$$

Proof: We compute the LST of Z as follows:

$$\begin{aligned}
\mathsf{E}(e^{-sZ}) &= \mathsf{E}(e^{-s(X_1 + X_2 + \cdots + X_n)}) \\
&= \mathsf{E}(e^{-sX_1})\mathsf{E}(e^{-sX_2}) \cdots \mathsf{E}(e^{-sX_n}) \\
&= \left(\frac{\lambda}{s + \lambda}\right)^n.
\end{aligned}$$

The result follows by taking the inverse LST from the table in Appendix F. ∎

Example 5.4 Suppose the times between consecutive births at a maternity ward in a hospital are iid exponential random variables with mean 1 day. What is the probability that the 10th birth in a calendar year takes place after Jan 15?

Note that it does not matter when the last birth in the previous year took place, since, due to strong memoryless property, the time until the first birth into the new year is exponentially distributed. Thus Z, the time of the tenth birth is a sum of 10 iid $\exp(1)$ random variables. Therefore the required probability is given by

$$P(Z > 15) = \sum_{r=0}^{9} e^{-15} \frac{15^r}{r!} = 0.0699.$$

We shall see later that the stochastic process of births is a Poisson process. ∎

5.1.7 Sum of Distinct Exponentials

In the previous subsection we computed the distribution of the sum of independent and identically distributed exponential random variables. In the next theorem we study the sum of independent but distinct exponential random variables.

Theorem 5.6 *Let $X_i \sim \exp(\lambda_i)$, $1 \le i \le n$, be independent random variables. Assume that all the λ_i's are distinct. Then the pdf of*

$$Z = X_1 + X_2 + \cdots + X_n$$

is given by

$$f_Z(z) = \sum_{i=1}^{n} \alpha_i \lambda_i e^{-\lambda_i z},$$

where

$$\alpha_i = \prod_{j=1, j \neq i}^{n} \frac{\lambda_j}{\lambda_j - \lambda_i}, \quad 1 \leq i \leq n.$$

Proof: The LST of Z is given by

$$
\begin{aligned}
\mathsf{E}(e^{-sZ}) &= \mathsf{E}(e^{-s(X_1 + X_2 + \cdots + X_n)}) \\
&= \mathsf{E}(e^{-sX_1})\mathsf{E}(e^{-sX_2}) \cdots \mathsf{E}(e^{-sX_n}) \\
&= \prod_{i=1}^{n} \frac{\lambda_i}{s + \lambda_i}.
\end{aligned}
$$

The result follows by using partial fraction expansion (see Appendix D) to compute the inverse LST. ∎

5.1.8 Random Sums of iid Exponentials

In this section we study the distribution of a sum of a random number of iid $\exp(\lambda)$ random variables. The main result is given in the next theorem.

Theorem 5.7 *Let $\{X_i, i \geq 1\}$ be a sequence of iid $\exp(\lambda)$ random variables and N be a G(p) random variable (i.e., geometric with parameter p), independent of the X's. Then the random sum*

$$Z = \sum_{i=1}^{N} X_i$$

is an $\exp(\lambda p)$ random variable.

Proof: We have

$$\mathsf{E}(e^{-sZ}|N = n) = \left(\frac{\lambda}{s + \lambda}\right)^n.$$

Hence we get

$$
\begin{aligned}
\mathsf{E}(e^{-sZ}) &= \sum_{n=1}^{\infty} \mathsf{E}(e^{-sZ}|N = n)\mathsf{P}(N = n) \\
&= \sum_{n=1}^{\infty} \left(\frac{\lambda}{s + \lambda}\right)^n (1 - p)^{n-1} p \\
&= \frac{\lambda p}{s + \lambda} \sum_{n=0}^{\infty} \left(\frac{\lambda(1 - p)}{s + \lambda}\right)^n \\
&= \frac{\lambda p}{s + \lambda} \cdot \frac{1}{1 - \lambda(1 - p)/(s + \lambda)}
\end{aligned}
$$

$$= \frac{\lambda p}{s + \lambda p}.$$

Hence, from Equation 5.2, Z must be an exp(λp) random variable. ∎

Example 5.5 A machine is subject to a series of randomly occurring shocks. The times between two consecutive shocks are iid exponential random variables with common mean of 10 hours. Each shock results in breaking the machine (if it is not already broken) with probability .3. Compute the distribution of the lifetime of the machine.

Suppose the Nth shock breaks the machine. Then N is G(.3) random variable. That is

$$P(N = k) = (.7)^{k-1}(.3), \quad k \geq 1.$$

Let X_i be the time between the $(i-1)$st and ith shock. We know that $\{X_i, i \geq 1\}$ is a sequence of iid exp(.1) random variables. The lifetime of the machine is given by

$$Z = \sum_{i=1}^{N} X_i.$$

Hence from Theorem 5.7, we see that $Z \sim \exp(.03)$. Thus the lifetime is exponentially distributed with mean $1/.03 = 33.33$ hours. ∎

5.2 Poisson Process: Definitions

A Poisson process is frequently used as a model for counting events occurring one at a time, such as the number of births in a hospital, the number of arrivals at a service system, the number of calls made, the number of accidents on a given section of a road, etc. In this section we give three equivalent definitions of a Poisson process and study its basic properties.

Let $\{X_n, n \geq 1\}$ be a sequence of non-negative random variables representing inter-event times. Define

$$S_0 = 0, \quad S_n = X_1 + X_2 + \cdots + X_n, \quad n \geq 1.$$

Then S_n is the time of occurrence of the nth event, $n \geq 1$. Now, for $t \geq 0$, define

$$N(t) = \max\{n \geq 0 : S_n \leq t\}.$$

Thus $N(t)$ is the number of events that take place over the time interval $(0, t]$, and $\{N(t), t \geq 0\}$ is called a *counting process* generated by $\{X_n, n \geq 1\}$. Poisson process is a special case of a counting process as defined below.

Definition 5.3 Poisson Process. *The counting process* $\{N(t), t \geq 0\}$ *generated by* $\{X_n, n \geq 1\}$ *is called a Poisson process with parameter (or rate) λ if* $\{X_n, n \geq 1\}$ *is a sequence of iid exp(λ) random variables.*

Thus the number of births in the maternity ward of Example 5.4 is a Poisson process. We denote a Poisson process with parameter (or rate) λ by the shorthand notation PP(λ). A typical sample path of a PP(λ) is shown in Figure 5.3. Note that

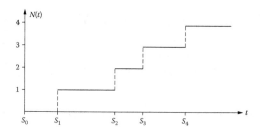

Figure 5.3 *A typical sample path of a Poisson process.*

$N(0) = 0$ and the process jumps up by one at $t = S_n$, $n \geq 1$. Thus it has piecewise constant, right-continuous sample paths. The next theorem gives the transient distribution of a Poisson process.

Theorem 5.8   Transient Distribution. *Let* $\{N(t), t \geq 0\}$ *be a PP(λ). Then*

$$P(N(t) = k) = e^{-\lambda t}\frac{(\lambda t)^k}{k!}, \quad k = 0, 1, 2, \cdots.$$

Proof: We have

$$N(t) \geq k \quad \Leftrightarrow \quad k \text{ or more events in } (0, t]$$
$$\Leftrightarrow \quad k\text{th event takes place at or before } t$$
$$\Leftrightarrow \quad S_k \leq t.$$

Hence

$$P(N(t) \geq k) = P(S_k \leq t).$$

From Theorem 5.5, S_k is an Erl(k, λ) random variable. Hence,

$$
\begin{aligned}
P(N(t) = k) &= P(N(t) \geq k) - P(N(t) \geq k+1) \\
&= P(S_k \leq t) - P(S_{k+1} \leq t) \\
&= \left[1 - \sum_{r=0}^{k-1} e^{-\lambda t}\frac{(\lambda t)^r}{r!}\right] - \left[1 - \sum_{r=0}^{k} e^{-\lambda t}\frac{(\lambda t)^r}{r!}\right] \\
&= e^{-\lambda t}\frac{(\lambda t)^k}{k!}
\end{aligned}
$$

as desired. ∎

Theorem 5.8 says that, for a fixed t, $N(t)$ is a Poisson random variable with parameter λt, denoted as P(λt). This provides one justification for calling $\{N(t), t \geq 0\}$ a Poisson process.

Example 5.6   Arrivals at a Post Office. Suppose customers arrive at a post office according to a PP with rate 10 per hour.

(i) Compute the distribution of the number of customers who use the post office during an 8-hour day.

Let $N(t)$ be the number of arrivals over $(0, t]$. We see that the arrival process is PP(λ) with $\lambda = 10$ per hour. Hence

$$N(8) \sim P(\lambda \cdot 8) = P(80).$$

Thus

$$P(N(8) = k) = e^{-80} \frac{(80)^k}{k!}, \quad k = 0, 1, 2, \cdots.$$

(ii) Compute the expected number of customers who use the post office during an 8-hour day.

Since $N(8) \sim P(80)$, the desired answer is given by $E(N(8)) = 80$. ∎

Next we compute the finite dimensional joint probability distributions of a Poisson process. The above theorem does not help us there. We develop a crucial property of a Poisson process that will help us do this. We start with a definition.

Definition 5.4   Shifted Poisson Process. *Let $\{N(t), t \geq 0\}$ be a PP(λ), and define, for a fixed $s \geq 0$,*
$$N_s(t) = N(t + s) - N(s), \quad t \geq 0.$$
The process $\{N_s(t), t \geq 0\}$ is called a shifted Poisson process.

Theorem 5.9   Shifted Poisson Process. *A shifted Poisson process $\{N_s(t), t \geq 0\}$ is a PP(λ), and is independent of $\{N(u), 0 \leq u \leq s\}$.*

Proof: It is clear from the definition of $N_s(t)$ that it equals the number of events in $(s, s + t]$. From Figure 5.4 we see that the first event after s occurs at time $S_{N(s)+1}$

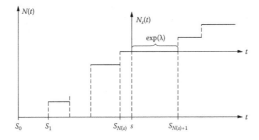

Figure 5.4 *A shifted Poisson process.*

and the last event at or before time s occurs at $S_{N(s)}$. We have

$$
\begin{aligned}
P(S_{N(s)+1} - s > y | N(s) &= k, S_{N(s)} = x, N(u) : 0 \le u \le s) \\
&= P(S_k + X_{k+1} - s > y | N(s) = k, S_k = x, N(u) : 0 \le u \le s) \\
&= P(X_{k+1} > s + y - x | X_{k+1} > s - x) \\
&= e^{-\lambda y}
\end{aligned}
$$

due to the memoryless property of X_{k+1}. Thus the first event in the counting process $\{N_s(t), t \ge 0\}$ takes place after an $\exp(\lambda)$ amount of time. The rest of the inter-event times are iid $\exp(\lambda)$. Hence the result follows. ∎

Next we introduce the concept of a stochastic process with stationary and independent increments.

Definition 5.5 Stationary and Independent Increments. *Let $\{X(t), t \ge 0\}$ be a continuous-time real-valued stochastic process. For given $s, t \ge 0$, $X(s+t) - X(s)$ is called the increment over the interval $(s, s + t]$. $\{X(t), t \ge 0\}$ is said to have stationary and independent increments if*

(i) the distribution of the increment over an interval $(s, s + t]$ is independent of s,

(ii) the increments over non-overlapping intervals are independent.

The next theorem states an important property of a Poisson process.

Theorem 5.10 Stationary and Independent Increments. *A Poisson process has stationary and independent increments.*

Proof: Let $\{N(t), t \ge 0\}$ be a PP(λ). From Theorems 5.8 and 5.9 it follows that

$$
P(N(t+s) - N(s) = k) = e^{-\lambda t} \frac{(\lambda t)^k}{k!}, \quad k = 0, 1, 2, \cdots \tag{5.10}
$$

which is independent of s. Thus $\{N(t), t \ge 0\}$ has stationary increments.

Now suppose $0 \le t_1 \le t_2 \le t_3 \le t_4$ are fixed. Then $N(t_2) - N(t_1)$ and $N(t_4) - N(t_3)$ are increments over non-overlapping intervals $(t_1, t_2]$ and $(t_3, t_4]$. From Theorem 5.9 $N_{t_3}(t_4 - t_3) = N(t_4) - N(t_3)$ is independent of $\{N(u), 0 \le u \le t_3\}$, and hence independent of $N(t_2) - N(t_1)$. This proves the independence of increments over non-overlapping intervals. This proves the theorem. ∎

The above theorem helps us compute the finite dimensional joint probability distributions of a Poisson process, as shown in the next theorem.

Theorem 5.11 Finite Dimensional Distributions. *Let $\{N(t), t \ge 0\}$ be a PP(λ) and $0 \le t_1 \le \cdots \le t_n$ be given real numbers and $0 \le k_1 \le \cdots \le k_n$ be given integers. Then*

$$
\begin{aligned}
P(N(t_1) &= k_1, N(t_2) = k_2, \cdots, N(t_n) = k_n) \\
&= e^{-\lambda t_n} \frac{(\lambda t_1)^{k_1}}{k_1!} \cdot \frac{(\lambda(t_2 - t_1))^{k_2 - k_1}}{(k_2 - k_1)!} \cdots \frac{(\lambda(t_n - t_{n-1}))^{k_n - k_{n-1}}}{(k_n - k_{n-1})!}.
\end{aligned}
$$

Proof:

$$P(N(t_1) = k_1, N(t_2) = k_2, \cdots, N(t_n) = k_n)$$
$$= P(N(t_1) = k_1, N(t_2) - N(t_1) = k_2 - k_1,$$
$$\cdots, N(t_n) - N(t_{n-1}) = k_n - k_{n-1})$$
$$= P(N(t_1) = k_1)P(N(t_2) - N(t_1) = k_2 - k_1) \cdots$$
$$P(N(t_n) - N(t_{n-1}) = k_n - k_{n-1})$$
$$\text{(by independent increments property)}$$
$$= e^{-\lambda t_1} \frac{(\lambda t_1)^{k_1}}{k_1!} \cdot e^{-\lambda(t_2 - t_1)} \frac{(\lambda(t_2 - t_1))^{k_2 - k_1}}{(k_2 - k_1)!}$$
$$\cdots e^{-\lambda(t_n - t_{n-1})} \frac{(\lambda(t_n - t_{n-1}))^{k_n - k_{n-1}}}{(k_n - k_{n-1})!},$$

where the last equation follows from Equation 5.10. Further simplification yields the desired result. ∎

The independent increments property is very useful in computing probabilistic quantities associated with a Poisson process as shown in the next two examples.

Example 5.7   Auto-Covariance Function. Let $\{N(t), t \geq 0\}$ be a PP(λ). Compute its auto-covariance function, namely, $\text{Cov}(N(s), N(s + t))$, for $t \geq 0, s \geq 0$.

We have

$$E(N(s)N(s + t)) = E(N(s)(N(s + t) - N(s) + N(s)))$$
$$= E(N(s)(N(s + t) - N(s))) + E(N(s)^2)$$
$$= E(N(s))E(N(s + t) - N(s)) + E(N(s)^2)$$
$$\text{(by independent increments property)}$$
$$= \lambda s \cdot \lambda t + (\lambda s)^2 + \lambda s.$$

Hence,

$$\text{Cov}(N(s), N(s + t)) = E(N(s)N(s + t)) - E(N(s))E(N(s + t))$$
$$= \lambda s \cdot \lambda t + (\lambda s)^2 + \lambda s - \lambda s(\lambda(s + t))$$
$$= \lambda s.$$

The fact that the auto-covariance function is independent of t is a result of the stationarity of increments. In general we can write

$$\text{Cov}(N(s), N(t)) = \lambda \min(s, t). \blacksquare$$

Example 5.8   Consider the post office of Example 5.6. What is the probability that one customer arrives between 1:00 pm and 1:06 pm, and two customers arrive between 1:03 pm and 1:12 pm?

Using time homogeneity and hours as units of time, we write the required probability as $P(N(0.1) = 1; N(0.2) - N(0.05) = 2)$. Using independence of increments over non-overlapping intervals $(0, 0.05]$, $(0.05, 0.1]$, and $(0.1, 0.2]$ we get

$$P(N(0.1) = 1; N(0.2) - N(0.05) = 2)$$

$$= \sum_{k=0}^{1} P(N(0.05) = k, N(0.1) - N(0.05) = 1 - k, N(0.2) - N(0.1) = 1 + k)$$

$$= \sum_{k=0}^{1} P(N(0.05) = k)P(N(0.1) - N(0.05) = 1 - k)P(N(0.2) - N(0.1) = 1 + k)$$

$$= \sum_{k=0}^{1} e^{-.5} \frac{(0.5)^k}{k!} e^{-.5} \frac{(0.5)^{1-k}}{(1 - k)!} e^{-1} \frac{(1)^{(1+k)}}{(1 + k)!}.$$

Numerical calculations yield the desired probability as 0.1015. ∎

The next theorem states that the properties stated in Theorems 5.8 and 5.10 in fact characterize a Poisson process.

Theorem 5.12   Alternate Characterization 1. *A stochastic process* $\{N(t), t \geq 0\}$ *is a PP(λ) if and only if*

(i) it has stationary and independent increments,

(ii) $N(t) \sim P(\lambda t)$, for all $t \geq 0$.

Proof: The "only if" part is contained in Theorems 5.8 and 5.10. Here we prove the "if " part. From (ii) it is clear that $N(0) = 0$ with probability 1. Also, since $N(s + t) - N(s)$ is a $P(\lambda t)$ random variable, it is clear that almost all sample paths of the process are piecewise constant with jumps of size 1. Let

$$S_1 = \inf\{t \geq 0 : N(t) = 1\}.$$

Now, by an argument similar to that in Theorem 5.8,

$$P(S_1 > t) = P(N(t) = 0) = e^{-\lambda t}.$$

Hence $S_1 \sim \exp(\lambda)$. Similarly, we can define

$$S_k = \inf\{t \geq 0 : N(t) = k\},$$

and, using stationary and independent increments, show that $\{X_k = S_k - S_{k-1}, k \geq 1\}$ (with $S_0 = 0$) is a sequence of iid $\exp(\lambda)$ random variables. Hence $\{N(t), t \geq 0\}$ is a PP(λ). This completes the proof. ∎

Since the conditions given in the above theorem are necessary and sufficient, they can be taken as an alternate definition of a Poisson process. Finally, we give yet another characterization of a Poisson process. We need the following definition.

Definition 5.6 $o(h)$ Functions. *A function $f : R \to R$ is said to be an $o(h)$ function (written $f(h) = o(h)$ and read "f is little o of h") if*

$$\lim_{h \to 0} \frac{f(h)}{h} = 0.$$

We illustrate the above definition with an example.

Example 5.9

(i) Let $f(x) = \sum_{i=0}^{\infty} a_i x^i$ be an absolutely convergent series with radius of convergence $R > 0$. f is an $o(h)$ function if and only if $a_0 = a_1 = 0$. This follows since

$$\lim_{h \to 0} \frac{1}{h} \sum_{i=2}^{\infty} a_i h^i = \lim_{h \to 0} \sum_{i=2}^{\infty} a_i h^{i-1} = 0.$$

(ii) The function $f(x) = e^{\lambda x} - 1 - \lambda x$ is $o(h)$.

(iii) If f and g are $o(h)$, then so is $f + g$. ∎

We use this definition to give the second alternate characterization of a Poisson process in the next theorem.

Theorem 5.13   Alternate Characterization 2. *A counting process* $\{N(t), t \geq 0\}$ *is a PP(λ) if and only if*

(i) it has stationary and independent increments,

(ii) $N(0) = 0$ and

$$\begin{aligned}
\mathsf{P}(N(h) = 0) &= 1 - \lambda h + o(h), \\
\mathsf{P}(N(h) = 1) &= \lambda h + o(h), \\
\mathsf{P}(N(h) = j) &= o(h), \quad j \geq 2.
\end{aligned}$$

Proof: Suppose $\{N(t), t \geq 0\}$ is a PP(λ). Thus it satisfies conditions (i) and (ii) of Theorem 5.12. Condition (i) above is the same as condition (i) of Theorem 5.12. We shall show that condition (ii) of Theorem 5.12 implies condition (ii) above. We have $N(0) \sim \mathsf{P}(0) = 0$ with probability 1. Furthermore,

$$\lim_{h \to 0} \frac{1}{h} \left(\mathsf{P}(N(h) = 0) - 1 + \lambda h \right) = \lim_{h \to 0} \frac{1}{h} \left(e^{-\lambda h} - 1 + \lambda h \right) = 0.$$

Hence $\mathsf{P}(N(h) = 0) - 1 + \lambda h$ is an $o(h)$ function. Thus we can write

$$\mathsf{P}(N(h) = 0) = 1 - \lambda h + o(h).$$

The other statements in (ii) can be proved similarly. This proves that conditions (i) and (ii) of Theorem 5.12 imply conditions (i) and (ii) of this theorem.

Next we assume conditions (i) and (ii) above, and derive condition (ii) of Theorem 5.12. Now, let

$$p_k(t) = \mathsf{P}(N(t) = k), \quad k \geq 0.$$

Then, for $k \geq 1$, we get

$$p_k(t + h) = \mathsf{P}(N(t + h) = k)$$

$$= \sum_{j=0}^{k} \mathsf{P}(N(t + h) = k | N(t) = j) \mathsf{P}(N(t) = j)$$

$$= \sum_{j=0}^{k} P(N(t+h) - N(t) = k - j | N(t) = j) p_j(t)$$

$$= \sum_{j=0}^{k} P(N(t+h) - N(t) = k - j) p_j(t)$$

(by independence of increments)

$$= \sum_{j=0}^{k} P(N(h) - N(0) = k - j) p_j(t)$$

(by stationarity of increments)

$$= \sum_{j=0}^{k} P(N(h) = k - j) p_j(t) \quad \text{(since } N(0) = 0)$$

$$= P(N(h) = 0) p_k(t) + P(N(h) = 1) p_{k-1}(t) + \sum_{j=2}^{k} P(N(h) = j) p_{k-j}(t)$$

$$= (1 - \lambda h + o(h)) p_k(t) + (\lambda h + o(h)) p_{k-1}(t) + \sum_{j=2}^{k} o(h) p_{k-j}(t)$$

$$= (1 - \lambda h) p_k(t) + \lambda h p_{k-1}(t) + \sum_{j=0}^{k} o(h) p_{k-j}(t).$$

Rearranging and dividing by h, we get

$$\frac{1}{h}(p_k(t+h) - p_k(t)) = -\lambda p_k(t) + \lambda p_{k-1}(t) + \frac{o(h)}{h} \sum_{j=0}^{k} p_{k-j}(t).$$

Letting $h \to 0$, we get

$$p_k'(t) = \frac{dp_k(t)}{dt} = -\lambda p_k(t) + \lambda p_{k-1}(t). \tag{5.11}$$

Proceeding in a similar fashion for the case $k = 0$ we get

$$p_0'(t) = -\lambda p_0(t). \tag{5.12}$$

Using the initial condition $p_0(0) = 1$, the above equation admits the following solution:

$$p_0(t) = e^{-\lambda t}, \quad t \geq 0.$$

Using the initial condition $p_k(0) = 0$ for $k \geq 1$, we can solve Equation 5.11 recursively to get

$$p_k(t) = e^{-\lambda t} \frac{(\lambda t)^k}{k!}, \quad t \geq 0, \tag{5.13}$$

which implies that $N(t) \sim P(\lambda t)$. Thus conditions (i) and (ii) of Theorem 5.13 imply conditions (i) and (ii) of Theorem 5.12. This proves the result. ∎

Since the conditions of the above theorem are necessary and sufficient, they can be taken as yet another definition of a Poisson process.

5.3 Event Times in a Poisson Process

In this section we study the joint distribution of the event times $S_1, S_2, \cdots, S_n$, given that $N(t) = n$. We begin with some preliminaries about order statistics of uniformly distributed random variables. (See Appendix C for more details.)

Let $U_1, U_2, \cdots, U_n$ be n iid random variables that are uniformly distributed over $[0, t]$, where $t > 0$ is a fixed number. Let $\tilde{U}_1, \tilde{U}_2, \cdots, \tilde{U}_n$ be the order statistics of $U_1, U_2, \cdots, U_n$, that is, a permutation of $U_1, U_2, \cdots, U_n$ such that

$$\tilde{U}_1 \leq \tilde{U}_2 \leq \cdots \leq \tilde{U}_n.$$

Thus

$$\begin{aligned} \tilde{U}_1 &= \min\{U_1,\ U_2,\ \cdots,\ U_n\}, \\ \tilde{U}_n &= \max\{U_1,\ U_2,\ \cdots,\ U_n\}. \end{aligned}$$

Let $f(t_1, t_2, \cdots, t_n)$ be the joint density of $\tilde{U}_1, \tilde{U}_2, \cdots, \tilde{U}_n$. Then one can show that (see Appendix C)

$$f(t_1, t_2, \cdots, t_n) = \begin{cases} \frac{n!}{t^n} & \text{if } 0 \leq t_1 \leq t_2 \leq \cdots \leq t_n \leq t. \\ 0 & \text{otherwise.} \end{cases} \tag{5.14}$$

From the above equation it is possible to derive the marginal density $f_k(u)$ of $\tilde{U}_k$ $(1 \leq k \leq n)$ as

$$f_k(u) = \frac{k}{t}\binom{n}{k}\left(\frac{u}{t}\right)^{k-1}\left(1 - \frac{u}{t}\right)^{n-k}, \quad 0 \leq u \leq t, \tag{5.15}$$

and the expected value of $\tilde{U}_k$ as

$$\mathsf{E}(\tilde{U}_k) = \frac{kt}{n+1}. \tag{5.16}$$

The main result about the joint distribution of $S_1, S_2, \cdots, S_n$ is given in the next theorem.

Theorem 5.14 Campbell's Theorem. *Let S_n be the nth event time in a PP(λ) $\{N(t), t \geq 0\}$. Given $N(t) = n$,*

$$(S_1, S_2, \cdots, S_n) \sim (\tilde{U}_1,\ \tilde{U}_2,\ \cdots, \tilde{U}_n).$$

Proof: Let $0 = t_0 \leq t_1 \leq t_2 \leq \cdots \leq t_n \leq t$. We have

$$\mathsf{P}(S_i \in (t_i, t_i + dt_i); 1 \leq i \leq n | N(t) = n)$$
$$= \frac{\mathsf{P}(S_i \in (t_i, t_i + dt_i); 1 \leq i \leq n, ; N(t) = n)}{\mathsf{P}(N(t) = n)}$$

$$= \frac{P(S_i \in (t_i, t_i + dt_i); 1 \le i \le n, ; S_{n+1} > t)}{P(N(t) = n)}$$

$$= \frac{P(X_i \in (t_i - t_{i-1}, t_i - t_{i-1} + dt_i); 1 \le i \le n, ; X_{n+1} > t - t_n)}{P(N(t) = n)}$$

$$= \frac{\left(\prod_{i=1}^{n} \lambda e^{-\lambda(t_i - t_{i-1})} dt_i\right) e^{-\lambda(t - t_n)}}{e^{-\lambda t} \frac{(\lambda t)^n}{n!}}$$

$$= \frac{e^{-\lambda t} \lambda^n}{e^{-\lambda t} \frac{(\lambda t)^n}{n!}} \cdot dt_1 dt_2 \cdots dt_n.$$

Hence the conditional joint density of $(S_1, S_2, \cdots, S_n)$, given $N(t) = n$, is given by

$$f(t_1, t_2, \cdots, t_n) = \begin{cases} \frac{n!}{t^n} & \text{if } 0 \le t_1 \le t_2 \le \cdots \le t_n \le t, \\ 0 & \text{otherwise,} \end{cases}$$

which is the joint density of $(\tilde{U}_1, \tilde{U}_2, \cdots, \tilde{U}_n)$. This proves the theorem. ∎

Note the curious fact that the joint distribution is independent of λ! This theorem can be interpreted as follows: Suppose $N(t) = n$ is given. Let $(U_1, U_2, \cdots, U_n)$ be n iid random variables uniformly distributed over $[0, t]$. The smallest among them can be thought of as S_1, the second smallest as S_2, and so on, with the largest as S_n. We give several applications of the above theorem.

Example 5.10

(i) Let $\{N(t), t \ge 0\}$ be a PP(λ), and let S_n be the time of the nth event. Compute $P(S_1 > s | N(t) = n)$. Assume $n \ge 1$ is a given integer.
Theorem 5.14 implies that, given $N(t) = n$, $S_1 \sim \min\{U_1, U_2, \cdots, U_n\}$. From Equation 5.15 we get

$$f_1(u) = \frac{n}{t}\left(1 - \frac{u}{t}\right)^{n-1}, \quad 0 \le u \le t.$$

Hence

$$P(S_1 > s | N(t) = n) = \int_s^t f_1(u) du = \left(1 - \frac{s}{t}\right)^n, \quad 0 \le s \le t.$$

(ii) Compute $E(S_k | N(t) = n)$.
From Theorem 5.14 and Equation 5.16, for $1 \le k \le n$, we get

$$E(S_k | N(t) = n) = E(\tilde{U}_k) = \frac{kt}{n+1}.$$

For $k > n$, using memoryless property of the exponentials, we get

$$E(S_k | N(t) = n) = t + \frac{k - n}{\lambda}. \quad ∎$$

Example 5.11 Suppose passengers arrive at a bus depot according to a PP(λ). Buses leave the depot every T time units. Assume that the bus capacity is sufficiently large so that when a bus leaves there are no more passengers left at the depot. What is the average waiting time of the passengers?

Let $\{N(t), t \geq 0\}$ be a PP(λ). Suppose a bus has just left the depot at time 0, so the bus depot is empty at time 0. Consider the time interval $(0, T]$. Number of passengers waiting for the bus at time T is $N(T)$.

Now suppose $N(T) = n > 0$. Let $S_1, S_2, \cdots, S_n$ be the arrival times of these n passengers. The waiting time of the ith passenger is $W_i = T - S_i$. Hence the average waiting time is

$$\overline{W} = \frac{1}{n} \sum_{i=1}^{n} W_i = \frac{1}{n} \sum_{i=1}^{n} (T - S_i) = T - \frac{1}{n} \sum_{i=1}^{n} S_i.$$

Now, using Theorem 5.14, we get

$$\sum_{i=1}^{n} S_i \sim \sum_{i=1}^{n} \tilde{U}_i = \sum_{i=1}^{n} U_i,$$

where U_i's are iid uniformly distributed uniformly over $[0, T]$. Hence

$$\mathsf{E}(\overline{W} | N(T) = n) = T - \mathsf{E}(\frac{1}{n} \sum_{i=1}^{n} U_i) = T - T/2 = T/2.$$

Hence

$$\mathsf{E}(\overline{W}) = \frac{T}{2}.$$

Thus the average waiting time is $T/2$, which makes intuitive sense. This is one more manifestation of the fact that the events in a Poisson process occur in a uniform, time-independent fashion. ∎

Example 5.12 Customers arrive at a service station according to a PP(λ). Each customer pays a service charge of \$1 when he enters the system. Suppose the service station manager discounts the revenues at a continuous discount factor $\alpha > 0$ so that a dollar earned at time t has a present value of $e^{-\alpha t}$ at time 0. Compute the expected present value of all the service charges earned until time t.

Let $N(t)$ be the number of arrivals over $(0, t]$. We are given that $\{N(t), t \geq 0\}$ is a PP(λ). Let $C(t)$ be the total discounted service charges accumulated over $(0, t]$. It is given by

$$C(t) = \sum_{i=1}^{N(t)} e^{-\alpha S_i},$$

where S_i is the arrival time of the ith customer. We have

$$\mathsf{E}(C(t) | N(t) = n) = \mathsf{E}\left(\sum_{i=1}^{n} e^{-\alpha S_i} \Big| N(t) = n \right)$$

$$= \mathsf{E}\left(\sum_{i=1}^{n} e^{-\alpha \tilde{U}_i} \right)$$

(from Theorem 5.14)

$$= \; \mathsf{E}\left(\sum_{i=1}^{n} e^{-\alpha U_i}\right)$$

$$= \; n\mathsf{E}(e^{-\alpha U}),$$

where U is uniformly distributed over $[0, t]$. We have

$$\mathsf{E}(e^{-\alpha U}) = \frac{1}{t}\int_0^t e^{-\alpha u} du = \frac{1 - e^{-\alpha t}}{\alpha t}.$$

Hence

$$\mathsf{E}(C(t)|N(t)) = N(t)\frac{1 - e^{-\alpha t}}{\alpha t}.$$

This yields

$$\mathsf{E}(C(t)) = \mathsf{E}(N(t))\frac{1 - e^{-\alpha t}}{\alpha t} = \frac{\lambda}{\alpha}(1 - e^{-\alpha t}).$$

The expected present value of all future revenue is

$$\lim_{t \to \infty} \mathsf{E}(C(t)) = \frac{\lambda}{\alpha}.$$

Note that this is a finite quantity although the total revenue is infinite. ∎

5.4 Superposition and Splitting of Poisson Processes

The superposition of two counting processes is a counting process that counts events in both the counting processes. Splitting of a counting process is the operation of generating two counting processes by classifying the events in the original counting process as belonging to one or the other counting process. We first study the superposition and then the splitting of Poisson processes.

5.4.1 Superposition

Superposition occurs naturally when two Poisson processes merge to generate a combined process. For example, a telephone exchange may get domestic calls and international calls, each forming a Poisson process. Thus the process that counts both calls is a superposition of the two processes. Figure 5.5 illustrates such a superposition. In this section we study the superposition of two or more independent Poisson processes.

Let $\{N_i(t), t \geq 0\}, (i = 1, 2, \cdots r)$, be *independent* Poisson processes. Define

$$N(t) = N_1(t) + N_2(t) + \cdots + N_r(t), \quad t \geq 0.$$

The process $\{N(t), t \geq 0\}$ is called the superposition of the r processes $\{N_i(t), t \geq 0\}, (i = 1, 2, \cdots r)$. The next theorem describes the superposed process.

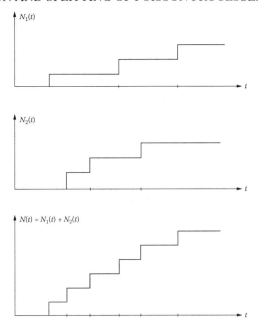

Figure 5.5 *Superposition of two Poisson processes.*

Theorem 5.15   Superposition of Poisson Processes. *Let $\{N_i(t), t \geq 0\}$, $(i = 1, 2, \cdots r)$, be independent Poisson processes. Let λ_i be the parameter of the ith process. Define*

$$N(t) = N_1(t) + N_2(t) + \cdots + N_r(t), \quad t \geq 0.$$

$\{N(t), t \geq 0\}$ is a Poisson process with parameter λ, where

$$\lambda = \lambda_1 + \lambda_2 + \cdots + \lambda_r.$$

Proof: Since the r independent processes $\{N_i(t), t \geq 0\}$, $(i = 1, 2, \cdots, r)$, have stationary and independent increments, it follows that $\{N(t), t \geq 0\}$ inherits this property. Thus in order to show that $\{N(t), t \geq 0\}$ is a PP(λ) it suffices to show that $N(t)$ is a $P(\lambda t)$ random variable. (See Theorem 5.12.) For a fixed t, we know that $\{N_i(t) \sim P(\lambda_i t), 1 \leq i \leq r\}$ are independent Poisson random variables. Hence their sum, $N(t)$, is a Poisson random variable with parameter $\lambda t = (\lambda_1 + \lambda_2 + \cdots + \lambda_r)t$. This proves the theorem. ∎

We illustrate the above result by an example below.

Example 5.13 Jobs submitted for execution on a central computer are divided into four priority classes, indexed 1, 2, 3, and 4. The inter-arrival times for the jobs of class i are exponential random variables with mean m_i minutes, with $m_1 = 10$, $m_2 = 15$, $m_3 = 30$, and $m_4 = 60$. Assume all arrival streams behave independently of each other. Let $N(t)$ be the total number of jobs of all classes that arrive during $(0, t]$. Characterize the stochastic process $\{N(t), t \geq 0\}$.

Let $N_i(t)$ be the number of jobs of class i that arrive during $(0, t]$. Due to the iid exponential inter-arrival times, we know that $\{N_i(t), t \geq 0\}$ is a PP(λ_i) where $\lambda_i = 60/m_i$ per hour, and the four arrival processes are independent. Now, we have

$$N(t) = N_1(t) + N_2(t) + N_3(t) + N_4(t).$$

Hence, from Theorem 5.15, it follows that $\{N(t), t \geq 0\}$ is a PP(λ) where, using the time units of hours,

$$
\begin{aligned}
\lambda &= \lambda_1 + \lambda_2 + \lambda_3 + \lambda_4 \\
&= 60/m_1 + 60/m_2 + 60/m_3 + 60/m_4 \\
&= 6 + 4 + 2 + 1 = 13 \text{ per hour.} \quad \blacksquare
\end{aligned}
$$

We now study how events from the individual processes $\{N_i(t), t \geq 0\}$, $(i = 1, 2, \cdots, r)$, are interleaved in the superposed process $\{N(t), t \geq 0\}$. Let $Z_n = i$ if the nth event in the superposed process belongs to the ith process. Thus, for the sample path shown in Figure 5.5, we have $Z_1 = 1$, $Z_2 = 2, Z_3 = 2, Z_4 = 1, Z_5 = 2, Z_6 = 1, \cdots$. The next theorem gives an interesting property of the sequence $\{Z_n, n \geq 1\}$.

Theorem 5.16 $\{Z_n, n \geq 1\}$ *is a sequence of iid random variables with*

$$\mathsf{P}(Z_n = i) = \frac{\lambda_i}{\lambda}, \quad 1 \leq i \leq r.$$

Proof: Let S_i be the time of occurrence of the first event in $\{N_i(t), t \geq 0\}$. Then $S_i \sim \exp(\lambda_i)$. Also, since the r processes are independent, $\{S_i, 1 \leq i \leq r\}$ are independent. Hence

$$\mathsf{P}(Z_1 = i) = \mathsf{P}(S_i = \min\{S_j, 1 \leq j \leq r\}) = \frac{\lambda_i}{\lambda},$$

where we have used the result from Equation 5.7. Now suppose the first event in the $\{N(t), t \geq 0\}$ process takes place at time s. Then, from Theorem 5.9 it follows that the shifted processes $\{N_i(t + s) - N_i(s), t \geq 0\}$, $(i = 1, 2, \cdots r)$, are independent Poisson processes with parameters λ_i, respectively. Hence Z_2, the type of the next event in $\{N(t), t \geq 0\}$, has the same distribution as Z_1, and is independent of it. Proceeding in this fashion we see that Z_n has the same distribution as Z_1 and is independent of $\{Z_k, 1 \leq k \leq n - 1\}$. This proves the theorem. $\blacksquare$

This is yet another indication that the events in a PP take place uniformly in time. That is why the probability of an event being from the ith process is proportional to the rate of the ith process, and is independent of when the event occurs.

Example 5.14 Customers arriving at a bank can be classified into three categories. Customers of category 1 deposit money, those of category 2 withdraw money, and those of category 3 do both. The deposit transaction takes 3 minutes, the withdrawals take 4 minutes, and the combined transaction takes 6 minutes. Customers of category i arrive according to PP(λ_i) with $\lambda_1 = 20$, $\lambda_2 = 15$, and $\lambda_3 = 10$ per hour. What is the average transaction time of a typical customer in the bank? Are the successive transaction times iid?

Let Z be the category of a typical customer. From Theorem 5.16 we get

$$P(Z = 1) = \frac{20}{45}, \ P(Z = 2) = \frac{15}{45}, \ P(Z = 3) = \frac{10}{45}.$$

Hence the average transaction time of a typical customer is

$$3 \cdot \frac{20}{45} + 4 \cdot \frac{15}{45} + 6 \cdot \frac{10}{45} = 4 \text{ minutes.}$$

The successive transaction times are iid since the categories of the successive customers are iid random variables, from Theorem 5.16. ∎

5.4.2 Splitting

Splitting is the opposite of superposition: we start with a single Poisson process and "split" it to create two or more counting processes. For example, the original process may count the number of arrivals at a store, while the split processes might count the male and female arrivals separately. The nature of these counting processes will depend on the rule used to split the original process. We use a special rule called the Bernoulli splitting, described below.

Let $\{N(t), t \geq 0\}$ be a Poisson process. Suppose each event is classified as a type i event ($1 \leq i \leq r$) with probability p_i, where $\{p_i, 1 \leq i \leq r\}$ are given numbers such that

$$p_i > 0, \ (1 \leq i \leq r), \ \sum_{i=1}^{r} p_i = 1.$$

The successive classifications are independent. This classification mechanism is called the *Bernoulli splitting mechanism*. Now let $N_i(t)$ be the number of events during $(0, t]$ that get classified as type i events. We say that the original process $\{N(t), t \geq 0\}$ is "split" into r processes $\{N_i(t), t \geq 0\}$, ($1 \leq i \leq r$). Clearly,

$$N(t) = N_1(t) + N_2(t) + \cdots + N_r(t), \quad t \geq 0,$$

so $\{N(t), t \geq 0\}$ is a superposition of $\{N_i(t), t \geq 0\}$, ($1 \leq i \leq r$).

The next theorem gives the probabilistic structure of the split processes.

Theorem 5.17 Bernoulli Splitting. *Let $\{N(t), t \geq 0\}$ be a PP(λ) and let $\{N_i(t), t \geq 0\}$, ($1 \leq i \leq r$), be generated by the Bernoulli splitting mechanism using splitting probabilities $[p_1, p_2, \cdots, p_r]$. Then $\{N_i(t), t \geq 0\}$ is a PP(λp_i), ($1 \leq i \leq r$), and the r processes are independent.*

Proof: We shall first show that $\{N_i(t), t \geq 0\}$ is a PP(λp_i) for a given $i \in$

$\{1, 2, \cdots, r\}$. There are many ways of showing this. We shall show it by using Theorem 5.7. Let $\{X_n, n \geq 1\}$ be the iid $\exp(\lambda)$ inter-event times in $\{N(t), t \geq 0\}$, and $\{Y_n, n \geq 1\}$ be the inter-event times in the $\{N_i(t), t \geq 0\}$ process. Let $\{R_n, n \geq 1\}$ be a sequence of iid geometric random variables with parameter p_i, i.e.,

$$P(R_n = k) = (1 - p_i)^{k-1} p_i, \quad k \geq 1.$$

Let $T_0 = 0$ and $T_n = R_1 + R_2 + \cdots + R_n$, $n \geq 1$. The Bernoulli splitting mechanism implies that

$$Y_n = \sum_{i=T_{n-1}+1}^{T_n} X_i, \quad n \geq 1.$$

Thus $\{Y_n, n \geq 1\}$ is a sequence of iid random variables. Now Y_1 is a geometric sum of iid exponential random variables. Hence, from Theorem 5.7, Y_1 is an $\exp(\lambda p_i)$ random variable. Thus the inter-event times in $\{N_i(t), t \geq 0\}$ are iid $\exp(\lambda p_i)$ random variables. Hence $\{N_i(t), t \geq 0\}$ is a $PP(\lambda p_i)$.

Next we show independence of the r processes. We treat the case $r = 2$, the general case follows similarly. To show that $\{N_1(t), t \geq 0\}$ and $\{N_2(t), t \geq 0\}$ are independent, it suffices to show that any increment in $\{N_1(t), t \geq 0\}$ is independent of any increment in $\{N_2(t), t \geq 0\}$. This reduces to showing

$$P(N_1(t) = i, N_2(t) = j) = P(N_1(t) = i)P(N_2(t) = j),$$

for all $i, j \geq 0$. We have

$$
\begin{aligned}
&P(N_1(t) = i, N_2(t) = j) \\
&= P(N_1(t) = i, N_2(t) = j | N(t) = i + j)P(N(t) = i + j) \\
&= \binom{i+j}{i} p_1^i p_2^j e^{-\lambda t} \frac{(\lambda t)^{i+j}}{(i+j)!} \\
&= e^{-\lambda p_1 t} \frac{(\lambda p_1 t)^i}{i!} e^{-\lambda p_2 t} \frac{(\lambda p_2 t)^j}{j!} \\
&= P(N_1(t) = i)P(N_2(t) = j).
\end{aligned}
$$

This proves the theorem. ∎

It should be noted that the independence of the r split processes is rather surprising, since they all arise out of the same original process. This theorem makes the Bernoulli splitting mechanism very attractive in applications. We illustrate with two examples.

Example 5.15 Geiger Counter. A geiger counter is a device to count the radioactive particles emitted by a source. Suppose the particles arrive at the counter according to a $PP(\lambda)$ with $\lambda = 1000$ per second. The counter fails to count a particle with probability .1, independent of everything else. Suppose the counter registers four particles in .01 seconds. What is the probability that at least six particles have actually arrived at the counter during this time period?

Let $N(t)$ be the number of particles that arrive at the counter during $(0, t]$, $N_1(t)$

the number of particles that are registered by the counter during $(0, t]$, and $N_2(t)$ the number of particles that go unregistered by the counter during $(0, t]$. Then $\{N(t), t \geq 0\}$ is a PP(1000), and Theorem 5.17 implies that $\{N_1(t), t \geq 0\}$ is a PP(900), and it is independent of $\{N_2(t), t \geq 0\}$, which is a PP(100). We are asked to compute $P(N(.01) \geq 6|N_1(.01) = 4)$. We have

$$
\begin{aligned}
P(N(.01) \geq 6|N_1(.01) = 4) &= P(N_2(.01) \geq 2|N_1(.01) = 4) \\
&= P(N_2(.01) \geq 2) = 0.264.
\end{aligned}
$$

Here we have used independence to get the second equality, and used $N_2(.01) \sim P(1)$ to compute the numerical answer. ∎

Example 5.16 Turnpike Traffic. Consider the toll highway from Orlando to Miami, with n interchanges where the cars can enter and exit, interchange 1 is at the start in Orlando, and n is at the end in Miami. Suppose the cars going to Miami enter at interchange i according to a PP(λ_i) and travel at the same speed, $(1 \leq i \leq n - 1)$. A car entering at interchange i will exit at interchange j with probability $p_{i,j}$, $1 \leq i < j \leq n$, independent of each other. Let $N_i(t)$ be the number of cars that cross a traffic counter between interchanges i and $i + 1$ during $(0, t]$. Show that $\{N_i(t), t \geq 0\}$ is a PP and compute its rate.

Let $N_{k,i}(t)$ be the number of cars that enter at interchange $k \leq i$ and cross the traffic counter between interchanges i and $i + 1$ during $(0, t]$. Assuming Bernoulli splitting, we see that

$$
\{N_{k,i}(t), t \geq 0\} \sim \text{PP}(\lambda_k \sum_{j>i} p_{k,j}).
$$

Now,

$$
N_i(t) = \sum_{k=1}^{i} N_{k,i}(t).
$$

Theorem 5.17 implies that $\{N_{k,i}(t), t \geq 0\}$, $1 \leq k \leq i$, are independent Poisson processes. Hence, from Theorem 5.15, we have

$$
\{N_i(t), t \geq 0\} \sim \text{PP}(\sum_{k=1}^{i} \lambda_k \sum_{j>i} p_{k,j}).
$$

Are $\{N_i(t), t \geq 0\}$, $1 \leq i \leq n - 1$, independent? ∎

Now we consider the *non-homogeneous* Bernoulli splitting. Let $p : [0, \infty) \to [0, 1]$ be a given function. Let $\{N(t), t \geq 0\}$ be a PP(λ). Under the non-homogeneous Bernoulli splitting mechanism, an event occurring at time s is registered with probability $p(s)$, independent of everything else. Let $R(t)$ be the number of registered events during $(0, t]$. The next theorem shows that $R(t)$ is a Poisson random variable.

Theorem 5.18 *Let p be an integrable function over $[0, t]$. Then $R(t)$ is a Poisson* *random variable with parameter*

$$\lambda \int_0^t p(s)ds.$$

Proof: Let $N(t) = m$ be given, and let S_i $(1 \le i \le m)$ be the occurrence time of the ith event in $\{N(t), t \ge 0\}$. Let $I(S_i)$ be 1 if the i-th event is registered and 0 otherwise. Then we can write

$$R(t) = \sum_{i=1}^m I(S_i),$$

where, given $(S_1, S_2, \cdots S_m)$, $\{I(S_i), 1 \le i \le m\}$ are independent random variables with distribution

$$I(S_i) = \begin{cases} 1 & \text{with probability } p(S_i) \\ 0 & \text{with probability } 1 - p(S_i). \end{cases}$$

Using Theorem 5.14 and the notation there, we get

$$\begin{aligned} R(t) &= \sum_{i=1}^m I(S_i) \\ &\sim \sum_{i=1}^m I(\tilde{U}_i) \\ &= \sum_{i=1}^m I(U_i). \end{aligned}$$

However, $\{I(U_i), 1 \le i \le m\}$ are iid Ber(α) random variables with

$$\alpha = \frac{1}{t} \int_0^t p(s)ds.$$

Hence

$$P(R(t) = k | N(t) = m) = \binom{m}{k} \alpha^k (1 - \alpha)^{m-k}.$$

Hence

$$\begin{aligned} P(R(t) = k) &= \sum_{m=0}^{\infty} P(R(t) = k | N(t) = m) P(N(t) = m) \\ &= \sum_{m=k}^{\infty} \binom{m}{k} \alpha^k (1 - \alpha)^{m-k} e^{-\lambda t} \frac{(\lambda t)^m}{m!} \\ &= e^{-\lambda t} \frac{\alpha^k}{k!} \sum_{m=k}^{\infty} (1 - \alpha)^{m-k} \frac{(\lambda t)^m}{(m - k)!} \\ &= e^{-\lambda t} \frac{(\lambda \alpha t)^k}{k!} \sum_{m=k}^{\infty} \frac{(\lambda(1 - \alpha)t)^{m-k}}{(m - k)!} \end{aligned}$$

$$\begin{aligned} &= e^{-\lambda t}\frac{(\lambda\alpha t)^k}{k!}e^{\lambda(1-\alpha)t} \\ &= e^{-\lambda\alpha t}\frac{(\lambda\alpha t)^k}{k!}. \end{aligned}$$

Thus $R(t)$ is Poisson random variable with parameter

$$\lambda\alpha t = \lambda\int_0^t p(s)ds.$$

This proves the theorem. ■

Note that although $R(t)$ is a Poisson random variable for each t, and $\{R(t), t \geq 0\}$ has independent increments, it is not a Poisson process since the increments are not stationary. We shall identify such a process as a non-homogeneous Poisson process, and study it in the next section. An interesting application of a non-homogeneous Bernoulli splitting is given in the next example.

Example 5.17 Infinite Server Queue. Users arrive at a public library according to a PP(λ). User i stays in the library for T_i amount of time and then departs. Assume that $\{T_i, i \geq 1\}$ are iid non-negative random variables with common cdf $G(\cdot)$ and mean τ. Suppose the library opens at time zero, with zero patrons in it. Compute the distribution of $R(t)$, the number of users in the library at time t.

Let $N(t)$ be the number of users who enter the library during $(0, t]$. If user i enters the library at time $s \leq t$, then he or she will be in the library at time t with probability

$$p(s) = \mathsf{P}(T_i > t - s) = 1 - G(t - s).$$

We can imagine that the user entering at time s is registered with probability $p(s)$. Thus $R(t)$, the number of users who are registered during $(0, t]$, is the same as the number of users in the library at time t. Since $p(\cdot)$ is a monotone bounded function, it is integrable, and we can apply Theorem 5.18 to see that $R(t)$ is a Poisson random variable with parameter

$$\begin{aligned} \lambda\int_0^t p(s)ds &= \lambda\int_0^t (1 - G(t - s))ds \\ &= \lambda\int_0^t (1 - G(s))ds. \end{aligned}$$

Thus the expected number of users in the library at time t is

$$\mathsf{E}(R(t)) = \lambda\int_0^t (1 - G(s))ds.$$

Thus, as $t \to \infty$, $R(t)$ converges in distribution to a Poisson random variable with parameter

$$\lambda\int_0^\infty (1 - G(s))ds = \lambda\tau.$$

We shall see later that what we analyzed here is called the infinite server queue. ■

5.5 Non-Homogeneous Poisson Process

In this section we study a generalization of the Poisson process by relaxing the requirement that the increments be stationary.

We begin with some preliminaries. Let $\lambda : [0, \infty) \to [0, \infty)$ be a given function. Assume that it is integrable over finite intervals and define

$$\Lambda(t) = \int_0^t \lambda(s)ds.$$

Definition 5.7 *A counting process $\{N(t), t \geq 0\}$ is said to be a non-homogeneous Poisson process with rate function $\lambda(\cdot)$ if*

(i) $\{N(t), t \geq 0\}$ has independent increments,

(ii) $N(t) \sim P(\Lambda(t))$, for $t \geq 0$.

We denote a non-homogeneous Poisson process with rate function $\lambda(\cdot)$ as $NPP(\lambda(\cdot))$. Note that when $\lambda(t) = \lambda$ for all $t \geq 0$, the $NPP(\lambda(\cdot))$ reduces to the standard $PP(\lambda)$. It is clear that if $\{N(t), t \geq 0\}$ is an $NPP(\lambda(\cdot))$,

(i) $N(0) = 0$,

(ii) $N(s + t) - N(s) \sim P(\Lambda(s + t) - \Lambda(s))$.

(See Conceptual Exercise 5.7.) Using the independence of increments one can derive the finite dimensional joint probability distributions of an NPP as in Theorem 5.11. We state the result in the next theorem and omit the proof.

Theorem 5.19   Finite Dimensional Distributions. *Let $\{N(t), t \geq 0\}$ be a $NPP(\lambda(\cdot))$ and $0 \leq t_1 \leq \cdots \leq t_n$ be given real numbers and $0 \leq k_1 \leq \cdots \leq k_n$ be given integers. Then*

$$P(N(t_1) = k_1, N(t_2) = k_2, \cdots, N(t_n) = k_n)$$
$$= e^{-\Lambda(t_n)} \frac{(\Lambda(t_1))^{k_1}}{k_1!} \cdot \frac{(\Lambda(t_2) - \Lambda(t_1))^{k_2-k_1}}{(k_2 - k_1)!} \cdots \frac{(\Lambda(t_n) - \Lambda(t_{n-1}))^{k_n-k_{n-1}}}{(k_n - k_{n-1})!}.$$

We illustrate with an example.

Example 5.18 A fast food restaurant is open from 6 am to midnight. During this time period customers arrive according to an $NPP(\lambda(\cdot))$, where the rate function $\lambda(\cdot)$ (in customers/hr) is as shown in Figure 5.6. The corresponding $\Lambda(\cdot)$ is shown in Figure 5.7.

(i) Compute the mean and variance of the number arrivals in one day (6 am to midnight).
   Let $N(t)$ be the number of arrivals in the first t hours after 6 am. The number of arrivals in one day is given by $N(18)$ which is a $P(\Lambda(18))$ random variable by definition. Hence the mean and the variance is given by $\Lambda(18) = 275$.

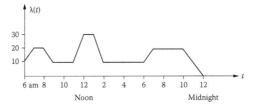

Figure 5.6 *Arrival rate function for the fast food restaurant.*

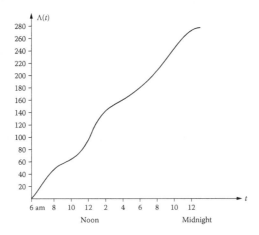

Figure 5.7 *Cumulative arrivals function for the fast food restaurant.*

(ii) Compute the mean and variance of the number arrivals from 6 pm to 10 pm.
The number of arrivals from 6 pm to 10 pm is given by $N(16) - N(12)$ which
is a $P(\Lambda(16) - \Lambda(12))$ random variable by definition. Hence the mean and the
variance is given by $\Lambda(16) - \Lambda(12) = 255 - 180 = 75$.

(iii) What is the probability that exactly one person arrives from 8 am to 8:06 am?
The number of arrivals from 8 am to 8:06 am is given by $N(2.1) - N(2)$ which is
a $P(\Lambda(2.1) - \Lambda(2)) = P(1.95)$ random variable by definition. Hence the required
probability is given by $e^{-1.95}(1.95)^1/1! = 0.2774$. ∎

The next theorem gives necessary and sufficient condition for a counting process
to be an NPP.

Theorem 5.20 *A counting process* $\{N(t), t \geq 0\}$ *is a* NPP$(\lambda(\cdot))$ *if and only if*

(i) $\{N(t), t \geq 0\}$ *has independent increments,*

(ii) $N(0) = 0$ *and*

$$
\begin{aligned}
P(N(t+h) - N(t) = 0) &= 1 - \lambda(t)h + o(h), \\
P(N(t+h) - N(t) = 1) &= \lambda(t)h + o(h),
\end{aligned}
$$

$$P(N(t+h) - N(t) = j) = o(h), \quad j \geq 2.$$

Proof: The "if" part is straightforward and we leave the details to the reader; see Conceptual Exercise 5.8. To establish the "only if" part it suffices to show that the conditions (i) and (ii) in this theorem imply that $N(t) \sim P(\Lambda(t))$. To do this we define

$$p_k(t) = P(N(t) = k), \quad k \geq 0.$$

We follow the proof of Theorem 5.13. For $k \geq 1$, we get

$$
\begin{aligned}
p_k(t+h) &= P(N(t+h) = k) \\
&= \sum_{j=0}^{k} P(N(t+h) = k | N(t) = j) P(N(t) = j) \\
&= \sum_{j=0}^{k} P(N(t+h) - N(t) = k - j | N(t) = j) p_j(t) \\
&= \sum_{j=0}^{k} P(N(t+h) - N(t) = k - j) p_j(t) \\
&\quad \text{(by independence of increments)} \\
&= P(N(t+h) - N(t) = 0) p_k(t) + P(N(t+h) - N(t) = 1) p_{k-1}(t) \\
&\quad + \sum_{j=2}^{k} P(N(t+h) - N(t) = j) p_{k-j}(t) \\
&= (1 - \lambda(t)h + o(h)) p_k(t) + (\lambda(t)h + o(h)) p_{k-1}(t) + \sum_{j=2}^{k} o(h) p_{k-j}(t) \\
&= (1 - \lambda(t)h) p_k(t) + \lambda(t)h p_{k-1}(t) + \sum_{j=0}^{k} o(h) p_{k-j}(t).
\end{aligned}
$$

Rearranging and dividing by h, we get

$$\frac{1}{h}(p_k(t+h) - p_k(t)) = -\lambda(t) p_k(t) + \lambda(t) p_{k-1}(t) + \frac{o(h)}{h} \sum_{j=0}^{k} p_{k-j}(t).$$

Letting $h \to 0$, we get

$$p_k'(t) = \frac{dp_k(t)}{dt} = -\lambda(t) p_k(t) + \lambda(t) p_{k-1}(t). \tag{5.17}$$

Proceeding in a similar fashion for the case $k = 0$ we get

$$p_0'(t) = -\lambda(t) p_0(t).$$

Using the initial condition $p_0(0) = 1$, the above equation admits the following solution:

$$p_0(t) = e^{-\Lambda(t)}, \quad t \geq 0.$$

Using the initial condition $p_k(0) = 0$ for $k \geq 1$, we can solve Equation 5.17 recursively to get

$$p_k(t) = e^{-\Lambda(t)} \frac{\Lambda(t)^k}{k!}, \quad t \geq 0,$$

which implies that $N(t) \sim P(\Lambda(t))$. This proves the theorem. ∎

Since the conditions in the above theorem are necessary and sufficient, we can use them as an alternate definition of an NPP.

5.5.1 Event Times in an NPP

Suppose the nth event in an NPP occurs at time S_n ($n \geq 1$). Define $X_n = S_n - S_{n-1}$ ($S_0 = 0$) be the nth inter-event time. In general the inter-event times are neither independent, nor identically distributed. We can compute the marginal distribution of X_n as given in the next theorem.

Theorem 5.21

$$P(X_{n+1} > t) = \begin{cases} e^{-\Lambda(t)} & \text{if } n = 0, \\ \int_0^\infty \lambda(s) e^{-\Lambda(t+s)} \frac{\Lambda(s)^{n-1}}{(n-1)!} ds & \text{if } n \geq 1. \end{cases}$$

Proof: Left as Conceptual Exercise 5.14. ∎

Next we generalize Theorem 5.14 to non-homogeneous Poisson processes. Let $t \geq 0$ be fixed and $\{U_i, 1 \leq i \leq n\}$ be iid random variables with common distribution

$$P(U_i \leq s) = \frac{\Lambda(s)}{\Lambda(t)}, \quad 0 \leq s \leq t.$$

Let $\tilde{U}_1 \leq \tilde{U}_2 \leq \cdots \leq \tilde{U}_n$ be the order statistics of $\{U_i, 1 \leq i \leq n\}$.

Theorem 5.22   Campbell's Theorem for NPPs. *Let S_n be the nth event time in an NPP$(\lambda(\cdot))$ $\{N(t), t \geq 0\}$. Given $N(t) = n$,*

$$(S_1, S_2, \cdots, S_n) \sim (\tilde{U}_1, \tilde{U}_2, \cdots, \tilde{U}_n).$$

Proof: Left as Conceptual Exercise 5.11. ∎

5.6 Compound Poisson Process

In this section we shall study another generalization of a Poisson process, this time relaxing the assumption that the events occur one at a time. Such a generalization is useful in practice where multiple events can occur at the same time. For example, customers arrive in batches to a restaurant, multiple cars are involved in an accident, demands occur in batches, etc. To model such situations we introduce the compound Poisson process (CPP) as defined below.

Definition 5.8 *Let $\{N(t), t \geq 0\}$ be a PP(λ) and $\{Z_n, n \geq 1\}$ be a sequence of iid random variables that is independent of the PP. Let*

$$Z(t) = \sum_{n=1}^{N(t)} Z_n, \quad t \geq 0. \tag{5.18}$$

Then $\{Z(t), t \geq 0\}$ is called a compound Poisson process.

Note $Z(t) = 0$ if $N(t) = 0$. Also, if $Z_n = 1$ for all $n \geq 1$, Equation 5.18 reduces to $Z(t) = N(t)$. Thus CPP is a generalization of a PP. We have not restricted Z_n's to be positive, or integer valued. It is convenient to think of Z_n as the cost incurred (or reward earned) when the nth event occurs in the PP. Then $Z(t)$ is the cumulative cost incurred (or reward earned) over $(0, t]$. Since Z_n's are allowed to be positive as well as negative, the sample paths of $\{Z(t), t \geq 0\}$ may jump up or down. A typical sample path is shown in Figure 5.8.

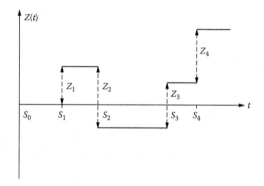

Figure 5.8 *A typical sample path of a compound Poisson process.*

Since $\{Z_n, n \geq 1\}$ is a sequence of iid random variables, and the PP has stationary and independent increments, it follows that $\{Z(t), t \geq 0\}$ also has stationary and independent increments. Thus if we know the marginal distribution of $Z(t)$, then all finite dimensional distributions of $\{Z(t), t \geq 0\}$ are known. Specifically, consider the case where Z_n is a non-negative integer valued random variable. Then the state-space of $\{Z(t), t \geq 0\}$ is $\{0, 1, 2, \cdots\}$. Let

$$p_k(t) = P(Z(t) = k), \quad k = 0, 1, 2, \cdots.$$

The next theorem gives the finite dimensional distribution.

Theorem 5.23 *Let $\{Z(t), t \geq 0\}$ be a CPP with state-space $\{0, 1, 2, \cdots\}$. Let $0 \leq t_1 \leq \cdots \leq t_n$ be given real numbers and $0 \leq k_1 \leq \cdots \leq k_n$ be given integers. Then*

$$P(Z(t_1) = k_1, Z(t_2) = k_2, \cdots, Z(t_n) = k_n)$$
$$= p_{k_1}(t_1)p_{k_2 - k_1}(t_2 - t_1) \cdots p_{k_n - k_{n-1}}(t_n - t_{n-1}).$$

Proof: See Conceptual Exercise 5.18. ∎

Thus the entire CPP is described by the marginal distributions of $Z(t)$ for all $t \geq 0$. In general, computing the distribution of $Z(t)$ is hard, but the Laplace Stieltjes Transform (LST) of $Z(t)$ can be computed easily, as shown in the next theorem.

Theorem 5.24 *Let* $\{Z(t), t \geq 0\}$ *be a CPP as defined by Equation 5.18. Let*

$$A(s) = \mathsf{E}(e^{-sZ_n}), \quad n \geq 1.$$

Then

$$\phi(s) = \mathsf{E}(e^{-sZ(t)}) = e^{-\lambda t(1 - A(s))}.$$

Proof: We have

$$\mathsf{E}(e^{-sZ(t)} | N(t) = n) = \mathsf{E}(e^{-s(Z_1 + Z_2 + \cdots + Z_n)}) = A(s)^n,$$

since $\{Z_n, n \geq 1\}$ are iid with common LST $A(s)$. Thus

$$\phi(s) = \mathsf{E}(e^{-sZ(t)}) = \mathsf{E}(\mathsf{E}(e^{-sZ(t)} | N(t))) = \mathsf{E}((A(s))^{N(t)}),$$

which is the generating function of $N(t)$ evaluated at $A(s)$. Since $N(t) \sim \mathrm{P}(\lambda t)$, we know that

$$\mathsf{E}(z^{N(t)}) = e^{-\lambda t(1 - z)}.$$

Hence

$$\phi(s) = e^{-\lambda t(1 - A(s))}.$$

This proves the theorem. ∎

In general it is difficult to invert $\phi(s)$ analytically to obtain the distribution of $Z(t)$. However, the moments of $Z(t)$ are easy to obtain. They are given in the following theorem.

Theorem 5.25 *Let* $\{Z(t), t \geq 0\}$ *be a CPP as defined by Equation 5.18. Let* τ *and* s^2 *be the mean and the second moment of* Z_n, $n \geq 1$. *Then*

$$\mathsf{E}(Z(t)) = \lambda \tau t,$$

$$\mathrm{Var}(Z(t)) = \lambda s^2 t.$$

Proof: See Conceptual Exercise 5.19. ∎

We end this section with an example.

Example 5.19 Suppose customers arrive at a restaurant in batches that are iid with the following common distribution

$$p_k = \mathrm{P}(\text{Batch Size} = k), \quad 1 \leq k \leq 6,$$

where

$$[p_1, \; p_2, \; p_3, \; p_4, \; p_5, \; p_6] = [.1, \; .25, \; .1, \; .25, \; .15, \; .15].$$

The batches themselves arrive according to a PP(λ). Compute the mean and variance of $Z(t)$, the number of arrivals during $(0, t]$.

$\{Z(t), t \geq 0\}$ is a CPP with batch arrival rate λ and the mean and second moment of batch sizes given by

$$\tau = \sum_{k=1}^{6} kp_k = 3.55,$$

$$s^2 = \sum_{k=1}^{6} k^2 p_k = 15.15.$$

Hence, using Theorem 5.25, we get

$$\mathsf{E}(Z(t)) = 3.55\lambda t, \quad \mathrm{Var}(Z(t)) = 15.15\lambda t. \quad \blacksquare$$

5.7 Computational Exercises

5.1 Consider a network with three arcs as shown in Figure 5.9. Let $X_i \sim \exp(\lambda_i)$ be the length of arc i, $(1 \leq i \leq 3)$. Suppose the arc lengths are independent. Compute the distribution of the length of the shortest path from node a to node c.

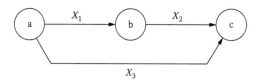

Figure 5.9 *The three arc network.*

5.2 Compute the probability that $a - b - c$ is the shortest path from node a to node c in the network in Figure 5.9.

5.3 Compute the distribution of the length of the longest path from node a to node c in the network in Figure 5.9.

5.4 Compute the probability that $a - b - c$ is the longest path from node a to node c in the network in Figure 5.9.

5.5 Consider a system consisting of n components in parallel, that is, the system fails when all of its components fail. The lifetimes of the components are iid $\exp(\lambda)$ random variables. Compute the cdf of the time when the system fails, assuming that all components are functioning at time 0.

5.6 A spaceship is controlled by three independent computers on board. The ship can function as long as at least two of these computers are functioning. Suppose the lifetimes of the computers are iid $\exp(\lambda)$ random variables. Assume that at time zero all computers are functional. What is the probability that the ship functions over the duration $(0, t]$?

5.7 The lifetime of a machine is an exp(λ) random variable. A maintenance crew inspects this machine every T time units, starting at time 0, where $T > 0$ is a fixed number. Find the expected time when the machine is discovered to have failed. What is the expected duration of time when the machine is down before it is discovered to be down?

5.8 There are n organisms in a colony at time 0. The lifetimes of these organisms are iid exp(λ) random variables. What is the probability that exactly k organisms are living at time t? What is the expected time by which the colony experiences its kth death? ($0 \leq k \leq n$).

5.9 A machine has two parts: A and B. There are two spares available for A and one for B. The machine needs both parts to function. As soon as a part fails, it is replaced instantaneously by its spare. The lifetimes of all parts are independent. The part A and its spares have iid exp(λ) lifetimes and B and its spare have iid exp(μ) lifetimes. Spares fail only while in service. What is the expected lifetime of the machine?

5.10 A rope consists of n strands. When a strand is carrying a load of x tons, its failure rate is λx. At time zero all strands are working and equally share a combined load of L tons. When a strand fails, the remaining strands share the load equally. This process continues until all strands break, at which point the rope breaks. Compute the distribution of the lifetime of the rope.

5.11 Two jobs are waiting to be processed by a single machine. The time required to complete the ith job is an exp(λ_i) random variable ($i = 1, 2$). The processing times are independent of each other, and of the sequence in which the jobs are processed. To keep the ith job in the machine shop costs C_i dollars per unit time. In what sequence should the jobs be processed so that the expected total cost is minimized?

5.12 Suppose customers arrive at a single-server queue according to a PP(λ), require iid exp(μ) service times, and are served in a first come first served fashion. At time 0 there is one customer in the system, and he is in service. What is the probability that there will be no customers in the system after the second customer completes his service?

5.13 Suppose there are $n \geq 3$ sticks whose lengths are iid exp(λ) random variables. Show that the probability that an n-sided polygon can be formed from these sticks is $n(1/2)^{n-1}$. Hint: A polygon can be formed if and only if the longest stick is shorter than the sum of the rest.

5.14 Let A, B, and C be iid exp(λ) random variables. Show that the probability that both the roots of $Ax^2 + Bx + C = 0$ are real is 1/3.

5.15 Suppose U is uniformly distributed over $(0, 1)$. Show that $-\ln(U)$ is an exp(1) random variable.

5.16 Mary arrives at a single server service facility to find a random number, N, of customers already in the system. The customers are served in a first come first served order. Service times of all the customers are iid $\exp(\lambda)$ random variables. The pmf of N is given by $P(N = n) = (1 - \rho)\rho^n$, $n = 0, 1, 2, \cdots$, where $\rho < 1$ is a given constant. Let W be the time when Mary completes her service. Compute the LST and the pdf of W.

5.17 A machine needs a single critical component to operate properly. When that component fails it is instantaneously replaced by a spare. There is a supply of k spares. The lifetimes of the original component and the k spares are iid $\exp(\lambda)$ random variables. What is the smallest number of spares that must be provided to guarantee that the system will last for at least T hours with probability α, a given constant? Assume that a component only fails while in service.

5.18 A system consists of two components in series. The lifetime of the first component is $\exp(\lambda)$ and that of the second component is $Erl(\mu, n)$. The system fails as soon any one of the two components fails. Assuming the components behave independently of each other, compute the expected lifetime of the system.

5.19 Let $X_i \sim \exp(\lambda_i)$ ($i = 1, 2$) be two independent random variables. Compute the distribution of X_1/X_2 and show that it has infinite expectation.

5.20 Let $\{N(t), t \geq 0\}$ be a PP(λ). Compute $P(N(t) = k | N(t + s) = k + m)$ for $s, t \geq 0$, $k, m \geq 0$.

5.21 Let X_i be a P(λ_i) random variable, and assume that $\{X_i, 1 \leq i \leq r\}$ are independent. Show that $X = X_1 + X_2 + \cdots + X_n$ is a P(λ) random variable, where $\lambda = \lambda_1 + \lambda_2 + \cdots + \lambda_n$.

5.22 Customers arrive according to a PP(λ) to a service facility with $s \geq 1$ identical servers in parallel. The service times are iid $\exp(\mu)$ random variables. At time 0, all the servers are busy serving one customer each, and no customers are waiting.

1. Compute the probability that the next arriving customer finds all servers busy.
2. Let N be the number of arrivals before the first service completion. Compute the pmf of N.
3. Compute the probability that the next arriving customer finds at least two servers idle.

5.23 Let S_i be the ith event time in a PP(λ) $\{N(t), t \geq 0\}$. Show that

$$E(S_{N(t)}) = t - \frac{1 - e^{-\lambda t}}{\lambda}.$$

5.24 Let $\{N(t), t \geq 0\}$ be a PP(λ). Suppose a Bernoulli splitting mechanism tags each event as type one with probability p, and type two with probability $1 - p$. Let $N_i(t)$ be the number of type i events over $(0, t]$. Let T_i be the time until the first event in the type i process. Compute the joint pdf of (T_1, T_2).

5.25 Let $\{N(t), t \geq 0\}$ be a PP(λ). Compute the probability that $N(t)$ is odd.

5.26 A system is subject to k types of shocks. Shocks of type i arise according to a PP(λ_i), $1 \leq i \leq r$. A shock of type i induces a system failure with probability p_i independently of everything else. Let T be the time when the system fails, and let S be the type of the shock that causes the failure. Compute $P(T > t, S = i)$, $t \geq 0, 1 \leq i \leq r$.

5.27 Let $\{N(t), t \geq 0\}$ be a PP(λ). The events in this process are registered by a counter that locks up for a fixed period of time $\tau > 0$ every time it registers an event. While it is locked, it cannot register any event. What is the probability that the counter is not locked at time t, assuming that it is not locked at time 0?

5.28 Suppose customers arrive at a bank according to a Poisson process with a rate of 8 per hour. Compute the following:

1. The mean and variance of the number of customers who enter the bank during an 8-hour day.

2. Probability that more than four customers enter the bank during an hour-long lunch break.

3. Probability than no customers arrive during the last 15 minutes of the day.

4. Correlation between the number of customers who enter the bank between 9 am and 11 am, and those who enter between 10 am and noon.

5.29 Consider a one-way road where the cars form a PP(λ) with rate λ cars/sec. The road is x feet wide. A pedestrian, who walks at a speed of u feet/sec, will cross the road if and only if she is certain that no cars will cross the pedestrian crossing while she is on it. Show that the expected time until she completes the crossing is $(e^{\lambda x/u} - 1)/\lambda$.

5.30 A machine is up at time zero. It then alternates between two states: up or down. (When an up machine fails it goes to the down state, and when a down machine is repaired it moves to the up state.) Let U_n be the nth up-duration, followed by the nth down-duration, D_n. Suppose $\{U_n, n \geq 0\}$ is a sequence of iid exp(λ) random variables. The nth down-duration is proportional to the nth up-duration, i.e., there is a constant $c > 0$ such that $D_n = cU_n$.

1. Let $F(t)$ be the number of failures up to time t. Is $\{F(t), t \geq 0\}$ a Poisson process? If yes, what is its rate parameter? If not, why not?

2. Let $R(t)$ be the number of repair completions up to time t. Is $\{R(t), t \geq 0\}$ a Poisson process? If yes, what is its rate parameter? If not, why not?

5.31 Suppose customers arrive at a system according to PP(λ). Every customer stays in the system for an exp(μ) amount of time and then leaves. Customers behave independently of each other. Show that the expected number of customers in the system at time t is $\frac{\lambda}{\mu}(1 - e^{-\mu t})$.

5.32 Two individuals, 1 and 2, need kidney transplants. Without a transplant the remaining lifetime of person i is an $\exp(\mu_i)$ random variable, the two lifetimes being independent. Kidneys become available according to a Poisson process with rate λ. The first available kidney is supposed to go to person 1 if she is still alive when the kidney becomes available; else it will go to person 2. The next kidney will go to person 2, if she is still alive and has not already received a kidney. Compute the probability that person i receives a new kidney ($i = 1, 2$).

5.33 It has been estimated that meteors entering the earth's atmosphere over a certain region form a Poisson process with rate $\lambda = 100$ per hour. About 1 percent of those are visible to the naked eye.

1. What is the probability that a person is unlucky enough not to see any shooting stars in a one-hour period?

2. What is the probability that a person will see two shooting stars in one hour? More than two in one hour?

5.34 A machine is subject to shocks that arrive according to a PP(λ). The strength of each shock is a non-negative random variable with cdf $G(\cdot)$. If the shock has strength x it causes the machine to fail with probability $p(x)$. Assuming the successive shock strengths are independent, what is the distribution of the lifetime of the machine?

5.35 Customers arrive at a bank according to a PP with rate of 10/hour. Forty percent of them are men, and the rest are women. Given that 10 men have arrived during the first hour, what is the expected number of women who arrived in the first hour?

5.36 Let $c > 0$ be a constant and define:
$$\lambda(t) = \begin{cases} c & 2n < t < 2n+1, \ n \geq 0 \\ 0 & 2n+1 \leq t \leq 2n+2, \ n \geq 0. \end{cases}$$
Let $\{N(t), t \geq 0\}$ be a NPP with the above rate function. Compute the distribution of S_1, the time of occurrence of the first event.

5.37 For the NPP of Computational Exercise 5.36, compute the distribution of $N(t)$.

5.38 For the NPP of Computational Exercise 5.36, compute $E(S_1 | N(t) = n)$, for $0 \leq t \leq 2$.

5.39 Redo Example 5.17 under the assumption that the arrivals to the library follow an NPP($\lambda(\cdot)$).

5.40 A factory produces items one at a time according to a PP(λ). These items are loaded onto trucks that leave the factory according to a PP(μ) and transport the items to a warehouse. The truck capacity is large enough so that all the items produced after the departure of a truck can be loaded onto the next truck. The travel time between the factory and the warehouse is a constant and can be ignored. Let $Z(t)$ be the number of items received at the warehouse during $(0, t]$. Is $\{Z(t), t \geq 0\}$ a PP, NPP, or CPP or none of the above? Show that $E(Z(t)) = \lambda t - \lambda(1 - e^{-\mu t})/\mu$.

5.41 A customer makes deposits in a bank according to a PP with rate λ_d per week. The sizes of successive deposits are iid random variables with mean τ_d and variance σ_d^2. Compute the mean and variance of the total amount deposited over $(0, t]$. Unknown to the customer, the customer's spouse makes withdrawals from the same account according to a PP with rate λ_w. The sizes of successive withdrawals are iid random variables with mean τ_w and variance σ_w^2. Assume that the deposit and withdrawal processes are independent of each other. Let $Z(t)$ be the account balance at time t. (The account balance is allowed to go negative.) Show that $\{Z(t), t \geq 0\}$ is a CPP. Compute the mean and variance of $Z(t)$.

5.42 Consider a CPP $\{Z(t), t \geq 0\}$ with batch size distribution

$$P(Z_n = k) = (1 - \alpha)^{k-1}\alpha, \quad k \geq 1.$$

Compute the LST of $Z(t)$. Hence or otherwise, compute $P(Z(t) = k)$.

5.43 Suppose the life time of a machine is a continuous random variable with hazard rate $r(x), \quad x \geq 0$. When the machine fails we perform a "minimal repair," i.e., the repair takes place instantaneously and after the repair the machine behaves as if it had never failed. (This is called minimal repair.) That is, if the repair takes place at time t, the failure rate of the machine immediately after repair is $r(t)$. Let $N(t)$ be the number of minimal repairs performed on the machine up to time t, assuming that the machine is new at time 0.

1. Show that $\{N(t), t \geq 0\}$ is a NPP$(r(\cdot))$.
2. Suppose the cost of the nth minimal repair is nc, where $c > 0$ is a fixed constant. Thus successive repairs get more and more expensive. Let $C(t)$ be the total cost incurred up to time t. Compute $E(C(t))$.
3. Suppose $r(t) = 2t$, i.e., the failure rate increases linearly with time. Compute $E(N(t))$ and $E(C(t))$.

5.44 Customers arrive at an infinite capacity shop according to a Poisson process with rate λ. Upon arrival each customer spends some time in the shop browsing, which is uniformly distributed between 0 and 1 independently of the other customers. If the customer spends x units of time browsing, the probability that she will decide to buy something and join the checkout queue is x; or she leaves without buying anything with probability $1 - x$.

1. What is the long-run average arrival rate to the checkout queue?
2. Let q_t denote the probability that a customer who entered the shop some time during $(0, t)$ joins the checkout queue by time t. Compute q_t.
3. Let $A(t)$ denote the number of arrivals to the check out queue over $(0, t]$. Determine the probability distribution of $A(t)$.
4. Is $\{A(t), t \geq 0\}$ a Poisson process? Is it a non-homogeneous Poisson process?

5.45 Let $N(t)$ be the total number of cameras sold by a store over $(0, t]$ (years). Suppose $\{N(t), t \geq 0\}$ is an NPP with rate function

$$\lambda(t) = 200(1 + e^{-t}), \quad t \geq 0.$$

The interval from $t = n$ to $t = n + 1$ is called year n. Suppose the price of the camera is $\$350(1.08)^n$ in year $n = 0, 1, 2, ...$ Thus sales rate changes continuously with time, but the price changes from year to year. Compute the mean and variance of the sales revenue in the nth year.

5.8 Conceptual Exercises

5.1 Derive Equation 5.4.

5.2 Let $X \sim \text{Erl}(n, \lambda)$ and $Y \sim \text{Erl}(m, \mu)$ be two independent Erlang random variables. Let

$$F(n, m) = P(X < Y).$$

Show that $F(n, m)$ satisfy the following recursion

$$F(n, m) = \frac{\lambda}{\lambda + \mu} F(n - 1, m) + \frac{\mu}{\lambda + \mu} F(n, m - 1), \quad n, m \geq 1,$$

with boundary conditions

$$F(n, 0) = 0, \ n \geq 1; \quad F(0, m) = 1, \ m \geq 1.$$

Hence or otherwise show that

$$F(n, m) = \sum_{k=n}^{n+m-1} \binom{n + m - 1}{k} \left(\frac{\lambda}{\lambda + \mu} \right)^k \left(\frac{\mu}{\lambda + \mu} \right)^{n+m-k-1}.$$

5.3 Let $X_i \sim \exp(\lambda_i)$, $i = 1, 2, \cdots, k$ be k independent random variables. Show that, for $1 \leq i \leq k$,

$$P(X_i < X_j, j \neq i) = \frac{\lambda_i}{\lambda},$$

where

$$\lambda = \sum_{j=1}^{k} \lambda_j.$$

5.4 Prove Equation 5.9.

5.5 Using Laplace transforms (LT) solve Equations 5.12 and 5.11 as follows: denote the LT of $p_k(t)$ by $p_k^*(s)$. Using appropriate initial conditions show that

$$
\begin{aligned}
(\lambda + s)p_0^*(s) &= 1, \\
(\lambda + s)p_k^*(s) &= \lambda p_{k-1}^*(s), \quad k \geq 1.
\end{aligned}
$$

Solve these to obtain

$$p_k^*(s) = \frac{\lambda^k}{(\lambda + s)^{k+1}}, \quad k \geq 0.$$

Invert this to obtain Equation 5.13.

5.6 Let $\{N_i(t), t \geq 0\}$ $(i = 1, 2)$ be two independent Poisson processes with rates λ_1 and λ_2, respectively. At time 0 a coin is flipped that turns up heads with probability p. Define

$$N(t) = \begin{cases} N_1(t) & \text{if the coin turns up heads,} \\ N_2(t) & \text{if the coin turns up tails.} \end{cases}$$

1. Is $\{N(t), t \geq 0\}$ a PP?

2. Does $\{N(t), t \geq 0\}$ have stationary increments?

3. Does $\{N(t), t \geq 0\}$ have independent increments?

5.7 Show that if $\{N(t), t \geq 0\}$ is an NPP($\lambda(\cdot)$),

1. $N(0) = 0$,

2. $N(s + t) - N(s) \sim P(\Lambda(s + t) - \Lambda(s))$.

Hint: Use generating functions to show part 2.

5.8 Prove the "if" part of Theorem 5.20 by using the independence of increments of $\{N(t), t \geq 0\}$ and that $N(t + h) - N(t) \sim P(\Lambda(t + h) - \Lambda(t))$.

5.9 Let $\{N_i(t), t \geq 0\}$ $(i = 1, 2)$ be two independent Poisson processes with rates λ_1 and λ_2, respectively. Let A_i be the number of events in the ith process before the first event in the jth process $(i \neq j)$.

1. Compute $P(A_i = k)$ for $k \geq 0$.

2. Are A_1 and A_2 independent random variables?

5.10 Prove Theorem 5.19.

5.11 Prove Theorem 5.22 by following the proof of Theorem 5.14 on page 173.

5.12 Let $\{N_i(t), t \geq 0\}$ $(i = 1, 2)$ be two independent Poisson processes with rates λ_1 and λ_2, respectively. Let T be a non-negative random variable that is independent of both these processes. Let F be its cdf and ϕ be its LST. Let $B_i = N_i(T)$.

1. Compute the generating function $\psi_i(z) = E(z^{B_i})$ in terms of ϕ, $i = 1, 2$.

2. Compute the joint generating function $\psi(z_1, z_2) = E(z_1^{B_1} z_2^{B_2})$ in terms of ϕ.

3. Are B_1 and B_2 independent? What if T is a constant?

5.13 This is an example of deterministic splitting. Let $\{N(t), t \geq 0\}$ be a PP(λ). Suppose all odd events (i.e., events occurring at time S_{2n+1} for $n \geq 0$) in this process are classified as type 1 events, and all even events are classified as type 2 events. Let $N_i(t)$ be the number of type i $(i = 1, 2)$ events that occur during $(0, t]$.

1. Is $\{N_1(t), t \geq 0\}$ a PP?

2. Is $\{N_2(t), t \geq 0\}$ a PP?

3. Are $\{N_i(t), t \geq 0\}$ $(i = 1, 2)$ independent of each other?

5.14 Prove Theorem 5.21 by conditioning on S_n.

5.15 Let $\{N_i(t), t \geq 0\}$ be an NPP($\lambda_i(\cdot)$), $(1 \leq i \leq r)$. Suppose they are independent and define

$$N(t) = N_1(t) + N_2(t) + \cdots + N_r(t), \quad t \geq 0.$$

Show that $\{N(t), t \geq 0\}$ is an NPP($\lambda(\cdot)$), where

$$\lambda(t) = \lambda_1(t) + \lambda_2(t) + \cdots + \lambda_r(t), \quad t \geq 0.$$

5.16 Show that the process $\{R(t), t \geq 0\}$ of Theorem 5.18 is an NPP with rate function $\lambda(t) = \lambda p(t)$, $t \geq 0$.

5.17 Let $\{N(t), t \geq 0\}$ be an NPP($\lambda(\cdot)$), and $p : [0, \infty) \to [0, 1]$ be a function. Suppose an event occurring in the NPP at time s is registered with probability $p(s)$, independent of everything else. Let $R(t)$ be the number of registered events during $(0, t]$. Show that $\{R(t), t \geq 0\}$ is an NPP with rate function $\lambda(t)p(t)$, $t \geq 0$.

5.18 Prove Theorem 5.23, by following the initial steps in the proof of Theorem 5.11.

5.19 Prove Theorem 5.25 by using the following identities:

$$\mathsf{E}(Z(t)) = -\phi'(0), \quad \mathsf{E}(Z(t)^2) = \phi''(0).$$

5.20 The process $\{Z(t), t \geq 0\}$ of Equation 5.18 is called a non-homogeneous compound Poisson process if $\{N(t), t \geq 0\}$ is an NPP($\lambda(\cdot)$). Compute the LST, the mean, and the variance of $Z(t)$.

5.21 Prove or disprove that the process $\{R(t), t \geq 0\}$ defined in Example 5.17 is a non-homogeneous Poisson process.

5.22 Derive Equations 5.14, 5.15, and 5.16.

5.23 A two-dimensional PP(λ) is a process of points occurring randomly in a Euclidean plane R^2 so that (i) the number of points occurring in a region of area A is P(λA) random variable, and (ii) the number of points in non-overlapping regions are independent. Let X_1 be the Euclidean distance from the origin to the point closest to the origin. Compute P($X_1 > x$).

5.24 Derive the variance-covariance matrix of $(S_1, S_2, \cdots, S_n)$ given $N(t) = n$, where $\{N(t), t \geq 0\}$ is a PP(λ).

5.25 Let $S_0 = 0$ and S_n be the time of occurrence of the nth event in an NPP$(\lambda(\cdot))$. Show that $\{S_n, n \geq 0\}$ is a DTMC with state-space $[0, \infty)$. Compute

$$P(S_{n+1} > s + t | S_n = s)$$

for $s, t \geq 0$.

CHAPTER 6

Continuous-Time Markov Chains

The amount of pain involved in solving a mathematical problem is independent of the route taken to solve it.

Proof: Follows from the general theorem, "there is no such thing as a free lunch."

6.1 Definitions and Sample Path Properties

In Chapters 2, 3, and 4 we studied DTMCs. They arose as stochastic models of systems with countable state-space that change their state at times $n = 1, 2, \cdots$, and have Markov property at those times. Thus, the probabilistic nature of the future behavior of these systems after time n depends on the past only through their state at time n.

In this chapter we study a system with a countable state-space that can change its state at any point in time. Let $S_n, n \geq 1$, be time of the nth change of state or *transition*, $Y_n = S_n - S_{n-1}$, (with $S_0 = 0$), be the nth sojourn time, and X_n be the state of the system immediately after the nth transition. Define

$$N(t) = \sup\{n \geq 0 : S_n \leq t\}, \quad t \geq 0.$$

Thus $N(t)$ is the number of transitions the system undergoes over $(0, t]$, and $\{N(t), t \geq 0\}$ is a counting process generated by $\{Y_n, n \geq 1\}$. It has piecewise constant sample paths that start with $N(0) = 0$ and jump up by $+1$ at times $S_n, n \geq 1$.

We make the following important assumption, called the "regularity assumption":

$$P(\lim_{n \to \infty} S_n = \infty) = 1. \tag{6.1}$$

This ensures that

$$P(N(t) < \infty) = 1, \tag{6.2}$$

for finite t. The regularity condition implies that the system undergoes finite number of transitions during finite intervals of time. This is a critical assumption, and we always assume that it holds.

Now let $X(0) = X_0$ be the initial state of the system, and $X(t)$ be the state of the system at time t. Under the regularity assumption of Equation 6.1, $N(t)$ is well defined for each $t \geq 0$, and hence we can write

$$X(t) = X_{N(t)}, \quad t \geq 0. \tag{6.3}$$

The continuous time stochastic process $\{X(t), t \geq 0\}$ has piecewise constant right-continuous sample paths. A typical sample path of such a system is shown in Figure 6.1. We see that the $\{X(t), t \geq 0\}$ process is in the initial state X_0 at time $t = 0$.

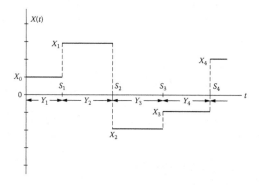

Figure 6.1 *A typical sample path of a CTMC.*

It stays there for a sojourn time Y_1 and then jumps to state X_1. In general it stays in state X_n for a duration given by Y_{n+1} and then jumps to state X_{n+1}, $n \geq 0$. Note that if $X_n = X_{n+1}$, there is no jump in the sample path of $\{X(t), t \geq 0\}$ at time S_{n+1}. Thus, without loss of generality, we can assume that $X_{n+1} \neq X_n$ for all $n \geq 0$.

In this chapter we study the case where $\{X(t), t \geq 0\}$ belongs to a particular class of stochastic processes called the continuous-time Markov chains (CTMC), defined below.

Definition 6.1   CTMC: Definition. *The stochastic process $\{X(t), t \geq 0\}$ as defined by Equation 6.3 is called a CTMC if it has a countable state-space S, and the sequence $\{X_0, (X_n, Y_n), n \geq 1\}$ satisfies*

$$\begin{aligned} \mathsf{P}(X_{n+1} = j&, Y_{n+1} > y | X_n = i, Y_n, X_{n-1}, Y_{n-1}, \cdots, X_1, Y_1, X_0) \\ &= \mathsf{P}(X_1 = j, Y_1 > y | X_0 = i) = p_{i,j} e^{-q_i y}, \quad i, j \in S, n \geq 0, \end{aligned} \tag{6.4}$$

where $P = [p_{i,j}]_{i,j \in S}$ is a stochastic matrix with $p_{i,i} = 0$ for all $i \in S$, and $0 \leq q_i < \infty, i \in S$.

Thus a system modeled by a CTMC $\{X(t), t \geq 0\}$ evolves as follows: it starts in state $X_0 = i_0$ (say). It stays in that state for a random duration of time $Y_1 \sim \exp(q_{i_0})$ and then jumps to state $X_1 = X(Y_1+) = i_1 \neq i_0$ with probability p_{i_0,i_1}, independent of the amount of time spent in state i_0. In general, after the nth transition at time S_n, it moves to state $X_n = X(S_n+) = i_n \neq i_{n-1}$. It stays in that state for a random duration of time $Y_{n+1} \sim \exp(q_{i_n})$ and then jumps at time S_{n+1} to state i_{n+1} with probability $p_{i_n, i_{n+1}}$, independent of the history over $[0, S_{n+1})$. In particular, given X_n, the random variables X_{n+1} and Y_{n+1} are independent of each other.

From Equation 6.4 it is clear that $\{X_n, n \geq 0\}$ is a DTMC on state-space S with transition probability matrix P. It is called the "embedded DTMC" corresponding to the CTMC. We illustrate with several examples below.

Example 6.1 Two-State Machine. Consider a machine that can be up or down at any time. If the machine is up, it fails after an $\exp(\mu)$ amount of time. If it is down, it is repaired in an $\exp(\lambda)$ amount of time. The successive up times are iid, and so are the successive down times, and the two are independent of each other. Define

$$X(t) = \begin{cases} 0 & \text{if the machine is down at time } t, \\ 1 & \text{if the machine is up at time } t. \end{cases}$$

Is $\{X(t), t \geq 0\}$ a CTMC?

The state-space of $\{X(t), t \geq 0\}$ is $\{0, 1\}$. Suppose the machine fails at time t, i.e., the process enters state 0 at time t. Then the process stays in state 0 for an $\exp(\lambda)$ amount of time and then jumps to state 1. Thus, $q_0 = \lambda$ and $p_{0,1} = 1$. Now suppose the machine repair completes at time t, i.e., the process enters state 1 at time t. Then the process stays in state 1 for an $\exp(\mu)$ amount of time and then jumps to state 0. Thus, $q_1 = \mu$ and $p_{1,0} = 1$. It is easy to see that the independence assumptions implicit in Equation 6.4 are satisfied. Thus $\{X(t), t \geq 0\}$ a CTMC. ∎

Example 6.2 Poisson Process. Show that a PP(λ) is a CTMC.

Let $\{X(t), t \geq 0\}$ be a PP(λ). (See Section 5.2.) Its state-space is $\{0, 1, 2, \cdots\}$. Suppose the process enters state i at time t, i.e., the ith event in the PP occurs at time t. Then the process stays there for an $\exp(\lambda)$ amount of time, independent of the history, and then jumps to state $i + 1$. This, along with the properties of the PP, implies that Equation 6.4 is satisfied with $q_i = \lambda$ and $p_{i,i+1} = 1$ for $i \in S$. Hence $\{X(t), t \geq 0\}$ is a CTMC. ∎

Example 6.3 Compound Poisson Process. Let $\{X(t), t \geq 0\}$ be a CPP with batch arrival rate λ and iid integer valued batch sizes with common pmf

$$\alpha_k = P(\text{Batch Size} = k), \quad k = 1, 2, 3, \cdots.$$

Is $\{X(t), t \geq 0\}$ a CTMC?

See Section 5.6 for the definition of the CPP. The state-space of $\{X(t), t \geq 0\}$ is $\{0, 1, 2, \cdots\}$. Suppose the process enters state i at time t, i.e., a batch arrives at time t and brings the total number of arrivals over $(0, t]$ to i. Then the process stays there for an $\exp(\lambda)$ amount of time (until the next batch arrives), independent of the history, and then jumps to state $j > i$ if the new batch is of size $j - i$. This, along with the properties of the CPP, implies that Equation 6.4 is satisfied with $q_i = \lambda$, $p_{i,j} = \alpha_{j-i}, j > i$. Hence $\{X(t), t \geq 0\}$ is a CTMC. ∎

We shall describe many more examples of CTMCs in the next section. In the remaining section we study general properties of a CTMC. The next theorem shows that a CTMC has Markov property at all times.

Theorem 6.1   Markov Property. *A CTMC* $\{X(t), t \geq 0\}$ *with state-space S has Markov property at each time point, i.e., for* $s, t \geq 0$, *and* $i, j \in S$,

$$P(X(t+s) = j|X(s) = i, X(u) : 0 \leq u < s) = P(X(t+s) = j|X(s) = i). \quad (6.5)$$

Furthermore, it is time homogeneous, i.e.,

$$P(X(t + s) = j|X(s) = i) = P(X(t) = j|X(0) = i). \quad (6.6)$$

Proof: Let $X(t)$ be as defined in Equation 6.3. Suppose $S_{N(s)} = \nu$, i.e., the last transition at or before s takes place at $\nu \leq s$. Then $X(s) = i$ implies that $X(\nu) = i$, and Y, the sojourn time in state i that started at ν, ends after s. Thus $Y > s - \nu$, which implies that the remaining sojourn time in state i at time s, given by $Y - (s - \nu)$, is exp(q_i). Also, Y depends on $\{X(u) : 0 \leq u \leq s\}$ only via $X(\nu)$ which equals $X(s)$. Also, the next state $X_{N(s)+1}$ depends on the history only via $X(\nu) = X(s)$. Thus all future evolution of the process after time s depends on the history only via $X(s) = i$. This gives Equation 6.5. The same argument yields Equation 6.6 since the q_i and $p_{i,j}$ do not depend on when the process enters state i. ∎

Intuitively, if a system is modeled by a CTMC the probabilistic nature of the future behavior of the system after time t depends on the past only through the state of the system at time t, for all $t \geq 0$. We shall concentrate on time-homogeneous CTMCs in this chapter. Thus, unless otherwise mentioned, when we use the term CTMC we mean a time-homogeneous CTMC.

Now consider a CTMC $\{X(t), t \geq 0\}$ with a countable state-space S, and let

$$p_{i,j}(t) = P(X(t) = j|X(0) = i), \quad i, j \in S, \ t \geq 0, \quad (6.7)$$

be its transition probabilities, and

$$P(t) = [p_{i,j}(t)]_{i,j \in S}, \quad t \geq 0, \quad (6.8)$$

be its transition probability matrix. Note that the notation $p_{i,j}$ used in Equation 6.4 implies that

$$p_{i,j} = P(X_1 = j|X_0 = i) = P(X(S_1) = j|X(0) = i).$$

Do not confuse this $p_{i,j}$ with $p_{i,j}(t)$ defined in Equation 6.7. Next, let

$$a_i = P(X(0) = i), \quad i \in S,$$

and

$$a = [a_i]_{i \in S},$$

be a row vector representing the initial distribution of the CTMC. The next theorem shows how to compute the finite dimensional distributions of a CTMC.

Theorem 6.2   Characterizing a CTMC. *A CTMC* $\{X(t), t \geq 0\}$ *is completely* *described by its initial distribution a and the set of transition probability matrices $\{P(t) : t \geq 0\}$.*

Proof: We shall prove the theorem by showing that all finite dimensional distributions of the CTMC are determined by a and $\{P(t) : t \geq 0\}$. Let $n \geq 1$, $0 \leq t_1 \leq t_2 \leq \cdots \leq t_n$ and $i_1, i_2, \cdots i_n \in S$ be given. We have

$$
\begin{aligned}
&\mathsf{P}(X(t_1) = i_1, X(t_2) = i_2, \cdots, X(t_n) = i_n) \\
&= \sum_{i_0 \in S} \mathsf{P}(X(t_1) = i_1, X(t_2) = i_2, \cdots, X(t_n) = i_n | X(0) = i_0) \mathsf{P}(X(0) = i_0) \\
&= \sum_{i_0 \in S} a_{i_0} \mathsf{P}(X(t_2) = i_2, \cdots, X(t_n) = i_n | X(0) = i_0, X(t_1) = i_1) \\
&\quad \cdot \mathsf{P}(X(t_1) = i_1 | X(0) = i_0) \\
&= \sum_{i_0 \in S} a_{i_0} p_{i_0, i_1}(t_1) \mathsf{P}(X(t_2) = i_2, \cdots, X(t_n) = i_n | X(t_1) = i_1)
\end{aligned}
$$

(from Markov property of the CTMC)

$$
= \sum_{i_0 \in S} a_{i_0} p_{i_0, i_1}(t_1) \mathsf{P}(X(t_2 - t_1) = i_2, \cdots, X(t_n - t_1) = i_n | X(0) = i_1)
$$

(from the time homogeneity of the CTMC).

Continuing in this fashion we get

$$
\begin{aligned}
&\mathsf{P}(X(t_1) = i_1, X(t_2) = i_2, \cdots, X(t_n) = i_n) \\
&= \sum_{i_0 \in S} a_{i_0} p_{i_0, i_1}(t_1) p_{i_1, i_2}(t_2 - t_1) \cdots p_{i_{n-1}, i_n}(t_n - t_{n-1}).
\end{aligned}
$$

This proves the theorem. ∎

Next we address the issue of how to describe a CTMC in a tractable fashion. Theorem 6.2 says that we can do this by specifying the initial distribution $a = [a_i]$, and the transition probability matrix $P(t)$ for each $t \geq 0$. However, the matrix $P(t)$ is hard to specify even for very simple CTMCs. Hence we look for alternative ways of describing a CTMC. We defined a CTMC (under regularity conditions) by using the parameters $q_i, (i \in S)$ and $p_{i,j}, (i, j \in S)$. If these parameters can determine $P(t)$ for all $t \geq 0$, then we can use them in place of $P(\cdot)$. The regularity condition implies that this is indeed the case. We refer the reader to Chung (1967) and Cinlar (1975) for further details on this topic. We shall provide a constructive proof in Section 6.4. Right now we shall assert that in all the applications in this book, the parameters $q_i, (i \in S)$ and $p_{i,j}, (i, j \in S)$, along with the initial distribution do describe the CTMCs completely. Hence we use this method in the rest of this book.

6.2 Examples

In this section we shall describe an often useful method of constructing a CTMC model to describe the stochastic evolution of a system. It is based on the following conceptualization of how the system evolves.

Consider a system with a countable state-space. An event $E_{i,j}$ is associated with each pair (i,j) of states $(i \neq j)$ so that if that event occurs while the system is in state i, the system jumps to state j. Which event actually occurs is determined as follows. Suppose the system enters state i at time t. Then event $E_{i,j}$ is scheduled to occur at time $t + T_{i,j}$, where $T_{i,j} \sim \exp(q_{i,j})$, where $q_{i,j} \geq 0$. If there is no event whose occurrence can cause the system to jump from state i to state j, we simply set $q_{i,j} = 0$. Furthermore the random variables $\{T_{i,j}, j \in S\}$ are mutually independent and also independent of the history of the system up to time t. Let

$$T_i = \min\{T_{i,j} : j \in S, j \neq i\}. \tag{6.9}$$

Then the system stays in state i until time $t+T_i$ and then jumps to state j if $T_i = T_{i,j}$, i.e., if $E_{i,j}$ is the first event that occurs among all the scheduled events. All the other scheduled events are now canceled, and new events $E_{j,k}$ $(k \neq j)$ are scheduled, and the process continues.

Now let $X(t)$ be the state of the system at time t, and S_n be the time of the nth transition. Let $X_n = X(S_n+)$ and $Y_n = S_n - S_{n-1}$ (with $S_0 = 0$), as before. From the distributional and independence assumptions about $\{T_{i,j}, j \neq i\}$, we see that T_i of Equation 6.9 is an $\exp(q_i)$ random variable, where

$$q_i = \sum_{j \neq i} q_{i,j}, \quad i \in S. \tag{6.10}$$

If $q_i = 0$ there are no events that will take the system out of state i, i.e., state i is an absorbing state. In this case we define $p_{i,i} = 1$, since this makes the state i absorbing in the embedded DTMC as well. If $q_i > 0$, we have

$$
\begin{aligned}
\mathsf{P}(X_{n+1} = j, Y_{n+1} > y | X_n &= i, Y_n, X_{n-1}, Y_{n-1}, \cdots, X_1, Y_1, X_0) \\
&= \mathsf{P}(T_{i,j} = T_i, T_i > y) \\
&= \frac{q_{i,j}}{q_i} e^{-q_i y}, \quad i, j \in S, i \neq j, y \geq 0.
\end{aligned}
$$

Now define

$$p_{i,j} = \frac{q_{i,j}}{q_i}, \quad i, j \in S, \ j \neq i. \tag{6.11}$$

From the above derivation it is clear that $\{X(t), t \geq 0\}$ is a CTMC with parameters $\{q_i\}$ and $\{p_{i,j}\}$ as defined in Equations 6.10 and 6.11. Thus we can describe a CTMC by specifying $q_{i,j}$, called the transition rate from state i to j, for all the pairs (i,j), with $i \neq j$. Note that the quantity $q_{i,i}$ is as yet undefined. For strictly technical reasons, we define

$$q_{i,i} = - \sum_{k:k \neq i} q_{i,k} = -q_i, \quad i \in S.$$

It is convenient to put all the $q_{i,j}$'s in a matrix form:

$$Q = [q_{i,j}]_{i,j \in S}.$$

This is called the infinitesimal generator or simply the generator matrix of the CTMC. It will become clear later that the seemingly arbitrary definition of $q_{i,i}$ makes it easy to write many equations of interest in matrix form. An important property of the generator matrix is that its row sums are zero, i.e.,

$$\sum_{j \in S} q_{i,j} = 0, \quad i \in S.$$

The generator matrix of a CTMC plays the same role as the one-step transition probability matrix of a DTMC.

Analogous to the DTMC case, a CTMC can also be represented graphically by what are known as rate diagrams. A rate diagram is a directed graph in which each node represents a state of the CTMC, and there is an arc from node i to node j if $q_{i,j} > 0$. We generally write $q_{i,j}$ on this arc. The rate diagram helps us visualize the dynamic evolution of a CTMC.

In the remaining section we present several examples of CTMCs. Modeling a system by a CTMC is an art, and like any other art, is perfected by practice. The first step in the modeling process is to identify the state-space. This step is guided by intuition and by what kind of information we want the model to produce. The next step involves identifying the triggering events $E_{i,j}$'s—events that trigger a transition from state i to j. The final step is to verify the distributional and independence assumptions about the $T_{i,j}$'s and obtain the transition rates $q_{i,j}$'s. If the $T_{i,j}$'s are not exponential, or the independence assumptions are violated, then either the system cannot be modeled by a CTMC or the state-space needs to modified.

In the examples that follow we do not explicitly verify the independence assumptions. However, we urge the students to do so for themselves.

Example 6.4   Two-State Machine. Let $X(t)$ be the state at time t of the machine of Example 6.1. The state-space of $\{X(t), t \geq 0\}$ is $\{0, 1\}$. In state 0 we have $E_{0,1}$ = machine repair completes, and $T_{0,1} \sim \exp(\lambda)$. Hence $q_{0,1} = \lambda$. Similarly, in state 1 we have $E_{1,0}$ = machine fails, and $T_{1,0} \sim \exp(\mu)$. Hence $q_{1,0} = \mu$. Thus $\{X(t), t \geq 0\}$ is a CTMC with generator matrix

$$Q = \begin{bmatrix} -\lambda & \lambda \\ \mu & -\mu \end{bmatrix}, \tag{6.12}$$

and the rate diagram as shown in Figure 6.2. ∎

Example 6.5   Two-Machine Two-Repairperson Workshop. Now consider a machine shop that has two machines that are independent, identical, and behave as described in Example 6.1. Each machine has its own repairperson. Let $X(t)$ be the number of working machines at time t. Is $\{X(t), t \geq 0\}$ a CTMC? If it is, what is its generator matrix?

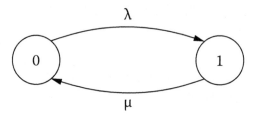

Figure 6.2 *The rate diagram of a two-state machine.*

The state-space of $\{X(t), t \geq 0\}$ is $\{0, 1, 2\}$. We analyze the states one by one.

State 0. $X(t) = 0$ implies that both the machines are down at time t. When either one of them gets repaired the state changes to 1. Hence we have $E_{0,1} =$ one of the two failed machines completes repair. Since the remaining repair times are exponential due to memoryless property, $T_{0,1}$ is the minimum of two independent $\exp(\lambda)$ random variables. Hence $T_{0,1} \sim \exp(2\lambda)$. Hence

$$q_0 = 2\lambda, \quad q_{0,1} = 2\lambda.$$

State 1. $X(t) = 1$ implies that one machine is up and the other is down at time t. Now there are two triggering events: $E_{1,0} =$ the working machine fails, and $E_{1,2} =$ the failed machine completes repair. If $E_{1,0}$ occurs before $E_{1,2}$, the process moves to state 0, else it moves to state 2. Since the remaining repair time and life time are exponential due to memoryless property, we see that $T_{1,0} \sim \exp(\mu)$ and $T_{1,2} \sim \exp(\lambda)$. Hence

$$q_1 = \lambda + \mu, \quad q_{1,2} = \lambda, \quad q_{1,0} = \mu.$$

State 2. $X(t) = 2$ implies that both the machines are up at time t. When either one of them fails the state changes to 1. Hence we have $E_{2,1} =$ one of the two working machines fails. Since the remaining life times are exponential due to memoryless property, $T_{2,1}$ is the minimum of two independent $\exp(\mu)$ random variables. Hence $T_{2,1} \sim \exp(2\mu)$. Hence

$$q_2 = 2\mu, \quad q_{2,1} = 2\mu.$$

Thus $\{X(t), t \geq 0\}$ is a CTMC with generator matrix

$$Q = \begin{bmatrix} -2\lambda & 2\lambda & 0 \\ \mu & -(\lambda + \mu) & \lambda \\ 0 & 2\mu & -2\mu \end{bmatrix},$$

and the rate diagram as shown in Figure 6.3. ∎

Example 6.6 Two-Machine One-Repairperson Workshop. Consider the machine shop of Example 6.5. Now suppose there is only one repairperson and the machines are repaired in the order in which they fail. The repair times are iid. Let $X(t)$ be the number of working machines at time t. Is $\{X(t), t \geq 0\}$ a CTMC? If it is, what is its generator matrix?

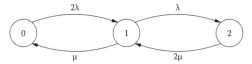

Figure 6.3 *The rate diagram for the two-machine, two-repairpersons workshop.*

The state-space of $\{X(t), t \geq 0\}$ is $\{0, 1, 2\}$. The analysis of states 1 and 2 is the same as in Example 6.5. In state 0, both machines are down, but only one is under repair, and its remaining repair time is $\exp(\lambda)$. Hence the triggering event is $E_{0,1} =$ machine under repair completes repair, and we have $T_{0,1} \sim \exp(\lambda)$. Thus $\{X(t), t \geq 0\}$ is a CTMC with generator matrix

$$Q = \begin{bmatrix} -\lambda & \lambda & 0 \\ \mu & -(\lambda + \mu) & \lambda \\ 0 & 2\mu & -2\mu \end{bmatrix},$$

and the rate diagram as shown in Figure 6.4. ∎

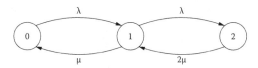

Figure 6.4 *The rate diagram for the two-machine, one-repairperson workshop.*

Example 6.7 Poisson Process. Suppose $\{X(t), t \geq 0\}$ is a PP(λ). We saw in Example 6.2 that it is a CTMC with state space $S = \{0, 1, 2, \cdots\}$. What is its generator matrix?

Consider state i. When the next event in the PP occurs, the process moves to state $j = i + 1$. Hence we have $E_{i,i+1} =$ the next event occurs, and $T_{i,i+1} \sim \exp(\lambda)$. Hence we have $q_i = \lambda$, and $q_{i,i+1} = \lambda$, for all $i \geq 0$. The generator matrix (where we shown only non-zero elements) is given below:

$$Q = \begin{bmatrix} -\lambda & \lambda & & \\ & -\lambda & \lambda & \\ & & -\lambda & \lambda \\ & & & \ddots & \ddots \end{bmatrix},$$

and the rate diagram as shown in Figure 6.5. ∎

Example 6.8 Pure Birth Process. An immediate extension of the PP(λ) is the pure birth process, which is a CTMC on $S = \{0, 1, 2, \cdots\}$ with the following generator

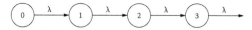

Figure 6.5 *The rate diagram of a PP(λ).*

matrix:

$$Q = \begin{bmatrix} -\lambda_0 & \lambda_0 & & & \\ & -\lambda_1 & \lambda_1 & & \\ & & -\lambda_2 & \lambda_2 & \\ & & & \ddots & \ddots \end{bmatrix}.$$

The rate diagram is shown in Figure 6.6. Such a process spends an $\exp(\lambda_i)$ amount

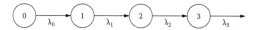

Figure 6.6 *The rate diagram for a pure birth process.*

of time in state i and then jumps to state $i+1$. In biological systems, $X(t)$ represents the number of organisms in a colony at time t, and the transition from i to $i+1$ represents a birth. Hence the name "pure birth process." The parameter λ_i is called the birth rate in state i. The PP(λ) is a special case of a pure birth process with all birth rates given by $\lambda_i = \lambda$. We illustrate with several situations where pure birth processes can arise.

Yule Process. Let $X(t)$ be the number of amoebae at time t in a colony of amoebae. Suppose each amoeba lives for an $\exp(\lambda)$ amount of time and then splits into two. All amoebae in the colony behave independently of each other. Suppose $X(t) = i$, $i \geq 1$. When one of the i amoebae splits, the process moves to state $i+1$. Hence $E_{i,i+1}$ = one of the i living amoebae splits. This will happen after an $\exp(i\lambda)$ amount of time, i.e., the minimum of i iid $\exp(\lambda)$ random variables. Thus $T_{i,i+1} \sim \exp(i\lambda)$. Thus we get $q_{i,i+1} = i\lambda$, and $q_{i,i} = -i\lambda$. State 0 is absorbing. Hence $\{X(t), t \geq 0\}$ is a pure birth process with birth rates $\lambda_i = i\lambda$, $i \geq 0$. Such a process is called the Yule process.

Yule Process with Immigration. Consider the above colony of amoebae. Suppose amoebae arrive at this colony from outside according to a PP(θ). All amoebae in the colony, whether native or immigrants, behave independently and identically. Let $X(t)$ be the number of amoebae in the colony at time t. As before, the state-space of $\{X(t), t \geq 0\}$ is $S = \{0, 1, 2, \cdots\}$. Suppose $X(t) = i$. The system moves to state $i+1$ if one of the existing amoebae splits, which happens after an $\exp(i\lambda)$ amount of time, or an amoeba arrive from outside, which happens after an $\exp(\theta)$ amount of time. Thus $T_{i,i+1} \sim \exp(i\lambda + \theta)$. Hence $q_{i,i+1} = i\lambda + \theta$ and $q_{i,i} = -(i\lambda + \theta)$. Hence $\{X(t), t \geq 0\}$ is a pure birth process with birth rates $\lambda_i = i\lambda + \theta$, $i \geq 0$.

Maintenance. Suppose a machine is brand new at time zero. It fails after an $\exp(\theta_0)$ amount of time, and is repaired instantaneously. The lifetime of a machine that has undergone $n \geq 1$ repairs is an $\exp(\theta_n)$ random variable, and is independent of how old the machine is. Let $X(t)$ be the number of repairs the machine has undergone over time $(0, t]$. Then $\{X(t), t \geq 0\}$ is a pure birth process with birth rates $\lambda_i = \theta_i$, $i \geq 0$.

Example 6.9   Pure Death Process. A CTMC $\{X(t), t \geq 0\}$ on $S = \{0, 1, 2, \cdots\}$ is called a pure death process if its generator matrix is given by

$$
Q = \begin{bmatrix}
0 & & & & \\
\mu_1 & -\mu_1 & & & \\
 & \mu_2 & -\mu_2 & & \\
 & & \mu_3 & -\mu_3 & \\
 & & & \ddots & \ddots
\end{bmatrix}.
$$

The rate diagram is shown in Figure 6.7. Such a process spends an $\exp(\mu_i)$ amount

Figure 6.7 *The rate diagram for a pure death process.*

of time in state $i \geq 1$ and then jumps to state $i - 1$. In biological systems, $X(t)$ represents the number of organisms in a colony at time t, and the transition from i to $i - 1$ represents death. Hence the name "pure death process." The parameter μ_i is called the death rate in state i. Note that state 0 is absorbing, with $q_{0,0} = 0$. Thus once the colony has no members left in it, it stays extinct forever. We discuss a special case below.

Multi-Component Machine. Consider a machine consisting of n parts. The life-times of these parts are iid $\exp(\mu)$ random variables. The failed parts cannot be re-paired. Once all parts fail, the machine fails, and stays that way forever. Let $X(t)$ be the number of functioning parts at time t. Suppose $X(t) = i \geq 1$. The system moves to state $i - 1$ if one of the i components fails, which happens after an $\exp(i\mu)$ time. Hence $T_{i,i-1} \sim \exp(i\mu)$. Hence $\{X(t), t \geq 0\}$ is a pure death process with death rates $\mu_i = i\mu$. The initial state is $X(0) = n$. ■

Example 6.10   Birth and Death Process. A CTMC $\{X(t), t \geq 0\}$ on $S = \{0, 1, 2, \cdots\}$ is called a birth and death process if its generator matrix is given by

$$
Q = \begin{bmatrix}
-\lambda_0 & \lambda_0 & & & \\
\mu_1 & -(\lambda_1 + \mu_1) & \lambda_1 & & \\
 & \mu_2 & -(\lambda_2 + \mu_2) & \lambda_2 & \\
 & & \mu_3 & -(\lambda_3 + \mu_3) & \lambda_3 \\
 & & & \ddots & \ddots
\end{bmatrix}.
$$

The rate diagram is shown in Figure 6.8. In biological systems, $X(t)$ represents the number of organisms in a colony at time t, and the transition from i to $i-1$ represents

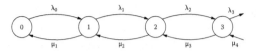

Figure 6.8 *The rate diagram for a birth and death process.*

a death and that from i to $i+1$ represents a birth. Hence the name "birth and death process." The parameter μ_i is called the death rate in state i and λ_i is called the birth rate in state i. We define $\mu_0 = 0$, since death cannot occur when there are no organisms to die. A birth and death process spends $\exp(\lambda_i + \mu_i)$ amount of time in state i, and then jumps to state $i - 1$ with probability $\mu_i/(\lambda_i + \mu_i)$, or to state $i + 1$ with probability $\lambda_i/(\lambda_i + \mu_i)$. A huge number of queuing models give rise to birth and death processes and we shall study them in detail in Chapter 7. Here we give a few examples. ∎

Example 6.11   Single-Server Queue. Consider a single-server service system where customers arrive according to a PP(λ) and request iid $\exp(\mu)$ service times. The customers wait in an unlimited waiting area and are served in the order of arrival. Let $X(t)$ be the number of customers in the system (waiting or in service) at time t. We shall show that $\{X(t), t \geq 0\}$ is a birth and death process.

The state-space is $S = \{0, 1, 2, \cdots\}$. Suppose $X(t) = 0$, that is, there are no customers in the system at time t. The triggering event is $E_{0,1} =$ arrival of a new customer. From the properties of the PP we have $T_{0,1} \sim \exp(\lambda)$. Thus $q_{0,1} = \lambda$. Now suppose $X(t) = i > 0$. Then one customer is in service and $i - 1$ are waiting at time t. Now there are two triggering events: $E_{i,i+1} =$ arrival of a new customer, and $E_{i,i-1} =$ departure of the customer in service. From the memoryless property of exponentials, we see that $T_{i,i+1} \sim \exp(\lambda)$ and $T_{i,i-1} \sim \exp(\mu)$. Hence $q_{i,i+1} = \lambda$ and $q_{i,i-1} = \mu$ for $i \geq 1$. Hence $\{X(t), t \geq 0\}$ is a birth and death process with birth parameters $\lambda_i = \lambda$ for $i \geq 0$, and death parameters $\mu_i = \mu$ for $i \geq 1$. ∎

Example 6.12   Infinite-Server Queue. Suppose customers arrive according to PP(λ) to a service system with infinite servers. The customer service times are iid $\exp(\mu)$ random variables. Since there are an infinite number of servers, there is no waiting and each arriving customer starts getting served at the time of arrival. Such systems arise as models of self service systems such as parking lots, cafeterias, reading rooms, etc. Let $X(t)$ be the number of customers in the system at time t. We shall show that $\{X(t), t \geq 0\}$ is a birth and death process.

The state-space is $S = \{0, 1, 2, \cdots\}$. Suppose $X(t) = i \geq 1$. Then i customers are in service at time t. Now there are two triggering events: $E_{i,i+1} =$ arrival of a new customer, and $E_{i,i-1} =$ departure of one of the i customers in service. From the memoryless property of exponentials, we see that $T_{i,i+1} \sim \exp(\lambda)$ and $T_{i,i-1} \sim \exp(i\mu)$. Hence $q_{i,i+1} = \lambda$ and $q_{i,i-1} = i\mu$ for $i \geq 1$. We can similarly show that $q_{0,1} = \lambda$. Hence $\{X(t), t \geq 0\}$ is a birth and death process with birth parameters $\lambda_i = \lambda$ for $i \geq 0$, and death parameters $\mu_i = i\mu$ for $i \geq 1$. ∎

Example 6.13   Linear Growth Model. Consider a colony of individuals whose lifetimes are independent $\exp(\mu)$ random variables. During its lifetime an organism produces offspring according to a $PP(\lambda)$, independent of everything else. Let $X(t)$ be the number of organisms in the colony at time t. We shall show that $\{X(t), t \geq 0\}$ is a birth and death process.

The state-space is $S = \{0, 1, 2, \cdots\}$. Suppose $X(t) = 0$, that is, there are no organisms in the colony at time t. Then $X(u) = 0$ for all $u \geq t$. Thus state 0 is absorbing. Next suppose $X(t) = i > 0$. Now there are two triggering events: $E_{i,i+1}$ = one of the i organisms gives birth to an individual, and $E_{i,i-1}$ = one of the i organisms dies. From the results about the superposition of Poisson processes, we see that $T_{i,i+1} \sim \exp(i\lambda)$ and from the memoryless properties of exponential random variables we get $T_{i,i-1} \sim \exp(i\mu)$. Hence $q_{i,i+1} = i\lambda$ and $q_{i,i-1} = i\mu$. Hence $\{X(t), t \geq 0\}$ is a birth and death process with birth parameters $\lambda_i = i\lambda$ and death parameters $\mu_i = i\mu$ for $i \geq 0$. Since the birth and death parameters are linear functions of the system state, this is called a linear growth model. ∎

Example 6.14 N-Machine Workshop. Consider an extension of Example 6.5 to N independent and identical machines, each with its own dedicated repairperson. Let $X(t)$ be the number of working machines at time t. We shall show that $\{X(t), t \geq 0\}$ is a birth and death process.

The state-space of $\{X(t), t \geq 0\}$ is $\{0, 1, \cdots, N\}$. Suppose $X(t) = i$, i.e., i machines are up and $N - i$ machines are down at time t. There are two possible triggering events: $E_{i,i-1}$ = one of the i functioning machines fails, and $E_{i,i+1}$ = one of the $N - i$ machines under repair completes its repair. We see that $T_{i,i-1}$ is the minimum of i iid $\exp(\mu)$ random variables, hence $T_{i,i-1} \sim \exp(i\mu)$. Similarly, $T_{i,i+1}$ is the minimum of $(N - i)$ iid $\exp(\lambda)$ random variables, hence $T_{i,i+1} \sim \exp((N - i)\lambda)$. Thus we get $q_{i,i-1} = i\mu$ and $q_{i,i+1} = (N - i)\lambda$, $0 \leq i \leq N$. Hence $\{X(t), t \geq 0\}$ is a birth and death process with birth parameters $\lambda_i = (N - i)\lambda$ and death parameters $\mu_i = i\mu$ for $0 \leq i \leq N$. ∎

Example 6.15   Retrial Queue. We describe a simple retrial queue here. Consider a single-server system where customers arrive according to a $PP(\lambda)$ and have iid $\exp(\mu)$ service times. Suppose the system capacity is 1. Thus if an arriving customer finds the server idle, he immediately starts getting served. On the other hand, if an arriving customer finds the server busy, he goes away (we say he joins an orbit) and tries his luck again after an $\exp(\theta)$ amount of time independent of everything else. He persists this way until he is served. All customers behave in this fashion. We model this as a CTMC.

Let $X(t)$ be the number of customers in service at time t. Obviously $X(t) \in \{0, 1\}$. Let $R(t)$ be the number of customers in the orbit. The reader should convince oneself that neither $\{X(t), t \geq 0\}$ nor $\{R(t), t \geq 0\}$ is a CTMC. We consider the bivariate process $\{(X(t), R(t)), t \geq 0\}$ with state-space $S = \{(i, k) : i = 0, 1; \ k = 0, 1, 2, \cdots\}$.

Suppose $(X(t), R(t)) = (1, k), k \geq 0$, i.e., the server is busy and there are k customers in the orbit at time t. There are two possible events that can lead to a state change: $E_{(1,k),(1,k+1)} =$ an external arrival occurs, and $E_{(1,k),(0,k)} =$ the customer in service departs. Notice that if a customer in orbit conducts a retrial, he simply rejoins the orbit, and hence there is no change of state. We have $T_{(1,k),(1,k+1)} \sim \exp(\lambda)$ and $T_{(1,k),(0,k)} \sim \exp(\mu)$. Thus $q_{(1,k),(1,k+1)} = \lambda$ and $q_{(1,k),(0,k)} = \mu$.

Next suppose $(X(t), R(t)) = (0, k), k \geq 1$, i.e., the server is idle and there are k customers in the orbit at time t. There are two possible events that can lead to a state change: $E_{(0,k),(1,k)} =$ an external arrival occurs, and $E_{(0,k),(1,k-1)} =$ one of the k customers in the orbit conducts a retrial, and finding the server idle, joins service. We see that $T_{(0,k),(1,k)} \sim \exp(\lambda)$ and $T_{(0,k),(1,k-1)}$, being the minimum of k iid $\exp(\theta)$ random variables, is an $\exp(k\theta)$ random variable. Thus $q_{(0,k),(1,k)} = \lambda$ and $q_{(0,k),(1,k-1)} = k\theta$.

Finally, suppose $(X(t), R(t)) = (0, 0)$. There is only one triggering event in this state, namely, arrival of an external customer, leading to the new state $(1, 0)$. Hence $T_{(0,0),(1,0)} \sim \exp(\lambda)$. Thus $q_{(0,0),(1,0)} = \lambda$ and $\{(X(t), R(t)), t \geq 0\}$ is a CTMC. The rate diagram is shown in Figure 6.9. ∎

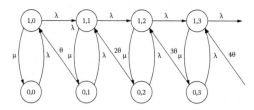

Figure 6.9 *The rate diagram for the retrial queue.*

6.3 CTMCs in Other Fields

The examples in the previous section may give an impression that CTMCs are mostly used to model queueing systems. In this section we describe CTMC models in other fields.

6.3.1 Organ Transplants

Organ transplants have become routine in the field of medicine since the discovery of immuno-suppressant drugs and enhanced cross-matching that reduce the probability of rejection of a transplanted organ by human immuno-response. Each year hundreds of thousands of hearts, lungs, and kidneys (among other organs) are transplanted. For ethical reasons there is no financial market in organs in the US, that is, one cannot buy or sell organs for transplant purposes. Also, typically the demand for organs always out-strips the supply. Hence there is a national list of patients waiting for organ transplants. In theory an incoming organ should be given to the patient who has the best expected outcome from the transplant operation, measured in QALYs (quality adjusted life-years). In practice the waiting patients are prioritized according to a fair and efficient rule and an incoming organ is first made available to the first patient on the list, and if that patient does not accept it (for whatever reason: not a good match, low quality of the organ, etc.), it is then made available to the next patient, and so on. If no patient accepts the organ, it is discarded.

A simple model of organ transplant assumes that the patients arrive according to a PP(λ) and organs arrive according an independent PP(μ). Organs cannot be stored, so if there is no patient waiting when an organ arrives, the organ is discarded. The patients are ordered according to a priority rule, with the highest priority patient first. A new organ is offered to the waiting patients in the order of their priority, until it is accepted by someone, or is discarded if it is rejected by all of them. If the organ is rejected by the first $i - 1$ patients, it is made available to the ith patient ($i \geq 1$), who accepts it with probability α_i and rejects it with probability $1 - \alpha_i$. Note that the acceptance probabilities depend only on the position in the queue, and not on the patient occupying this position. Thus, as a patient moves up in the queue, his acceptance probability will change. If there are n patients waiting when an organ arrives, it will be accepted by one of them with probability

$$\beta_n = 1 - \prod_{i=1}^{n}(1 - \alpha_i),$$

and rejected with probability $1 - \beta_n$. (Define $\beta_0 = 0$.) We assume that this activity of organ acceptance/rejection takes place instantaneously. A waiting patient may also leave the queue (mainly due to death) without getting a transplant. Assume that patient life-times are iid exp(θ) random variables.

Now let $X(t)$ be the number of patients waiting in the queue at time t. We can see that $\{X(t), t \geq 0\}$ is a birth and death process on $\{0, 1, 2, \cdots\}$ with birth rates $\lambda_n = \lambda$ and death rates $\mu_n = \mu\beta_n + n\theta$ for $n \geq 0$

Typical performance measures associated with an organ transplant system are: the long run fraction of the organs that are wasted (those that were accepted by no one), the long run fraction of the patients who get a transplant, expected time spent in the queue by an arriving customer, the long run reward earned (in terms of QALYs), etc. We shall develop methods in this chapter to compute such quantities.

6.3.2 Disease Progression

Disease progression models are motivated by the observation that we can identify different stages of a disease in a given patient based on symptoms and tests, for example, colorectal cancer, leukemia, renal diseases, AIDS, diabetes, breast cancer, etc. The particulars of the categorization of the stages of a disease depend on the disease. The disease progression models can be useful in devising public health policies about the frequency of screening for the disease, type of medical intervention, etc. For example, the current policy in the US recommends mammography (screening for breast cancer) every two years among women of age 50 to 74 years.

We denote the stage of the disease in a given patient at time t by $X(t)$. It is common to assume that $\{X(t), t \geq 0\}$ is a CTMC with state-space $\{0, 1, \cdots, S\}$, where higher stage indicates more serious level of the disease, with stage 0 indicating absence of the disease, and S indicating death. A simple model assumes that the transition rates of the CTMC are

$$q_{i,i+1} = \lambda_i, \; q_{i,S} = \theta, \; 0 \leq i < S.$$

Here, we can think of the patient's life span as an $\exp(\theta)$ random variable, independent of the disease stage. This patient, while alive, spends $\exp(\lambda_i)$ amount of time in disease stage i and then moves to stage $i+1$. This model assumes that the disease can only become worse over time, until the patient dies. We can make the model slightly more realistic by making the rate (θ) at which the patient dies dependent on the stage of the disease. That is, we can model the transition rates as

$$q_{i,i+1} = \lambda_i, \; q_{i,S} = \theta_i, \; 0 \leq i < S.$$

It is possible for the disease to be reversible, that is, the disease stage can become worse or better, with or without intervention. One way to model this is to assume the following transition rates:

$$q_{i,i+1} = \lambda_i, \; q_{i,S} = \theta_i, \; q_{i,i-1} = \mu_i, \; 0 \leq i < S,$$

with $\mu_0 = 0$. Thus the disease stage can improve from i to $i-1$ at rate μ_i.

The important questions are: how often should we screen for the disease, should the frequency of screening depend on the age of the patient or the stage of the disease, what interventions to do in what stage, etc. Some of these questions can be answered by using the tools developed later in this chapter, while other questions will require tools from Markov decision processes, which are beyond the scope of this book.

6.3.3 Epidemics

An epidemic occurs when an infectious disease enters a susceptible population from outside, spreads through the population infecting a significant fraction of it, and eventually dies out as the population becomes immune to the disease. The most common epidemics in humans are the annual flu epidemics, recent SARS and Ebola epidemics, West Nile virus, AIDS, malaria, plague, etc. The most common topics in epidemiology are: source of the epidemic, isolation and cure, controlling the spread, vaccination, quarantine policies, cost control, etc. The discussion of these issues benefits from building mathematical models of how epidemics evolve with time.

A typical infectious-disease progression model of an individual is as follows: An individual is initially susceptible (S) to a disease. At time t_1 he gets exposed (E), but does not exhibit any symptoms. At time $t_2 \geq t_1$ he becomes infectious (I), that is he can now transmit the disease to another susceptible individual by some means of contact. At time $t_3 \geq t_2$ the individual exhibits symptoms of the disease. Finally at time $t_4 \geq t_3$, the individual recovers (R) and becomes immune to the disease, or recovers but again becomes susceptible to the disease, or dies. It is common to describe an epidemic model by its disease progression model. An SI model assumes that a susceptible individual becomes infections and stays that way forever. Similarly we have the SIS model, the SIR model, the SIRS model, the SEIRS model, etc.

Next we describe an SIR model of an epidemic in a population of N individuals. At time zero, one of them becomes infected (catches the infection from outside the population) and the rest are susceptible. Suppose an infected individual stays infectious for an $\exp(\gamma)$ amount of time and then recovers and becomes immune to further infections by this disease. Any two individuals come into contact with each other after an $\exp(\beta)$ amount of time. If an infected individual comes in contact with a susceptible individual, the susceptible individual becomes infectious.

Let $S(t)$ be the number of susceptible individuals at time t, $I(t)$ be the number of infected individuals at time t and $R(t)$ be the recovered individuals at time t. Thus we have $S(0) = N - 1$, $R(0) = 0$, $I(0) = 1$ and

$$S(t) + I(t) + R(t) = N, \quad t \geq 0.$$

We see that $\{(S(t), I(t)), t \geq 0\}$ is a CTMC on state-space $\{(i, j) : 0 \leq i + j \leq N\}$ with transition rates as follow:

$$q_{(i,j),(i-1,j+1)} = \beta i j,$$

$$q_{(i,j),(i,j-1)} = \gamma j.$$

The states $(i, 0)$ are absorbing. It is of interest to study the transient behavior of $E(I(t))$, the expected size of the epidemic at time t, and see when the epidemic reaches its maximum. One way to control the epidemic is to quarantine or treat the infected individuals. One simple way to model such a program is to assume that the recovery rate under such a program increases from γ to $\gamma + \delta$. The cost of the control is an increasing function of δ. Another way to control an epidemic is to embark on a mass vaccination drive. This has an effect of reducing β (the spreading rate) to $\beta \alpha$, where $0 \leq \alpha < 1$ represents the effectiveness of the vaccine (the smaller the α, the better the vaccine). The cost of the vaccine is a decreasing function of α. The idea is to choose an optimal treatment/quarantine/vaccination policy that minimizes the social cost. The tools developed in this chapter are useful in solving such problems.

6.3.4 Order Books

The stock market forms the backbone of the modern capitalist system. An electronic exchange (or a broker) provides a service for orderly trading (buying and selling) of stocks in companies. The exchange maintains a limit order book, which is essentially a list of buy and sell orders that are currently awaiting execution. An order has the following information associated with it: the code for the trader who has placed the order, whether it is a buy order or sell order, the name of the stock that is being traded, the time when the order is placed, and the price at which the trader is willing to trade (it is called a bid price if it is a buy order, and ask price if it is a sell order), and the size of the order (that is, the number of shares that are put up for trade). The largest bid price is called the market bid price, and the smallest ask price is called a market ask price. Clearly, the market ask price is always above the market bid price. Orders arrive at random, and leave the order book when they are executed (that is, when a matching trader on the opposite side is found). Some orders may be cancelled and leave the order book before they get a chance at execution.

To understand the dynamics, consider the simple case where only stock is being traded, and all orders are for one share of the stock. When a buy order comes whose bid price is at or above the market ask price, it is matched with the oldest sell order at the market ask, and a transaction occurs, and both orders are removed from the book. Similarly, when a sell order comes whose ask price is at or below the market bid price, it is matched with the oldest buy order at the market bid, and a transaction occurs, and both orders are removed from the book. Orders arrive at random, and wait on the order book until they are executed, or cancelled. In general the arrival process will depend on the state of the order book, and can be quite complex.

Here we describe a contrived order book for a single stock trading at a single price. Let $S(t)$ be the number of sell orders on the order book at time t, and $B(t)$ be the number of buy orders at time t. Clearly we cannot have both $B(t) > 0$ and $S(t) > 0$, since in that case enough of the buy and sell orders will be matched and leave the system until there are either no buy orders left or no sell orders left on the order book. Thus the state of the order book is completely described by $X(t) = S(t) - B(t)$. If $X(t) > 0$, it means $S(t) = X(t)$ and $B(t) = 0$, and if $X(t) < 0$, it means $S(t) = 0$ and $B(t) = -X(t)$. Finally, $X(t) = 0$ means $S(t) = 0$ and $B(t) = 0$. Note that if $X(t) \geq 0$ it increases by one if a new sell order arrives and decreases by one if a new buy order arrives, or one of the existing sell orders is cancelled. Also if $X(t) \leq 0$ it increases by one if a new sell order arrives or one of the existing buy orders is cancelled, and decreases by one if a new buy order arrives. Suppose new sell orders arrive at rate λ_i^s and buy orders arrive at rate λ_i^b if $X(t) = i$. Also, each sell order on the book gets cancelled at rate α^s and each buy order gets cancelled at rate α^b. With this notation we see that $\{X(t), t \geq 0\}$ is a birth and death process on all integers with birth rates

$$\lambda_i = \begin{cases} \lambda_i^s & \text{if } i \geq 0 \\ -i\alpha^b + \lambda_i^s & \text{if } i < 0, \end{cases} \qquad (6.13)$$

and death rates

$$\mu_i = \begin{cases} i\alpha^s + \lambda_i^b & \text{if } i > 0 \\ \lambda_i^b & \text{if } i \leq 0. \end{cases} \qquad (6.14)$$

It is easy see how the state of the order book gets more complicated when there are multiple bid and ask prices. We shall not go into the details here.

6.3.5 Dealership Markets

In the previous subsection we discussed models of order books on electronic exchanges or brokers. There are other players in the market, called dealers, who also provide liquidity to the stock market. A dealer does this by posting a bid price and an ask price at all times: he is always ready to buy from the sellers at the posted bid price and sell to the buyers at the posted ask price from the inventory of shares that he holds. Clearly the bid price is always lower than the ask price, and the dealer makes money by buying low and selling high. The dealer varies the bid and ask prices to maximize his profits. We shall develop a simple CTMC model below of the dealer inventory of shares.

Let $X(t)$ be the number of shares of the stock held by the dealer at time t. Assume that $X(t) \in \{0, 1, \cdots, M\}$ for concreteness. That is, the dealer never lets the number of shares in the inventory fall below zero, or go above M. Assume that there are functions λ^s and λ^b with the following interpretation: if the ask price is a, buyers arrive at the dealer according to a $PP(\lambda^b(a))$, and if the bid price is b, sellers arrive at the dealer according to a $PP(\lambda^s(b))$. It makes sense to assume that $\lambda^b(a)$ is a decreasing function of a, and $\lambda^s(b)$ is an increasing function of b. The dealer changes the ask and bid prices according to his current inventory in order to maximize his profits.

Suppose the dealer uses bid price b_i and ask price a_i when his inventory is i. We assume that there is a large constant $A > 0$ such that $a_0 = A$ and $\lambda^b(A) = 0$, implying that there are no buyers when the inventory is zero, so it never goes negative. Similarly, assume that $b_M = 0$ and $\lambda^s(0) = 0$, so that there are no sellers when the inventory is M, so the inventory never goes above M. Let

$$\lambda_i = \lambda^s(b_i), \quad \mu_i = \lambda^b(a_i), \quad 0 \leq i \leq M.$$

Then it is clear that when the inventory is i it increases by one at rate λ_i and decreases by one at rate μ_i. Thus $\{X(t), t \geq 0\}$ is a birth and death process with birth rates λ_i and death rates μ_i ($0 \leq i \leq M$).

If the dealer sells a share when $X(t) = i$ his cash account increases by a_i dollars, and if he buys a share when $X(t) = i$, his cash account decreases by b_i dollars. Thus the expected net rate at which his cash account changes at time t is

$$r(t) = \sum_{i=0}^{M} (\mu_i a_i - \lambda_i b_i) P(X(t) = i).$$

The dealer chooses the a_i's and b_i's to maximize g, the long run net profit per unit time, where

$$g = \lim_{T \to \infty} \frac{1}{T} \int_0^T r(u) du.$$

We shall develop tools in this chapter that will help us compute such quantities.

6.3.6 Service and Production Systems

CTMCs have been used to develop models of service and production systems. We shall consider them in greater detail in Chapter 7, and hence we do not describe them here.

6.4 Transient Behavior: Marginal Distribution

In this section we study the transient behavior of a CTMC $\{X(t), t \geq 0\}$ on state-space $S = \{0, 1, 2, \cdots\}$ with initial distribution $a = [a_i]$ and generator matrix Q. To be specific, we study the pmf of $X(t)$ for a fixed t. Let

$$p_j(t) = \mathsf{P}(X(t) = j), \quad j \in S, \ t \geq 0,$$

and

$$p(t) = [p_j(t)]_{j \in S}$$

be the pmf of $X(t)$. Conditioning on $X(0)$ we get

$$
\begin{aligned}
p_j(t) &= \mathsf{P}(X(t) = j) \\
&= \sum_{i \in S} \mathsf{P}(X(t) = j | X(0) = i)\mathsf{P}(X(0) = i) \\
&= \sum_{i \in S} a_i p_{i,j}(t).
\end{aligned}
$$

In matrix form, we get

$$p(t) = aP(t), \quad t \geq 0.$$

Thus it suffices to study $P(t)$ to study $p(t)$. We start with the following theorem.

Theorem 6.3   Chapman–Kolmogorov Equations. *Let $P(t)$ be the transition probability matrix of a CTMC $\{X(t), t \geq 0\}$. Then*

(i) $p_{i,j}(t) \geq 0, \quad i, j \in S, \ t \geq 0.$

(ii) $\displaystyle\sum_{j \in S} p_{i,j}(t) = 1, \quad i \in S, \ t \geq 0.$ (6.15)

(iii) $p_{i,j}(s + t) = \displaystyle\sum_{k \in S} p_{i,k}(s) p_{k,j}(t), \quad i, j \in S, \ s, t \geq 0.$ (6.16)

Proof: Part (i) is obvious since $p_{i,j}(t)$'s are conditional probabilities. To show Equation 6.15, we assume the regularity condition in 6.1 and use the representation in Equation 6.3 to get

$$
\begin{aligned}
\sum_{j \in S} p_{i,j}(t) &= \sum_{j \in S} \mathsf{P}(X_{N(t)} = j | X_0 = i) \\
&= \sum_{j \in S} \sum_{n=0}^{\infty} \mathsf{P}(X_{N(t)} = j | X_0 = i, N(t) = n)\mathsf{P}(N(t) = n | X_0 = i)
\end{aligned}
$$

$$
\begin{aligned}
&= \sum_{j \in S} \sum_{n=0}^{\infty} \mathsf{P}(X_n = j | X_0 = i, N(t) = n) \mathsf{P}(N(t) = n | X_0 = i) \\
&= \sum_{n=0}^{\infty} \sum_{j \in S} \mathsf{P}(X_n = j | X_0 = i, N(t) = n) \mathsf{P}(N(t) = n | X_0 = i) \\
&= \sum_{n=0}^{\infty} \mathsf{P}(N(t) = n | X_0 = i) \\
&= \mathsf{P}(N(t) < \infty | X_0 = i) = 1.
\end{aligned}
$$

Here the second to last equality is a result of the fact that $X_n \in S$ for all $n \geq 0$, and the last equality is the result of Equation 6.2.

To show Equation 6.16, we condition on $X(s)$ to get

$$
\begin{aligned}
p_{i,j}(s+t) &= \mathsf{P}(X(s+t) = j | X(0) = i) \\
&= \sum_{k \in S} \mathsf{P}(X(s+t) = j | X(0) = i, X(s) = k) \mathsf{P}(X(s) = k | X(0) = i) \\
&= \sum_{k \in S} \mathsf{P}(X(s+t) = j | X(s) = k) p_{i,k}(s) \\
&\quad \text{(from Markov property, Equation 6.5)} \\
&= \sum_{k \in S} p_{i,k}(s) \mathsf{P}(X(t) = j | X(0) = k) \\
&\quad \text{(from time-homogeneity, Equation 6.6)} \\
&= \sum_{k \in S} p_{i,k}(s) p_{k,j}(t).
\end{aligned}
$$

This completes the proof. ∎

Equation 6.16 is called the Chapman–Kolmogorov equation for the CTMCs. It is analogous to the Chapman–Kolmogorov equation for the DTMCs, as given in Equation 2.22. Equation 6.16 can be written in matrix form as

$$
P(s+t) = P(s)P(t), \quad s, t \geq 0.
$$

By interchanging s and t we get

$$
P(s+t) = P(t)P(s), \quad s, t \geq 0.
$$

This shows a very important and unusual property of the matrices $\{P(t), t \geq 0\}$: they commute, i.e.,

$$
P(s)P(t) = P(t)P(s), \quad s, t \geq 0.
$$

We develop a set of differential equations satisfied by the transition probability matrix $P(t)$ in the next theorem. We say that $P(t)$ is differentiable with respect to t if $p_{i,j}(t)$ is differentiable with respect to t for all $t > 0$ and $i, j \in S$. We define the derivative of the matrix $P(t)$ with respect to t as the matrix of the derivatives of $p_{i,j}(t)$.

Theorem 6.4 Forward and Backward Equations. *Let $P(t)$ be the transition probability matrix of a CTMC with state-space $S = \{0, 1, 2, \cdots\}$ and generator matrix Q. Then $P(t)$ is differentiable with respect to t and satisfies*

$$\frac{d}{dt}P(t) = P'(t) = QP(t), \quad t \geq 0, \text{ (\textbf{Backward Equations})} \tag{6.17}$$

and

$$\frac{d}{dt}P(t) = P'(t) = P(t)Q, \quad t \geq 0, \text{ (\textbf{Forward Equations})} \tag{6.18}$$

with initial condition

$$P(0) = I,$$

where I is an identity matrix of appropriate size.

Proof: We shall first prove that

$$p_{i,j}(h) = \delta_{i,j} + q_{i,j}h + o(h), \quad i, j \in S, \tag{6.19}$$

where $\delta_{i,j} = 1$ if $i = j$ and 0 otherwise, and $o(h)$ is a function such that

$$\lim_{h \to 0} \frac{o(h)}{h} = 0.$$

From the properties of the sequence $\{X_0, (X_n, Y_n), n \geq 1\}$, we get

$$\begin{aligned}
P(N(h) = 0 | X_0 = i) &= P(Y_1 > h | X_0 = i) \\
&= e^{-q_i h} = 1 - q_i h + o(h) \\
&= 1 + q_{i,i}h + o(h).
\end{aligned}$$

Also,

$$\begin{aligned}
P(N(h) = 1 | X_0 = i) &= P(Y_1 \leq h, Y_1 + Y_2 > h | X_0 = i) \\
&= \sum_{j \in S, j \neq i} \int_0^h q_i e^{-q_i s} e^{-q_j (h-s)} p_{i,j} ds \\
&= \sum_{j \in S, j \neq i} q_i p_{i,j} e^{-q_j h} \frac{1 - e^{-(q_i - q_j)h}}{q_i - q_j} \quad \text{(assuming } q_i \neq q_j) \\
&= q_i h + o(h)
\end{aligned}$$

where the last equality follows after a bit of algebra. (The calculations in the case $q_i = q_j$ are similar and yield the same result.) Similar analysis yields

$$P(N(h) \geq 2 | X_0 = i) = o(h).$$

Using these we get

$$\begin{aligned}
p_{i,i}(h) &= P(X(h) = i | X(0) = i) \\
&= P(X_{N(h)} = i | X_0 = i) \\
&= P(X_{N(h)} = i | X_0 = i, N(h) = 0)P(N(h) = 0 | X_0 = i) \\
&\quad + P(X_{N(h)} = i | X_0 = i, N(h) = 1)P(N(h) = 1 | X_0 = i)
\end{aligned}$$

$$+\mathsf{P}(X_{N(h)} = i|X_0 = i, N(h) \geq 2)\mathsf{P}(N(h) \geq 2|X_0 = i)$$
$$= 1 \cdot (1 + q_{i,i}h + o(h)) + 0 \cdot (q_i h + o(h)) + o(h)$$
$$= 1 + q_{i,i}h + o(h).$$

Also, for $j \neq i$, we get

$$
\begin{aligned}
p_{i,j}(h) &= \mathsf{P}(X(h) = j|X(0) = i) \\
&= \mathsf{P}(X_{N(h)} = j|X_0 = i) \\
&= \mathsf{P}(X_{N(h)} = j|X_0 = i, N(h) = 0)\mathsf{P}(N(h) = 0|X_0 = i) \\
&\quad + \mathsf{P}(X_{N(h)} = j|X_0 = i, N(h) = 1)\mathsf{P}(N(h) = 1|X_0 = i) \\
&\quad + \mathsf{P}(X_{N(h)} = j|X_0 = i, N(h) \geq 2)\mathsf{P}(N(h) \geq 2|X_0 = i) \\
&= 0 \cdot (1 + q_{i,i}h + o(h)) + \mathsf{P}(X_1 = j|X_0 = i)(q_i h + o(h)) + o(h) \\
&= p_{i,j}(q_i h + o(h)) + o(h) = q_{i,j}h + o(h).
\end{aligned}
$$

This proves Equation 6.19. Using this we get

$$
\begin{aligned}
p_{i,j}(t + h) &= \sum_{k \in S} p_{i,k}(h)p_{k,j}(t) \\
&\qquad \text{(Chapman–Kolmogorov Equations 6.16)} \\
&= \sum_{k \in S} (\delta_{i,k} + q_{i,k}h + o(h))p_{k,j}(t) \quad \text{(Equation 6.19)}.
\end{aligned}
$$

Hence

$$p_{i,j}(t + h) - p_{i,j}(t) = \sum_{k \in S} q_{i,k}h p_{k,j}(t) + o(h).$$

Dividing by h yields

$$\frac{p_{i,j}(t + h) - p_{i,j}(t)}{h} = \sum_{k \in S} q_{i,k}p_{k,j}(t) + \frac{o(h)}{h}.$$

Letting $h \to 0$ we see that the right-hand side has a limit, and hence $p_{i,j}(t)$ is differentiable with respect to t and satisfies

$$p'_{i,j}(t) = \sum_{k \subset S} q_{i,k}p_{k,j}(t).$$

Writing this in matrix form we get the backward equations 6.17. The forward equations follow similarly by interchanging t and h in applying the Chapman–Kolmogorov equations. ∎

Thus one may solve forward or backward equations to obtain $P(t)$. Once we have the matrix $P(t)$, we have the distribution of $X(t)$. We illustrate by means of an example.

Example 6.16 Two-State Machine. Consider the CTMC of Example 6.4 with the rate matrix given in Equation 6.12. The forward equations are

$$p'_{0,0}(t) = -\lambda p_{0,0}(t) + \mu p_{0,1}(t), \quad p_{0,0}(0) = 1,$$

$$
\begin{aligned}
p'_{0,1}(t) &= \lambda p_{0,0}(t) - \mu p_{0,1}(t), \quad p_{0,1}(0) = 0,\\
p'_{1,0}(t) &= -\lambda p_{1,0}(t) + \mu p_{1,1}(t), \quad p_{1,0}(0) = 0,\\
p'_{1,1}(t) &= \lambda p_{1,0}(t) - \mu p_{1,1}(t), \quad p_{1,1}(0) = 1,
\end{aligned}
$$

and the backward equations are

$$
\begin{aligned}
p'_{0,0}(t) &= -\lambda p_{0,0}(t) + \lambda p_{1,0}(t), \quad p_{0,0}(0) = 1,\\
p'_{1,0}(t) &= \mu p_{0,0}(t) - \mu p_{1,0}(t), \quad p_{1,0}(0) = 0,\\
p'_{0,1}(t) &= -\lambda p_{0,1}(t) + \lambda p_{1,1}(t), \quad p_{0,1}(0) = 0,\\
p'_{1,1}(t) &= \mu p_{1,0}(t) - \mu p_{1,1}(t), \quad p_{1,1}(0) = 1.
\end{aligned}
$$

Note that we do not need to solve four equations in four unknowns simultaneously, but only two equations in two unknowns at a time. We solve the first two forward equations here. We have

$$
p_{0,0}(t) + p_{0,1}(t) = \mathsf{P}(X(t) = 0 \text{ or } 1 | X(0) = 0) = 1. \tag{6.20}
$$

Substituting for $p_{0,1}(t)$ in the first forward equation we get

$$
p'_{0,0}(t) = -\lambda p_{0,0}(t) + \mu(1 - p_{0,0}(t)) = -(\lambda + \mu)p_{0,0}(t) + \mu.
$$

This is a non-homogeneous first-order differential equation with constant coefficients, and can be solved by standard methods (see Appendix I) to get

$$
p_{0,0}(t) = \frac{\lambda}{\lambda + \mu} e^{-(\lambda+\mu)t} + \frac{\mu}{\lambda + \mu}.
$$

Using Equation 6.20 we get

$$
p_{0,1}(t) = \frac{\lambda}{\lambda + \mu}(1 - e^{-(\lambda+\mu)t}).
$$

Similarly we can solve the last two forward equations to get $p_{1,0}(t)$ and $p_{1,1}(t)$. Combining all these solutions we get

$$
P(t) = \begin{bmatrix} \frac{\lambda}{\lambda+\mu}e^{-(\lambda+\mu)t} + \frac{\mu}{\lambda+\mu} & \frac{\lambda}{\lambda+\mu}(1 - e^{-(\lambda+\mu)t}) \\ \frac{\mu}{\lambda+\mu}(1 - e^{-(\lambda+\mu)t}) & \frac{\mu}{\lambda+\mu}e^{-(\lambda+\mu)t} + \frac{\lambda}{\lambda+\mu} \end{bmatrix}. \tag{6.21}
$$

It is easy to check that $P(t)$ satisfies the Chapman–Kolmogorov Equations 6.16 and that $P(t)$ and $P(s)$ commute. Figure 6.10 shows graphically various $p_{i,j}(t)$ functions. ∎

6.5 Transient Behavior: Occupancy Times

Following the steps in the study of DTMCs, we now study the occupancy times in the CTMCs. Let $\{X(t), t \geq 0\}$ be a CTMC on state-space S with generator matrix Q. Let $V_j(t)$ be the amount of time the CTMC spends in state j over $(0, t]$. Thus $V_j(0) = 0$ for all $j \in S$. Define

$$
M_{i,j}(t) = \mathsf{E}(V_j(t) | X(0) = i), \quad i, j \in S, \ t \geq 0. \tag{6.22}
$$

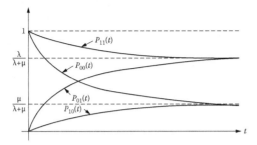

Figure 6.10 *The transition probability functions for the two-state CTMC.*

$M_{i,j}(t)$ is called the *occupancy time* of state j up to time t starting from state i. Define the occupancy times matrix as

$$M(t) = [M_{i,j}(t)].$$

The next theorem shows how to compute the occupancy times matrix $M(t)$.

Theorem 6.5   Occupancy Times Matrix. *We have*

$$M(t) = \int_0^t P(u)du, \quad t \geq 0, \tag{6.23}$$

where the integral of a matrix is defined to be the matrix of the integrals of its elements.

Proof: Fix a $j \in S$. Let $Z(u) = 1$ if $X(u) = j$, and zero otherwise. Then

$$V_j(t) = \int_0^t Z(u)du.$$

Hence we get

$$
\begin{aligned}
M_{i,j}(t) &= \mathsf{E}(V_j(t)|X(0) = i) \\
&= \mathsf{E}\left(\int_0^t Z(u)du \,\bigg|\, X(0) = i\right) \\
&= \int_0^t \mathsf{E}(Z(u)|X(0) = i)du \\
&= \int_0^t \mathsf{P}(X(u) = j|X(0) = i)du \\
&= \int_0^t p_{i,j}(u)du.
\end{aligned}
$$

Writing the above equation in a matrix form yields Equation 6.23. This proves the theorem. ∎

We illustrate with an example.

Example 6.17 Two-State Machine. Consider the CTMC of Example 6.4 with the rate matrix given in Equation 6.12. The transition probability matrix $P(t)$ of this CTMC is given in Equation 6.21. Using that, and carrying out the integration, we see that the occupancy matrix is given by

$$
\begin{bmatrix}
\frac{\mu}{\lambda+\mu}t + \frac{\lambda}{(\lambda+\mu)^2}(1 - e^{-(\lambda+\mu)t}) & \frac{\lambda}{\lambda+\mu}t - \frac{\lambda}{(\lambda+\mu)^2}(1 - e^{-(\lambda+\mu)t}) \\
\frac{\mu}{\lambda+\mu}t - \frac{\mu}{(\lambda+\mu)^2}(1 - e^{-(\lambda+\mu)t}) & \frac{\lambda}{\lambda+\mu}t + \frac{\mu}{(\lambda+\mu)^2}(1 - e^{-(\lambda+\mu)t})
\end{bmatrix}. \tag{6.24}
$$

Thus, if the CTMC starts in state 0, the expected time it spends in state 1 up to time t is given by $M_{0,1}(t)$. ∎

6.6 Computation of $P(t)$: Finite State-Space

In this section we present several methods of computing the transition matrix $P(t)$ of a CTMC with finite state-space $S = \{1, 2, \cdots, N\}$ and generator matrix Q. We shall illustrate the methods with two examples: the two-state machine of Example 6.4 which we analyze algebraically, and the two-machine workshop of Example 6.6 which we analyze numerically.

6.6.1 Exponential of a Matrix

Define the exponential of an $N \times N$ square matrix A as follows:

$$
e^A = I + \sum_{n=1}^{\infty} \frac{A^n}{n!}.
$$

Note that e^A is an $N \times N$ square matrix. One can show that the series on the right-hand side converges absolutely, and hence e^A is well defined. Note that if A is a diagonal matrix

$$
A = \text{diag}[a_1, a_2, \cdots, a_N],
$$

then

$$
e^A = \text{diag}[e^{a_1}, e^{a_2}, \cdots, e^{a_N}].
$$

In particular $e^0 = I$ where $\mathbf{0}$ is a square matrix with all elements equal to zero. The next theorem gives the main result. Following the discussion in Section 2.4, we say that Q is diagonalizable if there exists a diagonal matrix D and an invertible matrix X such that

$$
Q = XDX^{-1}. \tag{6.25}
$$

We saw in that section how to obtain the D and X if Q is diagonalizable.

Theorem 6.6 Exponential of Qt. *The transition probability matrix of a finite-state CTMC with generator matrix Q is given by*

$$
P(t) = e^{Qt}, \quad t \geq 0. \tag{6.26}
$$

Furthermore, if Q is diagonalizable and satisfies Equation 6.25

$$P(t) = X e^{Dt} X^{-1} = \sum_{i=1}^{N} e^{\lambda_i t} x_i y_i, \qquad (6.27)$$

where $\lambda_i = D_{i,i}$, x_i is the ith column of X, and y_i is the ith row of X^{-1}.

Proof: We have

$$e^{Qt} = I + \sum_{n=1}^{\infty} \frac{(Qt)^n}{n!}, \quad t \geq 0. \qquad (6.28)$$

Since the infinite series converges uniformly, we can take derivatives term by term to get

$$\frac{d}{dt} e^{Qt} = Q e^{Qt} = e^{Qt} Q.$$

Thus e^{Qt} satisfies the forward and backward equations. Since there is a unique solution to those equations in the finite state-space case, we get Equation 6.26. Now, from Theorem 2.7 we get

$$Q^n = X D^n X^{-1}, \quad n \geq 0.$$

Substituting in Equation 6.28, and factoring out X on the left and X^{-1} on the right, we get

$$e^{Qt} = X \left(I + \sum_{n=1}^{\infty} \frac{(Dt)^n}{n!} \right) X^{-1}, \quad t \geq 0,$$

which yields Equation 6.27. The last representation follows from a similar representation in Equation 2.34 on page 40. ∎

The next theorem is analogous to Theorem 2.8 on page 40.

Theorem 6.7 **Eigenvalues of Q.** *Let Q be an $N \times N$ generator matrix of a CTMC, with N eigenvalues λ_i, $1 \leq i \leq N$. Let*

$$q = \max\{-q_{i,i} : 1 \leq i \leq N\}.$$

Then

1. At least one of the eigenvalues is zero.
2. $|\lambda_i + q| \leq q$ for all $1 \leq i \leq N$.

Proof: Define

$$P = I + \frac{Q}{q}.$$

(We shall see this matrix again in the subsection on uniformization.) It is easy to verify that P is a stochastic matrix. Now it can be seen that the N eigenvalues of P are given by $1 + \lambda_i/q$, $(1 \leq i \leq N)$. From Theorem 2.8 we see that P has at least one eigenvalue equal to one, thus at least one $\lambda_i/q + 1$ equals 1. Thus at least one

λ_i is zero. Also, all eigenvalues of P lie in the complex plane within the unit circle centered at zero. Thus we must have

$$|1 + \frac{\lambda_i}{q}| \leq 1.$$

Thus each λ_i must lie in the complex plane within a circle of radius q with center at $-q$, which is the second part of the theorem. ∎

Many matrix-oriented software programs provide built-in ways to compute the matrix exponential function. For example, MATLAB® provides a function expm: we can compute $P(t)$ by the MATLAB® statement: $P(t) = \text{expm}(Q * t)$. Although simple to use for reasonable sized matrices with numerical entries, it does not provide any insight about the behavior of $P(t)$ as a function of t. We get this insight from Equation 6.27, which gives $P(t)$ as an explicit function of t. We illustrate with examples.

Example 6.18   Two-State Machine. Consider the two-state CTMC with generator matrix given in Equation 6.12. We have

$$Q = XDX^{-1},$$

where

$$X = \begin{bmatrix} 1 & \frac{\lambda}{\lambda+\mu} \\ 1 & -\frac{\mu}{\lambda+\mu} \end{bmatrix}, \quad D = \begin{bmatrix} 0 & 0 \\ 0 & -(\lambda+\mu) \end{bmatrix}, \quad X^{-1} = \begin{bmatrix} \frac{\mu}{\lambda+\mu} & \frac{\lambda}{\lambda+\mu} \\ 1 & -1 \end{bmatrix}.$$

Substituting in Equation 6.27 we get

$$P(t) = X \begin{bmatrix} 1 & 0 \\ 0 & e^{-(\lambda+\mu)t} \end{bmatrix} X^{-1}.$$

Straightforward calculations show that this reduces to the transition probability matrix given in Equation 6.21. ∎

Example 6.19   Two-Machine One-Repairperson Workshop. Consider the machine shop of Example 6.6. Suppose that the mean life time of a machine is 10 days, while the mean repair time is one day. Let $X(t)$ be the number of working machines at time t. Using days as the unit of time, we see that $\lambda = 1$ and $\mu = .1$, and thus $\{X(t), t \geq 0\}$ is a CTMC on state-space $\{0, 1, 2\}$ and generator matrix

$$Q = \begin{bmatrix} -1 & 1 & 0 \\ .1 & -1.1 & 1 \\ 0 & .2 & -.2 \end{bmatrix}.$$

Numerical calculations show that

$$Q = XDX^{-1},$$

where

$$X = \begin{bmatrix} 0.9029 & 0.9844 & 0.5774 \\ -0.4245 & 0.1675 & 0.5774 \\ 0.0668 & -0.0532 & 0.5774 \end{bmatrix}, \quad X^{-1} = \begin{bmatrix} 0.3178 & -1.4943 & 1.1765 \\ 0.7076 & 1.2041 & -1.9117 \\ 0.0284 & 0.2839 & 1.4197 \end{bmatrix},$$

and

$$D = \begin{bmatrix} -1.4702 & 0 & 0 \\ 0 & -0.8298 & 0 \\ 0 & 0 & 0 \end{bmatrix}.$$

Hence, from Theorem 6.6, we get

$$P(t) = e^{-1.4702t} \begin{bmatrix} 0.2870 & -1.3493 & 1.0623 \\ -0.1349 & 0.6344 & -0.4994 \\ 0.0212 & -0.0999 & 0.0786 \end{bmatrix} \quad (6.29)$$

$$+ e^{-0.8298t} \begin{bmatrix} 0.6966 & 1.1853 & -1.8820 \\ 0.1185 & 0.2017 & -0.3202 \\ -0.0376 & -0.0640 & 0.1017 \end{bmatrix} \quad (6.30)$$

$$+ \begin{bmatrix} 0.0164 & 0.1639 & 0.8197 \\ 0.0164 & 0.1639 & 0.8197 \\ 0.0164 & 0.1639 & 0.8197 \end{bmatrix}. \quad (6.31)$$

Thus if all machines are initially down, the probability that one machine is working at the end of the first day is given by

$$p_{0,1}(1) = -1.3493e^{-1.4702} + 1.1853e^{-0.8298} + 0.1639 = 0.3707.$$

From the explicit expression for $P(t)$ we see that

$$\lim_{t \to \infty} P(t) = \begin{bmatrix} 0.0164 & 0.1639 & 0.8197 \\ 0.0164 & 0.1639 & 0.8197 \\ 0.0164 & 0.1639 & 0.8197 \end{bmatrix}.$$

Thus $P(t)$ has a limit, and in the limit all its rows are identical. This is similar to the behavior we had encountered in the study of DTMCs. We shall study this in more detail in Section 6.11. ∎

6.6.2 Laplace Transforms

Define the Laplace transform (LT) (see Appendix F) of $p_{i,j}(t)$ as follows:

$$p_{i,j}^*(s) = \int_0^\infty e^{-st} p_{i,j}(t) dt, \quad Re(s) > 0,$$

where $Re(s)$ is the real part of the complex number s. Defining the LT of a matrix as the matrix of the LTs of its elements, we write

$$P^*(s) = [p_{i,j}^*(s)]_{i,j \in S}.$$

The next theorem gives the main result.

Theorem 6.8 *The LT of the transition probability matrix $P(t)$ is given by*

$$P^*(s) = [sI - Q]^{-1}, \quad Re(s) > 0.$$

Proof: Using the properties of the Laplace transforms, we get

$$\int_0^\infty e^{-st} p'_{i,j}(t)dt = s p^*_{i,j}(s) - p_{i,j}(0).$$

In matrix form, this yields that the LT of the derivative matrix $P'(t)$ is given by $sP^*(s) - I$. Now taking the LT on both sides of Equation 6.17 and 6.18, and using the initial condition $P(0) = I$, we get

$$sP^*(s) - I = QP^*(s) = P^*(s)Q. \tag{6.32}$$

This can be rearranged to get

$$(sI - Q)P^*(s) = P^*(s)(sI - Q) = I.$$

Since $sI - Q$ is invertible for $Re(s) > 0$, the theorem follows. ∎

Example 6.20 Two-State Machine. Consider the two-state CTMC with generator matrix given in Equation 6.12. From Theorem 6.8 we have

$$P^*(s) = \frac{1}{s(s + \lambda + \mu)} \begin{bmatrix} s + \mu & \lambda \\ \mu & s + \lambda \end{bmatrix}.$$

Using partial fraction expansion, we get, for example

$$p^*_{0,0}(s) = \frac{s + \mu}{s(s + \lambda + \mu)} = \frac{\mu}{\lambda + \mu} \cdot \frac{1}{s} + \frac{\lambda}{\lambda + \mu} \cdot \frac{1}{s + \lambda + \mu}.$$

Using the table of LTs (see Appendix F), we can invert the above transform to get

$$p_{0,0}(t) = \frac{\mu}{\lambda + \mu} + \frac{\lambda}{\lambda + \mu} e^{-(\lambda + \mu)t}.$$

We can compute other transition probabilities in a similar fashion to obtain the transition probability matrix given in Equation 6.21. ∎

Example 6.21 Two-Machine One-Repairperson Workshop. Consider the CTMC of Example 6.19. Using Theorem 6.8 we get

$$P^*(s) = \begin{bmatrix} s + 1 & -1 & 0 \\ -.1 & s + 1.1 & -1 \\ 0 & -.2 & s + .2 \end{bmatrix}^{-1}$$

$$= \frac{1}{D(s)} \begin{bmatrix} 50s^2 + 65s + 1 & 10(5s + 1) & 50 \\ 5s + 1 & 10(s + 1)(5s + 1) & 50(s + 1) \\ 1 & 10(s + 1) & 5*(10s^2 + 21s + 10) \end{bmatrix},$$

where

$$D(s) = s(50s^2 + 115s + 61) = s(s + 0.8298)(s + 1.4702).$$

Thus, using partial fractions, we get

$$p^*_{0,1}(s) = \frac{10(5s + 1)}{s(50s^2 + 115s + 61)} = \frac{0.1639}{s} + \frac{1.1853}{s + 0.8298} - \frac{1.3493}{s + 1.4702}.$$

This can be inverted easily to get

$$p_{0,1}(t) = 0.1639 + 1.1853e^{-0.8298t} - 1.3493e^{-1.4702t},$$

which matches the result in Example 6.19. ∎

6.6.3 Uniformization

We begin with the following theorem.

Theorem 6.9 *Let $\{X_n, n \geq 0\}$ be a DTMC on state-space $S = \{1, 2, \cdots, N\}$ with one-step transition probability matrix $P = [p_{i,j}]$. Let $\{N(t), t \geq 0\}$ be a $PP(\lambda)$ that is independent of $\{X_n, n \geq 0\}$. Define $X(t) = X_{N(t)}$. Then $\{X(t), t \geq 0\}$ is a CTMC on state-space S with generator matrix*

$$Q = \lambda(P - I). \tag{6.33}$$

Proof: Let $N_i = \min\{n \geq 0 : X_n \neq i\}$. Then

$$P(N_i = k | X_0 = i) = p_{i,i}^{k-1}(1 - p_{i,i}), \quad k \geq 1.$$

That is, given $X(0) = X_0 = i$, N_i is a geometric random variable with parameter $1 - p_{i,i}$. Now let $T_i = \min\{t \geq 0 : X(t) \neq i\}$. Thus T_i is a sum of a geometric number of iid $\exp(\lambda)$ random variables. Hence from Theorem 5.7 on page 164 it follows that $T_i \sim \exp(\lambda(1 - p_{i,i}))$. Also,

$$P(X_{N_i} = j | X_0 = i) = \begin{cases} \frac{p_{i,j}}{1 - p_{i,i}} & j \neq i, \\ 0 & j = i. \end{cases}$$

Now let

$$q_i = -q_{i,i} = \lambda(1 - p_{i,i}), \quad q_{i,j} = \lambda p_{i,j}, \quad i \neq j. \tag{6.34}$$

Due to the Markov properties of $\{X_n, n \geq 0\}$, and $\{N(t), t \geq 0\}$, we see that $\{X(t), t \geq 0\}$ stays in state i for an $\exp(q_i)$ amount of time and then jumps to state $j \neq i$ with probability $q_{i,j}/q_i$, independent of the length of the sojourn in state i. Hence $\{X(t), t \geq 0\}$ is a CTMC with generator matrix $Q = [q_{i,j}]$. Equation 6.34 implies that Q is given by Equation 6.33. ∎

We use the above theorem in reverse to compute $P(t)$ as shown in the following theorem.

Theorem 6.10   Uniformization. Let $\{X(t), t \geq 0\}$ be a CTMC on finite state-space $S = \{1, 2, \cdots, N\}$ with rate matrix Q. Let

$$\lambda = \max_{i \in S}\{-q_{i,i}\},$$

and define

$$P = I + \frac{Q}{\lambda}. \tag{6.35}$$

Then, $P(t)$, the transition probability matrix of $\{X(t), t \geq 0\}$ is given by

$$P(t) = \sum_{n=0}^{\infty} e^{-\lambda t} \frac{(\lambda t)^n}{n!} P^n, \quad t \geq 0. \tag{6.36}$$

Proof: First, Equation 6.35 implies that P is a stochastic matrix. Now let $\{X_n, n \geq 0\}$ be a DTMC with one-step transition matrix P and let $\{N(t), t \geq 0\}$ be a PP(λ) that is independent of the DTMC. From Theorem 6.9 it follows that $\{X_{N(t)}, t \geq 0\}$ is a CTMC with generator matrix Q. Hence, for $i, j \in S$ and $t \geq 0$, we get

$$
\begin{aligned}
p_{i,j}(t) &= \mathsf{P}(X(t) = j | X(0) = i) \\
&= \mathsf{P}(X_{N(t)} = j | X_0 = i) \\
&= \sum_{n=0}^{\infty} \mathsf{P}(X_{N(t)} = j | N(t) = n, X_0 = i)\mathsf{P}(N(t) = n | X_0 = i) \\
&= \sum_{n=0}^{\infty} \mathsf{P}(X_n = j | N(t) = n, X_0 = i)e^{-\lambda t}\frac{(\lambda t)^n}{n!} \\
&= \sum_{n=0}^{\infty} e^{-\lambda t}\frac{(\lambda t)^n}{n!}[P^n]_{i,j}.
\end{aligned}
$$

Putting the above equation in matrix form yields Equation 6.36. ∎

In practice one would compute $P(t)$ by truncating the series in Equation 6.36 after a finite number of terms. We discuss below how to decide when to truncate the series. For a given $\tau > 0$, and an $\epsilon > 0$, it is possible to find an $M < \infty$ such that

$$\sum_{n=M+1}^{\infty} e^{-\lambda \tau} \frac{(\lambda \tau)^n}{n!} < \epsilon.$$

Choose the smallest such M, and define

$$P_M(t) = \sum_{n=0}^{M} e^{-\lambda t}\frac{(\lambda t)^n}{n!}P^n.$$

We see that, for all $0 \leq t \leq \tau$,

$$
\begin{aligned}
\|P(t) - P_M(t)\| &= \max_{i,j \in S} |[P(t)]_{i,j} - [P_M(t)]_{i,j}| \\
&= \max_{i,j \in S} \sum_{n=M+1}^{\infty} e^{-\lambda t}\frac{(\lambda t)^n}{n!}[P^n]_{i,j} \\
&\leq \sum_{n=M+1}^{\infty} e^{-\lambda t}\frac{(\lambda t)^n}{n!} < \epsilon.
\end{aligned}
$$

Thus $P_M(t)$ is an ϵ lower bound on $P(t)$ for all $0 \leq t \leq \tau$. In our experience uniformization has proved to be an extremely stable and efficient numerical procedure for computing $P(t)$. A subtle numerical difficulty arises when λt is so large that

$e^{-\lambda t}$ is numerically computed to be zero. In this case we need more ingenious ways of computing the series. We shall not go into the details here.

Example 6.22 Two-State Machine. Consider the two-state CTMC with generator matrix given in Equation 6.12. To avoid confusing notation, we use q instead of λ in Equation 6.35 to get

$$q = \max(-q_{0,0}, -q_{1,1}) = \max(\lambda, \mu).$$

Note that any q larger than the right-hand side would do. Thus instead of choosing $q = \max(\lambda, \mu)$, we use $q = \lambda + \mu$. Then we get

$$P = I + \frac{Q}{q} = \frac{1}{\lambda + \mu} \begin{bmatrix} \mu & \lambda \\ \mu & \lambda \end{bmatrix}.$$

Clearly, we have $P^0 = I$ and $P^n = P$ for $n \geq 1$. Thus we have

$$P(t) = e^{-(\lambda + \mu)t} \left(I + P \sum_{n=1}^{\infty} \frac{((\lambda + \mu)t)^n}{n!} \right),$$

which reduces to Equation 6.21. ∎

Example 6.23 Two-Machine One-Repairperson Workshop. Consider the CTMC of Example 6.19. From Equation 6.35 we get

$$\lambda = \max(1, 1.1, .2) = 1.1.$$

Hence we get

$$P = I + \frac{Q}{1.1} = \frac{1}{1.1} \begin{bmatrix} .1 & 1 & 0 \\ .1 & 0 & 1 \\ 0 & .2 & .9 \end{bmatrix}.$$

Using Theorem 2.7 on page 40 we get

$$P^n = (-.3365)^n \begin{bmatrix} 0.2870 & -1.3493 & 1.0623 \\ -0.1349 & 0.6344 & -0.4994 \\ 0.0212 & -0.0999 & 0.0786 \end{bmatrix}$$

$$+ (.2456)^n \begin{bmatrix} 0.6966 & 1.1853 & -1.8820 \\ 0.1185 & 0.2017 & -0.3202 \\ -0.0376 & -0.0640 & 0.1017 \end{bmatrix}$$

$$+ (1)^n \begin{bmatrix} 0.0164 & 0.1639 & 0.8197 \\ 0.0164 & 0.1639 & 0.8197 \\ 0.0164 & 0.1639 & 0.8197 \end{bmatrix}.$$

Substituting this in Equation 6.36 and simplifying, we get Equation 6.29. ∎

It should be noted that the method of uniformization works even if the CTMC has infinite state-space as long as λ defined in Theorem 6.10 is finite. Such CTMCs are called uniformizable. It is easy to see that uniformizable CTMCs are automatically regular, since they can have at most $P(\lambda t)$ transitions over $(0, t]$.

Finally, one can always solve the backward or forward equations by standard numerical methods of solving differential equations. We refer the readers to any book on differential equations for more details.

6.7 Computation of $P(t)$: Infinite State-Space

In this section we discuss the computation of the transition matrix $P(t)$ for a CTMC $\{X(t), t \geq 0\}$ on infinite state-space $S = \{0, 1, 2, \cdots\}$. This can be done only when the CTMC has highly specialized structure. In our experience the transform methods are most useful in such cases. We illustrate with several examples.

Example 6.24 Poisson Process. Let $\{X(t), t \geq 0\}$ be a PP(λ). We have $X(0) = 0$. To simplify notation we write

$$p_i(t) = p_{0,i}(t) = P(X(t) = i | X(0) = 0), \quad i \geq 0, \ t \geq 0.$$

The forward equations are

$$
\begin{aligned}
p_0'(t) &= -\lambda p_0(t), \\
p_i'(t) &= -\lambda p_i(t) + \lambda p_{i-1}(t), \quad i \geq 1,
\end{aligned}
$$

with initial conditions

$$p_0(0) = 1, \quad p_i(0) = 0, \quad i \geq 1.$$

Taking Laplace transforms on both sides of the above differential equations, we get

$$
\begin{aligned}
sp_0^*(s) - 1 &= -\lambda p_0^*(s), \\
sp_i^*(s) &= -\lambda p_i^*(s) + \lambda p_{i-1}^*(s), \quad i \geq 1.
\end{aligned}
$$

These can be solved recursively to obtain

$$p_i^*(s) = \frac{\lambda^i}{(s+\lambda)^{i+1}}, \quad i \geq 0.$$

Inverting this using the table in Appendix F, we get

$$p_i(t) = e^{-\lambda t} \frac{(\lambda t)^i}{i!}, \quad i \geq 0,$$

which is as expected. ∎

Example 6.25 Pure Birth Process. Let $\{X(t), t \geq 0\}$ be a pure birth process as described in Example 6.8 with birth parameters $\lambda_i, i \geq 0$, and assume that $X(0) = 0$. As before we write

$$p_i(t) = p_{0,i}(t) = P(X(t) = i | X(0) = 0), \quad i \geq 0, \ t \geq 0.$$

The forward equations are

$$
\begin{aligned}
p_0'(t) &= -\lambda_0 p_0(t), \\
p_i'(t) &= -\lambda_i p_i(t) + \lambda_{i-1} p_{i-1}(t), \quad i \geq 1
\end{aligned}
$$

with initial conditions

$$p_0(0) = 1, \quad p_i(0) = 0, \quad i \geq 1.$$

Taking Laplace transforms of the above differential equations, we get

$$
\begin{aligned}
sp_0^*(s) - 1 &= -\lambda_0 p_0^*(s), \\
sp_i^*(s) &= -\lambda_i p_i^*(s) + \lambda_{i-1} p_{i-1}^*(s), \quad i \geq 1.
\end{aligned}
$$

These can be solved recursively to obtain

$$p_i^*(s) = \frac{1}{\lambda_i} \prod_{k=0}^{i} \frac{\lambda_k}{s + \lambda_k}, \quad i \geq 0.$$

Consider the case when all the λ_i's are distinct. Then we can use partial fractions expansion easily to get

$$p_i(t) = \sum_{k=0}^{i} A_{ki} \lambda_k e^{-\lambda_k t}, \quad i \geq 0,$$

where

$$A_{ki} = \frac{1}{\lambda_i} \prod_{r=0, r \neq k}^{i} \frac{\lambda_r}{\lambda_r - \lambda_k}, \quad 0 \leq k \leq i.$$

As a special case, suppose $\lambda_i = i\lambda$, $(i \geq 0)$, but with $X(0) = 1$. It can be shown that (see Computational Exercise 6.4) in this case we get

$$\mathsf{P}(X(t) = i | X(0) = 1) = e^{-\lambda t}(1 - e^{-\lambda t})^{i-1}.$$

Thus given $X(0) = 1$, $X(t)$ is geometrically distributed with parameter $e^{-\lambda t}$. ∎

Example 6.26 Pure Death Process. Let $\{X(t), t \geq 0\}$ be a pure death process as described in Example 6.9 with death parameters μ_i, $i \geq 1$, and assume that $X(0) = N$. We want to compute

$$p_i(t) = p_{N,i}(t) = \mathsf{P}(X(t) = i | X(0) = N), \quad 0 \leq i \leq N, \ t \geq 0.$$

Using the fact that $p_{N+1}(t) = p_{N,N+1}(t) = 0$, we can write the forward equations as

$$
\begin{aligned}
p_N'(t) &= -\mu_N p_N(t), \\
p_i'(t) &= -\mu_i p_i(t) + \mu_{i+1} p_{i+1}(t), \quad 1 \leq i \leq N - 1, \\
p_0'(t) &= \mu_1 p_1(t),
\end{aligned}
$$

with initial conditions

$$p_N(0) = 1, \quad p_i(0) = 0, \quad 0 \leq i \leq N - 1.$$

Taking Laplace transforms on both sides of the above differential equations, we get

$$
\begin{aligned}
sp_N^*(s) - 1 &= -\mu_N p_N^*(s), \\
sp_i^*(s) &= -\mu_i p_i^*(s) + \mu_{i+1} p_{i+1}^*(s), \quad 1 \leq i < N, \\
sp_0^*(s) &= \mu_1 p_1^*(s).
\end{aligned}
$$

These can be solved recursively to obtain

$$p_i^*(s) = \frac{1}{\mu_i} \prod_{k=i}^{N} \frac{\mu_k}{s + \mu_k}, \quad 1 \le i \le N,$$

and

$$p_0^*(s) = \mu_1 p_1^*(s).$$

When all the λ_i's are distinct we can use partial fractions expansion easily to get

$$p_i(t) = \sum_{k=i}^{N} B_{ki} \mu_k e^{-\mu_k t}, \quad 1 \le i \le N$$

where

$$B_{ki} = \frac{1}{\mu_i} \prod_{r=i, r \ne k}^{N} \frac{\mu_r}{\mu_r - \mu_k},$$

and

$$p_0(t) = 1 - \sum_{i=1}^{N} p_i(t).$$

As a special case, suppose $\mu_i = i\mu$, $(i \ge 0)$, with $X(0) = N$. It can be shown that (see Computational Exercise 6.5) given $X(0) = N$, $X(t)$ is binomially distributed with parameters N and $e^{-\mu t}$. ∎

Example 6.27 Linear Growth Model. Let $X(t)$ be the birth and death process of Example 6.13 with birth parameters $\lambda_i = i\lambda$ and death parameters $\mu_i = i\mu$, $i \ge 0$. Suppose $X(0) = 1$. We shall compute

$$p_i(t) = p_{1,i}(t) = P(X(t) = i | X(0) = 1), \quad i \ge 0, \ t \ge 0.$$

The forward equations are

$$
\begin{aligned}
p_0'(t) &= \mu p_1(t), \\
p_i'(t) &= (i-1)\lambda p_{i-1}(t) - i(\lambda + \mu)p_i(t) + (i+1)\mu p_{i+1}(t), \quad i \ge 1,
\end{aligned}
$$

with initial conditions

$$p_1(0) = 1, \quad p_i(0) = 0, \quad i \ne 1.$$

Now define the generating function

$$p(z, t) = \sum_{i=0}^{\infty} z^i p_i(t).$$

Multiplying the differential equation for $p_i(t)$ by z^i and summing over all i from 0 to ∞ we get

$$\sum_{i=0}^{\infty} z^i p_i'(t) = \sum_{i=1}^{\infty} (i-1)\lambda z^i p_{i-1}(t) - \sum_{i=1}^{\infty} i(\lambda + \mu)z^i p_i(t) + \sum_{i=0}^{\infty} (i+1)\mu z^i p_{i+1}(t),$$

which reduces to

$$\frac{\partial}{\partial t}p(z,t) = (\lambda z^2 - (\lambda + \mu)z + \mu)\frac{\partial}{\partial z}p(z,t).$$

This equation can be written in the canonical form as

$$\frac{\partial}{\partial t}p(z,t) - a(z)\frac{\partial}{\partial z}p(z,t) = 0,$$

where

$$a(z) = \lambda z^2 - (\lambda + \mu)z + \mu = (z-1)(\lambda z - \mu).$$

Thus we have a linear first-order partial differential equation, which can be solved by the method of characteristic functions (see Chaudhry and Templeton (1983)). We first form the total differential equations

$$\frac{dt}{1} = \frac{dz}{-a(z)} = \frac{dp}{0}.$$

The first equation can be integrated to yield

$$\int dt = \int \frac{dz}{-a(z)}$$

or

$$t - \frac{1}{\lambda - \mu}\ln\left(\frac{\lambda z - \mu}{z - 1}\right) = c,$$

where c is an arbitrary constant. The last equation gives

$$p = \text{ constant.}$$

Hence the general solution is

$$
\begin{aligned}
p(z,t) &= \hat{f}\left(t - \frac{1}{\lambda - \mu}\ln\left(\frac{\lambda z - \mu}{z - 1}\right)\right) \\
&= f\left((\lambda - \mu)t - \ln\left(\frac{\lambda z - \mu}{z - 1}\right)\right),
\end{aligned}
\tag{6.37}
$$

where f is a function to be determined by the boundary condition

$$p(z,0) = \sum_{i=0}^{\infty} z^i p_i(0) = z.$$

Hence

$$f\left(\ln\left(\frac{z-1}{\lambda z - \mu}\right)\right) = z.
\tag{6.38}$$

Now write

$$u = \ln\left(\frac{z-1}{\lambda z - \mu}\right)$$

and invert it to get

$$z = \frac{\mu e^u - 1}{\lambda e^u - 1}.$$

Substituting in Equation 6.38 we get

$$f(u) = \frac{\mu e^u - 1}{\lambda e^u - 1}.$$

Substituting in Equation 6.37 we get

$$p(z, t) = \frac{\mu \exp\left((\lambda - \mu)t - \ln\left(\frac{\lambda z - \mu}{z - 1}\right)\right) - 1}{\lambda \exp\left((\lambda - \mu)t - \ln\left(\frac{\lambda z - \mu}{z - 1}\right)\right) - 1}$$

which can be simplified to obtain

$$p(z, t) = \frac{\mu(1 - e^{(\lambda-\mu)t}) - (\lambda - \mu e^{(\lambda-\mu)t})z}{(\mu - \lambda e^{(\lambda-\mu)t}) - \lambda(1 - e^{(\lambda-\mu)t})z}.$$

The above expression can be expanded in a power series in z, and then the coefficient of z^i will give us $p_i(t)$. Doing this, and using $\rho = \lambda/\mu$, we get

$$p_0(t) = \frac{1 - e^{(\lambda-\mu)t}}{1 - \rho e^{(\lambda-\mu)t}},$$

$$p_i(t) = \rho^{i-1}(1 - \rho)e^{(\lambda-\mu)t}\frac{(1 - e^{(\lambda-\mu)t})^{i-1}}{(1 - \rho e^{(\lambda-\mu)t})^{i+1}}, \quad i \geq 1.$$

This completes the solution for the case $X(0) = 1$. The case $X(0) = i \geq 2$ can be analyzed by treating the linear growth process as the sum of i independent linear growth processes, each starting in state 1. This example shows that even for highly structured CTMCs computation of $P(t)$ is a formidable task.

Although computing $P(t)$ is hard, it is relatively easy to compute

$$m_i(t) = \mathsf{E}(X(t)|X(0) = i) = \sum_{k=0}^{\infty} k p_{i,k}(t), \quad i \geq 0, \ t \geq 0.$$

We have

$$m_i'(t) = \sum_{k=0}^{\infty} k p_{i,k}'(t).$$

By using the forward differential equations for $p_{i,k}(t)$ and carrying out the above sum we get

$$m_i'(t) = (\lambda - \mu)m_i(t).$$

Using the initial condition, $m_i(0) = i$, we can solve the above equation to get

$$m_i(t) = i e^{(\lambda-\mu)t}.$$

Thus if λ, the birth rate per individual, is greater than μ, the death rate per individual, then the mean population size explodes as $t \to \infty$. If it is less, the mean population size exponentially reduces to zero. If the two rates are equal, the mean population size is constant. All these conclusions confirm our intuition. ∎

6.8 First-Passage Times

We follow the developments of DTMCs and first study the first passage times in the CTMCs before we study their limiting behavior. Specifically, let $\{X(t), t \geq 0\}$ be a CTMC on $S = \{0, 1, 2, \cdots\}$ with generator matrix $Q = [q_{i,j}]$, and define

$$T = \min\{t \geq 0 : X(t) = 0\}. \tag{6.39}$$

We study the random variable T in this section. Since the first passage times in a CTMC have a lot of similarity to the first passage times in the embedded DTMC, many of the results follow from similar results in Chapter 3. Hence we do not spend as much time on this topic here.

6.8.1 Cumulative Distribution of T

Define

$$r_{i,j}(t) = P(T > t, X(t) = j | X(0) = i), \quad i, j \geq 1, \ t \geq 0.$$

Then the complementary cdf of T, conditioned on $X(0) = i$, is given by

$$r_i(t) = P(T > t | X(0) = i) = \sum_{j=1}^{\infty} r_{i,j}(t).$$

The next theorem gives the differential equations satisfied by

$$R(t) = [r_{i,j}(t)]_{i,j \geq 1}.$$

We shall need the following notation

$$B = [q_{i,j}]_{i,j \geq 1}.$$

Thus the matrix B is obtained by deleting the row and the column corresponding to the state 0.

Theorem 6.11 *The matrix $R(t)$ satisfies the following set of differential equations:*

$$R'(t) = BR(t) = R(t)B, \tag{6.40}$$

with the initial condition

$$R(0) = I.$$

Proof: Follows along the same lines as the proof of the backward and forward equations in Theorem 6.4. ∎

Thus we can use the methods described in the previous two sections to compute $R(t)$, and hence the complementary cdf of T.

The next theorem gives a more explicit expression for the cdf of T when the CTMC has a finite state-space.

Theorem 6.12 *Suppose the CTMC has state-space $\{0, 1, \cdots, K\}$ and initial distribution*

$$\alpha = [\alpha_1, \alpha_2, \cdots, \alpha_K],$$

with $\sum_{i=1}^{K} \alpha_i \leq 1$. Then

$$P(T \leq t) = 1 - \alpha e^{Bt} \mathbf{1}, \tag{6.41}$$

where $\mathbf{1}$ is a column vector of ones.

Proof: Since B is finite and $R(0) = I$, we can use Theorem 6.6 to obtain

$$R(t) = e^{Bt}.$$

Now,

$$
\begin{aligned}
P(T \leq t) &= 1 - P(T > t) \\
&= 1 - \sum_{i=1}^{K} \alpha_i P(T > t | X(0) = i) \\
&= 1 - \sum_{i=1}^{K} \sum_{j=1}^{K} \alpha_i P(T > t, X(t) = j | X(0) = i) \\
&= 1 - \sum_{i=1}^{K} \sum_{j=1}^{K} \alpha_i r_{i,j}(t) \\
&= 1 - \alpha R(t) \mathbf{1} \\
&= 1 - \alpha e^{Bt} \mathbf{1}
\end{aligned}
$$

as desired. ∎

Note that $P(T = 0) = 1 - \sum_{i=1}^{K} \alpha_i$ may be positive.

6.8.2 Absorption Probabilities

Define

$$v_i = P(T = \infty | X(0) = i), \quad i \geq 1. \tag{6.42}$$

Thus v_i is the probability that the CTMC never visits state zero starting from state i. The next theorem gives the main result.

Theorem 6.13 *Let $v = [v_1, v_2, \cdots]^{\top}$. Then v is given by the largest solution bounded above by 1 to*

$$Bv = 0. \tag{6.43}$$

Proof: Let $\{X_n, n \geq 0\}$ be the embedded DTMC in the CTMC $\{X(t), t \geq 0\}$. The transition probability matrix P of the embedded DTMC is related to the generator matrix Q of the CTMC by Equation 6.11. We have

$$
\begin{aligned}
v_i &= P(\{X(t), t \geq 0\} \text{ never visits state } 0 | X(0) = i) \\
&= P(\{X_n, n \geq 0\} \text{ never visits state } 0 | X_0 = i).
\end{aligned}
$$

Stating Equation 3.8 on page 62 in scalar form we get

$$v_i = \sum_{j=1}^{\infty} p_{i,j} v_j, \quad i \geq 1,$$

where $p_{i,j}$'s are the transition probabilities of the embedded DTMC. Substituting from Equation 6.11 we get

$$v_i = \sum_{j=1, j \neq i}^{\infty} \frac{q_{i,j}}{q_i} v_j.$$

Multiplying on both sides by q_i and using $q_i = -q_{i,i}$ we get

$$\sum_{j=1}^{\infty} q_{i,j} v_j = 0,$$

which, in matrix form, yields Equation 6.43. The maximality of the solution follows from Theorem 3.2. ∎

6.8.3 Moments and LST of T

In this subsection we develop methods of computing the moments of T. The next theorem gives a method of computing

$$m_i = \mathsf{E}(T|X(0) = i), \quad i \geq 1. \tag{6.44}$$

Obviously, $m_i = \infty$ if $v_i > 0$. Hence we consider the case $v_i = 0$ in the next theorem.

Theorem 6.14 *Suppose $v = 0$. Then $m = [m_1, m_2, \cdots]^\top$ is given by the smallest non-negative solution to*

$$Bm + 1 = 0. \tag{6.45}$$

Proof: We use the first-step analysis for CTMCs, which is analogous to the first-step analysis in the DTMCs as explained in Section 3.1. Let $X(0) = i > 0$ and Y_1 be the first sojourn time in state i. Then $Y_1 \sim \exp(q_i)$. Now condition on $X_1 = X(Y_1) = j$. Clearly, if $j = 0$, $T = Y_1$. If $j > 0$, then $T = Y_1 + T'$, where T' has the same distribution as T conditioned on $X_0 = j$. Hence we get

$$
\begin{aligned}
m_i &= \mathsf{E}(T|X(0) = i) \\
&= \sum_{j=0}^{\infty} \mathsf{E}(T|X(0) = i, X(Y_1) = j) \mathsf{P}(X(Y_1) = j|X(0) = i) \\
&= \mathsf{E}(Y_1|X(0) = i, X(Y_1) = 0) \mathsf{P}(X(Y_1) = 0|X(0) = i) \\
&\quad + \sum_{j=1}^{\infty} \mathsf{E}(Y_1|X(0) = i, X(Y_1) = j) \mathsf{P}(X(Y_1) = j|X(0) = i)
\end{aligned}
$$

$$+ \sum_{j=1}^{\infty} \mathsf{E}(T'|X(0) = i, X(Y_1) = j)\mathsf{P}(X(Y_1) = j|X(0) = i)$$

$$= \mathsf{E}(Y_1|X(0) = i) + \sum_{j=1}^{\infty} \mathsf{E}(T|X(0) = j)\mathsf{P}(X(Y_1) = j|X(0) = i)$$

$$= \frac{1}{q_i} + \sum_{j=1, j \neq i}^{\infty} \frac{q_{i,j}}{q_i} m_j.$$

Multiplying on both sides by q_i and using $q_i = -q_{i,i}$ we get

$$\sum_{j=1}^{\infty} q_{i,j} m_j + 1 = 0,$$

which, in matrix form, yields Equation 6.45. The minimality of the solution follows from Theorem 3.3. ∎

We next derive the Laplace Stieltjes transform (LST) of T. For $s \geq 0$, define

$$\phi_i(s) = \mathsf{E}(e^{-sT}|X(0) = i), \quad i \geq 1,$$

and

$$\phi(s) = [\phi_1(s), \phi_2(s), \cdots]^\top.$$

The main result is given in the next theorem.

Theorem 6.15 $\{\phi(s), s \geq 0\}$ *is given by the smallest non-negative solution to*

$$s\phi(s) = w + B\phi(s), \tag{6.46}$$

where $w = [q_{1,0}, q_{2,0}, \cdots]^\top$.

Proof: Use the terminology and the first-step analysis as described in the proof of Theorem 6.14. Conditioning on (X_1, Y_1) we get

$$\phi_i(s) = \mathsf{E}(e^{-sT}|X(0) = i)$$

$$= \sum_{j=0}^{\infty} p_{i,j} \mathsf{E}(e^{-sY_1}|X(0) = i)\mathsf{E}(e^{-sT}|X(0) = j)$$

$$= \frac{q_i}{s + q_i} \left(\sum_{j=1}^{\infty} p_{i,j} \phi_j(s) + p_{i,0} \right).$$

Multiplying by $(s + q_i)$ and using $q_i = -q_{i,i}$ and $q_{i,j} = q_i p_{i,j}$ $(i \neq j)$, we get Equation 6.46. The minimality follows along the same lines as in the proof of Theorem 3.2. ∎

We can use the above theorem to immediately derive the moments of T denoted by

$$m_i(k) = \mathsf{E}(T^k|X(0) = i), \quad i \geq 1, \ k \geq 1.$$

Theorem 6.16 Suppose $v = 0$. Then $m(k) = [m_1(k), m_2(k), \cdots]^\top$ is given by the smallest non-negative solution to

$$Bm(k) + km(k - 1) = 0, \quad k \geq 1, \tag{6.47}$$

with $m(0) = 1$.

Proof: Follows from taking successive derivatives of Equation 6.46 and using

$$m_i(k) = (-1)^k \frac{d^k}{ds^k} \phi_i(s) \Big|_{s=0} . \quad \blacksquare$$

We end this section with an example.

Example 6.28 Birth and Death Processes. Let $\{X(t), t \geq 0\}$ be the birth and death process of Example 6.10 with birth parameters $\lambda_i, i \geq 0$, and death parameters $\mu_i, i \geq 1$. Let T be as defined in Equation 6.39.

Let us first compute the quantities $\{v_i\}$ as defined in Equation 6.42. The Equation 6.43 can be written in scalar form as

$$\mu_i v_{i-1} - (\lambda_i + \mu_i)v_i + \lambda_i v_{i+1} = 0, \quad i \geq 1,$$

with boundary condition $v_0 = 0$. The above equations can be written as

$$v_i = \frac{\mu_i}{\lambda_i + \mu_i} v_{i-1} + \frac{\lambda_i}{\lambda_i + \mu_i} v_{i+1}, \quad i \geq 1.$$

But these equations are identical to Equations 3.14 on page 66 with

$$p_i = \frac{\lambda_i}{\lambda_i + \mu_i}, \quad q_i = \frac{\mu_i}{\lambda_i + \mu_i}. \tag{6.48}$$

Using the results of Example 3.10 on page 66 we get

$$v_i = \begin{cases} \dfrac{\sum_{j=0}^{i-1} \alpha_j}{\sum_{j=0}^{\infty} \alpha_j} & \text{if } \sum_{j=0}^{\infty} \alpha_j < \infty, \\[4mm] 0 & \text{if } \sum_{j=0}^{\infty} \alpha_j = \infty, \end{cases} \tag{6.49}$$

where $\alpha_0 = 1$ and

$$\alpha_i = \frac{\mu_1 \mu_2 \cdots \mu_i}{\lambda_1 \lambda_2 \cdots \lambda_i}, \quad i \geq 1. \tag{6.50}$$

Next we compute m_i, as defined in Equation 6.44, under the assumption that $\sum_{j=0}^{\infty} \alpha_j = \infty$, so that $v_i = 0$ for all $i \geq 0$. The Equation 6.45 can be written in scalar form as

$$\mu_i m_{i-1} - (\lambda_i + \mu_i)m_i + \lambda_i m_{i+1} + 1 = 0, \quad i \geq 1,$$

with boundary condition $m_0 = 0$. The above equation can be written as

$$m_i = \frac{1}{\lambda_i + \mu_i} + \frac{\mu_i}{\lambda_i + \mu_i} m_{i-1} + \frac{\lambda_i}{\lambda_i + \mu_i} m_{i+1}, \quad i \geq 1.$$

These equations are similar to Equations 3.24 on page 73 with p_i and q_i as defined in Equation 6.48. The only difference is that we have $\frac{1}{\lambda_i + \mu_i}$ on the right-hand side instead of 1. However, we can solve these equations using the same procedure as in Example 3.16 to get

$$m_i = \sum_{k=0}^{i-1} \alpha_k \sum_{j=k+1}^{\infty} \frac{1}{\lambda_j \alpha_j}. \quad \blacksquare$$

6.9 Exploring the Limiting Behavior by Examples

Let $\{X(t), t \geq 0\}$ be a CTMC on $S = \{0, 1, 2, \cdots\}$ with generator matrix Q. In Sections 6.4 and 6.5 we studied two main aspects of the transient behavior of the CTMCs: the transition probability matrix $P(t)$ and the occupancy matrix $M(t)$. Theorem 6.6 showed that

$$P(t) = e^{Qt}, \quad t \geq 0,$$

and Theorem 6.5 showed that

$$M(t) = \int_0^t P(u) du, \quad t \geq 0.$$

In the next several sections we study the limiting behavior of $P(t)$ as $t \to \infty$. Since the row sums of $P(t)$ are 1, it follows that the row sums of $M(t)$ are t. Hence we study the limiting behavior of $M(t)/t$ as $t \to \infty$. Note that $[M(t)]_{i,j}/t$ can be interpreted as the fraction of the time spent by the CTMC in state j during $(0, t]$ starting from state i. Hence studying this limit makes practical sense. We begin by some examples illustrating the types of limiting behavior that can arise.

Example 6.29   Two-State Example. Consider the CTMC of Example 6.16 with the $P(t)$ matrix as given in Equation 6.21. Hence we get

$$\lim_{t \to \infty} P(t) = \begin{bmatrix} \frac{\mu}{\lambda+\mu} & \frac{\lambda}{\lambda+\mu} \\ \frac{\mu}{\lambda+\mu} & \frac{\lambda}{\lambda+\mu} \end{bmatrix}.$$

Thus the limit of $P(t)$ exists and its row sums are one. As we had observed in the DTMCs, the rows of the limiting matrix are the same, implying that the limiting distribution of $X(t)$ does not depend upon the initial distribution of the CTMC.

Next, the occupancy matrix $M(t)$ for this CTMC is given in Equation 6.24. Hence, we get

$$\lim_{t \to \infty} \frac{M(t)}{t} = \begin{bmatrix} \frac{\mu}{\lambda+\mu} & \frac{\lambda}{\lambda+\mu} \\ \frac{\mu}{\lambda+\mu} & \frac{\lambda}{\lambda+\mu} \end{bmatrix}.$$

Thus the limit of $M(t)/t$ in this example is the same as that of $P(t)$. $\blacksquare$

Example 6.30   Three-state CTMC. Let $\{X(t), t \geq 0\}$ be a CTMC on state-space $\{0, 1, 2\}$ with generator matrix

$$Q = \begin{bmatrix} 0 & 0 & 0 \\ \mu & -(\lambda + \mu) & \lambda \\ 0 & 0 & 0 \end{bmatrix}.$$

Direct calculations show that

$$P(t) = \begin{bmatrix} 1 & 0 & 0 \\ \frac{\mu}{\lambda + \mu}(1 - e^{-(\lambda + \mu)t}) & e^{-(\lambda + \mu)t} & \frac{\lambda}{\lambda + \mu}(1 - e^{-(\lambda + \mu)t}) \\ 0 & 0 & 1 \end{bmatrix},$$

and

$$M(t) = \begin{bmatrix} t & 0 & 0 \\ \frac{\mu}{\lambda + \mu}t - \frac{p_{1,0}(t)}{\lambda + \mu} & \frac{1}{\lambda + \mu}(1 - e^{-(\lambda + \mu)t}) & \frac{\lambda}{\lambda + \mu}t - \frac{p_{1,2}(t)}{\lambda + \mu} \\ 0 & 0 & t \end{bmatrix}.$$

Hence we get

$$\lim_{t \to \infty} P(t) = \lim_{t \to \infty} \frac{M(t)}{t} = \begin{bmatrix} 1 & 0 & 0 \\ \frac{\mu}{\lambda + \mu} & 0 & \frac{\lambda}{\lambda + \mu} \\ 0 & 0 & 1 \end{bmatrix}.$$

Thus the limiting matrix has distinct rows, implying that the limiting behavior depends on the initial state. Furthermore, the row sums are one. ∎

Example 6.31   Linear Growth Model. Consider the CTMC of Example 6.27. Consider the case $\lambda > \mu$. In this case we can use the transition probabilities derived there to obtain

$$\lim_{t \to \infty} p_{i,j}(t) = \begin{cases} \left(\frac{\mu}{\lambda}\right)^i & \text{if } j = 0, \\ 0 & \text{if } j > 0. \end{cases}$$

Thus in this case the limits of $p_{i,j}(t)$ exist but depend on the initial state i. It is more tedious to show that $M_{i,j}(t)/t$ has the same limit as $p_{i,j}(t)$. Furthermore, the row sums of the limiting matrix are less than 1. ∎

Thus we have identified three cases:

Case 1: Limit of $P(t)$ exists, has identical rows, and each row sums to one.

Case 2: Limit of $P(t)$ exists, does not have identical rows, each row sums to one.

Case 3: Limit of $P(t)$ exists, but the rows may not sum to one.

We have also observed that limit of $M(t)/t$ always exists, and it equals the limit of $P(t)$. We have not seen any examples where the limit of $P(t)$ displays oscillatory behavior as in the case of periodic DTMCs. In the next section we develop the necessary theory to help us classify the CTMCs so we can understand their limiting behavior better.

6.10 Classification of States

Clearly, there is a close connection between the limiting behavior of a CTMC with state-space S and generator matrix Q and that of the corresponding embedded DTMC on the same state-space S and one-step transition matrix P that is related to Q as in Equation 6.11. Thus it seems reasonable that we will need to follow the same path as we did when we studied the limiting behavior of the DTMCs in Chapter 4.

6.10.1 Irreducibility

This motivates us to begin by introducing the concept of irreducibility for a CTMC. Note that a CTMC can visit state j from state i in a finite amount of time if and only if the corresponding embedded DTMC can visit state j from state i in a finite number of steps. Hence we can make the following definitions.

Definition 6.2   Communicating Class. *A set of states $C \subseteq S$ in a CTMC is said to be a (closed) communicating class if C is a (closed) communicating class of the corresponding embedded DTMC.*

Definition 6.3   Irreducibility. *A CTMC is called irreducible if the corresponding embedded DTMC is irreducible. Otherwise it is called reducible.*

Example 6.32   The CTMC of Example 6.1 is irreducible if $\lambda > 0$ and $\mu > 0$. The CTMC of Example 6.30 is reducible, with $\{0\}$ and $\{2\}$ as two closed communicating classes, and $\{1\}$ as a communicating class that is not closed. The linear growth process of Example 6.13, with $\lambda > 0$ and $\mu > 0$, has one closed communicating class, namely $\{0\}$. The set $\{1, 2, \cdots\}$ is a communicating class that is not closed. The CTMC is reducible. ∎

We see that the issue of periodicity does not arise in the CTMCs, since if a CTMC can go from state i to j at all, it can do so at any time. Thus, for the two-state CTMC of Example 6.1, the embedded DTMC is periodic. However, the $P(t)$ matrix for the two-state CTMC does not show a periodic behavior.

6.10.2 Transience and Recurrence

Following the developments in the DTMCs, we now introduce the concepts of recurrence and transience for the CTMCs. Let Y_1 be the first sojourn time in a CTMC $\{X(t), t \geq 0\}$ and define the first passage time (in a slightly modified form)

$$\tilde{T}_i = \inf\{t \geq Y_1 : X(t) = i\}, \quad i \in S. \tag{6.51}$$

This is a well-defined random variable if $Y_1 < \infty$ with probability 1, i.e., if $q_i > 0$. Note that if $q_i = 0$, i is an absorbing state in the CTMC. Let

$$\tilde{u}_i = \mathsf{P}(\tilde{T}_i < \infty | X(0) = i), \qquad (6.52)$$

and

$$\tilde{m}_i = \mathsf{E}(\tilde{T}_i | X(0) = i). \qquad (6.53)$$

When $\tilde{u}_i < 1$, $\tilde{m}_i = \infty$. However, as in the DTMCs, $\tilde{m}_i$ can be infinite even if $\tilde{u}_i = 1$. With this in mind, we make the following definition, which is analogous to the corresponding definition in the case of the DTMCs.

Definition 6.4   Transience and Recurrence. *A state i with $q_i > 0$ is said to be*

(i) recurrent if $\tilde{u}_i = 1$,

(ii) transient if $\tilde{u}_i < 1$.

If $q_i = 0$ we define i to be recurrent. This is consistent with the fact that an absorbing state in a DTMC is recurrent. The next theorem shows an easy way of establishing the transience and recurrence of a state in a CTMC.

Theorem 6.17   *A state i is recurrent (transient) in a CTMC if and only if it is recurrent (transient) in the corresponding embedded DTMC.*

Proof: Let $\{X(t), t \geq 0\}$ be a CTMC and $\{X_n, n \geq 0\}$ be the corresponding embedded DTMC. The implicit assumption of regularity of the CTMC implies that

$$\{X(t) = i \text{ for some } t \geq Y_1\} \Leftrightarrow \{X_n = i \text{ for some } n \geq 1\}.$$

Now,

$$\tilde{u}_i = 1 \Leftrightarrow \mathsf{P}(X(t) = i \text{ for some } t \geq Y_1 | X(0) = i) = 1.$$

Hence

$$\tilde{u}_i = 1 \Leftrightarrow \mathsf{P}(X_n = i \text{ for some } n \geq 1 | X_0 = i) = 1.$$

Hence it follows that a state i is recurrent in a CTMC if and only if it is recurrent in the corresponding embedded DTMC. The statement about the transient case follows by contraposition. ■

It follows from the above theorem that recurrence and transience of states in the CTMCs are class properties, just as they are in the DTMCs. This enables us to call an irreducible CTMC transient or recurrent if all its states are transient or recurrent, respectively. Since an irreducible CTMC is recurrent (transient) if and only if the corresponding embedded DTMC is recurrent (transient), we can use the criteria developed in Chapter 4 to establish the transience or recurrence of the CTMCs.

Next we define positive and null recurrence.

Definition 6.5   Null and Positive Recurrence. *A recurrent state i with $q_i > 0$ is said to be*

(i) *positive recurrent if* $\tilde{m}_i < \infty$,

(ii) *null recurrent if* $\tilde{m}_i = \infty$.

If $q_i = 0$ we define i to be positive recurrent. This is consistent with the fact that an absorbing state in a DTMC is positive recurrent. Establishing null and positive recurrence is more complicated than establishing recurrence and transience since a state i may be positive recurrent in the CTMC but null recurrent in the DTMC, and vice versa. Clearly, a CTMC with a single state is positive recurrent by definition. For CTMCs with two or more states we have the following theorem.

Theorem 6.18 *Let* $\{X(t), t \geq 0\}$ *be an irreducible CTMC on state-space S with at least two states, and let Q be its generator matrix. Let $\{X_n, n \geq 0\}$ be the corresponding embedded DTMC with transition probability matrix P. Suppose the DTMC is recurrent and π is a positive solution to*

$$\pi = \pi P.$$

The CTMC is positive recurrent if and only if

$$\sum_{i \in S} \frac{\pi_i}{q_i} < \infty. \tag{6.54}$$

Proof: Since the CTMC is irreducible and has at least two states, we see that none of the states can be absorbing. Thus $q_i > 0$ for all $i \in S$. We use the first-step analysis as in the proof of Theorem 6.14 to derive a set of equations for

$$\mu_{i,j} = \mathsf{E}(\tilde{T}_j | X(0) = i), \quad i, j \in S.$$

Let $X(0) = i$ and Y_1 be the first sojourn time in state i. Then $Y_1 \sim \exp(q_i)$. Now condition on $X_1 = X(Y_1) = k$. Clearly, if $k = j$, $\tilde{T}_j = Y_1$. If $k \neq j$, then $\tilde{T}_j = Y_1 + T'$, where T' has the same distribution as $\tilde{T}_j$ conditioned on $X(0) = k$. Hence we get

$$
\begin{aligned}
\mu_{i,j} &= \mathsf{E}(\tilde{T}_j | X(0) = i) \\
&= \sum_{k \in S} \mathsf{E}(\tilde{T}_j | X(0) = i, X(Y_1) = k) \mathsf{P}(X(Y_1) = k | X(0) = i) \\
&= \mathsf{E}(Y_1 | X(0) = i, X(Y_1) = j) \mathsf{P}(X(Y_1) = j | X(0) = i) \\
&\quad + \sum_{k \in S - \{j\}} \mathsf{E}(Y_1 | X(0) = i, X(Y_1) = k) \mathsf{P}(X(Y_1) = k | X(0) = i) \\
&\quad + \sum_{k \in S - \{j\}} \mathsf{E}(T' | X(0) = i, X(Y_1) = k) \mathsf{P}(X(Y_1) = k | X(0) = i) \\
&= \mathsf{E}(Y_1 | X(0) = i) + \sum_{k \in S - \{j\}} p_{i,k} \mathsf{E}(\tilde{T}_j | X(0) = k) \\
&= \frac{1}{q_i} + \sum_{k \in S - \{j\}} p_{i,k} \mu_{k,j}. \tag{6.55}
\end{aligned}
$$

Now multiply both sides by π_i and sum over all $i \in S$. We get

$$
\begin{aligned}
\sum_{i \in S} \pi_i \mu_{i,j} &= \sum_{i \in S} \frac{\pi_i}{q_i} + \sum_{i \in S} \pi_i \sum_{k \in S-\{j\}} p_{i,k} \mu_{k,j} \\
&= \sum_{i \in S} \frac{\pi_i}{q_i} + \sum_{k \in S-\{j\}} \sum_{i \in S} \pi_i p_{i,k} \mu_{k,j} \\
&= \sum_{i \in S} \frac{\pi_i}{q_i} + \sum_{k \in S-\{j\}} \pi_k \mu_{k,j}
\end{aligned}
$$

since

$$
\pi_k = \sum_{i \in S} \pi_i p_{i,k}.
$$

Hence, subtracting $\sum_{k \in S-\{j\}} \pi_k \mu_{k,j}$ from both sides, we get

$$
\pi_j \mu_{j,j} = \sum_{i \in S} \frac{\pi_i}{q_i},
$$

which yields

$$
\mu_{j,j} = \frac{1}{\pi_j} \left(\sum_{i \in S} \frac{\pi_i}{q_i} \right). \tag{6.56}
$$

Since $\pi_j > 0$, we see that $\tilde{m}_j = \mu_{j,j} < \infty$ if and only if Equation 6.54 is satisfied. The theorem then follows from the definition of positive recurrence. ∎

Theorem 6.18 also implies that null and positive recurrence in the CTMCs are class properties, just as in the DTMCs. We shall use this theorem to construct examples of null recurrent CTMCs with positive recurrent embedded DTMCs, and vice versa. However, note that if the CTMC is positive recurrent, but the embedded DTMC is null recurrent, the CTMC cannot be regular, since, this situation would imply that the CTMC makes an infinite number of transitions in a finite amount of time. Thus in our applications a positive recurrent CTMC will have a positive recurrent embedded DTMC.

Example 6.33   Success Runs. Consider a CTMC $\{X(t), t \geq 0\}$ on $S = \{0, 1, 2, \cdots\}$, with the corresponding embedded DTMC $\{X_n, n \geq 0\}$ with transition probabilities

$$
p_{i,i+1} = p_i, \quad p_{i,0} = 1 - p_i, \quad i \geq 0,
$$

with $p_0 = 1$. Thus the embedded DTMC is a success runs Markov chain similar to the one described in Example 4.15 on page 103. We consider two cases.

Case 1. Let $p_0 = 1$, $q_0 = -q_{0,0} = 1$, and $p_i = i/(i+1)$, and $q_i = -q_{i,i} = (i+1)$, for $i \geq 1$. From Example 4.8 we see that this DTMC is null recurrent, and a solution to

$$
\pi = \pi P
$$

is given by

$$
\pi_0 = 1, \pi_i = \frac{1}{i}, \quad i \geq 1.
$$

Now,

$$\sum_{i \in S} \frac{\pi_i}{q_i} = 1 + \sum_{i=1}^{\infty} \frac{1}{i(i+1)} = 2 < \infty.$$

Hence the CTMC is positive recurrent. Clearly such a CTMC cannot be regular.

Case 2. Let $p_0 = 1$, $q_0 = -q_{0,0} = 1$, and $p_i = p \in (0,1)$ and $q_i = -q_{i,i} = p^{i-1}$ for all $i \geq 1$. Then the DTMC is positive recurrent (see Example 4.15) with a positive solution to $\pi = \pi P$ given by

$$\pi_0 = 1, \quad \pi_i = p^{i-1}, \quad i \geq 1.$$

Then

$$\sum_{i \in S} \frac{\pi_i}{q_i} = 1 + \sum_{i=1}^{\infty} \frac{p^{i-1}}{p^{i-1}} = \infty.$$

Hence the CTMC is null recurrent. ∎

6.11 Limiting Behavior of Irreducible CTMCs

In this section we derive the main results regarding the limiting distribution of an irreducible CTMC $\{X(t), t \geq 0\}$ on state-space $S = \{0, 1, 2 \cdots\}$ with generator matrix Q. We treat the three types of irreducible CTMCs: transient, null recurrent, and positive recurrent. We treat the case of reducible CTMCs in the next section.

6.11.1 The Transient Case

We begin with the main result in the following theorem.

Theorem 6.19 *Let $\{X(t), t \geq 0\}$ be an irreducible transient CTMC. Then*

$$\lim_{t \to \infty} p_{i,j}(t) = 0, \quad i, j \in S. \tag{6.57}$$

Proof: Let $\{X_n, n \geq 0\}$ be the embedded DTMC corresponding to the irreducible transient CTMC $\{X(t), t \geq 0\}$. Theorem 6.17 implies that $\{X_n, n \geq 0\}$ is irreducible and transient. From Equation 4.21 we see that the expected number of visits by the DTMC to state j starting from state i over the infinite time horizon is finite. Every time the DTMC visits state j, the CTMC spends an average of $1/q_j$ time there. (Note that $q_j > 0$ since state j is transient.) Hence the total expected time spent by the CTMC over $[0, \infty)$ in state j, starting from state i is finite. Using the notation of occupancy times,

$$\lim_{t \to \infty} M_{i,j}(t) = M_{i,j}(\infty) < \infty.$$

From Theorem 6.5 we get

$$M_{i,j}(\infty) = \int_0^{\infty} p_{i,j}(t)dt < \infty.$$

Since $p_{i,j}(t) \geq 0$, we get Equation 6.57. ∎

As in the case of the transient DTMCs, a transient CTMC will eventually permanently exit any finite set with probability 1.

6.11.2 The Continuous Renewal Theorem

Following the development in Section 4.5, we start with the statement of the continuous renewal theorem, which is the continuous analogue of its discrete version in Theorem 4.15.

Theorem 6.20   Continuous Renewal Theorem. *Let f be a probability density function on $[0, \infty)$ with mean μ. Let $h : [0, \infty) \to (-\infty, \infty)$ be a monotone function with*

$$\int_0^\infty |h(u)|du < \infty. \tag{6.58}$$

Suppose a function $g : [0, \infty) \to (-\infty, \infty)$ satisfies

$$g(t) = h(t) + \int_0^t g(t - u)f(u)du, \quad t \geq 0. \tag{6.59}$$

Then

$$\lim_{t\to\infty} g(t) = \frac{1}{\mu} \int_0^\infty h(u)du. \tag{6.60}$$

If $\mu = \infty$, the limit on the right in the above equation is to be interpreted as 0.

Proof: As in the proof of Theorem 4.15, the hard part is to prove that $g(t)$ has a limit as $t \to \infty$. We refer the reader to Karlin and Taylor (1975) or Kohlas (1982) for the details. Here we assume that the limit exists and show that it is given as stated in Equation 6.60.

Define the Laplace transform (LT) as follows (see Appendix F). Here s is a complex number with $Re(s) \geq 0$.

$$f^*(s) = \int_0^\infty e^{-st} f(t)dt,$$

$$g^*(s) = \int_0^\infty e^{-st} g(t)dt,$$

$$h^*(s) = \int_0^\infty e^{-st} h(t)dt.$$

Multiplying Equation 6.59 by e^{-st} and integrating from 0 to ∞, and using the properties of LTs, we get

$$g^*(s) = h^*(s) + g^*(s)f^*(s),$$

which yields

$$g^*(s) = \frac{h^*(s)}{1 - f^*(s)}. \tag{6.61}$$

If $\lim_{t\to\infty} g(t)$ exists, we know that it is given by

$$\lim_{t\to\infty} g(t) = \lim_{s\to0} sg^*(s).$$

Using Equation 6.61 we get

$$
\begin{aligned}
\lim_{s\to0} sg^*(s) &= \lim_{s\to0} s\frac{h^*(s)}{1 - f^*(s)} \\
&= \lim_{s\to0} h^*(s)/ \lim_{s\to0} \frac{1 - f^*(s)}{s} \\
&= \frac{1}{\mu} \int_0^\infty h(u)du.
\end{aligned}
$$

Here the last equality follows because

$$\lim_{s\to0} \frac{1 - f^*(s)}{s} = \lim_{s\to0} \frac{f^*(0) - f^*(s)}{s} = -\lim_{s\to0} \frac{d}{ds}f^*(s) = \mu.$$

This proves the theorem. ∎

Equation 6.59 is sometimes called the continuous renewal equation, and is analogous to its discrete counterpart in Equation 4.24 on page 110. The next theorem shows that $\{p_{j,j}(t), t \geq 0\}$ satisfy a continuous renewal equation if $q_j > 0$.

Theorem 6.21   Renewal Equation for $p_{j,j}(t)$. *Suppose $q_j > 0$. Let $\tilde{T}_j$ be as defined in Equation 6.51. Let f_j be the density of $\tilde{T}_j$ given $X(0) = j$. Then $p_{j,j}(t)$ satisfies the continuous renewal equation*

$$p_{j,j}(t) = e^{-q_j t} + \int_0^t f_j(u)p_{j,j}(t - u)du, \quad t \geq 0. \tag{6.62}$$

Proof: Suppose $X(0) = j$ and $\tilde{T}_j = u$. If $u \leq t$, the CTMC starts all over again in state j at time u, due to Markov property. If $u > t$, then $X(t) = j$ if and only if $Y_1 > t$. Hence we have

$$P(X(t) = j|X(0) = j, \tilde{T}_j = u) = \begin{cases} P(Y_1 > t|\tilde{T}_j = u, X(0) = j) & \text{if } u > t, \\ P(X(t - u) = j|X(0) = j) & \text{if } u \leq t. \end{cases}$$

Thus we get

$$
\begin{aligned}
p_{j,j}(t) &= P(X(t) = j|X(0) = j) \\
&= \int_0^\infty P(X(t) = j|X(0) = j, \tilde{T}_j = u)f_j(u)du \\
&= \int_0^t P(X(t - u) = j|X(0) = j)f_j(u)du \\
&\quad + \int_t^\infty P(Y_1 > t|\tilde{T}_j = u, X(0) = j)f_j(u)du \\
&= \int_0^\infty P(Y_1 > t|\tilde{T}_j = u, X(0) = j)f_j(u)du + \int_0^t p_{j,j}(t - u)f_j(u)du
\end{aligned}
$$

$$\text{(since } \mathsf{P}(Y_1 > t | \tilde{T}_j = u, X(0) = j) = 0 \text{ if } u < t)$$

$$= \mathsf{P}(Y_1 > t | X(0) = j) + \int_0^t p_{j,j}(t - u) f_j(u) du$$

$$= e^{-q_j t} + \int_0^t f_j(u) p_{j,j}(t - u) du.$$

This proves the theorem. ∎

Using the above theorem we get the next important result.

Theorem 6.22 If $q_j = 0$

$$\lim_{t \to \infty} p_{j,j}(t) = 1.$$

If state j is recurrent with $q_j > 0$

$$\lim_{t \to \infty} p_{j,j}(t) = \frac{1}{q_j \tilde{m}_j}, \tag{6.63}$$

where $\tilde{m}_j$ is as defined in Equation 6.53.

Proof: If $q_j = 0$, we have $p_{j,j}(t) = 1$ for all $t \geq 0$. Hence the first equation in the theorem follows. If $q_j > 0$, we see from Theorem 6.21 that $p_{j,j}(t)$ satisfies the continuous renewal equation. Since j is recurrent, we have

$$\int_0^\infty f_j(t) dt = 1,$$

and

$$\int_0^\infty |h(t)| dt = \int_0^\infty e^{-q_j t} dt = \frac{1}{q_j}.$$

We also have

$$\mu = \int_0^\infty t f_j(t) dt = \mathsf{E}(\tilde{T}_j | X_0 = j) = \tilde{m}_j.$$

Hence we can apply Theorem 6.20 to get

$$\lim_{t \to \infty} p_{j,j}(t) = \frac{1}{\tilde{m}_j} \int_0^\infty e^{-q_j t} dt = \frac{1}{q_j \tilde{m}_j}.$$

This proves the theorem. ∎

6.11.3 The Null Recurrent Case

Now we study the limiting behavior of an irreducible null recurrent CTMC. The main result is given by

Theorem 6.23 The Null Recurrent CTMC. *For an irreducible null recurrent CTMC*

$$\lim_{t \to \infty} p_{i,j}(t) = 0.$$

Proof: Since the CTMC is null recurrent, we know that

$$q_j > 0, \text{ and } \tilde{m}_j = \infty, \quad j \in S.$$

Hence from Theorem **??** we see that

$$\lim_{t \to \infty} p_{j,j}(t) = 0.$$

Now let $i \neq j$, and let $f_{i,j}(\cdot)$ be the conditional probability density of $\tilde{T}_j$ given $X(0) = i$. Using the argument in the proof of Theorem 6.21 we see that

$$p_{i,j}(t) = \int_0^t f_{i,j}(u) p_{j,j}(t - u) du.$$

Since the CTMC is irreducible and recurrent it follows that

$$\mathsf{P}(\tilde{T}_j < \infty | X(0) = i) = 1.$$

Hence

$$\lim_{t \to \infty} p_{i,j}(t) = 0, \quad i, j \in S.$$

This proves the theorem. ∎

6.11.4 The Positive Recurrent Case

Now we study the limiting behavior of an irreducible positive recurrent CTMC. Such CTMCs are also called ergodic. If the state-space is a singleton, say $S = \{1\}$, then $q_1 = 0$, and $p_{1,1}(t) = 1$ for all $t \geq 0$. Hence the limiting behavior is trivial in this case. So suppose that S has at least two elements. Then q_j must be strictly positive and $\tilde{m}_j < \infty$ for all $j \in S$. Hence, from Theorem 6.22 we get

$$\lim_{t \to \infty} p_{j,j}(t) = \frac{1}{q_j \tilde{m}_j} > 0, \quad j \in S.$$

The next theorem yields the limiting behavior of $p_{i,j}(t)$ as $t \to \infty$.

Theorem 6.24   The Positive Recurrent CTMC. *For an irreducible positive recurrent CTMC*

$$\lim_{t \to \infty} p_{i,j}(t) = p_j > 0, \quad i, j \in S \tag{6.64}$$

where $p = [p_j, j \in S]$ is given by the unique solution to

$$pQ = 0, \tag{6.65}$$

$$\sum_{j \in S} p_j = 1. \tag{6.66}$$

Proof: The theorem is true if the CTMC has a single state. Hence assume that the CTMC has at least two states. Then $q_j > 0$ for all $j \in S$. Equation 6.63 implies that Equation 6.64 holds when $i = j$ with $p_j = 1/q_j \tilde{m}_j > 0$. Hence assume $i \neq j$. Following the proof of Theorem 6.21 we get

$$p_{i,j}(t) = \int_0^t f_{i,j}(u) p_{j,j}(t-u) du, \quad t \geq 0,$$

where $f_{i,j}(\cdot)$ is the density of $\tilde{T}_j$ conditioned on $X(0) = i$. Since $i \leftrightarrow j$ and the CTMC is positive recurrent, it follows that

$$\int_0^\infty f_{i,j}(u) du = P(X(t) = j \text{ for some } t \geq 0 \,|\, X(0) = i) = 1.$$

Now let $0 < \epsilon < 1$ be given. Thus it is possible to pick an N such that

$$\int_t^\infty f_{i,j}(u) du \leq \epsilon/2,$$

and

$$|p_{j,j}(t) - p_j| \leq \epsilon/2, \text{ for all } t \geq N.$$

Then, for $t \geq 2N$, we get

$$
\begin{aligned}
|p_{i,j}(t) - p_j| &= \left| \int_0^t f_{i,j}(u) p_{j,j}(t-u) du - p_j \right| \\
&= \left| \int_0^{t-N} f_{i,j}(u)(p_{j,j}(t-u) - p_j) du \right. \\
&\quad + \int_{t-N}^t f_{i,j}(u)(p_{j,j}(t-u) - p_j) du \\
&\quad \left. - \int_t^\infty f_{i,j}(u) p_j du \right| \\
&\leq \int_0^{t-N} f_{i,j}(u)|p_{j,j}(t-u) - p_j| du \\
&\quad + \int_{t-N}^t f_{i,j}(u)|p_{j,j}(t-u) - p_j| du \\
&\quad + p_j \int_t^\infty f_{i,j}(u) du \\
&\leq \int_0^{t-N} f_{i,j}(u) du \,\epsilon/2 + \int_{t-N}^t f_{i,j}(u) du + \int_t^\infty f_{i,j}(u) du \\
&\leq \epsilon/2 + \epsilon/2 \leq \epsilon.
\end{aligned}
$$

This proves Equation 6.64. Next we derive Equations 6.65 and 6.66. For any finite set $A \subset S$, we have

$$\sum_{j \in A} p_{i,j}(t) \leq \sum_{j \in S} p_{i,j}(t) = 1.$$

Letting $t \to \infty$ on the left-hand side, we get

$$\sum_{j \in A} p_j \leq 1.$$

Since the above equation holds for any finite A, we must have

$$\sum_{j \in S} p_j \leq 1. \tag{6.67}$$

Now let $a_j(t) = P(X(t) = j)$. Then Equation 6.64 implies

$$\lim_{t \to \infty} a_j(t) = p_j, \quad j \in S.$$

Now, the Chapman–Kolmogorov Equations 6.16 yield

$$a_j(s + t) = \sum_{i \in S} a_i(s) p_{i,j}(t), \quad s, t \geq 0.$$

Let $s \to \infty$ on both sides. The interchange of the limit and the sum on the right hand side is justified due to bounded convergence theorem. Hence we get

$$p_j = \sum_{i \in S} p_i p_{i,j}(t). \tag{6.68}$$

Replacing t by $t + h$ yields

$$p_j = \sum_{i \in S} p_i p_{i,j}(t + h). \tag{6.69}$$

Subtracting Equation 6.68 from 6.69 we get

$$\sum_{i \in S} p_i (p_{i,j}(t + h) - p_{i,j}(t)) = 0.$$

Dividing the above equation by h and letting $h \to 0$, the above equation in matrix form yields

$$pP'(t) = 0.$$

Using Equation 6.17, the above equation reduces to

$$pQP(t) = 0.$$

Substituting $t = 0$ in the above equation and using $P(0) = I$, we get Equation 6.65. Again, letting $t \to \infty$ in Equation 6.68 and using bounded convergence theorem to interchange the sum and the limit on the right-hand side we get

$$p_j = \left(\sum_{i \in S} p_i \right) p_j.$$

But $p_j > 0$. Hence we must have $\sum p_i = 1$, yielding Equation 6.66.

Now suppose $\{p'_i, i \in S\}$ is another solution to Equations 6.65 and 6.66. Using the same steps as before we get

$$p'_j = \sum_{i \in S} p'_i p_{i,j}(t), \quad t \geq 0.$$

Letting $t \to \infty$ we get

$$p'_j = \left(\sum_{i \in S} p'_i \right) p_j = p_j.$$

Thus Equations 6.65 and 6.66 have a unique solution. ∎

The next theorem removes the need to first check the positive recurrence of the CTMC before solving the balance and the normalizing equation. It is analogous to Theorem 4.22.

Theorem 6.25 *Let $\{X(t), t \geq 0\}$ be an irreducible CTMC. It is positive recurrent if and only if there is a positive solution to*

$$pQ = 0, \tag{6.70}$$

$$\sum_{j \in S} p_j = 1. \tag{6.71}$$

If there is a solution to the above equations, it is unique.

Proof: Let p be a positive solution to Equations 6.70 and 6.71. Then it is straightforward to verify that $\pi_j = p_j q_j$ solves the balance equations for the embedded DTMC. Substituting in Equation 6.56 we get

$$
\begin{aligned}
\mu_{j,j} &= \frac{1}{\pi_j} \left(\sum_{i \in S} \frac{\pi_i}{q_i} \right) \\
&= \frac{1}{p_j q_j} \left(\sum_{i \in S} p_i \right) = \frac{1}{p_j q_j} < \infty.
\end{aligned}
$$

Hence state j is positive recurrent, and hence the entire CTMC is positive recurrent. The uniqueness was already proved in Theorem 6.24. ∎

The vector $p = [p_j]$ is called the limiting distribution or the steady state distribution of the CTMC. It is also the stationary distribution or the invariant distribution of the CTMC, since if p is the pmf of $X(0)$ then it is also the pmf of $X(t)$ for all $t \geq 0$. See Conceptual Exercise 6.5.

The next theorem shows the relationship between the stationary distribution of the CTMC and that of the corresponding embedded DTMC.

Theorem 6.26 *Let $\{X(t), t \geq 0\}$ be an irreducible positive recurrent CTMC on state-space S with generator Q and stationary distribution $\{p_j, j \in S\}$. Let $\{X_n, n \geq 0\}$ be the corresponding embedded DTMC with transition probability matrix P and stationary distribution $\{\pi_j, j \in S\}$. Then*

$$p_j = \frac{\pi_j / q_j}{\sum_{i \in S} \pi_i / q_i}, \quad j \in S. \tag{6.72}$$

Proof: Follows by substituting $p_{i,j} = q_{i,j}/q_i$ in $\pi = \pi P$ and verifying that it reduces to $pQ = 0$. We leave the details to the reader. ∎

The last theorem of this section shows that the limiting occupancy distribution of a CTMC is always the same as its limiting distribution.

Theorem 6.27 Limiting Occupancy Distribution. *Let $\{X(t), t \geq 0\}$ be an irreducible CTMC. Then*

$$\lim_{t \to \infty} \frac{M_{i,j}(t)}{t} = \lim_{t \to \infty} p_{i,j}(t).$$

Proof: Follows from Equation 6.23 and the fact that $p_{i,j}(t) \geq 0$ and has a limit as $t \to \infty$. ∎

Thus, if Equations 6.70 and 6.71 have a solution p, it represents the limiting distribution, the stationary distribution as well as the limiting occupancy distribution of the CTMC. Equations 6.70 are called the *balance equations*. Sometimes they are called *global* balance equations to distinguish them from the local balance equations to be developed in Section 6.15. It is more instructive to write them in a scalar form:

$$\sum_{j:j \in S, j \neq i} p_i q_{i,j} = \sum_{j:j \in S, j \neq i} p_j q_{j,i}, \quad i \in S.$$

The left-hand side equals $p_i q_i$, the rate at which transitions take the system out of state i, and the right-hand side equals the rate at which transitions take the system into state i. In steady state the two rates must be equal. In practice it is easier to write these "rate out = rate in" equations by looking at the rate diagram, rather than by using the matrix Equation 6.70. We illustrate the theory developed in this section by several examples below.

Example 6.34 Two-State Machine. Consider the two-state CTMC of Example 6.29. The balance equations are

$$\lambda p_0 = \mu p_1,$$
$$\mu p_1 = \lambda p_0.$$

Thus there is only one independent balance equation. The normalizing equation is

$$p_0 + p_1 = 1.$$

Solving these we get

$$p_0 = \frac{\mu}{\lambda + \mu}, \quad p_1 = \frac{\lambda}{\lambda + \mu}.$$

This agrees with the result in Example 6.29. ∎

Example 6.35 Birth and Death Processes. Let $\{X(t), t \geq 0\}$ be the birth and death process of Example 6.10 with birth parameters $\lambda_i > 0$ for $i \geq 0$, and death parameter $\mu_i > 0$ for $i \geq 1$. This CTMC is irreducible. The balance equations are

$$\lambda_0 p_0 = \mu_1 p_1,$$
$$(\lambda_i + \mu_i) p_i = \lambda_{i-1} p_{i-1} + \mu_{i+1} p_{i+1}, \quad i \geq 1.$$

Summing the first $i + 1$ equations we get

$$\lambda_i p_i = \mu_{i+1} p_{i+1}, \quad i \geq 0. \tag{6.73}$$

These can be solved recursively to get

$$p_i = \rho_i p_0, \quad i \geq 0,$$

where $\rho_0 = 1$, and

$$\rho_i = \frac{\lambda_0 \lambda_1 \cdots \lambda_{i-1}}{\mu_1 \mu_2 \cdots \mu_i}, \quad i \geq 1. \tag{6.74}$$

Now, substituting in the normalizing equation we get

$$\sum_{i=0}^{\infty} p_i = \left(\sum_{i=0}^{\infty} \rho_i \right) p_0 = 1.$$

The above equation has a solution

$$p_0 = \left(\sum_{i=0}^{\infty} \rho_i \right)^{-1} \quad \text{if} \quad \sum_{i=0}^{\infty} \rho_i < \infty.$$

If the infinite sum diverges, there is no solution. Thus the CTMC is positive recurrent if and only if

$$\sum_{i=0}^{\infty} \rho_i < \infty$$

and, when it is positive recurrent, has the limiting distribution given by

$$p_j = \frac{\rho_j}{\sum_{i=0}^{\infty} \rho_i}, \quad j \geq 0. \tag{6.75}$$

If the CTMC is transient or null recurrent, $p_j = 0$ for all $j \geq 0$. We analyze two special birth and death processes in the next two examples. ■

Example 6.36   Single-Server Queue. Let $X(t)$ be the number of customers at time t in the single-server service system of Example 6.11. There we saw that $\{X(t), t \geq 0\}$ is a birth and death process with birth parameters $\lambda_i = \lambda$ for $i \geq 0$, and death parameters $\mu_i = \mu$ for $i \geq 1$. Thus we can use the results of Example 6.35 to study the limiting behavior of this system. Substituting in Equation 6.74 we get

$$\rho_i = \frac{\lambda_0 \lambda_1 \cdots \lambda_{i-1}}{\mu_1 \mu_2 \cdots \mu_i} = \rho^i, \quad i \geq 0$$

where

$$\rho = \frac{\lambda}{\mu}.$$

Now,

$$\sum_{i=0}^{\infty} \rho^i = \begin{cases} \frac{1}{1-\rho} & \text{if } \rho < 1, \\ \infty & \text{if } \rho \geq 1. \end{cases}$$

Thus the queue is stable (i.e., the CTMC is positive recurrent) if $\rho < 1$. In that case the limiting distribution can be obtained from Equation 6.75 as

$$p_j = \rho^j (1 - \rho), \quad j \geq 0. \tag{6.76}$$

Thus, in steady state, the number of customers in a stable single server queue is a modified geometric random variable with parameter $1 - \rho$. ■

Example 6.37 Infinite-Server Queue. Let $X(t)$ be the number of customers at time t in the infinite-server service system of Example 6.12. There we saw that $\{X(t), t \geq 0\}$ is a birth and death process with birth parameters $\lambda_i = \lambda$ for $i \geq 0$, and death parameters $\mu_i = i\mu$ for $i \geq 1$. Thus we can use the results of Example 6.35 to study the limiting behavior of this system. Substituting in Equation 6.74 we get

$$\rho_i = \frac{\lambda_0 \lambda_1 \cdots \lambda_{i-1}}{\mu_1 \mu_2 \cdots \mu_i} = \frac{\rho^i}{i!}, \quad i \geq 0$$

where

$$\rho = \frac{\lambda}{\mu}.$$

Now,

$$\sum_{i=0}^{\infty} \frac{\rho^i}{i!} = e^\rho.$$

Thus the queue is stable (i.e., the CTMC is positive recurrent) if $\rho < \infty$. In that case the limiting distribution can be obtained from Equation 6.75 as

$$p_j = e^{-\rho} \frac{\rho^j}{j!}, \quad j \geq 0.$$

Thus, in steady state, the number of customers in a stable infinite-server queue is a Poisson random variable with parameter ρ. ∎

Example 6.38 Retrial Queue. Consider the bivariate CTMC $\{(X(t), R(t)), t \geq 0\}$ of the retrial queue of Example 6.15. From of the rate diagram in Figure 6.9 we get the following balance equations:

$$\begin{aligned}
(\lambda + n\theta)p_{0,n} &= \mu p_{1,n}, \quad n \geq 0, \\
(\lambda + \mu)p_{1,0} &= \lambda p_{0,0} + \theta p_{0,1}, \\
(\lambda + \mu)p_{1,n} &= \lambda p_{0,n} + (n+1)\theta p_{0,n+1} + \lambda p_{1,n-1}, \quad n \geq 1.
\end{aligned}$$

These can be rearranged to yield

$$\begin{aligned}
(\lambda + n\theta)p_{0,n} &= \mu p_{1,n}, \quad n \geq 0, \\
\lambda p_{1,n} &= (n+1)\theta p_{0,n+1}, \quad n \geq 0.
\end{aligned}$$

Thus

$$\frac{(n+1)\theta}{\lambda} p_{0,n+1} = \frac{\lambda + n\theta}{\mu} p_{0,n}, \quad n \geq 0,$$

or

$$p_{0,n+1} = \frac{\lambda}{\mu} \frac{\lambda + n\theta}{(n+1)\theta} p_{0,n}, \quad n \geq 0.$$

This can be solved recursively to obtain

$$p_{0,n} = \frac{1}{n!} \left(\frac{\lambda}{\mu}\right)^n \prod_{k=0}^{n-1} \left(\frac{\lambda}{\theta} + k\right) p_{0,0}, \quad n \geq 1. \tag{6.77}$$

Then substituting in the equation for $p_{1,n}$ we get

$$p_{1,n} = \frac{1}{n!}\frac{\theta}{\lambda}\left(\frac{\lambda}{\mu}\right)^{n+1}\prod_{k=0}^{n}\left(\frac{\lambda}{\theta}+k\right)p_{0,0}, \quad n \geq 0. \tag{6.78}$$

Now using

$$\sum_{n=0}^{\infty}p_{0,n} + \sum_{n=0}^{\infty}p_{1,n} = 1$$

we get

$$\left[1 + \sum_{n=1}^{\infty}\frac{1}{n!}\left(\frac{\lambda}{\mu}\right)^{n}\prod_{k=0}^{n-1}\left(\frac{\lambda}{\theta}+k\right) + \sum_{n=0}^{\infty}\frac{1}{n!}\frac{\theta}{\lambda}\left(\frac{\lambda}{\mu}\right)^{n+1}\prod_{k=0}^{n}\left(\frac{\lambda}{\theta}+k\right)\right]p_{0,0} = 1.$$

It can be shown that the infinite sums converge if

$$\frac{\lambda}{\mu} < 1, \tag{6.79}$$

and, in that case

$$p_{0,0} = \left(1 - \frac{\lambda}{\mu}\right)^{\frac{\lambda}{\theta}+1}.$$

Using this in Equations 6.77 and 6.78 we get the complete limiting distribution of the CTMC. Thus the condition in Equation 6.79 is the condition of positive recurrence.

∎

Example 6.39 Batch Arrival Queue. Consider a single server queue where customers arrive in batches. The successive batch sizes are iid with common pmf

$$P(\text{Batch Size} = k) = \alpha_k, \quad k \geq 1.$$

The batches themselves arrive according to a PP(λ). Thus the customer arrival process is a CPP. Customers are served one at a time, service times being iid exp(μ) random variables. Assume that there is an infinite waiting room. Let $X(t)$ be the number of customers in the system at time t.

Using the methodology developed in Section 6.2 we can show that $\{X(t), t \geq 0\}$ is a CTMC with state-space $\{0, 1, 2, \cdots\}$ and the following transition rates:

$$q_{i,i-1} = \mu \quad i \geq 1,$$
$$q_{i,i+k} = \lambda\alpha_k, \quad k \geq 1, i \geq 0.$$

Thus we have $q_{0,0} = -\lambda$ and $q_{i,i} = -(\lambda+\mu)$ for $i \geq 1$. Hence the balance equations for this system are

$$\lambda p_0 = \mu p_1,$$
$$(\lambda + \mu)p_i = \lambda\sum_{r=0}^{i-1}\alpha_{i-r}p_r + \mu p_{i+1}, \quad i \geq 1.$$

Now define the generating function of $\{p_i, i \geq 0\}$ as

$$\phi(z) = \sum_{i=0}^{\infty} p_i z^i.$$

Multiplying the balance equation for state i by z^i and summing over all i, we get

$$\lambda p_0 + (\lambda + \mu) \sum_{i=1}^{\infty} p_i z^i = \mu p_1 + \lambda \sum_{i=1}^{\infty} z^i \sum_{r=0}^{i-1} \alpha_{i-r} p_r + \mu \sum_{i=1}^{\infty} z^i p_{i+1}.$$

Adding μp_0 on both sides, interchanging the i and r sums on the right-hand side, and regrouping terms, we get

$$
\begin{aligned}
(\lambda + \mu) \sum_{i=0}^{\infty} p_i z^i &= \mu p_0 + \frac{\mu}{z} \sum_{i=1}^{\infty} p_i z^i + \lambda \sum_{r=0}^{\infty} \left(\sum_{i=r+1}^{\infty} z^i \alpha_{i-r} \right) p_r \\
&= \mu \left(1 - \frac{1}{z} \right) p_0 + \frac{\mu}{z} \sum_{i=0}^{\infty} p_i z^i + \left(\lambda \sum_{r=0}^{\infty} z^r p_r \right) \left(\sum_{i=1}^{\infty} z^i \alpha_i \right).
\end{aligned}
$$

This can be rewritten as

$$(\lambda + \mu)\phi(z) = \mu \left(1 - \frac{1}{z} \right) p_0 + \frac{\mu}{z} \phi(z) + \lambda \phi(z) \psi(z),$$

where

$$\psi(z) = \sum_{r=1}^{\infty} z^r \alpha_r.$$

This yields

$$\phi(z) = \frac{\mu(1 - 1/z)}{\lambda(1 - \psi(z)) + \mu(1 - 1/z)} p_0. \tag{6.80}$$

Note that $\psi(z)$ is known. Thus the only unknown is p_0, which can be computed by using

$$\lim_{z \to 1} \phi(z) = 1.$$

Using L'Hopital's rule to compute the above limit, we get

$$\lim_{z \to 1} \phi(z) = \frac{\mu}{\mu - \lambda c} p_0$$

where

$$c = \lim_{z \to 1} \frac{d}{dz} \psi(z) = \sum_{i=1}^{\infty} i \alpha_i$$

is the expected batch size. Writing

$$\rho = \frac{\lambda c}{\mu},$$

we get

$$p_0 = 1 - \rho.$$

Thus the queue is stable (i.e., the CTMC is positive recurrent) if

$$\rho < 1.$$

If it is stable, the generating function in Equation 6.80 reduces to

$$\phi(z) = \frac{(1-\rho)(1-z)}{(1-z) - \rho z(1 - \psi(z))/c}.$$

The expected number of customers in the system in steady state is given by

$$L = \lim_{z \to 1} \frac{d}{dz} \phi(z) = \frac{\rho}{1-\rho} \cdot \frac{s^2 + c}{2c},$$

where s^2 is the second moment of the batch size. One needs to apply L'Hopital's rule twice to get the above expression. ∎

6.12 Limiting Behavior of Reducible CTMCs

In this section we derive the main results regarding the limiting distribution of a reducible CTMC $\{X(t), t \geq 0\}$ on state-space $S = \{0, 1, 2 \cdots\}$ and generator matrix Q. Assume that there are k closed communicating classes $C_i, 1 \leq i \leq k$, and T is the set of states that do not belong to any closed communicating class. Now, relabel the states in S by non-negative integers such that $i \in C_r$ and $j \in C_s$ with $r < s$ implies that $i < j$, and $i \in C_r$ and $j \in T$ implies that $i < j$. With this relabeling, the generator matrix Q has the following canonical block structure:

$$Q = \begin{bmatrix} Q(1) & 0 & \cdots & 0 & 0 \\ 0 & Q(2) & \cdots & 0 & 0 \\ \vdots & \vdots & \ddots & \vdots & \vdots \\ 0 & 0 & \cdots & Q(k) & 0 \\ & R & & & Q(T) \end{bmatrix}. \tag{6.81}$$

Here $Q(i)$ $(1 \leq i \leq k)$ is a generator matrix of an irreducible CTMC with state-space C_i, $Q(T)$ is a $|T| \times |T|$ sub-stochastic generator matrix (i.e., all row sums of $Q(T)$ being less than or equal to zero, with at least one being strictly less than zero), and R is a $|T| \times |S - T|$ matrix. Then the transition probability matrix P has the same block structure:

$$P(t) = \begin{bmatrix} P(1)(t) & 0 & \cdots & 0 & 0 \\ 0 & P(2)(t) & \cdots & 0 & 0 \\ \vdots & \vdots & \ddots & \vdots & \vdots \\ 0 & 0 & \cdots & P(k)(t) & 0 \\ & P_R(t) & & & P(T)(t) \end{bmatrix}.$$

Since $P(r)(t)$ $(1 \leq r \leq k)$ is a transition probability matrix of an irreducible CTMC with state-space C_r, we already know how $P(r)(t)$ behaves as $t \to \infty$. Similarly, since all states in T are transient, we know that $P(T)(t) \to 0$ as $t \to \infty$. Thus the study of the limiting behavior of $P(t)$ reduces to the study of the limiting behavior of $P_R(t)$ as $t \to \infty$. This is what we proceed to do.

Let $T(r)$ be the first passage time to visit the set C_r, i.e.,

$$T(r) = \min\{t \geq 0 : X(t) \in C_r\}, \quad 1 \leq r \leq k.$$

Let

$$u_i(r) = P(T(r) < \infty | X(0) = i), \quad 1 \leq r \leq k, \ i \in T. \tag{6.82}$$

The next theorem gives a method of computing the above probabilities.

Theorem 6.28   Absorption Probabilities. *The quantities* $\{u_i(r), i \in T, 1 \leq r \leq k\}$ *are given by the smallest non-negative solution to*

$$u_i(r) = \sum_{j \in C_r} \frac{q_{i,j}}{q_i} + \sum_{j \in T-\{i\}} \frac{q_{i,j}}{q_i} u_j(r), \quad i \in T. \tag{6.83}$$

Proof: Equation 6.83 can be derived as in the proof of Theorem 6.13. The rest of the proof is similar to the proof of Theorem 3.2 on page 62. ∎

Using the quantities $\{u_i(r), i \in T, 1 \leq r \leq k\}$ we can describe the limit of $P_R(t)$ as $t \to \infty$. This is done in the theorem below.

Theorem 6.29   Limit of $P_R(t)$. *Let* $\{u_i(r), i \in T, 1 \leq r \leq k\}$ *be as defined in Equation 6.82. Let* $i \in T$ *and* $j \in C_r$.

(i) If C_r *is transient or null recurrent,*

$$\lim_{t \to \infty} p_{i,j}(t) = 0. \tag{6.84}$$

(ii) If C_r *is positive recurrent,*

$$\lim_{t \to \infty} p_{i,j}(t) = u_i(r)p_j,$$

where $\{p_j, j \in C_r\}$ *is the unique solution to*

$$\sum_{m \in C_r} p_m q_{m,j} = 0, \quad j \in C_r,$$

$$\sum_{m \in C_r} p_m = 1.$$

Proof: Follows along the same lines as the proof of Theorem 4.24. ∎

We discuss two examples of reducible CTMCs below.

Example 6.40   Let $\{X(t), t \geq 0\}$ be the CTMC of Example 6.30. This is a reducible CTMC with two closed communicating classes $C_1 = \{0\}$ and $C_2 = \{2\}$.

The set $T = \{1\}$ is not closed. We do not do any relabeling. Equations 6.83 yield:

$$u_1(1) = \frac{\mu}{\lambda + \mu}, \quad u_1(2) = \frac{\lambda}{\lambda + \mu}.$$

We also have

$$p_0 = 1, \quad p_2 = 1$$

since these are absorbing states. Then Theorem 6.29 yields

$$\lim_{t \to \infty} P(t) = \begin{bmatrix} 1 & 0 & 0 \\ \frac{\mu}{\lambda + \mu} & 0 & \frac{\lambda}{\lambda + \mu} \\ 0 & 0 & 1 \end{bmatrix}.$$

This matches the result in Example 6.30. ■

Example 6.41 Linear Growth Model. Consider the CTMC of Example 6.27. This is a reducible CTMC with one closed class $C_1 = \{0\}$. Hence $p_0 = 1$. The rest of the states form the set T. Thus we need to compute $u_i(1)$ for $i \geq 1$. From the results of Example 6.28 we get

$$u_i(1) = \begin{cases} 1 & \text{if } \lambda \leq \mu, \\ \left(\frac{\mu}{\lambda}\right)^i & \text{if } \lambda > \mu. \end{cases}$$

Hence, for $i \geq 1$,

$$\lim_{t \to \infty} p_{i,0}(t) = \begin{cases} 1 & \text{if } \lambda \leq \mu, \\ \left(\frac{\mu}{\lambda}\right)^i & \text{if } \lambda > \mu. \end{cases} \tag{6.85}$$

which agrees with Example 6.31.

Recall that the linear growth model represents a colony of organisms where each organism produces new ones at rate λ, and each organism dies at rate μ. State 0 is absorbing: once all organisms die, the colony is permanently extinct. Equation 6.85 says that extinction is certain if the birth rate is no greater than the death rate. On the other hand, even if the birth rate is greater than the death rate, there is a positive probability of extinction in the long run, no matter how large the colony is to begin with. Of course, in this case, there is also a positive probability, $1 - (\mu/\lambda)^i$, that the size of the colony of initial size i will become infinitely large in the long run. ■

6.13 CTMCs with Costs and Rewards

As in the case of DTMCs, now we study CTMCs with costs and rewards. Let $X(t)$ be the state of a system at time t. Suppose $\{X(t), t \geq 0\}$ is a CTMC with state-space S and generator matrix Q. Consider a simple cost model where the system incurs cost at a rate of $c(i)$ per unit time it spends in state i. For other cost models, see Conceptual Exercises 6.6 and 6.7. Rewards can be thought of as negative costs. We consider costs incurred over an infinite horizon. For the analysis of costs over finite horizon, see Conceptual Exercise 6.8.

6.13.1 Discounted Costs

Suppose the costs are discounted continuously at rate α, where $\alpha > 0$ is a fixed (continuous) discount factor. Thus if the system incurs a cost of d at time t, its present value at time 0 is $e^{-\alpha t}d$, i.e., it is equivalent to incurring a cost of $e^{-\alpha t}d$ at time zero. Let C be the total discounted cost over the infinite horizon, i.e.,

$$C = \int_0^\infty e^{-\alpha t} c(X(t))dt.$$

Let $\phi(i)$ be the expected total discounted cost (ETDC) incurred over the infinite horizon starting with $X(0) = i$. That is,

$$\phi(i) = \mathsf{E}(C|X(0) = i).$$

The next theorem gives the main result regarding the ETDC. We introduce the following column vectors

$$c = [c(i)]_{i \in S}, \quad \phi = [\phi(i)]_{i \in S}.$$

Theorem 6.30 ETDC. *Suppose $\alpha > 0$, and c is a bounded vector. Then ϕ is given by*

$$\phi = (\alpha I - Q)^{-1}c. \tag{6.86}$$

Proof: We have

$$\begin{aligned}
\phi(i) &= \mathsf{E}(C|X(0) = i) \\
&= \mathsf{E}\left(\int_0^\infty e^{-\alpha t} c(X(t))dt \,\Big|\, X(0) = i\right) \\
&= \int_0^\infty e^{-\alpha t}\mathsf{E}(c(X(t))|X(0) = i)dt \\
&= \int_0^\infty e^{-\alpha t} \sum_{j \in S} p_{i,j}(t)c(j)dt.
\end{aligned}$$

The interchanges of integrals and expectations are justified since c is bounded. In matrix form the above equation becomes

$$\phi = \left(\int_0^\infty e^{-\alpha t} P(t)dt\right)c. \tag{6.87}$$

The right hand-side is the Laplace transform of $P(t)$ evaluated at α. From Theorem 6.8, we see that it is given by $(\alpha I - Q)^{-1}$. Hence we get Equation 6.86. ∎

Note that there is no assumption of irreducibility or transience or recurrence behind the above theorem. Equation 6.80 is valid for any generator matrix Q. Note that the matrix $\alpha I - Q$ is invertible for $\alpha > 0$. Also, the theorem remains valid if the c vector is bounded from below or above. It does not need to be bounded from both sides.

Example 6.42 Two-State Machine. Consider the two-state machine of Example 6.4 on page 207. It was modeled by a CTMC $\{X(t), t \geq 0\}$ with state-space $\{0, 1\}$ (0 being down, and 1 being up), and generator matrix

$$Q = \begin{bmatrix} -\lambda & \lambda \\ \mu & -\mu \end{bmatrix},$$

where $\lambda, \mu > 0$. Now suppose the machine produces a revenue of r per day when it is up, and it costs d in repair costs per day when the machine is down. Suppose a new machine in working order costs m. Is it profitable to purchase it if the continuous discount factor is $\alpha > 0$?

Let $c(i)$ be the expected cost incurred per unit time spent in state i. We have

$$c = [c(0) \ c(1)]^\top = [d \ -r]^\top.$$

Then, using Theorem 6.30 we get

$$\phi = [\phi(0) \ \phi(1)]^\top = (\alpha I - Q)^{-1} c.$$

Direct calculations yield

$$\phi = \frac{1}{\alpha(\alpha + \lambda + \mu)} \begin{bmatrix} d(\alpha + \mu) - r\lambda \\ d\mu - r(\alpha + \lambda) \end{bmatrix}.$$

Thus it is profitable to buy a new machine if the expected total discounted net revenue from a new machine over the infinite horizon is greater than the initial purchase price of m, i.e., if

$$m \leq \frac{r(\alpha + \lambda) - d\mu}{\alpha(\alpha + \lambda + \mu)}.$$

How much should you be willing to pay for a machine in down state? ∎

6.13.2 Average Costs

The discounted costs have the disadvantage that they depend upon the discount factor and the initial state, thus making decision making more complicated. These issues are addressed by considering the long run cost per unit time, called the average cost. The expected total cost up to time t, starting from state i, is given by $E(\int_0^t c(X(u)) du | X(0) = i)$. Dividing it by t gives the cost per unit time. Hence the long run expected cost per unit time is given by:

$$g(i) = \lim_{t \to \infty} \frac{1}{t} E\left(\int_0^t c(X(u)) du \,\Big|\, X(0) = i \right),$$

assuming that the above limit exists. To keep the analysis simple, we will assume that the CTMC is irreducible and positive recurrent with limiting distribution given by $\{p_j, j \in S\}$, which is also the limiting occupancy distribution. Intuitively, it makes sense that the long run cost per period should be given by $\sum p_j c(j)$, independent of the initial state i. This intuition is formally proved in the next theorem:

Theorem 6.31 Average Cost. Suppose $\{X(t), t \geq 0\}$ is an irreducible positive recurrent CTMC with limiting occupancy distribution $\{p_j, j \in S\}$. Suppose

$$\sum_{j \in S} p_j |c(j)| < \infty.$$

Then

$$g(i) = g = \sum_{j \in S} p_j c(j).$$

Proof: Let $M_{i,j}(t)$ be the expected time spent in state j over $(0, t]$ starting from state i. See Section 6.5. Then, we see that

$$
\begin{aligned}
g(i) &= \lim_{t \to \infty} \frac{1}{t} \sum_{j \in S} M_{i,j}(t) c(j) \\
&= \lim_{t \to \infty} \sum_{j \in S} \frac{M_{i,j}(t)}{t} c(j) \\
&= \sum_{j \in S} \lim_{t \to \infty} \frac{M_{i,j}(t)}{t} c(j) \\
&= \sum_{j \in S} p_j c(j).
\end{aligned}
$$

Here the last interchange of sum and the limit is allowed because the CTMC is positive recurrent. The last equality follows from Theorem 6.27. ∎

We illustrate with two examples.

Example 6.43 Two-State Machine. Consider the cost structure regarding the two-state machine of Example 6.42. Compute the long run cost per unit time.

The steady state distribution of the two-state machine is given by (see Example 6.34)

$$p_0 = \frac{\mu}{\lambda + \mu}, \quad p_1 = \frac{\lambda}{\lambda + \mu}.$$

From Theorem 6.31 we see that the expected cost per unit time in the long run is given by

$$g = c(0)p_0 + c(1)p_1 = \frac{d\mu - r\lambda}{\lambda + \mu}.$$

Thus it is profitable to operate this machine if $r\lambda > d\mu$. ∎

Example 6.44 Single-Server Queue. Let $X(t)$ be the number of customers in the single-server queue with arrival rate λ and service rate μ as described in Example 6.11. Now suppose the cost of keeping customers in the system is \$c per customer per hour. The entry fee is \$d per customer, paid upon entry. Compute the long run net revenue per unit time.

We saw in Example 6.36 that $\{X(t), t \geq 0\}$ is positive recurrent if $\rho = \lambda/\mu < 1$ and in that case the limiting distribution is given by

$$p_j = \rho^j(1 - \rho), \quad j \geq 0.$$

The cost structure implies that $c(j) = jc$, $j \geq 0$. Using Theorem 6.31 we see that the long run expected cost per unit time is given by

$$\sum_{j=0}^{\infty} p_j c(j) = \sum_{j=0}^{\infty} j\rho^j(1 - \rho)c = \frac{c\rho}{1 - \rho}.$$

Let $N(t)$ be the number of arrivals over $(0, t]$. Since the arrival process is PP(λ), we get r, the long run fees collected per unit time as

$$r = d\frac{\mathsf{E}(N(t))}{t} = d\frac{\lambda t}{t} = \lambda d.$$

Thus the net expected revenue per unit time is given by

$$\lambda d - \frac{c\rho}{1 - \rho}.$$

A bit of algebra shows that the entry fee must be at least $\frac{c}{\mu - \lambda}$ in order to break even.

∎

It is possible to use the results in Section 6.12 to extend this analysis to reducible CTMCs. However, the long run cost rate may depend upon the initial state in that case.

6.14 Phase Type Distributions

The distribution given in Equation 6.41 is called a *phase type distribution* with parameters (α, B). Any random variable whose distribution can be represented as in Equation 6.41 for a valid α and B is called a *phase type random variable*. (By "valid" we mean that α is a pmf, and B is obtained from a generator matrix of an irreducible CTMC by deleting rows and columns corresponding to a non-empty subset of states.) It is denoted by PH(α, B), and the size of B is called the number of phases in the random variable. Many well-known distributions are phase type distributions, as is demonstrated in the next example.

Example 6.45 Examples of Phase Type Random Variables.

1. **Exponential.** The exp(λ) is a PH(α, B) random variable with

$$\alpha = [1], \quad B = [-\lambda].$$

2. **Sums of Exponentials.** Let $X_i \sim \exp(\lambda_i)$, $1 \leq i \leq K$, be independent random variables. Then

$$T = X_1 + X_2 + \cdots + X_K$$

is a PH(α, B) random variable with

$$\alpha = [1, 0, 0, \cdots, 0],$$

and

$$B = \begin{bmatrix} -\lambda_1 & \lambda_1 & & & \\ & -\lambda_2 & \lambda_2 & & \\ \vdots & \vdots & \ddots & \vdots & \vdots \\ & & & -\lambda_{K-1} & \lambda_{K-1} \\ & & & & -\lambda_K \end{bmatrix}.$$

As a special case we see that Erl(λ, K) is a phase type random variable.

3. **Mixtures of Exponentials.** Let $X_i \sim \exp(\lambda_i)$, $1 \leq i \leq K$, be independent random variables. Let

$$\alpha = [\alpha_1, \alpha_2, \cdots, \alpha_K]$$

be such that

$$\alpha_i \geq 0, \quad \sum_{i=1}^{K} \alpha_i = 1.$$

Let

$$T = X_i \text{ with probability } \alpha_i, \quad 1 \leq i \leq K.$$

Thus T is a mixture of exponentials. It is also a PH(α, B) random variable with α as above, and

$$B = \begin{bmatrix} -\lambda_1 & 0 & \cdots & 0 & 0 \\ 0 & -\lambda_2 & \cdots & 0 & 0 \\ \vdots & \vdots & \ddots & \vdots & \vdots \\ 0 & 0 & \cdots & -\lambda_{K-1} & 0 \\ 0 & 0 & \cdots & 0 & -\lambda_K \end{bmatrix}. \quad \blacksquare$$

Examples 2 and 3 are special cases of the general theorem given below.

Theorem 6.32 *Sums and mixtures of a finite number of independent phase type random variables are phase type random variables.*

Proof: We treat the case of two independent random variables, $T_i \sim PH(\alpha_i, B_i)$ $(i = 1, 2)$. The general case follows similarly. Let $T = T_1 + T_2$. Let e_1 be a column vector of ones of appropriate length, and define

$$G = -B_1 e_1 \alpha_2.$$

Then it is possible to show that (see Conceptual Exercise 6.13) T is a PH(α, B) random variable with

$$\alpha = [\alpha_1, 0 \cdot \alpha_2], \tag{6.88}$$

and

$$B = \begin{bmatrix} B_1 & G \\ 0 & B_2 \end{bmatrix}. \tag{6.89}$$

Next, let β_1, β_2 be two scalars such that $\beta_i \geq 0$, and $\beta_1 + \beta_2 = 1$. Let

$$T = \begin{cases} T_1 & \text{with probability } \beta_1, \\ T_2 & \text{with probability } \beta_2. \end{cases} \tag{6.90}$$

Thus T is a $[\beta_1, \beta_2]$ mixture of $\{T_1, T_2\}$. Then it is possible to show that (see Conceptual Exercise 6.14) T is a PH(α, B) random variable with

$$\alpha = [\beta_1 \alpha_1, \beta_2 \alpha_2], \tag{6.91}$$

and

$$B = \begin{bmatrix} B_1 & 0 \\ 0 & B_2 \end{bmatrix}. \tag{6.92}$$

This establishes the theorem. ∎

The phase type distributions form a versatile family of distributions. We refer the reader to the book by Neuts (1981) for further details. It can be shown that the set of phase type distributions is dense in the set of all continuous distributions over $[0, \infty)$. Thus, any continuous distribution over $[0, \infty)$ can be approximated arbitrarily closely by a phase type distribution. However, this denseness is of little practical value if the approximating distribution has a very large number of phases. Use of PH type distributions does help in developing tractable algorithms for the performance evaluation of many stochastic systems.

6.15 Reversibility

In Section 4.9 we studied reversible DTMCs. In this section we study the reversible CTMCs. Intuitively, if we watch a movie of a reversible CTMC we will not be able to tell whether the time is running forward or backward. We begin with the definition.

Definition 6.6 *A CTMC is called reversible if the corresponding embedded DTMC is reversible.*

Using the definition of reversible DTMCs, we immediately get the following theorem.

Theorem 6.33 *A CTMC with state-space S and generator matrix Q is reversible if and only if for every $r \geq 1$, and $i_0 \neq i_1 \neq \cdots \neq i_r$,*

$$q_{i_0,i_1} q_{i_1,i_2} \cdots q_{i_{r-1},i_r} q_{i_r,i_0} = q_{i_0,i_r} q_{i_r,i_{r-1}} q_{i_{r-1},i_{r-2}} \cdots q_{i_1,i_0}. \tag{6.93}$$

Proof: Suppose the CTMC is reversible. Then, by definition, so is the corresponding embedded DTMC. Now, if $q_{i_k} = 0$ for any $0 \leq k \leq r$ the Equation 6.93 holds trivially since both sides are zero. So assume that $q_{i_k} > 0$ for all $0 \leq k \leq r$. Since the DTMC is reversible, we have, from Definition 4.11,

$$p_{i_0,i_1} p_{i_1,i_2} \cdots p_{i_{r-1},i_r} p_{i_r,i_0} = p_{i_0,i_r} p_{i_r,i_{r-1}} p_{i_{r-1},i_{r-2}} \cdots p_{i_1,i_0}. \tag{6.94}$$

Using

$$p_{i,j} = q_{i,j}/q_i, \quad i \neq j,$$

the above equation reduces to Equation 6.93.

Next suppose Equation 6.93 holds. Then, using the transition probabilities of the embedded DTMC we see that Equation 6.94 holds. Hence the embedded DTMC is reversible. Hence the CTMC is reversible. ■

The next theorem is analogous to Theorem 4.27.

Theorem 6.34 *An irreducible, positive recurrent CTMC with state-space S, generator matrix Q and stationary distribution $\{p_i, i \in S\}$ is reversible if and only if*

$$p_i q_{i,j} = p_j q_{j,i}, \quad i, j \in S, \ i \neq j. \tag{6.95}$$

Proof: Suppose the CTMC is irreducible, positive recurrent, and reversible. Since we implicitly assume that the CTMC is regular, the embedded DTMC is irreducible, positive recurrent, and reversible. Let $\{\pi_i, i \in S\}$ be the stationary distribution of the DTMC. Since it is reversible, Theorem 4.26 yields

$$\pi_i p_{i,j} = \pi_j p_{j,i}, \quad i \neq j \in S,$$

which implies that

$$\pi_i \frac{q_{i,j}}{q_i} = \pi_j \frac{q_{j,i}}{q_j}, \quad i \neq j \in S.$$

From Theorem 6.26, we know that $p_j = C\pi_j/q_j$, for some constant C. Hence the above equation reduces to Equation 6.95.

Now suppose the CTMC is regular, irreducible, positive recurrent, and Equation 6.95 holds. This implies that

$$p_i q_i p_{i,j} = p_j q_j p_{j,i}.$$

Then, using Theorem 6.26, we see that the above equation reduces to

$$\pi_i p_{i,j} = \pi_j p_{j,i}.$$

Thus the embedded DTMC is reversible. Hence the CTMC is reversible, and the theorem follows. ■

The Equations 6.95 are called the *local balance* or *detailed balance* equations, as opposed to Equations 6.70, which are called the *global balance* equations. Intuitively, the local balance equations say that, in steady state, the rate of transitions from state i to j is the same as the rate of transitions from j to i. This is in contrast to stationary CTMCs that are not reversible: for such CTMCs the global balance equations imply that the rate of transitions out of a state is the same as the rate of transitions into that state. It can be shown that the local balance equations imply global balance equations, but not the other way. See Conceptual Exercise 6.16.

Example 6.46   Birth and Death Processes. Consider a positive recurrent birth and death process as described in Example 6.35. Show that it is a reversible CTMC.

From Equation 6.73 we see that the stationary distribution satisfies the local balance equations

$$p_i q_{i,i+1} = p_{i+1} q_{i+1,i}.$$

Since the only transitions in a birth and death process are from i to $i + 1$, and i to $i - 1$, we see that Equations 6.95 are satisfied. Hence the CTMC is reversible. ∎

Reversible CTMCs arise in population models, networks of queues, and a variety of other applications. An excellent account is given in the books by Kelly (1979) and Whittle (1986). The steady state analysis of reversible CTMCs is usually very simple due to the simplicity of the local balance equations. In fact, one method of establishing reversibility is to guess the limiting distribution, and then verify that it satisfies the local balance equations. In many cases, Equation 6.95 itself is useful in coming up with a good guess. We illustrate with two examples.

Example 6.47   Database Locking. A database is a collection of information records stored on a computer. The records are indexed 1 through N. A user can access multiple records. We say that the user belongs to class A if he wishes to access records with indices in the set $A \subseteq \{1, 2, \cdots, N\}$. When a user of class A arrives, he checks to see if all the records in A are available. If they are, the user gets access to them and he immediately locks the records. If a record is locked, it is unavailable to any other user. When the user is done, he simultaneously unlocks all the records he was using. If an arriving user finds that some of the records that he needs to access are locked (and hence unavailable to him), he leaves immediately.

(In practice there are two classes of users: readers (who read the records) or writers (who alter the records), and correspondingly, the locks are of two types: shared and exclusive. When a writer wants to access records in the set A, he gets the access if none of the records in A have any kind of lock from other users, in which case he puts an exclusive lock on each of them. When a reader wants to access records in the set A, he gets it if none of the records in A have an exclusive lock on them, and then puts a shared lock on each of them. This ensures that many readers can simultaneously read a record, but if a record is being modified by a writer no one else can access a record. We do not consider this further complication here.)

Now suppose users of type A arrive according to a $PP(\lambda(A))$, and the arrival processes are independent of each other. A user of type A needs an $\exp(\mu(A))$ amount of time to process these records. The processing times are independent of each other and the arrival processes. We say that the database is in state $(A_1, A_2, \cdots, A_k)$ if there is exactly one user of type A_i ($1 \leq i \leq k$) in the system at time t. This implies that the sets $A_1, A_2, \cdots, A_k$ must be mutually disjoint. When there are no users in the system, we denote that state by ϕ.

Let $X(t)$ be the state of the database at time t. The process $\{X(t), t \geq 0\}$ is a CTMC with state-space

$$\begin{aligned} S \ = \ & \{\phi\} \cup \{(A_1, A_2, \cdots, A_k) : k \geq 1, A_i \neq \phi, A_i \subseteq \{1, 2, \cdots, N\}, \\ & A_i A_j = \phi \text{ if } i \neq j, 1 \leq i, j \leq k\}, \end{aligned}$$

and transition rates given below (we write $q(i \to j)$ instead of $q_{i,j}$ for ease of reading):

$$q(\phi \to (A)) = \lambda(A),$$
$$q((A_1, A_2, \cdots, A_k) \to (A_1, A_2, \cdots, A_k, A_{k+1})) = \lambda(A_{k+1}),$$
$$q((A_1, A_2, \cdots, A_k) \to (A_1, \cdots, A_{i-1}, A_{i+1}, \cdots, A_k)) = \mu(A_i),$$
$$q((A) \to \phi) = \mu(A).$$

The above rates are defined whenever $(A_1, A_2, \cdots, A_k) \in S$ and $(A_1, A_2, \cdots, A_k, A_{k+1}) \in S$. Thus the state-space is finite and the CTMC is irreducible and positive recurrent. Let

$$p(\phi) = \lim_{t \to \infty} \mathsf{P}(X(t) = \phi),$$
$$p(A_1, A_2, \cdots, A_k) = \lim_{t \to \infty} \mathsf{P}(X(t) = (A_1, A_2, \cdots, A_k)),$$

for $(A_1, A_2, \cdots, A_k) \in S$. Now suppose, for $(A_1, A_2, \cdots, A_k) \in S$,

$$p(A_1, A_2, \cdots, A_k) = \frac{\lambda(A_1)\lambda(A_2) \cdots \lambda(A_k)}{\mu(A_1)\mu(A_2) \cdots \mu(A_k)} p(\phi). \tag{6.96}$$

It is straightforward to verify that this satisfies the local balance equations:

$$\lambda(A)p(\phi) = \mu(A)p(A), \quad (A) \in S - \phi,$$
$$\lambda(A_{k+1})p(A_1, A_2, \cdots, A_k) = \mu(A_{k+1})p(A_1, A_2, \cdots, A_k, A_{k+1}),$$

for $(A_1, A_2, \cdots, A_{k+1}) \in S$. The above equations imply that $\{X(t), t \geq 0\}$ is a reversible CTMC with the limiting distribution given in Equation 6.96. Using the normalizing equation we get

$$p(\phi) = \left\{ 1 + \sum_{(A_1, A_2, \cdots, A_k) \in S} \frac{\lambda(A_1)\lambda(A_2) \cdots \lambda(A_k)}{\mu(A_1)\mu(A_2) \cdots \mu(A_k)} \right\}^{-1}.$$

Of course, the hard part here is to compute the sum in the above expression. ∎

Example 6.48 Restaurant Process. Consider a restaurant with seating capacity N. Diners arrive in batches (parties) and leave in the same party that they arrived in. Suppose parties of size i arrive according to $\mathrm{PP}(\lambda_i)$, independently of each other. When a party of size i arrives and the restaurant has i or more seats available, the party is seated immediately. (We assume that the seating is flexible enough so that no seats are wasted.) If not enough seats are available, the party leaves immediately. A party of size i that gets seated stays in the restaurant for an $\exp(\mu_i)$ amount of time, independent of everything else. This implies that the restaurant has ample service capacity so that the service is not affected by how busy the restaurant is.

We say that the restaurant is in state $(i_1, i_2, \cdots, i_k)$ if there are k parties in the restaurant and the size of the rth party is i_r $(1 \leq r \leq k)$. The empty restaurant is said to be in state ϕ. Let $X(t)$ be the state of the restaurant at time t. The process

$\{X(t), t \geq 0\}$ is a CTMC with state-space

$$S = \{\phi\} \cup \{(i_1, i_2, \cdots, i_k) : k \geq 1, i_r > 0, 1 \leq r \leq k, \sum_{r=1}^{k} i_r \leq N\},$$

and transition rates given below (we write $q(i \to j)$ instead of $q_{i,j}$ for ease of reading):

$$q(\phi \to (i)) = \lambda_i, \quad 1 \leq i \leq N,$$
$$q((i_1, i_2, \cdots, i_k) \to (i_1, i_2, \cdots, i_k, i_{k+1})) = \lambda_{i_{k+1}}, \quad (i_1, i_2, \cdots, i_{k+1}) \in S,$$
$$q((i_1, i_2, \cdots, i_k) \to (i_1, \cdots, i_{r-1}, i_{r+1}, \cdots, i_k)) = \mu_{i_r}, \quad (i_1, i_2, \cdots, i_k) \in S,$$
$$q((i) \to \phi) = \mu_i \quad 1 \leq i \leq N.$$

Thus the state-space is finite and the CTMC is irreducible and positive recurrent. Let

$$p(\phi) = \lim_{t \to \infty} \mathsf{P}(X(t) = \phi),$$
$$p(i_1, i_2, \cdots, i_k) = \lim_{t \to \infty} \mathsf{P}(X(t) = (i_1, i_2, \cdots, i_k)), \quad (i_1, i_2, \cdots, i_k) \in S.$$

Now suppose, for $(i_1, i_2, \cdots, i_k) \in S$,

$$p(i_1, i_2, \cdots, i_k) = \frac{\lambda_{i_1} \lambda_{i_2} \cdots \lambda_{i_k}}{\mu_{i_1} \mu_{i_2} \cdots \mu_{i_k}} p(\phi). \tag{6.97}$$

It is straightforward to verify that this satisfies the local balance equations:

$$\lambda_i p(\phi) = \mu_i p(i), \quad 1 \leq i \leq N,$$
$$\lambda_{i_{k+1}} p(i_1, i_2, \cdots, i_k) = \mu_{i_{k+1}} p(i_1, i_2, \cdots, i_k, i_{k+1}),$$

for $(i_1, i_2, \cdots, i_k, i_{k+1}) \in S$. The above equations imply that $\{X(t), t \geq 0\}$ is a reversible CTMC with the limiting distribution given in Equation 6.97 . Using the normalizing equation we get

$$p(\phi) = \left\{ 1 + \sum_{(i_1, i_2, \cdots, i_k) \in S} \frac{\lambda_{i_1} \lambda_{i_2} \cdots \lambda_{i_k}}{\mu_{i_1} \mu_{i_2} \cdots \mu_{i_k}} \right\}^{-1}.$$

As in Example 6.47, the hard part here is to compute the sum in the above expression since the number of terms in the sum grows exponentially in N. ∎

6.16 Modeling Exercises

6.1 Consider a workshop with k machines, each with its own repairperson. The k machines are identical and behave independently of each other as described in Examples 6.1 and 6.4. Let $X(t)$ be the number of working machines at time t. Show that $\{X(t), t \geq 0\}$ is a CTMC. Display its generator matrix and the rate diagram.

6.2 Do Modeling Exercise 6.1 assuming that there are r repairpersons ($1 \leq r < k$). The machines are repaired in the order of failure.

6.3 Consider the two-machine one repairperson workshop of Example 6.6. Suppose the repair time of the ith machine is $\exp(\lambda_i)$ and the lifetime is $\exp(\mu_i)$, $i = 1, 2$. The repairs take place in the order of failure. Construct an appropriate CTMC to model this system. Assume that all life times and repair times are independent.

6.4 Do Modeling Exercise 6.3 with three distinct machines. Care is needed to handle the "order of failure" service.

6.5 A metal wire, when subject to a load of L kilograms, breaks after an $\exp(\mu L)$ amount of time. Suppose k such wires are used in parallel to hold a load of L kilograms. The wires share the load equally, and their failure times are independent. Let $X(t)$ be the number of unbroken wires at time t, with $X(0) = k$. Formulate $\{X(t), t \geq 0\}$ as a CTMC.

6.6 A bank has five teller windows, and customers wait in a single line and are served by the first available teller. The bank uses the following operating policy: if there are k customers in the bank, it keeps one teller window open for $0 \leq k \leq 5$, two windows for $6 \leq k \leq 8$, three for $9 \leq k \leq 12$, four for $13 \leq k \leq 15$, and all five are kept open for $k \geq 16$. The service times are iid $\exp(\mu)$ at any of the tellers, and the arrival process is PP(λ). Let $X(t)$ be the number of customers in the system at time t. Show that $\{X(t), t \geq 0\}$ is a birth and death process and find its parameters.

6.7 Consider a two-server system where customers arrive according to a PP(λ). Those served by server i require $\exp(\mu_i)$ amount of service time, $i = 1, 2$. A customer arriving at an empty system goes to server 1 with probability α and to server 2 with probability $1 - \alpha$. Otherwise a customer goes to the first available server. Formulate this as a CTMC. Is it a birth and death process?

6.8 Consider a two-server queue that operates as follows: two different queues are formed in front of the two servers. The arriving customer joins the shorter of the two queues (the customer in service is counted as part of that queue). If the two queues are equal, he joins either one with probability .5. Queue jumping is not allowed. Let $X_i(t)$ be the number of customers in the ith queue, including any in service with server i, $i = 1, 2$. Assume that the arrival process is PP(λ) and the service times are iid $\exp(\mu)$ at either server. Model $\{(X_1(t), X_2(t)), t \geq 0\}$ as a CTMC and specify its transition rates and draw the rate diagram.

6.9 Consider a system consisting of n components in series, i.e., it needs all components in working order in order to function properly. The lifetime of the ith component is $\exp(\lambda_i)$ random variable. At time 0 all the components are up. As soon as any of the components fails, the system fails, and the repair starts immediately. The repair time of the ith component is an $\exp(\mu_i)$ random variable. When the system is down, no more failures occur. Let $X(t) = 0$ if the system is functioning at time t, and $X(t) = i$ if component i is down (and hence under repair) at time t. Show that $\{X(t), t \geq 0\}$ is a CTMC and display its rate diagram.

6.10 Consider an infinite server queue of Example 6.12. Suppose the customers arrive according to a CPP with batch arrival rate λ and the successive batch sizes are iid geometric with parameter $1 - p$. Each customer is served by a different server in an independent fashion, and the service times are iid $\exp(\mu)$ random variables. Model the number of customers in the system by a CTMC. Compute the transition rates.

6.11 Consider a queueing system with s servers. The incoming customers belong to two classes: 1 and 2. Class 2 customers are allowed to enter the system if and only if there is a free server immediately available upon their arrival, otherwise they are turned away. Class 1 customers always enter the system and wait in an infinite capacity waiting room for service if necessary. Assume that class i customers arrive according to independent $PP(\lambda_i)$, $i = 1, 2$. The service times are iid $\exp(\mu)$ for both classes. Let $X(t)$ be the number of customers in the system at time t. Show that $\{X(t), t \geq 0\}$ is a birth and death process and find its parameters.

6.12 Consider a bus depot where customers arrive according to a $PP(\lambda)$, and the buses arrive according to a $PP(\mu)$. Each bus has capacity $K > 0$. The bus departs with $\min(x, K)$ passengers, where x is the number of customers waiting when the bus arrives. Loading time is insignificant. Let $X(t)$ be the number of customers waiting at the depot at time t. Show that $\{X(t), t \geq 0\}$ is a CTMC and compute its generator matrix.

6.13 Customers arrive at a single-server facility according to a $PP(\lambda)$. An arriving customer, independent of everything else, belongs to class 1 with probability α and class 2 with probability $\beta = 1 - \alpha$. The service times of class i customers are iid $\exp(\mu_i)$, $i = 1, 2$, with $\mu_1 \neq \mu_2$. The customers form a single queue. Let $X(t)$ be the number of customers in the system at time t and $Y(t)$ be the class of the customer in service. Define $Y(t) = 0$ if $X(t) = 0$. Model $\{(X(t), Y(t)), t \geq 0\}$ as a CTMC and specify its transition rates and draw the rate diagram.

6.14 Consider the single-server queue of Example 6.11. Suppose that every customer has "patience time" such that if his wait for service to begin exceeds the patience time, he leaves the system without service. Suppose the patience times of customers are iid $\exp(\theta)$ random variables. Let $X(t)$ be the number of customers in the system at time t. Show that $\{X(t), t \geq 0\}$ is a birth and death process and find its parameters.

6.15 Genetic engineers have come up with a super-amoeba with $\exp(\mu)$ lifetimes and the following property: at the end of its lifetime it ceases to exist with probability 0.3, or splits into two clones of itself with probability .4, or three clones of itself with probability .3. Let $X(t)$ be the number of super-amoebae in a colony at time t. Suppose all the existing amoebae behave independently of each other. Show that $\{X(t), t \geq 0\}$ is a CTMC and compute its generator matrix.

6.16 Consider a single-server queue with the following operating policy: Once the server becomes idle, it stays idle as long there are less than N (a given positive integer) customers in the system. As soon as there are N customers in the system, the server starts serving them one by one and continues until there is no one left in the system, at which time the server becomes idle. The arrival process is a PP(λ) and the service times are iid exp(μ). Model this as a CTMC by describing its state-space and the generator matrix. Draw its rate diagram.

6.17 Consider a computer system with five independent and identical central processing units (CPUs). The lifetime of each CPU is exp(μ). When a CPU fails, an automatic mechanism instantaneously isolates the failed CPU, and the system continues in a reduced capacity with the remaining working CPUs. If this automatic mechanism fails to isolate the failed CPU, which can happen with probability $1 - c$, the system crashes. If no working CPUs are left, then also the system crashes. Since the system is aboard a spaceship, once the system crashes it cannot be repaired. Model the system by a CTMC by giving its state-space, generator matrix, and the rate diagram.

6.18 Consider the system in Modeling Exercise 6.17. Now suppose there is a robot aboard the space ship that can repair the failed CPUs one at a time, the repair time for each CPU being exp(λ). However, the robot uses the computer system for the repair operation, and hence requires a non-crashed system. That is, the robot works as long as the system is working. Model this modified system by a CTMC by giving its state-space, generator matrix, and the rate diagram.

6.19 Consider the s-server system of Modeling Exercise 6.11. Suppose the service times of the customers of class i are iid exp(μ_i), $i = 1, 2$. Model this modified system by a CTMC by giving its state-space and the generator matrix.

6.20 Unslotted Aloha. Consider a communications system where the messages arrive according to a PP(λ). As soon as a message arrives it attempts transmission. The message transmission times are iid exp(μ). If no other message tries to transmit during this time, the message transmission is successful. Otherwise a collision results, and both the messages involved in the collision are terminated instantaneously. All messages involved in a collision are called backlogged, and are forced to retransmit. All backlogged messages wait for an exp(θ) amount of time (independently of each other) before starting a retransmission attempt. Let $X(t)$ be the number of backlogged messages at time t, $Y(t) = 1$ if a message is under transmission at time t and zero otherwise. Model $\{(X(t), Y(t)), t \geq 0\}$ as a CTMC and specify its transition rates and draw the rate diagram.

6.21 Leaky Bucket. Packets arrive for transmission at a node according to a PP(λ) and join the packet queue. Tokens (permission slips to transmit) arrive at that node according to a PP(μ) and join the token pool. The node uses a "leaky bucket" transmission protocol to control the entry of the packets into the network, and it operates as follows: if the token pool is empty, the arriving packets wait in the packet queue. Otherwise, an incoming packet removes a token from the token pool and is instantaneously transmitted. Thus the packet queue and the token pool cannot simultaneously be non-empty. The token pool size is M, and any tokens generated when the token pool is full are discarded. Model this system by a CTMC by giving its state-space, generator matrix, and the rate diagram.

6.22 An inventory system is under continuous control as follows. Demands arrive according to a PP(λ) and are satisfied instantly if there are items in the warehouse. If the warehouse is empty when a demand occurs, the demand is lost. As soon as the inventory level drops to R (a given positive integer), an order is placed for K (another given positive integer $> R$) items from the supplier. It takes an $\exp(\theta)$ amount of time for the supplier to deliver the order to the warehouse. Let $X(t)$ be the number of items in the inventory at time t, and assume that $R < X(0) \leq K + R$. Model $\{X(t), t \geq 0\}$ as a CTMC by giving its state-space and the generator matrix.

6.23 Consider the following model of the growth of a macromolecule in a chemical solution. A chemical solution has two types of molecules, labeled A and T, suspended in it. The A molecules are active, while the T molecules are terminal. The A molecules can latch onto each other and form linear macromolecules, e.g., $AAA \cdots AA$. Such a macromolecule can grow by adding an A or a T molecule at either end. Once there is a T molecule at an end, no more molecules can be added to that end. Thus a macromolecule $TAAA$ can grow only at one end, while the macromolecule $TAAT$ cannot grow any more. Suppose that an A molecule can get attached to another A molecule at rate λ, while a T molecule can attach to an A with rate θ. The bond between two A molecules can come unglued at rate μ, while the bond between an A and a T molecule is permanent. Assume that the length of a macromolecule can only increase or decrease by one unit at a time. Let $X(t)$ be the state of a macromolecule at time t. Assume that at time zero a macromolecule consists of a single A molecule. Model $\{X(t), t \geq 0\}$ as a CTMC by giving its state-space and the rate diagram.

6.24 There are K parking spaces, indexed 1 through K, in front of a retail store. Space 1 is the closest (and the most preferred) and space K is the farthest (and the least preferred). Customers arrive according to a PP(λ) and occupy the closest available space. If all spaces are occupied, the customer goes away and is lost. Each customer spends an $\exp(\mu)$ amount of time in the store, independent of the others. Let $X_i(t) = 1$ ($1 \leq i \leq K$) if the i-th space is occupied at time t and 0 otherwise. Also define $Z_k(t)$ to be the number of spaces among $\{1, 2, ..., k\}$ that are occupied at time t, ($1 \leq k \leq K$). Is $\{X_i(t), t \geq 0\}$ a CTMC ($1 \leq i \leq K$)? If yes, show its generator matrix, or the rate diagram. Is $\{Z_k(t), t \geq 0\}$ a CTMC ($1 \leq k \leq K$)? If yes, show its generator matrix, or the rate diagram.

6.25 Customers arrive according to a PP(λ) to a queueing system with two servers. The ith server ($i = 1, 2$) needs $\exp(\mu_i)$ amount of time to serve one customer. Each incoming customer is routed either to server 1 with probability p_1 or to server 2 with probability $p_2 = 1 - p_1$, independently. Queue jumping is not allowed, thus there is a separate queue in front of each server. Let $X_i(t)$ be the number of customers in the i-th queue at time t (including any in service with the ith server). Show that $\{X_i(t), t \geq 0\}$ is a birth and death process, $i = 1, 2$. Are the two processes independent?

6.26 Jobs arrive at a central computer according to a PP(λ). The job processing times are iid exp(μ). The computer processes them one at a time in the order of arrival. The computer is subject to failures and repairs as follows: It stays functional for an exp(α) amount of time and then fails. It takes an exp(β) amount of time to repair it back to a fully functional state. Successive up and down times are iid and independent of the number of jobs in the system. When the computer fails, all the jobs in the system are lost. All jobs arriving at the computer while it is down are also lost. Model this system as a CTMC by giving its state-space and the rate diagram.

6.27 A single server queue serves K types of customers. Customers of type k arrive according to a PP(λ_k), and have iid exp(μ_k) service times. An arriving customer enters service if the server is idle, and leaves immediately if the server is busy at the time of arrival. Let $X(t)$ be 0 if the server is idle at time t, and $X(t) = k$ if the server is serving a customer of type k at time t. Model $X(t)$ as a CTMC by giving its state-space and the rate diagram.

6.28 Life time of a machine is exponentially distributed with parameter μ. A repairperson visits this machine periodically, the inter-visit times being iid exp(λ). If the machine is discovered to be in working condition, the repairperson leaves it alone; otherwise he repairs it. Repair takes exp(θ) amount of time and makes the machine as good as new. Repair visits continue after repair completion as before. Model this system as a CTMC by giving its state-space and the rate diagram.

6.29 Consider the following modification of the machine shop of Example 6.5. When both machines are down each machine is repaired by one repairperson. However, if only one machine is down, both repairpersons work on it together so that the repair occurs at twice the speed. Model this system as a CTMC by giving its state-space and the rate diagram.

6.30 A machine produces items according to a PP(λ). The produced items are stored in a buffer. Demands for these items occur according to a PP(μ). If an item is available in the buffer when a demand occurs, the demand is satisfied immediately. If the buffer is empty when a demand occurs, the demand is lost. The machine is turned off when the number of items in the buffer reaches K, a fixed positive number, and is turned on when the number of items in the buffer falls to a pre-specified number $k < K$. Once it is turned on it stays on until the number of items in the buffer reaches K. Model this system as a CTMC. State the state-space and show the rate diagram.

6.31 Customers of two classes arrive at a single server service station with infinite waiting room. Class i customers arrive according to a PP(λ_i), $i = 1, 2$. Customers of class 1 are always allowed to enter the system, while those of class 2 can enter the system if and only if the total number of customers in the system (before this customer joins) is less than K, where $K > 0$ is a fixed integer. The service times are iid exp(μ) for both the classes. Let $X(t)$ be the total number of customers in the system at time t. Show that $\{X(t), t \geq 0\}$ is a birth and death process and state its parameters.

6.32 Customers arrive at a service station according to a PP(λ). Each customer brings with him two tasks that can be done in parallel. Each task takes an $\exp(\mu)$ amount of time, and the two tasks are independent. There are two servers in the system. When a customer enters service, the servers begin working on one task each. When one of two tasks finishes, the freed up server helps the other server so that the remaining task is processed at twice the rate. (i.e., the remaining service time of the task is now $\exp(2\mu)$ instead of $\exp(\mu)$.) Model this system as a CTMC. State the state-space and show the rate diagram.

6.33 Consider a gas station with two pumps and three car-spaces as shown in Figure 6.11. Potential customers arrive at the gas station according to a PP(λ). An in-

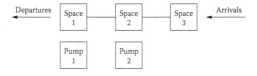

Figure 6.11 *The 2-pump gas station.*

coming customer goes to pump 1 if both pumps are idle. If pump 1 is busy but pump 2 is idle, she goes to pump 2. If pump 2 is occupied she waits in the space 3 (regardless of whether pump 1 occupied or not. If space 3 is occupied she goes away. It takes an $\exp(\mu)$ amount of time to fill gas. After finishing filling gas, the customer at pump 1 leaves, the customer at pump 2 leaves if space 1 is vacant, else she has to wait until the customer in space 1 is done (due to one way rules and space restrictions). In that case both customers leave simultaneously when the customer in space 1 is done. Model this system as a CTMC. State the state-space and show the rate diagram.

6.34 A machine shop has three identical machines, at most two of which can be in use at any time. Initially all three machines are in working condition. The policy of the shop is to keep a machine in standby mode if and only if there are two working machines in use. Each machine in use fails after an $\exp(\lambda)$ amount of time, whereas the standby machine (if in working condition) fails after an $\exp(\theta)$ amount of time. The failure of a standby machine can be detected only after it is put in use. There is a single repair person who repairs the machines in $\exp(\mu)$ amount of time. Model this system as a CTMC. Describe the state-space and show the rate diagram. Assume independence as needed.

6.35 Consider the following model of an epidemic: a population consists of N individuals, of whom one is infected and $N-1$ are susceptible initially. Any two individuals come in contact with each other after an $\exp(\beta)$ amount of time. If a susceptible individual comes in contact with an infected individual, he gets infected. An infected individual stays infectious forever. Let $I(t)$ be the number of infected individuals in the system at time t. Model $\{I(t), t \geq 0\}$ as a CTMC. Give its state-space and transition rates.

6.36 Consider an epidemic model for a society with N individuals, of which K are carriers of an infectious disease. The remaining $N - K$ individuals are susceptible. The carriers display no symptoms. However, when they come in contact with a susceptible individual, the susceptible individual gets infected, shows symptoms and is immediately quarantined, so he cannot infect any more individuals. Any two individuals come in contact with each other after an $\exp(\beta)$ amount of time. A carrier ceases to be a carrier after an $\exp(\gamma)$ amount of time, after which it becomes immune. Let $C(t)$ be the number of carriers and $S(t)$ be the number of susceptibles in the society at time t. Model $\{(S(t), C(t)), t \geq 0\}$ as a CTMC. Give its state-space and transition rates.

6.37 Consider the SIR epidemic model of Section 6.3.3. Suppose we institute a vaccination program, identified by a parameter ν as follows. Under this program every susceptible person gets vaccinated after an $\exp(\nu)$ amount of time. Thus, the higher the ν, the quicker the program takes effect. Once an individual is vaccinated, he/she is no longer susceptible, and behaves like a recovered individual. Let $S(t)$, $I(t)$ and $R(t)$ be as in Section 6.3.3. Model $\{(S(t), I(t)), t \geq 0\}$ as a CTMC. Give its state-space and transition rates.

6.38 Consider an order book for a single stock but with two prices $p_1 < p_2$. Buy (sell) orders arrive at bid (ask) price p_i according to a PP(λ_i^b) (PP(λ_i^s)) ($i = 1, 2$). A buy order waits in the system if there is no sell order at a price at or below its bid price. Similarly, a sell order waits in the system if there is no buy order at a price at or above its ask price. An arriving buy order that sees a sell order on the book at or below its bid price immediately removes the most favorable sell order and leaves the system. For example, if there are sell orders waiting at prices p_1 and p_2, and a buy order comes in at p_2, it is matched against the sell order at p_1. Similarly, an arriving sell order that sees a buy order on the book at or above its ask price immediately removes the most favorable buy order and leaves the system. There can be at most K orders waiting in the system at any given price (thus allowing a maximum of total $2K$ orders on the book). Once an order enters a system, it stays there until execution. Model this order book as a CTMC. Give its state-space and the transition diagram.

6.39 A buy order is called a market order if its bid price is equal to the market ask price (that is, the lowest ask price among all the sell orders on the order book). Consider a one-sided sell book in which all the buy orders are market orders. Suppose sell orders arrive according to a PP(λ). The market buy orders arrive according to a PP(μ). If there are no sell orders on the book when a market buy order arrives, it goes away and is lost. Each sell order on the book gets cancelled at rate θ. Let $S(t)$ be the number of sell orders on the order book at time t. Model $\{S(t), t \geq 0\}$ as a CTMC.

6.40 Consider an organ transplant system with J types of patients and I types of kidneys. Suppose type j patients arrive according to a PP(λ_j) ($1 \leq j \leq J$) and wait for kidneys. Type i kidneys arrive according to a PP(μ_i) ($1 \leq i \leq I$) and are allocated to patients of type j with probability $\beta_{i,j}$, where $\sum_{j=1}^{J} \beta_{i,j} = 1$. A patient always accepts a kidney that is allocated to him/her. If there are no patients of type j

then the kidneys allocated to those patients are lost. (Thus this is a state-independent allocation rule.) Kidneys do not wait for patients. Let $X_j(t)$ be the number of patients of type j waiting for kidneys. Model $\{X_j(t), t \geq 0\}$ as a CTMC.

6.41 Consider a hospital with B beds. There are two types of patients that use this hospital. Patients of type i ($i = 1, 2$) arrive according to a PP(λ_i) and require iid exp(μ_i) service. The hospital uses the following policy: it reserves b beds for patients of type 1. Thus when a type 1 patient arrives, she is admitted to the hospital as long as any bed is available. However, a patient of type 2 is admitted if and only if the number of type 2 patients currently occupying beds is less than $B-b$. Let $X_i(t)$ be the number of patients of type i in the hospital at time t ($i = 1, 2$). Is $\{(X_1(t), X_2(t)), t \geq 0\}$ a CTMC? If it is, give its state-space and transition rates.

6.42 Here is a simple model of a hospital that incorporates the observed phenomenon that the longer it takes for a service to begin, the longer is the service time. Model the hospital as a single server queueing system where customers arrive according to a PP(λ). Customers who find the system empty upon arrival are called type 0 customers and require exp(μ_0) service times, while the others are called type 1 customers and require exp(μ_1) service times. Assume that $\mu_1 < \mu_0$, and all service times are independent. Let $X(t)$ be the number of customers in the system at time t and $Y(t)$ be 0 if the server is idle or serving a type 0 customer at time t, and 1 otherwise. Show that $\{(X(t), Y(t)), t \geq 0\}$ is a CTMC. Give its state-space and transition rates.

6.17 Computational Exercises

6.1 Consider a pure birth process $\{X(t), t \geq 0\}$ with birth parameters

$$\lambda_{2n} = \alpha > 0, \quad \lambda_{2n+1} = \beta > 0, \quad n \geq 0.$$

Compute $\mathsf{P}(X(t) \text{ is odd} | X(0) = 0)$ for $t \geq 0$.

6.2 Let $m_i(t) = \mathsf{E}(X(t)|X(0) = i)$, where $X(t)$ is the number of customers at time t in an infinite-server queue of Example 6.12. Derive the differential equations for $m_i(t)$ ($i \geq 0$) by using the forward equations for $p_{i,j}(t)$. Solve for $m_0(t)$.

6.3 Let $\{X(t), t \geq 0\}$ be the CTMC of Example 6.5. Compute its transition probability matrix $P(t)$ by

1. exploiting the independence of the machines,
2. using the matrix exponential technique,
3. using the Laplace transform technique.

6.4 Consider the Yule process $\{X(t), t \geq 0\}$ with $\lambda_n = n\lambda$ for $n \geq 0$. Compute $\mathsf{P}(X(t) = j | X(0) = i)$.

6.5 For the pure death process $\{X(t), t \geq 0\}$ with $\mu_n = n\mu$ for $n \geq 0$, compute $P(X(t) = j | X(0) = i)$.

6.6 Consider the CTMC $\{X(t), t \geq 0\}$ of Modeling Exercise 6.15. Let $m_i(t) = E(X(t) | X(0) = i)$. Show that

$$m_i'(t) = 0.7\mu m_i(t), \quad m_i(0) = i.$$

Solve for $m_i(t)$.

6.7 Let $\{X(t), t \geq 0\}$ be the pure birth process of Computational Exercise 6.1. Compute $E(X(t) | X(0) = 0)$.

6.8 Customers arrive at a service center according to a PP(λ), and demand iid exp(μ) service. There is a single service person and waiting room for only one customer. Thus if an incoming customer finds the server idle he immediately enters service, or else he leaves without service. A special customer, who knows how the system works, wants to use it in a different way. He inspects the server at times $0, T, 2T, 3T, \ldots$ (here $T > 0$ is a fixed real number) and enters the system as soon as he finds the server idle upon inspection. Compute the distribution of the number of visits the special customer makes until he enters, assuming that he finds the server busy at time 0. Also compute the expected time that the special customer has to wait until he enters.

6.9 A machine is either up or down. The up machine stays up for an exp(μ) amount of time and then fails. The repairperson takes an exp(λ) amount of time to fix the machine to a good-as-new state. A repair person is on duty from 8:00 am to 4:00 pm every day, and works on the machine if it fails. A working machine can fail at any time, whether the repair person is on duty or not. If the machine is under repair at 4:00 pm on a given day, or it fails while the repair person is off-duty, it stays down until 8:00 am the next day, when the repairperson resumes (or starts) the repair work. Compute the steady state probability that the machine is working when the repairperson reports to work.

6.10 Compute the steady state distribution of the CTMC of Modeling Exercise 6.1.

6.11 Compute the steady state distribution of the CTMC of Modeling Exercise 6.6. What is the condition of stability?

6.12 Compute the steady state distribution of the CTMC of Modeling Exercise 6.7. What is the condition of stability?

6.13 Compute the steady state distribution of the CTMC of Modeling Exercise 6.9.

6.14 Compute the steady state distribution of the CTMC of Modeling Exercise 6.4.

6.15 Consider Modeling Exercise 6.10. Let p_k be the limiting probability that there are k customers in the systems. Define

$$G(z) = \sum_{k=0}^{\infty} p_k z^k.$$

Using the balance equations derive the following differential equation for $G(z)$:

$$G'(z) = \frac{\lambda}{\mu} \cdot \frac{1}{1 - pz} \cdot G(z).$$

Solve for $G(z)$.

6.16 Compute the steady state distribution of the CTMC of Modeling Exercise 6.11. What is the condition of stability?

6.17 Consider Modeling Exercise 6.13. Let

$$p_{i,k} = \lim_{t \to \infty} \mathsf{P}(X(t) = i, Y(t) = k), \quad i \geq 0, \ k = 1, 2,$$

and define

$$\phi_k(z) = \sum_{i=0}^{\infty} p_{i,k} z^i, \quad k = 1, 2.$$

Show that

$$\phi_1(z) = \frac{\alpha\lambda(\lambda(1 - z) + \mu_2)p_{0,0}}{\mu_1\mu_2/z - \lambda\mu_1(1 - \alpha/z) - \lambda\mu_2(1 - \beta/z) - \lambda^2(1 - z)}.$$

(Expression for ϕ_2 is obtained by interchanging α with β and μ_1 with μ_2.) Compute $p_{0,0}$ and show that the condition of stability is

$$\lambda\left(\frac{\alpha}{\mu_1} + \frac{\beta}{\mu_2}\right) < 1.$$

6.18 Compute the steady state distribution of the CTMC of Modeling Exercise 6.14. What is the condition of stability?

6.19 Consider the parking lot of Modeling Exercise 6.24. Compute the long run probability that the i-th space is occupied. Also compute the long run fraction of the customers that are lost.

6.20 Consider the queueing system of Modeling Exercise 6.25. When is $\{X_i(t), t \geq 0\}$ stable, $i = 1, 2$? Assuming it is stable, what is the mean and the variance of the number of customers in the entire system in steady state?

6.21 Consider the computer system of Modeling Example 6.26. When is this system stable? What is its limiting distribution, assuming stability? What fraction of the incoming jobs are completed successfully in steady state?

6.22 Compute the limiting distribution of the number of items in the inventory system described in Modeling Exercise 6.22. In steady state, what fraction of the demands are lost?

6.23 Consider the system described in Modeling Exercise 6.27. What is the limiting probability that the server is idle? What is the limiting probability that it is serving a customer of type k, $1 \leq k \leq K$?

6.24 Compute the steady state distribution of the system described in Modeling Exercise 6.28.

6.25 Consider the system of Modeling Exercise 6.31. What is the condition of stability for this process? Compute its limiting distribution assuming it is stable.

6.26 Consider the single server queue of Example 6.11. Now suppose the system capacity is K, i.e., an arriving customer who finds K customers already in the system is turned away. Compute the limiting distribution of the number of customers in such a system.

6.27 Consider the system described in Modeling Exercise 6.32. State the condition of stability and compute the expected number of customers in steady state assuming that the system is stable.

6.28 Consider the gas station of Modeling Exercise 6.33. Compute the limiting distribution of the state of the system. In steady state, compute the fraction of the potential customers who actually enter the gas station.

6.29 Consider the three-machine workshop of Modeling Exercise 6.34. Compute the limiting distribution of the state of the system. In steady state, what fraction of the time is the repairperson idle?

6.30 Consider the model of macromolecular growth as described in Modeling Exercise 6.23. Compute the limiting distribution of the length of the molecule.

6.31 A finite state CTMC with generator matrix Q is called doubly stochastic if the row as well column sums of Q are zero. Find the limiting distribution of an irreducible doubly stochastic CTMC with state-space $\{1, 2, \cdots, N\}$.

6.32 Let $\{N(t), t \geq 0\}$ be a PP(λ). Compute

$$\lim_{t \to \infty} P(N(t) \text{ is divisible by 3 or 7}).$$

You may use the result of Computational Exercise 6.31.

6.33 Consider a stochastic process on N nodes arranged in a circle. The process takes a clockwise step with rate λ and a counter-clockwise step with rate μ. Display the rate diagram of the CTMC. Compute its limiting distribution.

6.34 Consider the inventory system described in Modeling Exercise 6.22. Compute the expected time between two consecutive orders placed with the supplier.

6.35 Consider the machine shop of Modeling Exercise 6.29. Suppose machine 1 fails at time zero, and machine 2 is working, so that both repairpersons are available to repair machine 1 at that time. Let T be the time when machine 1 becomes operational again. Compute $E(T)$.

6.36 Consider the computer system described in Modeling Exercise 6.17. Suppose at time 0 all CPUs are functioning. Compute the expected time until the system crashes.

6.37 Consider the machine shop described in Modeling Exercise 6.1. Suppose one machine is functioning at time 0. Show that the expected time until all machines are down is given by

$$\frac{1}{k\lambda}\left(\left(\frac{\lambda+\mu}{\mu}\right)^k - 1\right).$$

6.38 Do Computational Exercise 6.36 for the computer system in Modeling Exercise 6.18.

6.39 Consider the finite capacity single server queue of Computational Exercise 6.26. Suppose at time zero there is a single customer in the system. Let T be the arrival time of the first customer who sees an empty system. Compute $E(T)$.

6.40 Consider the computer system of Modeling Exercise 6.26. Suppose initially the computer is up and there are i jobs in the system. Compute the expected time until there are no jobs in the system.

6.41 Consider the parking lot of Modeling Exercise 6.24. Suppose initially there is one car in the lot. What is the expected time until the lot becomes empty?

6.42 Consider the queueing system of Modeling Exercise 6.25. Suppose keeping a customer for one unit of time in the i-th queue costs h_i dollars $(i = 1, 2)$, including the time spent in service. Find the optimum routing probabilities that will minimize the sum of the expected total holding cost per unit time in the two queues in steady state.

6.43 Consider the system described in Modeling Exercise 6.28. Suppose the machine generates revenue at a rate of $30 per hour while it is operating. Each visit of the repairperson costs $100, and the repair itself costs $20 per hour. Compute the long run net income (revenue - cost) per hour if the mean machine lifetime is 80 hrs and mean repair time is 3 hrs. Find the optimal rate at which the repairperson should visit this machine so as to maximize the net income rate in steady state.

6.44 Consider the inventory system described in Modeling Exercise 6.22. Suppose each item sold produces a profit of r, while it costs h to hold one item in the inventory for one unit of time. Compute the long run net profit per unit time.

6.45 Consider the k machine workshop of Modeling Exercise 6.1. Suppose each working machine produces revenues at rate r per unit time. Compute the expected total discounted revenue over infinite horizon starting with k operating machines. Hint: Use independence of the machines.

6.46 A machine shop consists of K independent and identical machines. Each machine works for an $\exp(\mu)$ amount of time before it fails. During its lifetime each machine produces revenue at rate R per unit time. When the machine fails it needs to be replaced by a new and identical machine at a cost of C_m per machine. Any number of machines can be replaced simultaneously. The replacement operation is instantaneous. The repairperson charges C_v per visit, regardless of how many machines are replaced on a given visit. Consider the following policy: Wait until the number of working machines falls below k, (where $0 < k \leq K$ is a fixed integer) and then replace all the failed machines simultaneously. Compute $g(k)$, the long run cost rate of following this policy. Compute the optimal value of the parameter k that minimizes this cost rate for the following data: Number of machines in the machine shop is 10, mean lifetime of the machine is 10 days, the revenue rate is $100 per day, the replacement cost is $750 per machine, and the visit charge is $250.

6.47 Consider the computer system described in Modeling Exercise 6.17. Suppose at time 0 all CPUs are functioning. Suppose each functioning CPU executes r instructions per unit time. Compute the expected total number of instructions executed before the system crashes.

6.48 Consider the machine in Modeling Exercise 6.30. Suppose it costs h dollars to keep an item in the buffer for one unit of time. Each item sells for r dollars. Compute the long run net income (revenue-holding cost) per unit time, as a function of k and K.

6.49 Consider the system in Computational Exercise 6.8. Suppose it costs the special customer c per unit time to wait for service, and d to inspect the server. Compute the total expected cost to the special customer until he enters as a function of T. Suppose $\lambda = 1/\text{hr}$, $\mu = 2/\text{hr}$, $c = \$10/\text{hr}$ and $d = \$20$. Compute the value of T that minimizes the expected total cost.

6.50 A small local car rental company has a fleet of K cars. We are interested in deciding how many of them should be large cars and how many should be mid-size. The demands for large cars occur according to a PP with rate λ_l per day, while those for the mid-size cars occur according to a PP with rate λ_m. Any demand that can not be met immediately is lost. The rental durations (in days) for the large (mid-size) cars are iid exponential random variables with parameter μ_l (μ_m). The rental net revenue for the large (mid-size) cars is r_l (r_m) per day. Compute $g(k)$, the long run average net rental revenue per day, if the fleet has k large cars and $K - k$ mid-size cars. Compute the optimal fleet mix if the following data are known: the fleet size is fixed at 10, demand is 2 cars per day for large cars, 3 per day for mid-size cars, the mean rental duration for large cars is 2 days, while that for the mid-size cars is 3 days. The large cars generate a net revenue of $60 per day, while that for the mid-size cars is $40 per day.

6.51 Customers arrive at a single server service facility according to PP(λ) and request iid exp(μ) service times. Let V_i be the value of the service (in dollars) to the i-th customer. Suppose $\{V_i, i \geq 1\}$ are iid $U(0, 1)$ random variables. A customer has to pay a price p to enter service. An arriving customer enters the system if and only if the server is idle when he arrives *and* the service is worth more to him than the service charge p. Otherwise, the customer leaves without service and is permanently lost. Find the optimal service charge p that will maximize the long run expected revenue per unit time to the system. What fraction of the arriving customers join the system when the optimal charge is used?

6.52 Consider the single server system of Computational Exercise 6.51. An arriving customer enters the system if and only if the service is worth more to him than the service charge p, even if the server is busy. The customer incurs zero cost for waiting. However, the service provider incurs a holding cost of $h per customer per hour of staying in the system. Find the optimal service charge p that will maximize the long run expected revenue per unit time to the system.

6.53 Consider the system of Modeling Exercise 6.31. Class i customers pay c_i to enter the system, $c_1 > c_2$. Assuming the process is stable, compute the long run rate at which the system collects revenue.

6.54 Let $\{X_i(t), t \geq 0\}$ ($1 \leq i \leq K$) be K independent reversible, irreducible, positive recurrent CTMCs. The ith CTMC has state-space S_i and generator matrix Q_i. Let $X(t) = (X_1(t), X_2(t), \cdots, X_K(t))$. Show that $\{X(t), t \geq 0\}$ is a reversible CTMC.

6.55 Consider the following special case of the database locking model described in Example 6.47: $N = 3$ items in the database, $\lambda(1) = \lambda(2) = \lambda(3) = \lambda > 0$, $\lambda(123) = \theta > 0$. All other $\lambda(A)$'s are zero. Assume $\mu(A) = \mu$ for all A's with $\lambda(A) > 0$. Compute the steady state distribution of the state of the database.

6.56 Consider a warehouse that gets shipments from k different sources. Source i sends items to the warehouse according to a PP(λ_i). The items from source i need M_i (a fixed positive integer) amount of space. The capacity of the warehouse is $B > \max(M_1, M_2, \cdots, M_k)$. If there is not enough space in the warehouse for an incoming item, it is sent somewhere else. An item from source i stays in the warehouse for an exp(μ_i) amount of time before it is shipped off. The sojourn times are independent. Model this system by a CTMC and show that it is reversible. What is the probability that a warehouse has to decline an item from source i, in steady state?

6.57 A particle moves on an undirected connected network of N nodes as follows: it stays at the ith node for an $\exp(q_i)$ amount of time and then moves to any one of its immediate neighbors with equal probability. Let $X(t)$ be the index of the node occupied by the particle at time t. Show that $\{X(t), t \geq 0\}$ is a reversible CTMC, and compute its limiting distribution.

6.58 Is the retrial queue of Example 6.15 reversible?

6.59 A system consists of N urns containing k balls. Each ball moves independently among these urns according to a reversible CTMC with rate matrix Q. Let $X_i(t)$ be the number of balls in the ith urn at time t. Show that $\{X(t) = (X_1(t), X_2(t), \cdots, X_N(t)), t \geq 0\}$ is a reversible CTMC.

6.60 Consider the epidemic model of Modeling Exercise 6.35. Compute the expected time when every individual in the colony is infected.

6.61 Consider the epidemic model of Modeling Exercise 6.36. Compute the probability that the epidemic terminates (that is all carriers die out) before anyone gets infected.

6.62 Consider the epidemic model of Modeling Exercise 6.36. Suppose $K = 1$. Compute the probability that there are m survivors (that is, those susceptibles who never get infected) when the epidemic terminates. That is, compute

$$p_{m,0} = \lim_{t \to \infty} P(S(t) = m, C(t) = 0 | S(0) = N - 1, C(0) = 1), \quad 0 \leq k \leq N - 1.$$

What happens to the above limiting probabilities when $\beta = \gamma$?

6.63 Consider the epidemic model of Modeling Exercise 6.37. Fix a $0 \leq k \leq N$ and define

$$p_{i,j} = \lim_{t \to \infty} P(S(t) = k, I(t) = 0 | S(0) = i, I(0) = j).$$

Derive a recursive equation for $p_{i,j}$ in terms of $p_{i-1,j+1}$ and $p_{i,j-1}$. What are the boundary conditions? Obtain an explicit expression for $p_{k,j}$. Show how to obtain $p_{i,j}$'s recursively.

6.64 Acquired immunodeficiency syndrome (AIDS) is a disease of the human immune system caused by the human immunodeficiency virus (HIV). AIDS can be thought of as progressing over five states: (1) infected but antibody-negative; (2) antibody-positive but asymptomatic; (3) pre-AIDS symptoms and/or abnormal hematologic indicator; (4) clinical AIDS, and (5) death. τ_i, the mean time (in years) until death starting from state i, is estimated to be

$$\tau_1 = 11.8, \quad \tau_2 = 11.6, \quad \tau_3 = 7.2, \quad \tau_4 = 2.0.$$

Model the disease progression as a pure birth CTMC, and give the transition rates. Compute the $F(t)$, the probability that an individual in stage 1 of the AIDS disease will be dead by time t.

6.65 Consider the AIDS model of Computational Exercise 6.64. Suppose a patient is in stage 1. We want to test him at time T to see if he is in stage 2. Find the optimum value of T that would maximize the probability of the patient being in state 2 at the time of testing.

6.66 Consider the single-price, single-stock order book of Section 6.3.4. Compute the limiting probability distribution of the state of the order book.

6.67 Consider the single-price, single-stock order book of Section 6.3.4. Compute the limiting fraction of the buy orders that are executed.

6.68 Consider the two-prices, single-stock order book of Modeling Exercise 6.38, but with the condition that there can be at most one order on the book (at any price, buy, or sell). Compute the limiting probability distribution of the state of the order book.

6.69 Consider the one-sided order book of Modeling Exercise 6.39. Compute the limiting distribution of the number of sell order on the order book in steady state.

6.70 Consider the one-sided order book of Modeling Exercise 6.39. Suppose the sell orders get executed in a first-come first-served order. Compute the probability that a sell order with n orders ahead of it gets executed before it gets cancelled.

6.71 Consider the organ transplant system of Modeling Exercise 6.40. Let $\alpha_{i,j}$ be the probability that a transplant is successful if type i kidney is given to type j patient $(1 \leq i \leq I, 1 \leq j \leq J)$. Compute the long run fraction of kidneys that lead to a successful transplant. One could choose the allocation probabilities $\beta_{i,j}$'s to maximize this fraction.

6.72 Consider a call center with s servers. The service times at the ith server are iid exponential random variables with parameter μ_i $(1 \leq i \leq s)$. Without loss of generality, assume that

$$\mu_1 \geq \mu_2 \geq \cdots \geq \mu_s.$$

The customers arrive according to a PP(λ). An arriving customer waits in the queue if all servers are busy, else he goes to the fastest idle server. When a server completes a service he starts serving a waiting customer, if one is waiting. If there are no waiting customers, he takes over the service of a customer who is being served by the slowest busy server that is slower than him. If there are no waiting customers and there is no busy server that is slower than him, he stays idle, waiting for a new customer. Let $X(t)$ be the number of customers in the system at time t.

1. Is $\{X(t), t \geq 0\}$ a birth and death process? If so, give its birth and death parameters and draw its rate diagram.

2. What is the condition of positive recurrence? Assuming the system is positive recurrent, compute
$$p_i = \lim_{t \to \infty} P(X(t) = i), \quad i \geq 0$$

3. What is the probability that server i is busy?

6.73 Suppose batches of jobs arrive to a system according to a PP(λ). Assume that the batch sizes are i.i.d. G($1 - \alpha$), where $0 < \alpha < 1$ is a given constant. Let $X(t)$ be the maximum batch size that has arrived up to time $t \geq 0$.

1. Justify that $\{X(t), t \geq 0\}$ is a CTMC and specify its transition rate matrix, $Q = [q_{ij}]$.
2. Classify the states of $\{X(t), t \geq 0\}$ (which ones are recurrent, transient, positive recurrent).
3. Find a closed form expression for the transition probabilities $p_{i,j}(t)$ of $\{X(t), t \geq 0\}$.

6.74 Consider a tandem line of two stations. There is an infinite supply of jobs in front of station 1 and hence station 1 is always busy. The jobs that are processed at station 1 are stored in a buffer with a finite capacity b. Station 2 draws jobs from this buffer for processing. After a job completes processing at station 2, it leaves the system. Station 1 is blocked when there are b jobs in the buffer and station 1 completes the processing of another job. (Note that station 1 can start processing a job even if its output buffer is full.) The blockage is removed as soon as a job leaves station 2. We have two servers. Server i takes iid exp(μ_i) ($i = 1, 2$) amounts of time to complete jobs, with $\mu_1 < \mu_2$. Suppose server i is assigned to station i.

1. Model this system as a CTMC.
2. Obtain the long-run average throughput of this system, i.e., the long-run average number of jobs that leave the system per unit time.
3. Obtain the long-run average level of work-in-process (WIP), i.e., the number of jobs in stations 1 and 2, and in the buffer between these two stations.
4. Do part 3 assuming the server 2 is at station 1 and server 1 is at station 2. Which order will maximize the long-run average throughput?

6.75 Consider a service system with a single server, capacity K, and PP(λ) arrivals. The service starts as soon as the system is full and lasts for an exp(μ) amount of time. At the end of service all customers leave the system. No new customers can join when the service is in progress, due to finite capacity. Let $X(t)$ be the number of customers in the system at time t.

1. Is $\{X(t), t \geq 0\}$ a CTMC? If so, give its state-space and transition rates.
2. Compute the limiting probability that the server is busy.
3. Compute the expected long-run average number of customers in the queue (excluding the ones in service).
4. Compute the expected long-run average number of customers in the system.

6.76 Suppose customers arrive at a single server service station according to a PP(λ). The service times are iid exp(μ). When the number in the system reaches a fixed number $K > 0$, all the customers are shipped off instantly to another location and the system becomes empty, and the cycle repeats. Model this as a CTMC by giving its state-space and generator matrix.

1. Compute the limiting distribution of the state of the system.

2. Suppose the cost of keeping a customer in the system is h per person per unit time, while the cost of shipping off the customers to another location is s per customer. Compute the long run cost per unit time as a function of K.

3. Compute the expected time between two consecutive epochs when the customers are shipped off to the other location.

6.77 Customers of two types arrive to a system according to independent Poisson processes with rates λ_A and λ_B, respectively. The system has two servers, server A serves customers of type A with rates μ_A, and server B serves customers of type B with rate μ_B. There is waiting room for only one customer to wait in a queue which is shared by the two servers. When the queue position is occupied, arriving customers are lost. (Thus there can be at most three customers in the system at any time.) Model this system by a CTMC.

1. Define the state-space.

2. Write the balance equations. Do not solve them.

3. What is the long run utilization of server A in terms of the limiting distribution?

6.78 Consider the two-state machine of Example 6.4 with state-space $\{0, 1\}$. It produces items according to a PP(α_i) while it is in state i, $i = 0, 1$. Compute the expected number of items produced over $(0, t]$ if the machine starts in state 1 at time 0.

6.79 Passengers arrive at a bus depot according to a PP(λ). An officer visits the bus depot at time epochs that follow an independent PP(μ). If the officer finds that there are at least k passengers waiting at the bus depot, he immediately calls enough buses to transport all the waiting customers. Here $k > 0$ is a fixed integer. Buses are assumed to arrive and leave with the passengers instantaneously.

1. Model this system as a continuous-time Markov chain.

2. Find the necessary and sufficient conditions under which the Markov chain is positive recurrent. Then, assuming positive recurrence, find its limiting distribution.

3. Consider a passenger who arrives to an empty bus depot. What is the expected time until this passenger leaves on a bus?

6.80 What is the condition of stability of the CTMC developed in Modeling Exercise 6.42. Assume stability and compute

$$p_{(i,j)} = \lim_{t \to \infty} P(X(t) = i, Y(t) = j), \quad i \geq 0, \ j = 0, 1.$$

6.18 Conceptual Exercises

6.1 Let $\{X(t), t \geq 0\}$ be an irreducible CTMC on state-space $S = \{1, 2, ..., N\}$ with generator matrix Q. Let

$$T = \min\{t \geq 0 : X(t) = N\}.$$

Derive a set of simultaneous linear equations for

$$M_{i,j} = \mathsf{E}\{\text{time spent in state } j \text{ during } [0, T) | X(0) = i\}, \quad i, j \in S.$$

6.2 Complete the proof of Theorem 6.16.

6.3 Let $\{X(t), t \geq 0\}$ be a CTMC on state-space S and generator matrix Q. Let T be the first time the CTMC undergoes a transition from state i to j (here i and j are fixed). Derive a method to compute $\mathsf{E}(T)$.

6.4 Let $\{X(t), t \geq 0\}$ be a CTMC on state-space $\{0, 1, 2, ...\}$. Suppose the system earns rewards at rate r_i per unit time when the system is in state i. Let Z be the total reward earned by the system until it reaches state 0, and let $g(i) = \mathsf{E}(Z|X(0) = i)$. Using first-step analysis derive a set of equations satisfied by $\{g(i), i = 1, 2, 3, ...\}$.

6.5 Let $\{X(t), t \geq 0\}$ be an irreducible and positive recurrent CTMC with state-space S and limiting distribution $p = [p_j, j \in S]$. Suppose the initial distribution is given by

$$P(X(0) = j) = p_j, \quad j \in S.$$

Show that

$$P(X(t) = j) = p_j, \quad j \in S, \ t \geq 0.$$

6.6 Consider a CTMC with state-space S and generator matrix Q. Suppose the CTMC earns rewards at rate r_i per unit time it spends in state i ($i \in S$). Furthermore it earns a lump sum reward $r_{i,j}$ whenever it undergoes a transition from state i to state j. Let $g_\alpha(i)$ be the total expected discounted (with continuous discount factor $\alpha > 0$) reward earned over an infinite horizon if the initial state is i. Show that the vector $g_\alpha = [g_\alpha(i), i \in S]$ satisfies

$$[\alpha I - Q]g_\alpha = r,$$

where $r = [r(i), i \in S]$ is given by

$$r(i) = r_i + \sum_{j \neq i} q_{i,j} r_{i,j}.$$

Note that no lump sum reward is obtained at time 0.

6.7 Consider Conceptual Exercise 6.6 and assume that the CTMC is irreducible and positive recurrent. Let g be the long run average reward per unit time. Show that

$$g = \sum_{i \in S} r(i) p_i,$$

where p_i is the limiting probability that the CTMC is in state i, and $r(i)$ is as given in Conceptual Exercise 6.6.

6.8 Consider the cost structure of Section 6.13. Let $g_i(t)$ be the expected total cost incurred over $(0, t]$ by the CTMC starting from state i at time 0. Let $g(t) = [g_i(t), i \in S]$. Show that

$$g(t) = M(t)c,$$

where $M(t)$ is the occupancy times matrix, and c is the cost rate vector.

6.9 In Conceptual Exercise 6.4, suppose $r(i) > 0$ for all $1 \leq i < N$. Construct a new CTMC $\{Y(t), t \geq 0\}$ on $\{0, 1, 2, \cdots, N\}$ with the same initial distribution as that of $X(0)$, such that the reward Z equals the first passage time until the $\{Y(t), t \geq 0\}$ process visits state N.

6.10 Let $\{X(t), t \geq 0\}$ be an irreducible positive recurrent CTMC on state-space S, generator matrix Q, and limiting distribution $p = [p_j, j \in S]$. Let $N_{i,j}(t)$ be the number of times the CTMC undergoes a transition from state i to j during $(0, t]$. Show that

$$\lim_{t \to \infty} \frac{E(N_{i,j}(t))}{t} = p_i q_{i,j}.$$

6.11 Consider a computer system modeled by a CTMC $\{X(t), t \geq 0\}$ on state-space $\{1, 2, \cdots, N\}$ and generator matrix Q. Whenever the system is in state i it executes $r_i > 0$ instructions/time. A programming job requires x instructions to be executed. Since the numbers involved are in the millions, we shall treat r_i's and x as real numbers. Suppose the job starts executing at time 0, with the system in state i. Let T be the time when the job completes. Let

$$\phi_i(s, x) = E(e^{-sT}|X(0) = i), \quad \phi_i^*(s, w) = \int_0^\infty e^{-wx} \phi_i(s, x) dx.$$

Show that

$$\phi_i^*(s, w) = \frac{r_i}{s + q_i + r_i w} + \sum_{j=1, j \neq i}^N \frac{q_{i,j}}{s + q_i + r_i w} \phi_j^*(s, w).$$

6.12 Consider the computer system described in Conceptual Exercise 6.11, with the following modification: whenever the system changes state, all the work done on the job is lost, and the job starts from scratch in the new state. Now show that

$$\phi_i(s, x) = e^{-(s+q_i)x/r_i} + \sum_{j=1, j \neq i}^N \frac{q_{i,j}}{s + q_i} (1 - e^{-(s+q_i)x/r_i} \phi_j(s, x)).$$

6.13 Let $T_i \sim PH(\alpha_i, M_i)$, $i = 1, 2$, be two independent phase type random variables. Show that $T = T_1 + T_2$ is a $PH(\alpha, M)$ random variable with α and M as given in Equations 6.88 and 6.89.

6.14 Let $T_i \sim \text{PH}(\alpha_i, M_i)$, $i = 1, 2$, be two independent phase type random variables. Let T be as defined in Equation 6.90. Show that T is a $\text{PH}(\alpha, M)$ random variable with α and M as given in Equations 6.91 and 6.92.

6.15 Let $T_i \sim \text{PH}(\alpha_i, M_i)$, $i = 1, 2$, be two independent phase type random variables. Show that $T = \min(T_1, T_2)$ is a phase type random variable. Find its parameters.

6.16 Suppose a vector of probabilities p satisfies the local balance Equation 6.95. Show that it satisfies the global balance Equation 6.70.

6.17 Show that an irreducible CTMC on $\{1, 2, \cdots, N\}$ with a symmetric generator matrix is reversible.

6.18 Let $M(t)$ be the occupancy matrix of a CTMC as in Section 6.5. Show that $M(t)$ satisfies the following matrix differential equation:

$$\frac{d}{dt}M(t) = I + QM(t),$$

where Q is the identity matrix. Hint: Follow the infinitesimal analysis as in the proof of Theorem 6.4.

CHAPTER 7

Queueing Models

An operations research professor gets an offer from another university that gives him better research opportunities, better students, and better pay, but it means uprooting his family and moving to a new place. When he describes this quandary to his friend, the friend says, "Well, why don't you use the methodology of operations research? Set up an optimization model with an objective function and constraints to maximize your expected utility." The professor objects, saying, "No, no! This is serious!"

7.1 Introduction

Queues are an unavoidable aspect of modern life. We do not like queues because of the waiting involved. However, we like the fair service policies that a queueing system imposes. Imagine what would happen if an amusement park did not enforce a first-come first-served queueing discipline at its attractions!

Knowingly or unknowingly, we face queues every day. We stand in queues at grocery store checkout counters, for movie tickets, at the baggage claim areas in airports. Many times we may be in a queue without physically being there — as when we are put on hold for the "next available service representative" when we call the customer service number during peak times. Many times we do not even know that we are in a queue: when we pick up the phone we are put in a queue to get a dial tone. Since we generally get the dial tone within a fraction of a second, we do not realize that we went through a queue. But we did, nonetheless.

Finally, waiting in queues is not a uniquely human fate. All kinds of systems enforce queueing for all kinds of non-human commodities. For example, a modern computer system manages queues of computer programs at its central processing unit, its input/output devices, etc. A telephone system maintains a queue of calls and serves them by assigning circuits to them. A digital communication network transmits packets of data in a store-and-forward fashion, i.e., it maintains a queue of packets at each node before transmitting them further toward their destinations. In manufacturing settings queues are called inventories. Here the items are produced at a factory and stored in a warehouse, i.e., they form a queue in the warehouse. The items are removed from the warehouse whenever a demand occurs.

Managing a queue — whether human or non-human — properly is critical to a smooth (and profitable) operation of any system. In a manufacturing system, excessively long queues (i.e., inventories) of manufactured products are expensive — one needs larger warehouses to store them, enormous capital is idled in the inventory, there are costs of deterioration, spoilage, etc. In a computer setting, building long queues of programs inevitably means slower response times. Slow response times from the central computers can have a disastrous effect on modern banks, and stock exchanges, for example. The modern communication networks can carry voice and video traffic efficiently only if the queues of data packets are managed so that delays do not exceed a few milliseconds.

For human queues there is an additional cost to queueing — the psychological stress generated by having to wait, the violent road rage being the most extreme manifestation of this. Industries that manage human queues (airlines, retail stores, amusement parks, etc.) employ several methods to reduce the stress. For example, knowing beforehand that there is a half-hour wait in a queue for a ride in an amusement park helps reduce the stress. Knowing the reason for delay also reduces stress — hence the airline pilot's announcement about being tenth in a queue for takeoff after the plane leaves from the gate but sits on the tarmac for 20 minutes. Diverting the attention of the waiting public can make the wait more bearable — this explains the presence of tabloids and TVs near the grocery store checkout counters. Airports use an ingenious method. There is generally a long walk from the gate to the baggage claim area. This makes the total wait seem smaller. Finally, there is the famous story of the manager of a skyscraper who put mirrors in the elevator lobbies and successfully reduced the complaints about slow elevators!

Another aspect of queues mentioned earlier is the "fair service discipline." In human queues this generally means first-come first-served (or first-in, first-out, or head of the line). In non-human queues many other disciplines can be used. For example, blood banks may manage their blood inventory by last-in first-out policy. This ensures better quality blood for most clients. The bank may discard blood after it stays in the bank for longer than a certain period. Generally, queues (or inventories) of perishable commodities use last-in first-out systems. Computers use many specialized service disciplines. A common service discipline is called time sharing, under which each job in the system gets Δ milliseconds from the CPU in a round-robin fashion. Thus all jobs get some service in reasonable intervals of time. In the limit, as $\Delta \to 0$, all jobs get served continuously in parallel, each job getting a fraction of the CPU processing capacity. This limiting discipline is called *processor sharing*. As a last example, the service discipline may be random: the server picks one of the waiting customers at random for the next service. Such a discipline is common in statistical experiments to avoid bias. It is also common to have priorities in service, although we do not consider such systems in this chapter.

A block diagram of a simple queueing system is shown in Figure 7.1. There are several key aspects to describing a queueing system. We discuss them below.

Figure 7.1 *A typical queueing system.*

1. *The Arrival Process.* The simplest model is when the customers arrive one at a time and the times between two consecutive arrivals are iid non-negative random variables. We use special symbols to denote these inter-arrival times as follows: M for exponential (stands for memoryless or Markovian), E_k for Erlang with k phases, D for deterministic, PH for phase type, G for general (sometimes we use GI to emphasize independence). This list is by no means exhaustive, and new notation keeps getting introduced as newer applications demand newer arrival characteristics. For example, the applications in telecommunication systems use what are known as the Markovian arrival processes, or Markov modulated Poisson processes, denoted by MAP or MMPP.

2. *The Service Times.* The simplest model assumes that the service times are iid non-negative random variables. We use the notation of the inter-arrival times for the service times as well.

3. *The Number of Servers.* It is typically denoted by s (for servers) or c (for channels in telecommunication systems). It is typically assumed that all servers are identical and each server serves one customer at a time.

4. *The Holding Capacity.* This is the maximum number of customers that can be in the system at any time. We also call it the system capacity, or just capacity. If the capacity is K, an arriving customer who sees K customers in the system is permanently lost. If no capacity is mentioned, it is assumed to be infinity.

5. *The Service Discipline.* This describes the sequence in which the waiting customers are serviced. As described before, the possible disciplines are FCFS (first-come first-served), LCFS (last-come first-served), random, PS (processor sharing), etc.

We shall follow the symbolic representation introduced by G. Kendall to represent a queueing system as inter-arrival time distribution/service time distribution/number of servers/capacity/service discipline. Thus $M/G/3/15/LCFS$ represents a queueing system with Poisson arrivals, generally distributed service times, three servers, a capacity to hold 15 customers, and last-come first-served service discipline. If the capacity and the service discipline are not mentioned, we assume infinite capacity and FCFS discipline. Thus an $M/M/s$ queue has Poisson arrivals, exponential service times, s servers, infinite waiting room and FCFS discipline.

Several quantities are of interest in the study of a queueing system. We introduce the relevant notation below:

$X(t)$ = the number of customers in the system at time t,

X_n = the number of customers in the system just after the n-th customer departs,

X_n^* = the number of customers in the system just before the n-th customer enters,

$\hat{X}_n$ = the number of customers in the systems just before the n-th customer arrives,

W_n = time spent in the system by the n-th customer.

Note that we distinguish between an arriving customer and an entering customer, since an arriving customer may not enter because the system is full, or the customer may decide not to join if the system is too congested. We are interested in

$$p_j = \lim_{t \to \infty} P(X(t) = j),$$

$$\pi_j = \lim_{n \to \infty} P(X_n = j),$$

$$\pi_j^* = \lim_{n \to \infty} P(X_n^* = j),$$

$$\hat{\pi}_j = \lim_{n \to \infty} P(\hat{X}_n = j),$$

$$F(x) = \lim_{n \to \infty} P(W_n \leq x),$$

$$L = \lim_{t \to \infty} E(X(t)),$$

$$W = \lim_{n \to \infty} E(W_n).$$

Note that we can also interpret the above limits as long run averages. For example, we can think of $\hat{\pi}_j$ as the long run fraction of arrivals who see j customers ahead of them in the system when they arrive. Many times the limits defined above may not exist, but the long run averages exist. (See Computational Exercise 7.79.) The theorems in the next section remain valid for such long run averages as well.

We shall also find it useful to study the customers in the queue (i.e., those who are in the system but not in service). We define

$X^q(t)$ = the number of customers in the queue at time t,

X_n^q = the number of customers in the queue just after the n-th customer departs,

X_n^{*q} = the number of customers in the queue just before the n-th customer enters,

$\hat{X}_n^q$ = the number of customers in the queue just before the n-th customer arrives,

W_n^q = time spent in the queue by the n-th customer,

and

$$p_j^q = \lim_{t \to \infty} P(X^q(t) = j),$$

$$\pi_j^q = \lim_{n \to \infty} P(X_n^q = j),$$

$$\pi_j^{*q} = \lim_{n \to \infty} P(X_n^{*q} = j),$$

$$\hat{\pi}_j^q = \lim_{n \to \infty} P(\hat{X}_n^q = j),$$

$$F^q(x) = \lim_{n \to \infty} P(W_n^q \leq x),$$

$$L^q = \lim_{t \to \infty} \mathsf{E}(X^q(t)),$$
$$W^q = \lim_{n \to \infty} \mathsf{E}(W_n^q).$$

With this introduction we are ready to apply the theory of DTMCs and CTMCs to queueing systems. In particular we shall study queueing systems in which $\{X(t), t \geq 0\}$ or $\{X_n, n \geq 0\}$ or $\{X_n^*, n \geq 0\}$ is a Markov chain. We shall also study queueing systems modeled by multidimensional Markov chains. There is an extremely large and growing literature on queueing theory and this chapter is by no means an exhaustive treatment of even the Markovian queueing systems. Readers are encouraged to refer to one of the several excellent books that are devoted entirely to queueing theory.

It is obvious that $\{X(t), t \geq 0\}$, $\{X_n, n \geq 0\}$, $\{X_n^*, n \geq 0\}$, and $\{\hat{X}_n, n \geq 0\}$ are related processes. The exact relationship between these processes is studied in the next section.

7.2 Properties of General Queueing Systems

In this section we shall study several important properties of general queueing systems. They are discussed in Theorems 7.1, 7.2, 7.3, and 7.5.

In Theorem 7.1 we state the connection between the state of the system as seen by a potential arrival and an entering customer. In Theorem 7.2 we show that under mild conditions on the sample paths of $\{X(t), t \geq 0\}$, the limiting distributions of X_n and X_n^*, if they exist, are identical. Thus, in steady state, the state of the system as seen by an entering customer is identical to that as seen by a departing customer. In Theorems 7.3 and 7.4, we prove that, if the arrival process is Poisson, and certain other mild conditions are satisfied, the limiting distributions of $X(t)$ and $\hat{X}_n$ (if they exist) are identical. Thus in steady state the state of the system as seen by an arriving customer is the same as the state of the system at an arbitrary point of time. This property is popularly known as PASTA – Poisson Arrivals See Time Averages. The last result (Theorem 7.5) is called Little's Law, and it relates limiting averages L and W defined in the last section. All these results are very useful in practice, but their proofs are rather technical. We suggest that the reader should first read the statements of the theorems, and understand their implications, before reading the proofs.

7.2.1 Relationship between π_j^* and $\hat{\pi}_j$

As pointed out in the last section, we make a distinction between the arriving customers and the entering customers. One can think of the arriving customers as the potential customers and the entering customers as the actual customers. The potential customers become actual customers when they decide to enter. The relationship between the state of the system as seen by an arrival (potential customer) and as seen by an entering customer depends on the decision rule used by the arriving customer to actually enter the system. To make this more precise, let $I_n = 1$ if the n-th arriving customer enters, and 0 otherwise. Suppose the following limits exist

$$\lim_{n \to \infty} P(I_n = 1) = \alpha, \tag{7.1}$$

and

$$\lim_{n \to \infty} P(I_n = 1 | \hat{X}_n = j) = \alpha_j, \quad j \geq 0. \tag{7.2}$$

Note that α is the long run fraction of the arriving customers that enter the system. The next theorem gives a relationship between the state of the system as seen by an arrival and by an entering customer.

Theorem 7.1   Arriving and Entering Customers. *Suppose $\alpha > 0$, and one of the two limiting distributions $\{\pi_j^*, j \geq 0\}$ and $\{\hat{\pi}_j, j \geq 0\}$ exist. Then the other limiting distribution exists, and the two are related to each other by*

$$\pi_j^* = \frac{\alpha_j}{\alpha} \hat{\pi}_j, \quad j \geq 0. \tag{7.3}$$

Proof: Define

$$N(n) = \sum_{i=1}^{n} I_i, \quad n \geq 1.$$

Thus $N(n)$ is the number of customers who join the system from among the first n arrivals. This implies that

$$P(\hat{X}_n = j | I_n = 1) = P(X_{N(n)}^* = j | I_n = 1).$$

The assumption $\alpha > 0$ implies that $N(n) \to \infty$ as $n \to \infty$ with probability 1. Now

$$
\begin{aligned}
P(I_n = 1 | \hat{X}_n = j) P(\hat{X}_n = j) &= P(\hat{X}_n = j, I_n = 1) \\
&= P(\hat{X}_n = j | I_n = 1) P(I_n = 1) \\
&= P(X_{N(n)}^* = j | I_n = 1) P(I_n = 1).
\end{aligned}
$$

Letting $n \to \infty$ on both sides, and using Equations 7.1 and 7.2, we get

$$\alpha_j \hat{\pi}_j = \pi_j^* \alpha,$$

if either $P(\hat{X}_n = j)$ or $P(X_n^* = j)$ has a limit as $n \to \infty$. The theorem follows from this. ∎

Note that if we further know that

$$\sum_{j=0}^{\infty} \pi_j^* = 1$$

we can evaluate α as

$$\alpha = \sum_{i=0}^{\infty} \alpha_i \hat{\pi}_i.$$

Hence Equation 7.3 can be written as

$$\pi_j^* = \frac{\alpha_j \hat{\pi}_j}{\sum_{i=0}^{\infty} \alpha_i \hat{\pi}_i}, \quad j \geq 0.$$

Finally, if every arriving customer enters, we have $\alpha_j = 1$ for all $j \geq 0$ and $\alpha = 1$. Hence the above equation reduces to $\pi_j^* = \hat{\pi}_j$ for all $j \geq 0$, as expected.

Example 7.1 $M/M/1/K$ **System.** Consider an $M/M/1/K$ system. An arriving customer enters the system if he finds less than K customers ahead of him, else he leaves. Thus we have

$$P(I_n = 1 | \hat{X}_n = j) = \begin{cases} 1 & \text{if } 0 \leq j < K, \\ 0 & \text{if } j \geq K. \end{cases}$$

From Equation 7.2

$$\alpha_j = \begin{cases} 1 & \text{if } 0 \leq j < K, \\ 0 & \text{if } j \geq K. \end{cases}$$

Now suppose the limiting distribution $\{\pi_j^*, 0 \leq j \leq K - 1\}$ exists. Then clearly we must have

$$\sum_{j=0}^{K-1} \pi_j^* = 1.$$

The limiting distribution $\{\hat{\pi}_j, 0 \leq j \leq K\}$ exists, and we must have

$$\sum_{j=0}^{K} \hat{\pi}_j = 1.$$

Hence we get

$$\pi_j^* = \frac{\hat{\pi}_j}{\sum_{i=0}^{K-1} \hat{\pi}_i} = \frac{\hat{\pi}_j}{1 - \hat{\pi}_K}, \quad 0 \leq j \leq K - 1. \quad \blacksquare$$

Also see Computational Exercise 7.79.

7.2.2 Relationship between π_j^* and π_j

The next theorem gives the relationship between the state of the system as seen by an entering customer and a departing customer.

Theorem 7.2 Entering and Departing Customers. *Suppose the customers enter and depart a queueing system one at a time, and one of the two limiting distributions $\{\pi_j^*, j \geq 0\}$ and $\{\pi_j, j \geq 0\}$ exists. Then the other limiting distribution exists, and*

$$\pi_j = \pi_j^*, \quad j \geq 0. \tag{7.4}$$

Proof: We follow the proof as given in Cooper [1981]. Suppose that there are i customers in the system at time 0. We shall show that

$$\{X_{n+i} \leq j\} \Leftrightarrow \{X_{n+j+1}^* \leq j\}, \quad j \geq 0. \tag{7.5}$$

First suppose $X_{n+i} = k \le j$. This implies that there are exactly $k + n$ entries before the $(n + i)$th departure. Thus there can be only departures between the $(n + i)$th departure and $(n + k + 1)$st entry. Hence

$$X^*_{n+k+1} \le k.$$

Using $k = j$ we see that

$$\{X_{n+i} \le j\} \Rightarrow \{X^*_{n+j+1} \le j\}, \quad j \ge 0.$$

To go the other way, suppose $X^*_{n+j+1} = k \le j$. This implies that there are exactly $n + i + j - k$ departures before the $(n + j + 1)$st entry. Thus there are no entries between the $(n + i + j - k)$th departure and $(n + j + 1)$st entry. Hence

$$X_{n+i+j-k} \le X^*_{n+j+1} = k \le j, \quad j \ge 0.$$

Setting $k = j$ we get $X_{n+i} \le j$. Hence

$$\{X^*_{n+j+1} \le j\} \Rightarrow \{X_{n+i} \le j\}, \quad j \ge 0.$$

This proves the equivalence in Equation 7.5. Hence we have

$$P(X_{n+i} \le j) = P(X^*_{n+j+1} \le j), \quad j \ge 0.$$

Letting $n \to \infty$, and assuming one of the two limits exist, the theorem follows. ∎

Two comments are in order at this point. First, the above theorem is a sample path result and does not require any probabilistic assumptions. As long as the limits exist (or the long run averages exist), they are equal. Of course, we need probabilistic assumption to assert that the limits exist. Second, the above theorem can be applied even in the presence of batch arrivals and departures, as long as we sequence the entries (or departures) in the batch and observe the system after every entry and every departure in the batch. Thus, suppose n customers have entered so far, and a new batch of size 2 enters when there are i customers in the system. Then we treat this as two single entries occurring one after another, thus yielding $X^*_{n+1} = i$ and $X^*_{n+2} = i + 1$.

Example 7.2   Entering and Departing Customers. The $\{X(t), t \ge 0\}$ processes in the queueing systems $M/M/1$, $M/M/s$, $M/G/1$, $G/M/1$, etc., satisfy the hypothesis of Theorem 7.2. Since every arrival enters the system, we can combine the result of Theorem 7.1 with that of Theorem 7.2, to get

$$\pi_j = \pi^*_j = \hat{\pi}_j, \quad j \ge 0,$$

if any one of the three limiting distributions exist. At this point we do not know how to prove that they exist. ∎

Also see Computational Exercise 7.79.

7.2.3 Relationship between $\hat{\pi}_j$ and p_j

Now we discuss the relationship between the limiting distribution of the state of the system as seen by an arrival and that of the state of the system at any time point. This will lead us to an important property called PASTA — Poisson Arrivals See Time Averages. Roughly speaking, we shall show that, the limiting probability that an arriving customer sees the system in state j is the same as the limiting probability that the system is in state j, if the arrival process is Poisson. One can think of p_j as the time average (that is, the occupancy distribution): the long run fraction of the time that the system spends in state j. Similarly, we can interpret $\hat{\pi}_j$ as the long run fraction of the arriving customers that see the system in state j. PASTA says that, if the arrival process is Poisson, these two averages are identical.

We shall give a proof of this under the restrictive setting when the stochastic process $\{X(t), t \geq 0\}$ describing the system state is a CTMC on $\{0, 1, 2, \cdots\}$. However, PASTA is a very general result that applies even though the process is not Markovian. For example, it applies to the queue length process in an $M/G/1$ queue, even if that process is not a CTMC. However, its general proof is rather technical and will not be presented here. We refer the reader to Wolff [1989] or Heyman and Sobel [1982] for the general proof.

Let $X(t)$ be the state of queueing system at time t. Let $N(t)$ be the number of customers who arrive at this system up to time t, and assume that $\{N(t), t \geq 0\}$ is a PP(λ), and S_n is the time of arrival of the nth customer. Assume that $\{X(t), S_n \leq t < S_{n+1}\}$ is a CTMC with state-space S and rate matrix $G = [g_{ij}], n \geq 0$. When the nth arrival occurs at time S_n, it causes an instantaneous transition in the X process from i to j with probability r_{ij}, regardless of the past history up to time t. That is,

$$P(X(S_n+) = j | X(S_n-) = i, (X(u), N(u)), 0 \leq u < S_n) = r_{i,j}, \quad i, j \geq 0.$$

Now the nth arrival sees the system in state $\hat{X}_n = X(S_n-)$. Hence we have, assuming the limits exist,

$$\hat{\pi}_j = \lim_{n \to \infty} P(\hat{X}_n = j) = \lim_{n \to \infty} P(X(S_n-) = j), \quad j \geq 0 \qquad (7.6)$$

and

$$p_j = \lim_{t \to \infty} P(X(t) = j), \quad j \geq 0. \qquad (7.7)$$

The next theorem gives the main result.

Theorem 7.3 PASTA. *If the limits in Equations 7.6 and 7.7 exist,*

$$\hat{\pi}_j = p_j, \quad j \geq 0.$$

The proof proceeds via several lemmas, and is completely algebraic. We provide the following intuition to strengthen the feel for the result. Figure 7.2(a) shows a sample path $\{X(t), t \geq 0\}$ of a three-state system interacting with a PP(λ) $\{N(t), t \geq 0\}$.

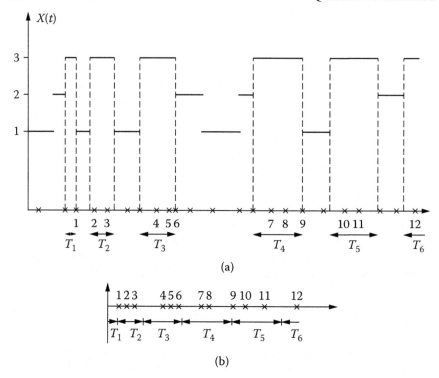

Figure 7.2 *(a) Sample path of a CTMC interacting with a PP. (b) Reduced sample path of a CTMC interacting with a PP.*

In the figure, the T_is are the intervals of time (open on the left and closed on the right) when the system is in state 3. The events in the PP $\{N(t), t \geq 0\}$ that see the system in state 3 are numbered consecutively from 1 to 12. Figure 7.2(b) shows the T_is spliced together, essentially deleting all the intervals of time when the system is not in state 3. Thus we count the Poisson events that trigger the transitions out of state 3 to states 1 and 2, but not those that trigger a transition into state 3 from state 1 and 2. We also count all the Poisson events that do not generate any transitions.

Now, the times between consecutive events in Figure 7.2(a) are iid $\exp(\lambda)$, due to the model of the interaction that we have postulated. Hence the process of events in Figure 7.2(b) is a PP(λ). Following the notation of Section 6.5, let $V_j(t)$ be the amount of time the system spends in state j over $(0, t]$. Hence the expected number of Poisson events up to time t that see the system in state 3 is $\lambda E(V_3(t))$. By the same argument, the expected number of Poisson events up to time t that see the system in state j is $\lambda E(V_j(t))$. The expected number of events up to time t in a PP(λ) is λt.

Hence the fraction of the Poisson events that see the system in state j is

$$\frac{\lambda E(V_j(t))}{\lambda t} = \frac{E(V_j(t))}{t}.$$

If the $\{X(t), t \geq 0\}$ process is a CTMC, the above fraction, as $t \to \infty$, goes to p_j, the long run fraction of the time the CTMC spends in state j. Hence, the limiting probability (if the limits exist) that the system is in state j just before an event in the PP, is the same as the long run probability that the system is in state j. The hard part is to prove that the limits exist.

We now continue with the proof of Theorem 7.3. First we study the structure of the process $\{X(t), t \geq 0\}$.

Lemma 7.1 $\{X(t), t \geq 0\}$ *is a CTMC on S with rate matrix Q given by*

$$Q = G + \lambda(R - I),$$

where $R = [r_{i,j}]$, and I is the identity matrix.

Proof: That $\{X(t), t \geq 0\}$ is a CTMC is a consequence of how the system interacts with the Poisson process, and the system behavior between the Poisson events. The rate at which it moves to state j from state $i \neq j$ is given by

$$q_{i,j} = \lambda r_{i,j} + g_{i,j}.$$

Hence we have

$$q_{ii} = g_{ii} - \lambda(1 - r_{ii}).$$

The above two relations imply Equation 7.8. ∎

The next lemma describes the probabilistic structure of the $\{\hat{X}_n, n \geq 0\}$ process.

Lemma 7.2 $\{\hat{X}_n, n \geq 0\}$ *is a DTMC on S with transition probability matrix $P = [p_{i,j}]$ given by*

$$P = RB \tag{7.8}$$

where $R = [r_{i,j}]$ and $B = [b_{i,j}]$ where

$$b_{i,j} = P(X(S_{n+1}-) = j | X(S_n+) = i), \quad i, j \in S.$$

Proof: That $\{\hat{X}_n, n \geq 0\}$ is a DTMC is clear. We have

$$
\begin{aligned}
p_{i,j} &= P(\hat{X}_{n+1} = j | \hat{X}_n = i) \\
&= P(X(S_{n+1}-) = j | X(S_n-) = i) \\
&= \sum_{k \in S} P(X(S_{n+1}-) = j | X(S_n-) = i, X(S_n+) = k) \\
&\quad \cdot P(X(S_n+) = k | X(S_n-) = i) \\
&= \sum_{k \in S} P(X(S_n+) = k | X(S_n-) = i) P(X(S_{n+1}-) = j | X(S_n+) = k)
\end{aligned}
$$

$$= \sum_{k \in S} r_{i,k} b_{k,j},$$

which yields Equation 7.8 in matrix form. ∎

The next lemma relates the B and G matrices in an algebraic fashion.

Lemma 7.3 *The matrix B satisfies*

$$(\lambda I - G)B = \lambda I. \tag{7.9}$$

Proof: Let $\{Y(t), t \geq 0\}$ be a CTMC with generator matrix G, and let

$$a_{i,j}(t) = \mathsf{P}(Y(t) = j | Y(0) = i).$$

Let $a_{i,j}^*(s)$ be the Laplace transform of $a_{i,j}(\cdot)$. From Equation 6.32 on page 230 we see that $A^*(s) = [a_{i,j}^*(s)]$ satisfies

$$sA^*(s) - I = GA^*(s).$$

Since the $\{X(t), S_n \leq t < S_{n+1}\}$ is a CTMC with generator G, and $S_{n+1} - S_n \sim \exp(\lambda)$, we get

$$
\begin{aligned}
b_{i,j} &= \mathsf{P}(X(S_{n+1}-) = j | X(S_n+) = i) \\
&= \int_0^\infty \lambda e^{-\lambda t} a_{i,j}(t) dt \\
&= \lambda a_{i,j}^*(\lambda).
\end{aligned}
$$

Thus

$$B = \lambda A^*(\lambda) = I + GA^*(\lambda) = I + GB/\lambda,$$

which yields Equation 7.9. ∎

With these three lemmas we can now give the proof.

The Proof of Theorem 7.3: Suppose $\{X(t), t \geq 0\}$ is an irreducible positive recurrent CTMC with limiting distribution $p = [p_j, j \in S]$. Then, from Theorems 4.22 and 6.25, we get

$$pQ = 0, \quad \sum p_j = 1, \quad \hat{\pi}P = \hat{\pi}, \quad \sum \hat{\pi}_j = 1.$$

This yields

$$
\begin{aligned}
0 &= pQ \\
&= pQB \\
&= p(G + \lambda(R - I))B \quad \text{(from Lemma 7.1)} \\
&= p((G - \lambda I)B + \lambda RB) \\
&= p(-\lambda I + \lambda P) \quad \text{(from Lemmas 7.3 and 7.2)} \\
&= -\lambda(p - pP).
\end{aligned}
$$

Hence we have

$$p = pP, \quad \sum p_j = 1.$$

Thus p is the stationary distribution of the DTMC $\{\hat{X}_n, n \geq 0\}$. However, since $\{X(t), t \geq 0\}$ is an irreducible CTMC, $\{\hat{X}_n, n \geq 0\}$ is irreducible and aperiodic. Hence it has a unique limiting distribution $\hat{\pi}$. Hence, we must have $p = \hat{\pi}$, as desired. ∎

We explain with three examples.

Example 7.3 $M/M/1/1$ **Queue.** Verify Theorem 7.3 directly for the $M/M/1/1$ queue.

Let $X(t)$ be the number of customers at time t in an $M/M/1/1$ queue with arrival rate λ and service rate μ. Let $N(t)$ be the number of arrivals over $(0, t]$. Then $\{N(t), t \geq 0\}$ is a PP(λ). Let S_n be the nth arrival epoch. We see that $\{X(t), S_n \leq t < S_{n+1}\}$ is a CTMC with generator matrix

$$G = \begin{bmatrix} 0 & 0 \\ \mu & -\mu \end{bmatrix}.$$

If the system is empty at an arrival instant, the arrival enters, else the arrival leaves. Thus the R matrix is given by

$$R = \begin{bmatrix} 0 & 1 \\ 0 & 1 \end{bmatrix}.$$

From Lemma 7.1 we see that $\{X(t), t \geq 0\}$ is a CTMC with generator matrix

$$Q = G + \lambda(R - I) = \begin{bmatrix} -\lambda & \lambda \\ \mu & -\mu \end{bmatrix}.$$

This is as expected. Next we compute the B matrix. We have

$$\begin{aligned} b_{10} &= \mathrm{P}(X(S_{n+1}-) = 0 | X(S_n+) = 1) \\ &= \mathrm{P}(\text{next departure occurs before next arrival}) \\ &= \frac{\mu}{\lambda + \mu}. \end{aligned}$$

Similar calculations yield

$$B = \begin{bmatrix} 1 & 0 \\ \frac{\mu}{\lambda+\mu} & \frac{\lambda}{\lambda+\mu} \end{bmatrix}.$$

Using Lemma 7.2 we see that $\{\hat{X}_n, n \geq 0\}$ is a DTMC on $\{0, 1\}$ with transition probability matrix

$$P = RB = \begin{bmatrix} \frac{\mu}{\lambda+\mu} & \frac{\lambda}{\lambda+\mu} \\ \frac{\mu}{\lambda+\mu} & \frac{\lambda}{\lambda+\mu} \end{bmatrix}.$$

Hence its limiting distribution is given by

$$\hat{\pi}_0 = \frac{\mu}{\lambda + \mu}, \quad \hat{\pi}_1 = \frac{\lambda}{\lambda + \mu}.$$

Using the results of Example 6.34 on page 258 we see that the limiting distribution of the CTMC $\{X(t), t \geq 0\}$ is given by

$$p_0 = \frac{\mu}{\lambda + \mu}, \quad p_1 = \frac{\lambda}{\lambda + \mu}.$$

Thus Theorem 7.3 is verified. ∎

Example 7.4   PASTA for $M/M/1$ and $M/M/s$ Queues. Let $X(t)$ be the number of customers at time t in an $M/M/1$ queue with arrival rate λ and service rate $\mu > \lambda$. Let $N(t)$ be the number of arrivals over $(0, t]$. Then $\{N(t), t \geq 0\}$ is a PP(λ). We see that X and N processes satisfy the assumptions described in this section. Hence we can apply PASTA (Theorem 7.3) to get

$$\hat{\pi}_j = p_j, \quad j \geq 0.$$

Furthermore, the X process satisfies the conditions for Theorem 7.2. Thus

$$\pi_j = \pi_j^*, \quad j \geq 0.$$

Finally, every arriving customer enters the system, hence

$$\hat{\pi}_j = \pi_j^*, \quad j \geq 0.$$

From Example 6.36 we see that this queue is stable, and using Equation 6.76 and the above equations, we get

$$\hat{\pi}_j = \pi_j^* = \pi_j = p_j = (1 - \rho)\rho^j, \quad j \geq 0, \tag{7.10}$$

where $\rho = \lambda/\mu < 1$.

Similar analysis for the $M/M/s$ queue shows that

$$\hat{\pi}_j = \pi_j^* = \pi_j = p_j, \quad j \geq 0.$$

We shall give the limiting distribution $\{p_j, j \geq 0\}$ for the $M/M/s$ queue in Section 7.3.3. ∎

Example 7.5 $M/M/1/K$ System. Let $X(t)$ be the number of customers at time t in an $M/M/1/K$ queue with arrival rate λ and service rate μ. Let $N(t)$ be the number of arrivals over $(0, t]$. An arriving customer enters if and only if the number in the system is less than K when he arrives at the system. Then $\{N(t), t \geq 0\}$ is a PP(λ). We see that X and N processes satisfy the assumptions described in this section. Hence we can apply PASTA (Theorem 7.3) to get

$$\hat{\pi}_j = p_j, \quad 0 \leq j \leq K.$$

Furthermore, the X process satisfies the conditions for Theorem 7.2. Thus

$$\pi_j = \pi_j^*, \quad 0 \leq j \leq K - 1.$$

Finally, from Example 7.1 we get

$$\pi_j^* = \frac{\hat{\pi}_j}{1 - \hat{\pi}_K}, \quad 0 \leq j \leq K - 1.$$

Thus the arriving customers see the system in steady state, but not the entering customers. This is because the arriving customers form a PP, but not the entering customers. We shall compute the limiting distribution $\{p_j, 0 \leq j \leq K\}$ for this system in Section 7.3.2. ∎

We have proved PASTA in Theorem 7.3 for an $\{X(t), t \geq 0\}$ process that interacts with a PP in a specific way, and behaves like a CTMC between the events of the PP. However, PASTA is a very general property. We state the general result here, but omit the proof. In almost all applications the version of PASTA given here is sufficient, since almost all the processes we study in this book can be turned into CTMCs by using phase-type distributions as we shall see in Section 7.6.

Let $X(t)$ be the state of a system at time t, and let $\{N(t), t \geq 0\}$ be a PP that may depend on the $\{X(t), t \geq 0\}$ process. However, we assume the following:

"Lack of Anticipation" Property: $\{N(t+s) - N(s), t \geq 0\}$ and $\{X(u), 0 \leq u \leq s\}$ are independent.

Note that $\{N(t+s) - N(s), t \geq 0\}$ is independent of $\{N(u), 0 \leq u \leq s\}$ due to the independence of increments property of a PP. Thus the lack of anticipation property says that the future arrivals after time s are independent of the system history up to time s.

Now let B be a given set of states and let $V_B(t)$ be the time spent by the system in the set B over $(0, t]$, and $A_B(t)$ be the number of arrivals over $(0, t]$ that see the system in the set B,

$$A_B(t) = \sum_{n=1}^{N(t)} 1_{\{X(S_n-) \in B\}}, \quad t \geq 0,$$

where S_n is the time of the nth arrival.

Now suppose the sample paths of the $\{X(t), t \geq 0\}$ are almost surely piecewise continuous and have finite numbers of jumps in finite intervals of time. Then the following theorem gives the "almost sure" version of PASTA.

Theorem 7.4 PASTA. $\frac{V_B(t)}{t}$ converges almost surely if and only if $\frac{A_B(t)}{N(t)}$ converges almost surely, and both have the same limit.

Proof: See Wolff (1989), or Heyman and Sobel (1982). ∎

Note that PASTA, which equates the limit of $\frac{V_B(t)}{t}$ to that of $\frac{A_B(t)}{N(t)}$, holds whenever the convergence holds. We use the tools developed in this book to show that

$$\lim_{t \to \infty} \frac{V_B(t)}{t} = \lim_{t \to \infty} \mathsf{P}(X(t) \in B),$$

and

$$\lim_{t \to \infty} \frac{A_B(t)}{N(t)} = \lim_{n \to \infty} \mathsf{P}(X(S_n-) \in B).$$

Thus according to PASTA, whenever the limits exist,

$$\lim_{t \to \infty} P(X(t) \in B) = \lim_{n \to \infty} P(X(S_n-) \in B),$$

or, when $B = \{j\}$,

$$p_j = \hat{\pi}_j.$$

Example 7.6   PASTA and the $M/G/1$ and $G/M/1$ Queues. Theorems 7.2 and 7.4 imply that for the $M/G/1$ queue

$$\hat{\pi}_j = \pi_j^* = \pi_j = p_j, \quad j \geq 0.$$

On the other hand, for the $G/M/1$ queue, we have

$$\hat{\pi}_j = \pi_j^* = \pi_j, \quad j \geq 0.$$

However, PASTA does not apply unless the inter-arrival times are exponential. Thus in general, for a $G/M/1$ queue, we do not have $\hat{\pi}_j = p_j$. ∎

7.2.4 Little's Law

Recall the definitions of $X(t)$, W_n, L, and W on page 299. In this subsection we study another important theorem of queueing theory: The Little's Law. It states that L, the average number of customers in the system, λ, the average arrival rate of customers to the system, and W, the average time a customer spends in the system (all in steady state), are related to each other by the following simple relation:

$$L = \lambda W. \tag{7.11}$$

The above relation is in fact a sample path property, rather than a probabilistic one. We need probabilistic assumption to assert that the above averages exist. An intuitive explanation of Little's Law is as follows: Suppose each arriving customer pays the system $1 per unit time the customer spends in the system. The long run rate at which the system earns revenue can be computed in two equivalent ways. First, the nth customer pays the system $\$W_n$, the time spent by that customer in the system. Hence the average amount paid by a customer in steady state is $\$W$. Since the average arrival rate is λ, the system earns $\$\lambda W$ per unit time in the long run. Second, since each customer in the system pays at rate $1 per unit time, the system earns revenue at the instantaneous rate of $\$X(t)$ per unit time at time t. Hence the long run rate of revenue is seen to be L, the steady state expected value of $X(t)$. Since these two calculations must provide the same answer, Equation 7.11 follows.

In the rest of the section we shall make the statement of Little's Law more precise. Consider a general queueing system where the customers arrive randomly, get served, and then depart. Let $X(t)$ be the number of customers in the system at time t, $A(t)$ be the number of arrivals over $(0, t]$, and W_n be the time spent in the system by the nth customer. We also refer to it as the waiting time or sojourn time. Now define the following limits, whenever they exist. Note that these are defined for every sample path of the queueing system.

$$L = \lim_{t\to\infty} \frac{1}{t} \int_0^t X(u)du, \tag{7.12}$$

$$\lambda = \lim_{t\to\infty} \frac{A(t)}{t}, \tag{7.13}$$

$$W = \lim_{n\to\infty} \frac{\sum_{k=1}^n W_k}{n}. \tag{7.14}$$

The next theorem states the relationship that binds the three limits defined above.

Theorem 7.5 Little's Law. *Suppose that for a fixed sample path the limits in Equations 7.13 and 7.14 exist and are finite. Then the limit in Equation 7.12 exists, and is given by*

$$L = \lambda W.$$

Proof: Let $S_0 = 0$, $D_0 = 0$, and S_n be the arrival time and $D_n \geq S_n$ be the departure time of the nth customer, $n \geq 1$, and assume that $0 \leq S_1 \leq S_2 \leq \cdots$. $A(t)$ is already defined to be the number of arrivals over $(0, t]$. Define $D(t)$ to be the number of departures over $(0, t]$. Without loss of generality we assume that $X(0) = 0$, since we can always assume that $X(0)$ customers arrived at time $0+$. Then we have

$$W_n = D_n - S_n, \quad n \geq 1,$$

$$X(t) = \sum_{n=1}^\infty 1_{\{S_n \leq t < D_n\}}, \quad t \geq 0,$$

$$A(t) = \sup\{n \geq 0 : S_n \leq t\}, \quad t \geq 0,$$

$$X(t) = A(t) - D(t), \quad t \geq 0.$$

Now the existence of the limits implies that (we omit the details of the proof of this assertion, and refer the readers to El-Taha and Stidham (1998))

$$\lim_{t\to\infty} \frac{X(t)}{t} = 0. \tag{7.15}$$

This implies that

$$\frac{D(t)}{t} = \frac{A(t) - X(t)}{t} \to \lambda.$$

The above relations also imply that

$$\sum_{n=1}^{D(t)} W_n \leq \int_0^t X(u)du \leq \sum_{n=1}^{A(t)} W_n.$$

Dividing by t on both sides we get

$$\frac{1}{t} \sum_{n=1}^{D(t)} W_n \leq \frac{1}{t} \int_0^t X(u)du \leq \frac{1}{t} \sum_{n=1}^{A(t)} W_n,$$

which can be written as

$$\frac{D(t)}{t} \frac{1}{D(t)} \sum_{n=1}^{D(t)} W_n \leq \frac{1}{t} \int_0^t X(u)du \leq \frac{A(t)}{t} \frac{1}{A(t)} \sum_{n=1}^{A(t)} W_n. \qquad (7.16)$$

Now, assume that $A(t) \to \infty$ as $t \to \infty$. Then $D(t) \to \infty$ as $t \to \infty$, and we have

$$\lim_{t \to \infty} \frac{1}{D(t)} \sum_{n=1}^{D(t)} W_n = \lim_{t \to \infty} \frac{1}{A(t)} \sum_{n=1}^{A(t)} W_n = W.$$

Now let $t \to \infty$ in Equation 7.16. We get

$$\lambda W \leq L \leq \lambda W.$$

If $A(t)$ remains bounded as $t \to \infty$, we necessarily have $L = \lambda = W = 0$. The theorem follows. ∎

There are many results that prove Little's Law under less restrictive conditions. We refer the readers to Wolff (1989) and Heyman and Sobel (1982) and El-Taha and Stidham (1998). The condition in Equation 7.15 usually holds since we concentrate on stable queueing systems, where the queue length has non-defective limiting distribution.

Note that as long as the service discipline is independent of the service times, the $\{X(t), t \geq 0\}$ process does not depend on the service discipline. Hence L is also independent of the service discipline. On the other hand, the $\{W_n, n \geq 0\}$ process does depend upon the service discipline. However, Little's Law implies that the average wait is independent of the service discipline, since the quantities L and λ are independent of the service discipline.

Example 7.7 The $M/G/\infty$ Queue. Consider the infinite server queue of Example 5.17 on page 183. In the queueing nomenclature introduce in Section 7.1 this is an $M/G/\infty$ queue. Verify Little's Law for this system.

From Example 5.17, we see that in steady state the number of customers in this system is a $P(\lambda \tau)$ random variable, where λ is the arrival rate of customers, and τ is the mean service time. Thus

$$L = \lambda \tau.$$

Since the system has infinite number of servers, W_n, the time spent by the nth customer in the system equals his service time. Hence

$$W = E(W_n) = \tau.$$

Thus Equation 7.11 is satisfied. ∎

Also see Computational Exercises 7.80 and 7.81.

7.3 Birth and Death Queues

Many queueing systems where customers arrive one at a time, form a single queue, and get served one at a time, can be modeled by birth and death processes. See Example 6.10 on page 211 for the definition, and Example 6.35 on page 258 for the limiting behavior of birth and death processes. We shall use these results in the rest of this section.

7.3.1 $M/M/1$ Queue

Consider an $M/M/1$ queue. Such a queue has a single server, infinite capacity waiting room, and the customers arrive according to a PP(λ) and request iid exp(μ) service times. Let $X(t)$ be the number of customers in the system at time t. We have seen in Example 6.11 on page 212 that $\{X(t), t \geq 0\}$ is a birth and death process with birth rates

$$\lambda_n = \lambda, \quad n \geq 0$$

and death rates

$$\mu_n = \mu, \quad n \geq 1.$$

We saw in Example 6.36 on page 259 that this queue is stable if

$$\rho = \lambda/\mu < 1.$$

The parameter ρ is called the traffic intensity of the queue, and it can be interpreted as the expected number of arrivals during one service time. The system serves one customer during one service time, and gets ρ new customers during this time on the average. Thus if $\rho < 1$, the system can serve more customers than it gets, so it should be stable. The stability condition can also be written as $\lambda < \mu$. In this form it says that the rate at which customers arrive is less than the rate at which they can be served, and hence the system should be stable. Note that the system is perfectly balanced when $\lambda = \mu$, but it is unstable, since it has no spare service capacity to handle random variation in arrivals and services. Example 6.36 shows that the limiting distribution of $X(t)$ in a stable $M/M/1$ queue is given by

$$p_j = (1 - \rho)\rho^j, \quad j \geq 0. \tag{7.17}$$

From Example 7.4 we get

$$\hat{\pi}_j = \pi_j^* = \pi_j = p_j = (1 - \rho)\rho^j, \quad j \geq 0.$$

We have

$$L = \sum_{j=0}^{\infty} jp_j = \frac{\rho}{1 - \rho}. \tag{7.18}$$

Thus as $\rho \to 1$, $L \to \infty$. This is a manifestation of increasing congestion as $\rho \to 1$.

Next we shall compute $F(\cdot)$, the limiting distribution of W_n, assuming that the service discipline is FCFS. We have

$$
\begin{aligned}
F(x) &= \lim_{n\to\infty} \mathsf{P}(W_n \le x) \\
&= \lim_{n\to\infty} \sum_{j=0}^{\infty} \mathsf{P}(W_n \le x | X_n^* = j) \mathsf{P}(X_n^* = j) \\
&= \sum_{j=0}^{\infty} \lim_{n\to\infty} \mathsf{P}(X_n^* = j) \lim_{n\to\infty} \mathsf{P}(W_n \le x | X_n^* = j).
\end{aligned}
$$

From Equation 7.10 we see that

$$
\lim_{n\to\infty} \mathsf{P}(X_n^* = j) = \pi_j^* = p_j = (1-\rho)\rho^j, \quad j \ge 0.
$$

Now suppose an entering customer sees j customers ahead in the system. Due to the FCFS discipline, these customers will be served before his service starts. The remaining service time of the customer in service (if any) is an $\exp(\mu)$ random variable. Hence this customer's time in the system is the sum of $j+1$ iid $\exp(\mu)$ random variables. Thus, for all $n \ge 1$,

$$
\mathsf{P}(W_n \le x | X_n^* = j) = 1 - \sum_{r=0}^{j} e^{-\mu x} \frac{(\mu x)^r}{r!}.
$$

Substituting, we get

$$
F(x) = \sum_{j=0}^{\infty} (1-\rho)\rho^j \left(1 - \sum_{r=0}^{j} e^{-\mu x} \frac{(\mu x)^r}{r!} \right),
$$

which, after some algebra, reduces to

$$
F(x) = 1 - e^{-(\mu-\lambda)x}, \quad x \ge 0.
$$

Thus the steady state waiting time is an $\exp(\mu - \lambda)$ random variable. Thus we have

$$
W = \mathsf{E}(\text{waiting time in steady state}) = \frac{1}{\mu - \lambda}.
$$

Using Equation 7.18 we see that $L = \lambda W$. Thus Little's Law is verified for the $M/M/1$ queue under FCFS discipline.

7.3.2 $M/M/1/K$ Queue

In an $M/M/1/K$ system, customers arrive according to a $PP(\lambda)$ and receive iid $\exp(\mu)$ service times from a single server. If an arriving customer finds K persons in the system, he leaves immediately without service. Let $X(t)$ be the number of customers in the system at time t. One can show that $\{X(t), t \ge 0\}$ is a birth and death process on $\{0, 1, 2, \cdots, K\}$ with birth rates

$$
\lambda_n = \lambda, \quad 0 \le n < K
$$

and death rates

$$\mu_n = \mu, \quad 1 \le n \le K.$$

Note that $\lambda_K = 0$ implies that the number in the system will not increase from K to $K + 1$. We can use the results of Example 6.35 on page 258 to compute the limiting distribution of the $X(t)$. Substituting in Equation 6.74 we get

$$\rho_j = \rho^j, \quad 0 \le j \le K,$$

where $\rho = \lambda/\mu$ is the traffic intensity. We have

$$\sum_{j=0}^{K} \rho_j = \begin{cases} \frac{1-\rho^{K+1}}{1-\rho} & \text{if } \rho \ne 1, \\ K+1 & \text{if } \rho = 1. \end{cases}$$

This is always finite, hence the queue is always stable. Substituting in Equation 6.75 we get, for $0 \le j \le K$,

$$p_j = \begin{cases} \frac{1-\rho}{1-\rho^{K+1}} \rho^j & \text{if } \rho \ne 1, \\ \frac{1}{K+1} & \text{if } \rho = 1. \end{cases} \tag{7.19}$$

Finally, from Example 7.5 we have

$$\hat{\pi}_j = p_j, \quad 0 \le j \le K$$

and

$$\pi_j = \pi_j^* = \frac{p_j}{1 - p_K}, \quad 0 \le j \le K - 1.$$

The mean number of customers in the system in steady state is given by

$$L = \sum_{j=0}^{K} j p_j = \begin{cases} \frac{\rho}{1-\rho} [1 - (K+1)p_K] & \text{if } \rho \ne 1, \\ \frac{K}{2} & \text{if } \rho = 1. \end{cases} \tag{7.20}$$

7.3.3 M/M/s Queue

Consider an $M/M/s$ queue. Such a queue has s identical servers, infinite capacity waiting room, the customers arrive according to a PP(λ) and request iid exp(μ) service times. The customers form a single line and the customer at the head of the line is served by the first available server. If more than one server is idle when a customer arrives, he goes to any one of the available servers. Let $X(t)$ be the number of customers in the system at time t. One can show that $\{X(t), t \ge 0\}$ is a birth and death process with birth rates

$$\lambda_n = \lambda, \quad n \ge 0$$

and death rates

$$\mu_n = \min(n, s)\mu, \quad n \ge 0.$$

We can use the results of Example 6.35 on page 258 to compute the limiting distribution of the $X(t)$. Now let

$$r = \frac{\lambda}{\mu},$$

and

$$\rho = \frac{\lambda}{s\mu}.$$

Equation 6.74 yields

$$\rho_n = \begin{cases} \frac{r^n}{n!} & \text{for } 0 \leq n < s, \\ \frac{s^s}{s!}\rho^n & \text{for } n \geq s. \end{cases}$$

We have

$$\sum_{n=0}^{\infty} \rho_n = \begin{cases} \sum_{n=0}^{s-1} \rho_n + \frac{s^s}{s!}\frac{\rho^s}{1-\rho} & \text{if } \rho < 1, \\ \infty & \text{if } \rho \geq 1. \end{cases}$$

Hence the stability condition for the $M/M/s$ queue is

$$\rho < 1.$$

This condition says that the queue is stable if the arrival rate λ is less than the maximum service rate $s\mu$, which makes intuitive sense. From now on we assume that the queue is stable. Using Equation 6.75 we get

$$p_0 = \left(\sum_{n=0}^{\infty} \rho_n\right)^{-1} = \left[\sum_{n=0}^{s-1} \rho_n + \frac{s^s}{s!}\frac{\rho^s}{1-\rho}\right]^{-1}$$

and

$$p_n = \rho_n p_0 = \begin{cases} \frac{r^n}{n!}p_0 & \text{for } 0 \leq n < s, \\ \frac{s^s}{s!}\rho^n p_0 & \text{for } n \geq s. \end{cases}$$

The limiting probability that all servers are busy is given by

$$\sum_{n=s}^{\infty} p_n = \frac{s^s}{s!}\frac{\rho^s}{1-\rho}p_0 = \frac{p_s}{1-\rho}.$$

The above probability, called the blocking probability, is commonly parameterized by r rather than ρ, and is denoted by $C(s,r)$ as given below

$$C(s,r) = \frac{\frac{r^s}{s!}\frac{s}{s-r}}{\sum_{j=0}^{s-1}\frac{r^j}{j!} + \frac{r^s}{s!}\frac{s}{s-r}}. \tag{7.21}$$

The above equation is called the Erlang-C formula (in contrast to the Erlang-B formula; see Computational Exercises 7.24 and 7.25). Many other performance measures in the $M/M/s$ queue can be written using this quantity. For example, the expected number of customers waiting for service can be shown to be

$$L_q = \frac{\rho}{1-\rho}C(s,r),$$

and the expected queueing time (time to start service) is

$$W_q = \frac{L_q}{\lambda} = \frac{1}{\mu}\frac{1}{s-r}C(s,r). \tag{7.22}$$

We leave it to the reader (see Computational Exercise 7.17) to show that the expected number of busy servers equals r. Hence the mean number of customers in the system is given by

$$L = r + L_q,$$ (7.23)

and the expected time in the system is given by

$$W = \frac{1}{\mu} + W_q.$$

In practice the $M/M/s$ queue is often used as a simple model of many multiserver service systems such as call-centers, hospitals, emergency rooms, to name a few. A typical call center can have several hundred to several thousand servers handling service calls from thousands of customers. A large hospital may have several hundred beds serving the patients. The main issue in designing such service systems is to size them correctly so that customers get high-quality service, and the servers get utilized efficiently. A design that achieves both these criteria is called a quality and efficiency driven design, or QED for short.

The quality of a queueing system is quantified by $C(s, r)$, the blocking probability in steady state. The efficiency of a queueing system is quantified by r/s, the fraction of the servers that are busy in steady state, or equivalently, the fraction of the time any one server is busy. Thus we are interested in designing queueing systems with $C(s, r)$ close to zero and r/s close to one. We describe here the Halfin–Whitt design that achieves this for systems with large number of servers.

It is intuitively clear that if we fix λ and increase s, the blocking probability (as given by Equation 7.21) will go to zero, and if we fix s and increase λ (or equivalently, increase the ratio λ/μ), the blocking probability will go to one. Thus, if we want to study large systems with the blocking probability strictly between zero and one, say α, we must have large λ/μ and large s. Such an asymptotic regime is called Halfin–Whitt regime, first studied by Halfin and Whitt. We summarize their results below.

Let ϕ and Φ be the pdf and cdf of a standard normal random variable and let

$$h(x) = \frac{\phi(x)}{1 - \Phi(x)}, \quad -\infty < x < \infty,$$ (7.24)

be its hazard rate. We call the asymptotic regime

$$HW = \{r \to \infty, \ s \to \infty, \ s = r + \beta\sqrt{r} + o(\sqrt{r})\},$$ (7.25)

where β is a constant, the Halfin–Whitt asymptotic regime. The importance of this regime is shown in the next theorem.

Theorem 7.6 In the Halfin–Whitt asymptotic regime,

$$\lim_{HW} C(s, r) = \frac{h(-\beta)}{\beta + h(-\beta)}.$$

Proof: Let X be a P(r) random variable. Then, in the Halfin-Whitt asymptotic regime we have

$$
\begin{aligned}
e^{-r}\frac{r^s}{s!} &= \mathrm{P}(s-1 < X \le s) \\
&= \mathrm{P}((s-1-r)/\sqrt{r} < (X-r)/\sqrt{r} \le (s-r)/\sqrt{r}) \\
&\to \frac{1}{\sqrt{r}}\phi(\beta).
\end{aligned}
$$

Similarly

$$
\begin{aligned}
\sum_{j=0}^{s-1} e^{-r}\frac{r^j}{j!} &= \mathrm{P}(X < s-1) \\
&= \mathrm{P}((X-r)/\sqrt{r} \le (s-r-1)/\sqrt{r}) \\
&\to \Phi(\beta).
\end{aligned}
$$

Using these limits we get

$$
\begin{aligned}
C(s,r) &= \frac{\frac{r^s}{s!}\frac{s}{s-r}}{\sum_{j=0}^{s-1}\frac{r^j}{j!} + \frac{r^s}{s!}\frac{s}{s-r}} \\
&= \frac{e^{-r}\frac{r^s}{s!}\frac{s}{s-r}}{\sum_{j=0}^{s-1}e^{-r}\frac{r^j}{j!} + e^{-r}\frac{r^s}{s!}\frac{s}{s-r}} \\
&\to \frac{\frac{1}{\sqrt{r}}\phi(\beta)\frac{\sqrt{r}}{\beta}}{\Phi(\beta) + \frac{1}{\sqrt{r}}\phi(\beta)\frac{\sqrt{r}}{\beta}} \\
&= \frac{\phi(\beta)}{\beta\Phi(\beta) + \phi(\beta)} \\
&= \frac{h(-\beta)}{\beta + h(-\beta)},
\end{aligned}
$$

where the last equality follows from $\phi(\beta) = \phi(-\beta)$ and $\Phi(\beta) = 1 - \Phi(-\beta)$. ∎

Note that the hazard rate function $h(x)$ of a standard normal random variable is an increasing function of x, increasing from 0 at $x = -\infty$ to $\sqrt{2/\pi}$ at $x = 0$, and growing to $+\infty$ as $x \to \infty$. Hence $\frac{h(-\beta)}{\beta+h(-\beta)}$ decreases from 1 at $\beta = 0$ to zero as $\beta \to \infty$. Hence there is a unique $\beta > 0$ that satisfies

$$
\frac{h(-\beta)}{\beta + h(-\beta)} = \alpha, \tag{7.26}
$$

for any given $\alpha \in (0,1)$. The solution $\beta = \beta(\alpha)$ to the above equation is a decreasing function of α. In practice, $\beta(\alpha)$ has to be computed numerically. Thus to design an $M/M/s$ queue with a given large $r = \lambda/\mu$ so as to achieve a given blocking probability α, we first compute $\beta(\alpha)$ that satisfies Equation 7.26 and use the number of servers given by

$$
s = r + \beta(\alpha)\sqrt{r}. \tag{7.27}
$$

The formula given in the above equation is sometimes called the (Halfin–Whitt) square-root staffing formula.

Note that α represents the desired blocking probability. Thus as α decreases, the service requirement becomes more stringent, the $\beta(\alpha)$ increases, and hence the recommended s in Equation 7.27 increases. This makes intuitive sense. This system is highly utilized, since the fraction of busy servers is given by

$$r/s = r/(r + \beta\sqrt{r}) \approx 1 - \beta/\sqrt{r}$$

which is close to one, since r is large. Thus almost all servers are busy in steady state. Furthermore, in Halfin–Whitt regime, the expected queueing time of Equation 7.22 reduces to

$$W_q = \frac{\alpha}{\mu} \frac{1}{\beta\sqrt{r}},$$

which is also quite small, since r is large. Thus Equation 7.27 gives us an efficient system (most servers busy) that gives high quality of service (low blocking probability). We have achieved a QED design!

How is it possible that in the Halfin–Whitt regime, all servers are almost always busy, but the queueing time is not overly long? This seems to go against the folk theorem that if you keep the servers very busy, the wait will be long (efficiency vs. throughput tradeoff). The answer is: the tradeoff exists in a system with finite fixed number of servers. However, the Halfin–Whitt regime holds when the number of servers is allowed to grow without bound with the arrival rate.

7.3.4 $M/M/\infty$ Queue

The $M/M/\infty$ queue is the limit of the $M/M/s$ queue as $s \to \infty$. It arises as a model of self-service queues and was modeled in Example 6.12 as a birth and death process with birth parameters

$$\lambda_n = \lambda, \quad n \geq 0$$

and death parameters

$$\mu_n = n\mu, \quad n \geq 0.$$

This queue is stable as long as

$$r = \frac{\lambda}{\mu} < \infty.$$

Its limiting distribution is given in Example 6.37 as

$$p_j = e^{-r}\frac{r^j}{j!}, \quad j \geq 0.$$

As in the case of the $M/M/1$ queue, we have

$$\hat{\pi}_j = \pi_j^* = \pi_j = p_j = e^{-r}\frac{r^j}{j!}, \quad j \geq 0.$$

The transient analysis of the $M/M/\infty$ queue was done in Example 5.17, from which we see that if $X(0) = 0$, then $X(t)$ is a Poisson random variable with mean $r(1 - e^{-\mu t})$. From this we get

$$\mathsf{E}(X(t)|X(0) = 0) = r(1 - e^{-\mu t}).$$

In comparison, the transient analysis of the $M/M/1$ and $M/M/s$ queues is quite messy. We refer the readers to one of the books on queueing theory for details: Saaty (1961) or Gross and Harris (1985).

7.3.5 Queues with Finite Populations

In Example 6.6 on page 208 we modeled a workshop with two machines and one repairperson. Here we consider a more general workshop with N machines and s repairpersons. The life times of the machines are iid $\exp(\mu)$ random variables, and the repair times are iid $\exp(\lambda)$ random variables, and are independent of the life times. The machines are as good as new after repairs. The machines are repaired in the order in which they fail. Let $X(t)$ be the number of working machines at time t. One can show that $\{X(t), t \geq 0\}$ is a birth and death process on $\{0, 1, 2, \cdots, N\}$ with birth parameters

$$\lambda_n = \min(N - n, s)\lambda, \quad 0 \leq n \leq N - 1$$

and death parameters

$$\mu_n = n\mu, \quad 0 \leq n \leq N.$$

Since this is a finite state queue, it is always stable. One can compute the limiting distribution using the results of Example 6.35 on page 258.

7.3.6 $M/M/1$ Queue with Balking and Reneging

Consider an $M/M/1$ queue of Section 7.3.1. Now suppose an arriving customer who sees n customers in the system ahead of him joins the system with probability α_n. This is called the balking behavior. Once he joins the system, he displays reneging behavior as follows: He has a patience time that is an $\exp(\theta)$ random variable. If his service does not begin before his patience time expires, he leaves the system without getting served (i.e., he reneges); else he completes his service and then departs. All customer patience times are independent of each other.

Let $X(t)$ be the number of customers in the system at time t. We can show that $\{X(t), t \geq 0\}$ is a birth and death process with birth rates

$$\lambda_n = \alpha_n \lambda, \quad n \geq 0$$

and death rates

$$\mu_n = \mu + (n - 1)\theta, \quad n \geq 1.$$

(Why do we get $(n-1)\theta$ and not $n\theta$ in the death parameters?) If $\theta > 0$, such a queue is always stable and its steady state distribution can be computed by using the results of Example 6.35 on page 258.

The above examples illustrate the usefulness of the birth and death processes in modeling queues. Further examples are given in Modeling Exercises 7.1 through 7.5.

7.4 Open Queueing Networks

The queueing models in Section 7.3 are for single-station queues, i.e., there is a single place where the service is rendered. Customers arrive at this facility, which may have more than one server, get served once, and then depart.

In this and the next section we consider more complicated queueing systems called queueing networks. A typical queueing network consists of several service stations, or nodes. Customers form queues at each of these queueing stations. After completing service at a station a customer may depart from the system or join a queue at some other service station. In open queueing networks, customers arrive at the nodes from outside the system, visit the nodes in some order, and then depart. Thus the total number of customers in an open queueing network varies over time. In closed queueing networks there are no external arrivals or departures, thus the total number of customers in a closed queueing network remains constant. We shall study open queueing networks in this section and closed queueing networks in the next.

Open queueing networks arise in all kinds of situations: hospitals, computers, telecommunication networks, assembly lines, and supply chains, to name a few. Consider a simple model of patient flow in a hospital. Patients arrive from outside at the admitting office or the emergency ward. Patients from the admitting office go to various clinics, which we have lumped into one node for modeling ease. In the clinics the patients are diagnosed and routed to the intensive care unit, or are discharged, or are given reappointment for a follow-up visit. Patients with reappointments go home and return to the admitting office at appropriate times. Patients from the emergency ward are either discharged after proper care, or are sent to the intensive care unit. From the intensive care unit they are either discharged or given reappointments for follow-up visits.

This simple patient flow model can be set up as a queueing network with five nodes as shown in Figure 7.3. The arrows interconnecting the nodes show the patient routing pattern. In this figure the customers can arrive at the system at two nodes: admissions and emergency. Customers can depart the system from three nodes: clinics, emergency, and intensive care. Note that a customer can visit a node a random number of times before departing the system, i.e., a queueing network can have cycles.

It is extremely difficult to analyze a queueing network in all its generality. In this section we shall concentrate on a special class of queueing networks called the Jackson networks. Jackson introduced this class in a seminal paper in 1957, and it has become a standard queueing network model since then.

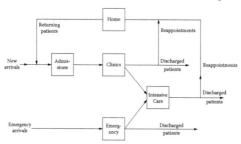

Figure 7.3 *Schematic diagram of patient flow in a hospital.*

A queueing network is called a Jackson network if it satisfies the following assumptions:

A1. It has N service stations (nodes).

A2. There are s_i servers at node i, $1 \leq s_i \leq \infty$, $1 \leq i \leq N$. Service times of customers at node i are iid $\exp(\mu_i)$ random variables. They are independent of service times of customers in other nodes.

A3. There is an infinite waiting room at each node.

A4. External arrivals at node i form a $PP(\lambda_i)$. All the arrival processes are independent of each other and the service times.

A5. After completing service at node i, a customer departs the system with probability r_i, or joins the queue at node j with probability $r_{i,j}$, independent of the number of customers at any node in the system. $r_{i,i}$ can be positive. We have

$$r_i + \sum_{j=1}^{N} r_{i,j} = 1, \quad 1 \leq i \leq N. \tag{7.28}$$

A6. The routing matrix $R = [r_{i,j}]$ is such that $I - R$ is invertible.

All the above assumptions are crucial to the analysis of Jackson networks. Some assumptions can be relaxed, but not the others. For example, in practice we would like to consider finite capacity waiting rooms (assumption 3) or state-dependent routing (assumption 5). However such networks are very difficult to analyze. On the other hand certain types of state-dependent service and arrival rates can be handled fairly easily. See Sections 7.4.1 and 7.4.2.

Now let us study a Jackson network described above. Let $X_i(t)$ be the number of customers at node i at time t, $(1 \leq i \leq N, \ t \geq 0)$, and let

$$X(t) = [X_1(t), X_2(t), \cdots, X_N(t)]$$

be the state of the queueing network at time t. The state-space of the $\{X(t), t \geq 0\}$ process is $S = \{0, 1, 2, \cdots\}^N$. To understand the transitions in this process we need some notation.

Let e_i be an N-vector with a 1 in the ith coordinate and 0 in all others. Suppose the system is in state $x = [x_1, x_2, \cdots, x_N] \in S$. If an external arrival takes place at node i, the system state changes to $x + e_i$. If a customer completes service at node i (this can happen only if $x_i > 0$) and departs the system, the system state changes from x to $x - e_i$, and if the customer moves to state j, the system state changes from x to $x - e_i + e_j$. It can be seen that $\{X(t), t \geq 0\}$ is a multidimensional CTMC with the following transition rates:

$$
\begin{aligned}
q(x, x + e_i) &= \lambda_i, \\
q(x, x - e_i) &= \min(x_i, s_i)\mu_i r_i, \\
q(x, x - e_i + e_j) &= \min(x_i, s_i)\mu_i r_{i,j}, \quad i \neq j.
\end{aligned}
$$

Hence we get

$$
q(x, x) = -q(x) = -\sum_{i=1}^{N} \lambda_i - \sum_{i=1}^{N} \min(x_i, s_i)\mu_i(1 - r_{i,i}).
$$

Now let

$$
\begin{aligned}
p(x) &= \lim_{t \to \infty} \mathsf{P}(X(t) = x) \\
&= \lim_{t \to \infty} \mathsf{P}(X_1(t) = x_1, X_2(t) = x_2, \cdots, X_N(t) = x_N)
\end{aligned}
$$

be the limiting distribution, assuming it exists.

We now study node j in isolation. The input to node j consists of two parts: the external input that occurs at rate λ_j, and the internal input originating from other nodes (including node j) in the network. Let a_j be the total arrival rate (external + internal) to node j. In steady state (assuming it exists) the departure rate from node i must equal the total input rate a_i to node i. A fraction $r_{i,j}$ of the departing customers go to node j. Thus the internal input from node i to node j is $a_i r_{i,j}$. Thus in steady state we must have

$$
a_j = \lambda_j + \sum_{i=1}^{N} a_i r_{i,j}, \quad 1 \leq j \leq N, \tag{7.29}
$$

which can be written in matrix form as

$$
a(I - R) = \lambda,
$$

where $a = [a_1, a_2, \cdots, a_N]$ and $\lambda = [\lambda_1, \lambda_2, \cdots, \lambda_N]$. Equation 7.29 is called the *traffic equation* of the Jackson network. Now, from assumption 6, $I - R$ is invertible. Hence we have

$$
a = \lambda(I - R)^{-1}. \tag{7.30}
$$

Note that the invertibility of $I - R$ implies that no customer stays in the network indefinitely.

Now consider an $M/M/s_i$ queue with arrival rate a_i and service rate μ_i. From the results of Section 7.3.3 we see that such a queue is stable if $\rho_i = a_i/s_i\mu_i < 1$, and

$\phi_i(n)$, the steady state probability that there are n customers in the system is given by

$$\phi_i(n) = \frac{s_i^{\min(n,s_i)}}{\min(n,s_i)!} \rho_i^n \phi_i(0), \quad n \geq 0,$$

where

$$\phi_i(0) = \left[\sum_{n=0}^{s_i-1} \frac{s_i^n}{n!} \rho_i^n + \frac{s_i^{s_i}}{s_i!} \frac{\rho_i^{s_i}}{1-\rho_i}\right]^{-1}.$$

With these preliminaries we are ready to state the main result about the Jackson networks below.

Theorem 7.7 Open Jackson Networks. *The CTMC $\{X(t), t \geq 0\}$ is positive recurrent if and only if*

$$a_i < s_i \mu_i, \quad 1 \leq i \leq N,$$

where $a = [a_1, a_2, \cdots, a_N]$ is given by Equation 7.30. When it is positive recurrent, its steady state distribution is given by

$$p(x) = \prod_{i=1}^{N} \phi_i(x_i). \tag{7.31}$$

Proof: The balance equations for $\{X(t), t \geq 0\}$ are

$$
\begin{aligned}
q(x)p(x) &= \sum_{i=1}^{N} \lambda_i p(x - e_i) + \sum_{i=1}^{N} r_i \mu_i \min(x_i + 1, s_i) p(x + e_i) \\
&+ \sum_{j=1}^{N} \sum_{i:i\neq j} r_{i,j} \mu_i \min(x_i + 1, s_i) p(x + e_i - e_j), \quad x \in S.
\end{aligned}
$$

Assume that $p(x) = 0$ if x has any negative coordinates. Substitute $p(x)$ of Equation 7.31 in the above equation to get

$$
\begin{aligned}
q(x) \prod_{k=1}^{N} \phi_k(x_k) &= \sum_{i=1}^{N} \lambda_i \prod_{k=1}^{N} \phi_k(x_k) \frac{\phi_i(x_i - 1)}{\phi_i(x_i)} \\
&+ \sum_{i=1}^{N} r_i \mu_i \min(x_i + 1, s_i) \prod_{k=1}^{N} \phi_k(x_k) \frac{\phi_i(x_i + 1)}{\phi_i(x_i)} \\
&+ \sum_{j=1}^{N} \sum_{i:i\neq j} r_{i,j} \mu_i \min(x_i + 1, s_i) \prod_{k=1}^{N} \phi_k(x_k) \frac{\phi_i(x_i + 1)}{\phi_i(x_i)} \frac{\phi_j(x_j - 1)}{\phi_j(x_j)}.
\end{aligned}
$$

Canceling $\prod_{k=1}^{N} \phi_k(x_k)$ from both sides and using the identity

$$\frac{\phi_i(x_i - 1)}{\phi_i(x_i)} = \frac{\mu_i \min(x_i, s_i)}{a_i}$$

the above equality reduces to

$$q(x) = \sum_{i=1}^{N} \lambda_i \frac{\mu_i \min(x_i, s_i)}{a_i} + \sum_{i=1}^{N} r_i a_i$$
$$+ \sum_{j=1}^{N} \sum_{i:i\neq j} (a_i r_{i,j}) \frac{\mu_j \min(x_j, s_j)}{a_j}.$$

Using Equation 7.29 and simplifying the above equation we get

$$\sum_{i=1}^{N} \lambda_i = \sum_{i=1}^{N} r_i a_i.$$

However, the above equation holds in steady state since the left-hand side is the rate at which the customers enter the network, and the right-hand side is the rate at which they depart the network. We can also derive this equation from Equations 7.29 and 7.28. Thus the solution in Equation 7.31 satisfies the balance equation. Since $\phi_i(0) > 0$ if and only if $a_i < s_i \mu_i$, the condition of positive recurrence follows. Also, the CTMC is irreducible. Hence there is a unique limiting distribution. Hence the theorem follows. ■

The form of the distribution in Equation 7.31 is called the product form, for the obvious reason. Theorem 7.7 says that, in steady state, the queue lengths at the N nodes are independent random variables. Furthermore, node i behaves as if it is an $M/M/s_i$ queue with arrival rate a_i and service rate μ_i. The phrase "behaves as if" is important, since the process $\{X_i(t), t \geq 0\}$ is not a birth and death process of an $M/M/s_i$ queue. For example, in general, the total arrival process to node i is not a PP(a_i). It just so happens that the steady distribution of $\{X_i(t), t \geq 0\}$ is the same as that of an $M/M/s_i$ queue. We illustrate the result of Theorem 7.7 with a few examples.

Example 7.8 Single Queue with Feedback. The simplest queueing network is a single station queue with feedback as shown in Figure 7.4. Customers arrive from

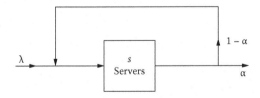

Figure 7.4 *An s-server queue with Bernoulli feedback.*

outside at this service station according to a PP(λ) and request iid exp(μ) services. The service station has s identical servers. When a customer completes service, he leaves the system with probability α. With probability $1 - \alpha$ he rejoins the queue and behaves as a new arrival.

This is a Jackson network with $N = 1, s_1 = s, \mu_1 = \mu, \lambda_1 = \lambda, r_1 = \alpha, r_{1,1} = 1 - \alpha$. From Equation 7.29, we get

$$a_1 = \lambda + (1 - \alpha)a_1,$$

which yields

$$a_1 = \lambda/\alpha.$$

Thus the queue is stable if

$$\rho = \frac{\lambda}{\alpha s \mu} < 1.$$

The steady state distribution of $X(t) = X_1(t)$ is given by

$$p_n = \lim_{t \to \infty} \mathsf{P}(X(t) = n) = \frac{s^{\min(n,s)}}{\min(n, s)!} \rho^n p_0, \quad n \geq 0,$$

where

$$p_0 = \left[\sum_{n=0}^{s-1} \frac{s^n}{n!} \rho^n + \frac{s^s}{s!} \frac{\rho^s}{1 - \rho} \right]^{-1}. \quad \blacksquare$$

Example 7.9 Tandem Queue. Consider N single server queues in tandem as shown in Figure 7.5. External customers arrive at node 1 according to a PP(λ). The

Figure 7.5 *A tandem queueing network.*

service times at the ith node are iid $\exp(\mu_i)$ random variables. Customers completing service at node i join the queue at node $i + 1, 1 \leq i \leq N - 1$. Customers leave the system after completing service at node N.

This is a Jackson network with N nodes and $\lambda_1 = \lambda, \lambda_i = 0$, for $2 \leq i \leq N$, $r_{i,i+1} = 1$ for $1 \leq i \leq N - 1, r_i = 0$ for $1 \leq i \leq N - 1$, and $r_N = 1$. The traffic Equations 7.29 yield

$$a_i = \lambda, \quad 1 \leq i \leq N.$$

Hence the tandem queueing system is stable if $\lambda < \mu_i$ for all $1 \leq i \leq N$, i.e.,

$$\lambda < \min\{\mu_1, \mu_2, \cdots, \mu_N\}.$$

Thus the slowest server determines the traffic handling capacity of the tandem network. Assuming stability, the limiting distribution is given by

$$\lim_{t \to \infty} \mathsf{P}(X_i(t) = x_i, \ 1 \leq i \leq N) = \prod_{i=1}^{N} \left(\frac{\lambda}{\mu_i} \right)^{x_i} \left(1 - \frac{\lambda}{\mu_i} \right). \quad \blacksquare$$

Example 7.10 Patient Flow. Consider the queueing network model of patient flow as shown in Figure 7.3. Suppose external patients arrive at the admitting ward at a rate of 4 per hour and at the emergency ward at a rate of 1/hr. The admissions desk is managed by one secretary who processes an admission in five minutes on the average. The clinic is served by k doctors, (here k is to be decided on), and the average consultation with a doctor takes 15 minutes. Generally, one out of every four patients going through the clinic is asked to return for another check up in two weeks (336 hours). One out of every ten is sent to the intensive care unit from the clinic. The rest are dismissed after consultations. Patients arriving at the emergency room require on the average one hour of consultation time with a doctor. The emergency room is staffed by m doctors, where m is to be decided on. One out of two patients in the emergency ward goes home after treatment, whereas the other is admitted to the intensive care unit. The average stay in the intensive care unit is four days, and there are n beds available, where n is to be decided on. From the intensive care unit, 20% of the customers go home, and the other 80% are given reappointments for follow-up in two weeks. Analyze this system assuming that the assumptions of Jackson networks are satisfied.

We model this as a Jackson network with five nodes as follows: Node 1 is Admissions, Node 2 is Emergency, Node 3 is Clinics, Node 4 is Intensive Care Unit, and Node 5 is Home. The parameters of the network (using time units of hours) are

$$\lambda_1 = 4, \ \lambda_2 = 1, \ \lambda_3 = 0, \ \lambda_4 = 0, \ \lambda_5 = 0,$$
$$s_1 = 1, \ s_2 = m, \ s_3 = k, \ s_4 = n, \ s_5 = \infty,$$
$$\mu_1 = 12, \ \mu_2 = 1, \ \mu_3 = 4, \ \mu_4 = 1/96, \ \mu_5 = 1/336,$$
$$r_1 = 0, \ r_2 = 0.5, \ r_3 = 0.65, \ r_4 = 0.2, \ r_5 = 0.$$

The routing matrix is given by

$$R = \begin{bmatrix} 0 & 0 & 1 & 0 & 0 \\ 0 & 0 & 0 & 0.5 & 0 \\ 0 & 0 & 0 & 0.1 & 0.25 \\ 0 & 0 & 0 & 0 & 0.80 \\ 1 & 0 & 0 & 0 & 0 \end{bmatrix}.$$

Equation 7.29 yields

$$\begin{aligned} a_1 &= 4 + a_5 \\ a_2 &= 1 \\ a_3 &= a_1 \\ a_4 &= .5a_2 + .1a_3 \\ a_5 &= .25a_3 + .8a_4. \end{aligned}$$

Solving the above equations we get

$$a_1 = 6.567, \ a_2 = 1, \ a_3 = 6.567, \ a_4 = 1.157, \ a_5 = 2.567.$$

We use Theorem 7.7 to establish the stability of the network. Note that $a_1 < s_1\mu_1$ and $a_5 < s_5\mu_5$. We must also have

$$1 = a_2 < s_2\mu_2 = m,$$
$$6.567 = a_3 < s_3\mu_3 = 4k,$$

$$1.157 = a_4 < s_4\mu_4 = n/96.$$

These are satisfied if we have

$$m > 1, \quad k > 1.642, \quad n > 111.043.$$

Thus the hospital must have at least two doctors in the emergency room, at least two in the clinics, and have at least 112 beds in the intensive care unit. So let us assume the hospital uses two doctors each in the emergency room and the clinics, and has 120 beds. With these parameters, the steady state analysis of the queueing network can be done by treating (1) the admissions queue as an $M/M/1$ with arrival rate 6.567 per hour, and service rate of 12 per hour; (2) the emergency ward queue as an $M/M/2$ with arrival rate 1 per hour, and service rate of 1 per hour; (3) the clinic queue as an $M/M/2$ queue with arrival rate 6.567 per hour and service rate of 4 per hour; (4) the intensive care queue as an $M/M/120$ with arrival rate 1.157 per hour, and service rate of 1/96 per hour; and (5) the home queue as an $M/M/\infty$ with arrival rate 2.567 per hour, and service rate of 1/336 per hour. Furthermore these five queues are independent of each other in steady state. ∎

Next we discuss two important generalizations of the Jackson networks.

7.4.1 State-Dependent Service

In the Jackson network model we had assumed that the service rate at node i, when there are n customers at that node, is given by $\min(s_i, n)\mu_i$. We define Jackson networks with state-dependent service by replacing assumption A2 by A2' as follows:

A2'. The service rate at node i, when there are n customers at that node, is given by $\mu_i(n)$, with $\mu_i(0) = 0$ and $\mu_i(n) > 0$ for $n \geq 0$, $1 \leq i \leq N$.

Note that the service rate at node i is not allowed to depend on the state of node $j \neq i$. Now define

$$\phi_i(0) = 1, \quad \phi_i(n) = \prod_{j=1}^{n} \left(\frac{a_i}{\mu_i(j)} \right), \quad n \geq 1, 1 \leq i \leq N \qquad (7.32)$$

where a_i is the total arrival rate to node i as given by Equation 7.30. Jackson networks with state-dependent service also admit a product form solution as shown in the next theorem.

Theorem 7.8 Jackson Networks with State-Dependent Service. *A Jackson network with state-dependent service rates is stable if and only if*

$$c_i = \sum_{n=0}^{\infty} \phi_i(n) < \infty \quad \text{for all} \;\; 1 \leq i \leq N.$$

If the network is stable, the limiting state distribution is given by

$$p(x) = \prod_{i=1}^{N} \frac{\phi_i(x_i)}{c_i}, \quad x \in S.$$

Proof: Follows along the same lines as the proof of Theorem 7.7. ∎

Thus, in steady state, the queues at various nodes in a Jackson network with state-dependent service are independent.

7.4.2 State-Dependent Arrivals and Service

It is also possible to further generalize the model of Section 7.4.1 by allowing the external arrival rate to the network to depend on the total number of customers in the network. Specifically, we replace assumption A4 by A4′ as follows:

A4′. External customers arrive at the network at rate $\lambda(n)$ when the total number of customers in the network is n. An arriving customer joins node i with probability u_i, where

$$\sum_{i=1}^{N} u_i = 1.$$

The above assumption implies that the external arrival rate to node i is $u_i\lambda(n)$ if the total number of customers in the network is n. To keep the $\{X(t), t \geq 0\}$ process irreducible, we assume that there is a $K \leq \infty$ such that $\lambda(n) > 0$ for $0 \leq n < K$, and $\lambda(n) = 0$ for $n \geq K$. We call the Jackson networks with assumptions A2 and A4 replaced by A2′ and A4′ "Jackson networks with state-dependent arrivals and service."

We shall see that such Jackson networks with state-dependent arrival and service rates continue to have a kind of product form limiting distribution. However, the queue at various nodes are not independent any more. The results are given in the next theorem. First, we need the following notation.

Let $\{a_i, 1 \leq i \leq N\}$ be the unique solution to

$$a_j = u_j + \sum_{i=1}^{N} a_i r_{i,j}, \quad 1 \leq j \leq N.$$

Let $\phi_i(n)$ be as defined in Equation 7.32 using the above $\{a_i, 1 \leq i \leq N\}$. Also, for $x = [x_1, x_2, \cdots, x_N] \in S$, let

$$|x| = \sum_{i=1}^{N} x_i.$$

Thus if the state of the network is $X(t)$, the total number of customers in it at time t is $|X(t)|$.

Theorem 7.9   Jackson Networks with State-Dependent Arrivals and Service.
The limiting state distribution in a Jackson network with state-dependent arrivals

and service is given by

$$p(x) = c \cdot \prod_{i=1}^{N} \phi_i(x_i) \cdot \prod_{j=1}^{|x|} \lambda(j), \quad x \in S,$$

where c is the normalizing constant given by

$$c = \left(\sum_{x \in S} \prod_{i=1}^{N} \phi_i(x_i) \cdot \prod_{j=1}^{|x|} \lambda(j) \right)^{-1}.$$

The network is stable if and only if $c > 0$.

Proof: Follows along the same lines as that of Theorem 7.7. ∎

Computation of the constant c is the hard part. There is a large literature on "product form" queueing networks, and it is more or less completely understood now as to what enables a network to have "product form" solution. See Kelly (1979) and Walrand (1988).

7.5 Closed Queueing Networks

In this section we consider the closed queueing networks. In these networks there are no external arrivals to the network, and there are no departures from the network. Thus the total number of customers in the network is constant. Closed queueing networks have been used to study population dynamics, multiprogrammed computer systems, telecommunication networks with window flow control, etc. We start with the definition.

A queueing network is called a closed Jackson network if it satisfies the following assumptions:

B1. It has N service stations (nodes) and a total of K customers.

B2. The service rate at node i, when there are n customers at that node, is given by $\mu_i(n)$, with $\mu_i(0) = 0$ and $\mu_i(n) > 0$ for $1 \leq n \leq K, 1 \leq i \leq N$.

B3. After completing service at node i, a customer joins the queue at node j with probability $r_{i,j}$, independent of the number of customers at any node in the system. $r_{i,i}$ can be positive.

B4. The routing matrix $R = [r_{i,j}]$ is a transition probability matrix of an irreducible DTMC.

Now let us study a closed Jackson network described above. Let $X_i(t)$ be the number of customers at node i at time t, $(1 \leq i \leq N, \ t \geq 0)$, and let

$$X(t) = [X_1(t), X_2(t), \cdots, X_N(t)]$$

be the state of the queueing network at time t. As in the case of open Jackson networks, we see that $\{X(t), t \geq 0\}$ is a CTMC on state-space

$$S = \{x = [x_1, x_2, \cdots, x_N] : x_i \geq 0, \sum_{i=1}^{N} x_i = K\}$$

with transition rates given by

$$q(x, x - e_i + e_j) = \mu_i(x_i) r_{i,j}, \quad i \neq j, \ x \in S.$$

Hence we get

$$q(x, x) = -q(x) = -\sum_{i=1}^{N} \mu_i(x_i)(1 - r_{i,i}), \quad x \in S.$$

Since the CTMC has finite state-space and is irreducible, it is positive recurrent. Let

$$\begin{aligned} p(x) &= \lim_{t \to \infty} \mathsf{P}(X(t) = x) \\ &= \lim_{t \to \infty} \mathsf{P}(X_1(t) = x_1, X_2(t) = x_2, \cdots, X_N(t) = x_N) \end{aligned}$$

be the limiting distribution. We need the following notation before we give the result about $p(x)$. Let $\pi = [\pi_1, \pi_2, \cdots, \pi_N]$ be the limiting distribution of the DTMC with transition matrix R. Since R is assumed to be irreducible, π is the unique solution to

$$\pi = \pi R, \quad \sum_{i=1}^{N} \pi_i = 1. \tag{7.33}$$

Next, define

$$\phi_i(0) = 1, \quad \phi_i(n) = \prod_{j=1}^{n} \left(\frac{\pi_i}{\mu_i(j)} \right), \quad 1 \leq n \leq K, \ 1 \leq i \leq N. \tag{7.34}$$

Theorem 7.10 Closed Jackson Networks. *The limiting distribution of the CTMC* $\{X(t), t \geq 0\}$ *is given by*

$$p(x) = G_N(K) \prod_{i=1}^{N} \phi_i(x_i), \tag{7.35}$$

where the normalizing constant $G_N(K)$ *is chosen so that*

$$\sum_{x \in S} p(x) = 1.$$

Proof: Follows by verifying that the solution in Equation 7.35 satisfies the balance equation

$$q(x)p(x) = \sum_{y \in S: y \neq x} p(y)q(y, x).$$

The verification proceeds along the same lines as that in the proof of Theorem 7.7.

∎

Thus the closed Jackson network has a "product form" limiting distribution. The hard part is the evaluation of $G_N(K)$, the normalizing constant. The computation is difficult since the size of the state-space grows exponentially in N and K: it has $\binom{N+K-1}{K}$ elements. A recursive method of computing $G_N(K)$ for closed Jackson networks of single-server queues is described in the next example.

Example 7.11 Tandem Closed Network. Consider a closed Jackson network of N single server nodes as shown in Figure 7.6. The service rate at node i is $\mu_i(n) = \mu_i$

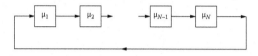

Figure 7.6 *A tandem closed network.*

for $n \geq 1$. The routing probabilities are $r_{i,i+1} = 1$ for $1 \leq i \leq N-1$, and $r_{N,1} = 1$. Thus the solution to Equation 7.33 is given by

$$\pi_i = \frac{1}{N}, \quad 1 \leq i \leq N.$$

We have

$$\phi_i(n) = \rho_i^n, \quad n \geq 0,$$

where

$$\rho_i = \frac{\pi_i}{\mu_i} = \frac{1}{N\mu_i}.$$

We introduce the notation

$$H_N(K) = \frac{1}{G_N(K)},$$

and use Theorem 7.10 to get

$$p(x) = \frac{1}{H_N(K)} \prod_{i=1}^{N} \rho_i^{x_i}, \quad x \in S.$$

We leave it to the readers to verify that the generating function of $H_N(K)$ is given by

$$\tilde{H}_N(z) = \sum_{K=0}^{\infty} H_N(K) z^K = \prod_{i=1}^{N} \frac{1}{1 - \rho_i z}.$$

Now we see that $\tilde{H}_0(z) = 1$ and

$$\tilde{H}_N(z)(1 - \rho_N z) = \tilde{H}_{N-1}(z), \quad N \geq 1,$$

which can be written as

$$\tilde{H}_N(z) = \rho_N z \tilde{H}_N(z) + \tilde{H}_{N-1}(z).$$

From this we can derive the following recursion

$$H_N(K) = H_{N-1}(K) + \rho_N H_N(K-1), \quad N \geq 1, K \geq 1,$$

with boundary conditions

$$H_0(0) = 1, \quad H_N(0) = 1, \quad N \geq 1, \quad H_0(K) = 0, \quad K \geq 1.$$

We can use this recursion to compute $H_N(K)$ (and hence $G_N(K)$) in $O(NK)$ steps.
■

Example 7.12   MultiProgramming Systems. Consider the following model of a multiprogramming computer system. It consists of a central processing unit (CPU) (node 1), a printer (node 2), and a disk drive (node 3). A program starts in the CPU. When the computing part is done, it goes to the printer with probability α or the disc drive with probability $1 - \alpha$. From the printer the program terminates with probability β or goes back to the CPU for further computing with probability $1 - \beta$. After completing the operation at the disc drive the program returns to the CPU with probability 1. Suppose the time required at the CPU phase is $\exp(\mu_1)$, the printer stage is $\exp(\mu_2)$, and the disc drive stage is $\exp(\mu_3)$. Suppose these times are independent. When a program departs the system from the printer queue, a new program is instantaneously admitted to the CPU queue, so that the total number of programs in the system remains constant, say K. The parameter K is called the degree of multiprogramming.

This system can be modeled by a closed Jackson network as shown in Figure 7.7. The parameters of this network are

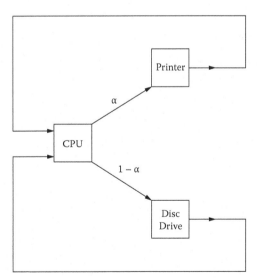

Figure 7.7 *Multi-programming system.*

$$\mu_i(n) = \mu_i, \quad i = 1, 2, 3; \; n \geq 1,$$

$$R = \begin{bmatrix} 0 & \alpha & 1-\alpha \\ 1 & 0 & 0 \\ 1 & 0 & 0 \end{bmatrix}.$$

Thus the solution to Equation 7.33 is given by

$$\pi_1 = 0.5, \quad \pi_2 = 0.5\alpha, \quad \pi_3 = 0.5(1-\alpha).$$

Using

$$\rho_1 = \frac{1}{2\mu_1}, \quad \rho_2 = \frac{\alpha}{2\mu_2}, \quad \rho_3 = \frac{1-\alpha}{2\mu_3},$$

we get

$$\phi_i(n) = \rho_i^n, \quad i = 1, 2, 3; \ 0 \le n \le K.$$

Thus

$$p(x_1, x_2, x_3) = G_3(K)\rho_1^{x_1}\rho_2^{x_2}\rho_3^{x_3}, \quad x \in S.$$

The constant $G_3(K)$ can be computed by using the method of Example 7.11.

The throughput of the system is defined as the rate at which jobs get completed in steady state. In our system, if the system is in state (x_1, x_2, x_3) with $x_2 > 0$, jobs get completed at rate $\mu_2\beta$. Hence we have

$$\text{throughput} = \mu_2\beta \sum_{x \in S: x_2 > 0} p(x_1, x_2, x_3).$$

The closed queueing systems have been found to be highly useful models for computing systems and there is a large literature in this area. See Gelenbe and Pujolle (1987) and Saur and Chandy (1981). ∎

7.6 Single Server Queues

So far we have studied queueing systems that are described by CTMCs. In this section we study single server queues where either the service times or the interarrival times are non-exponential, making the queue length process non-Markovian.

7.6.1 M/G/1 Queue

We study an $M/G/1$ queue where customers arrive according to a PP(λ) and form a single queue in an infinite waiting room in front of a single server and demand iid service times with common cdf $G(\cdot)$, with mean τ and variance σ^2. Let $X(t)$ be the number of customers in the system at time t. The stochastic process $\{X(t), t \ge 0\}$ is a CTMC if and only if the service times are exponential random variables. Thus in general we cannot use the theory of CTMCs to study

$$p_j = \lim_{t \to \infty} P(X(t) = j), \quad j \ge 0,$$

in an $M/G/1$ queue.

Recall the definitions of X_n, X_n^*, $\hat{X}_n$, π_j, π_j^*, and $\hat{\pi}_j$ from Section 7.1. From Example 7.6 we have

$$\hat{\pi}_j = \pi_j = \pi_j^* = p_j, \quad j \geq 0.$$

Thus we can compute the limiting distribution of $\{X(t), t \geq 0\}$ by studying the limiting distribution of $\{X_n, n \geq 0\}$. This is possible to do, since the next theorem shows that $\{X_n, n \geq 0\}$ is a DTMC.

Theorem 7.11   Embedded DTMC in an $M/G/1$ Queue. $\{X_n, n \geq 0\}$ is an irreducible and aperiodic DTMC on $S = \{0, 1, 2, \cdots\}$ with one-step transition probability matrix

$$P = \begin{bmatrix} \alpha_0 & \alpha_1 & \alpha_2 & \alpha_3 & \cdots \\ \alpha_0 & \alpha_1 & \alpha_2 & \alpha_3 & \cdots \\ 0 & \alpha_0 & \alpha_1 & \alpha_2 & \cdots \\ 0 & 0 & \alpha_0 & \alpha_1 & \cdots \\ 0 & 0 & 0 & \alpha_0 & \cdots \\ \vdots & \vdots & \vdots & \vdots & \ddots \end{bmatrix}, \tag{7.36}$$

where

$$\alpha_i = \int_0^\infty e^{-\lambda t} \frac{(\lambda t)^i}{i!} dG(t), \quad i \geq 0. \tag{7.37}$$

Proof: Let A_n be the number of arrivals to the queueing system during the nth service time. Since the service times are iid random variables with common distribution $G(\cdot)$, and the arrival process is PP(λ), we see that $\{A_n, n \geq 1\}$ is a sequence of iid random variables with common pmf

$$\begin{aligned} \mathsf{P}(A_n = i) &= \mathsf{P}(i \text{ arrivals during a service time}) \\ &= \int_0^\infty \mathsf{P}(i \text{ arrivals during a service time of duration } t) dG(t) \\ &= \int_0^\infty e^{-\lambda t} \frac{(\lambda t)^i}{i!} dG(t) \\ &= \alpha_i. \end{aligned}$$

Now, if $X_n > 0$, the $(n+1)$st service time starts immediately after the nth departure, and during that service time A_{n+1} customers join the system. Hence after the $(n+1)$st departure $X_n + A_{n+1} - 1$ customers are left in the system. On the other hand, if $X_n = 0$, the $(n+1)$st service time starts immediately after the $(n+1)$st arrival, and during that service time A_{n+1} customers join the system. Hence after the $(n+1)$st departure A_{n+1} customers are left in the system. Combining these two observations, we get

$$X_{n+1} = \begin{cases} A_{n+1} & \text{if } X_n = 0, \\ X_n - 1 + A_{n+1} & \text{if } X_n > 0. \end{cases} \tag{7.38}$$

This is identical to Equation 2.10 derived in Example 2.16 on page 20 if we define $Y_n = A_n$. The result then follows from the results in Example 2.16. The DTMC is irreducible and aperiodic since $\alpha_i > 0$ for all $i \geq 0$. ∎

The next theorem gives the result about the limiting distribution of $\{X_n, n \geq 0\}$.

Theorem 7.12   Limiting Distribution of an $M/G/1$ Queue. *The DTMC $\{X_n, n \geq 0\}$ is positive recurrent if and only if*

$$\rho = \lambda\tau < 1.$$

If it is positive recurrent, its limiting distribution has the generating function given by

$$\phi(z) = \sum_{j=0}^{\infty} \pi_j z^j = (1 - \rho)\frac{(1 - z)\tilde{G}(\lambda - \lambda z)}{\tilde{G}(\lambda - \lambda z) - z}, \qquad (7.39)$$

where

$$\tilde{G}(s) = \int_0^{\infty} e^{-st} dG(t).$$

Proof: Since $\{X_n, n \geq 0\}$ is the DTMC studied in Example 2.16 on page 20, we can use the results about its limiting distribution from Example 4.25 on page 124. From there we see that the DTMC is positive recurrent if and only if

$$\sum_{k=0}^{\infty} k\alpha_k < 1.$$

Substituting from Equation 7.37 we get

$$\begin{aligned}
\sum_{k=0}^{\infty} k\alpha_k &= \sum_{k=0}^{\infty} k \int_0^{\infty} e^{-\lambda t} \frac{(\lambda t)^k}{k!} dG(t) \\
&= \int_0^{\infty} e^{-\lambda t} \left(\sum_{k=0}^{\infty} k \frac{(\lambda t)^k}{k!} \right) dG(t) \\
&= \int_0^{\infty} \lambda t\, dG(t) = \lambda\tau = \rho.
\end{aligned}$$

Thus the DTMC is positive recurrent if and only if $\rho < 1$. From Equation 4.44 (using ρ in place of μ) we get

$$\phi(z) = (1 - \rho)\frac{\psi(z)(1 - z)}{\psi(z) - z}, \qquad (7.40)$$

where

$$\psi(z) = \sum_{k=0}^{\infty} \alpha_k z^k.$$

Substituting for α_k from Equation 7.37 in the above equation

$$\begin{aligned}
\psi(z) &= \sum_{k=0}^{\infty} z^k \int_0^{\infty} e^{-\lambda t} \frac{(\lambda t)^k}{k!} dG(t) \\
&= \int_0^{\infty} e^{-\lambda t} \left(\sum_{k=0}^{\infty} z^k \frac{(\lambda t)^k}{k!} \right) dG(t) \\
&= \int_0^{\infty} e^{-\lambda t} e^{\lambda z t} dG(t)
\end{aligned}$$

$$= \int_0^\infty e^{-\lambda(1-z)t} dG(t) = \tilde{G}(\lambda - \lambda z).$$

Substituting in Equation 7.40 we get Equation 7.39. This proves the theorem. ∎

One immediate consequence of Equation 7.39 is that the probability that the server is idle in steady state can be computed as

$$p_0 = \pi_0 = \phi(0) = 1 - \rho. \tag{7.41}$$

Also, since $p_j = \pi_j$ for all $j \geq 0$, $\phi(z)$ in Equation 7.39 is also the generating function of the limiting distribution of the $\{X(t), t \geq 0\}$ process. Using Equation 7.39 we can compute the expected number of customers in the system in steady state as given in the following theorem.

Theorem 7.13   Expected Number in an $M/G/1$ Queue. *The expected number in steady state in a stable $M/G/1$ queue is given by*

$$L = \rho + \frac{1}{2} \cdot \frac{\rho^2}{1 - \rho} \left(1 + \frac{\sigma^2}{\tau^2}\right) = \rho + \frac{\lambda^2 s^2}{2(1 - \rho)}, \tag{7.42}$$

where τ, σ^2, and s^2 are the mean, variance, and the second moment of the service time.

Proof: We have

$$
\begin{aligned}
L &= \lim_{t \to \infty} \mathsf{E}(X(t)) \\
&= \sum_{j=0}^\infty j p_j \\
&= \sum_{j=0}^\infty j \pi_j \\
&= \left. \frac{d\phi(z)}{dz} \right|_{z=1}.
\end{aligned}
$$

The theorem follows after evaluating the last expression in a straight forward fashion. This involves using L'Hopital's rule twice. ∎

The Equation 7.42 is called the Pollaczec–Khintchine formula. It is interesting to note that the first moment of the queue length depends on the second moment (or equivalently, the variance) of the service time. This has an important implication. We can decrease the queue length by making the server more consistent, that is, by reducing the variability of the service times. Since the variance is zero for constant service times, it follows that among all service times with the same mean, the deterministic service time will minimize the expected number in the system in steady state!

Example 7.13   The $M/M/1$ Queue. If the service time distribution is

$$G(x) = 1 - e^{-\mu x}, \quad x \geq 0$$

the $M/G/1$ queue reduces to $M/M/1$ queue with $\tau = 1/\mu$ and $\rho = \lambda\tau = \lambda/\mu$. In this case we get

$$\tilde{G}(s) = \frac{\mu}{s + \mu}.$$

Substituting in Equation 7.39 we get

$$
\begin{aligned}
\phi(z) &= (1-\rho)\frac{(1-z)\tilde{G}(\lambda - \lambda z)}{\tilde{G}(\lambda - \lambda z) - z} \\
&= (1-\rho)\frac{(1-z)\mu/(\lambda + \mu - \lambda z)}{\mu/(\lambda + \mu - \lambda z) - z} \\
&= (1-\rho)\frac{(1-z)\mu}{(1-z)(\mu - \lambda z)} \\
&= \frac{1-\rho}{1-\rho z}.
\end{aligned}
$$

By expanding the last expression in a power series in z we get

$$\phi(z) = \sum_{j=0}^{\infty} \pi_j z^j = \sum_{j=0}^{\infty} p_j z^j = (1-\rho)\sum_{j=0}^{\infty} \rho^j z^j.$$

Hence we get

$$p_j = (1-\rho)\rho^j, \quad j \geq 0.$$

This matches with the result in Example 6.36 on page 259, as expected. ∎

Example 7.14   The $M/E_k/1$ Queue. Suppose the service times are iid Erl(k, μ). Then the $M/G/1$ queue reduces to $M/E_k/1$ queue. In this case we get

$$\tau = \frac{k}{\mu}, \quad \sigma^2 = \frac{k}{\mu^2}.$$

The queue is stable if

$$\rho = \lambda\tau = \frac{k\lambda}{\mu} < 1.$$

Assuming the queue is stable, the expected number in steady state can be computed by using Equation 7.42 as

$$L = \rho + \frac{1}{2} \cdot \frac{\rho^2}{1-\rho} \frac{k+1}{k}. \quad ∎$$

A large number of variations of the $M/G/1$ queue have been studied in literature. See Modeling Exercises 7.11 and 7.13.

Next we study the waiting times (this includes time in service) in an $M/G/1$ queue assuming FCFS service discipline. Let $F_n(\cdot)$ be the cdf of W_n, the waiting time of the nth customer. Let

$$\tilde{F}_n(s) = E(e^{-sW_n}).$$

The next theorem gives the Laplace Stieltjes transform (LST) of the waiting time in steady state, defined as

$$\tilde{F}(s) = \lim_{n \to \infty} \tilde{F}_n(s).$$

Theorem 7.14 Waiting Times in an $M/G/1$ **Queue.** *The LST of the waiting time in steady state in a stable* $M/G/1$ *queue with FCFS service discipline is given by*

$$\tilde{F}(s) = (1 - \rho)\frac{s\tilde{G}(s)}{s - \lambda(1 - \tilde{G}(s))}. \tag{7.43}$$

Proof: Let A_n be the number of arrivals during the nth customer's waiting time in the system. Since X_n is the number of customers left in the system after the nth departure, the assumption of FCFS service discipline implies that $X_n = A_n$. The Poisson assumption implies that (see the derivation of $\psi(z)$ in the proof of Theorem 7.12)

$$\mathsf{E}(z^{A_n}) = \tilde{F}_n(\lambda - \lambda z).$$

Hence

$$\phi(z) = \lim_{n \to \infty} \mathsf{E}(z^{X_n}) = \lim_{n \to \infty} \mathsf{E}(z^{A_n}) = \lim_{n \to \infty} \tilde{F}_n(\lambda - \lambda z) = \tilde{F}(\lambda - \lambda z).$$

Substituting $\lambda - \lambda z = s$ we get Equation 7.43. ∎

Equation 7.43 is also known as the Pollaczec–Khintchine formula. Using the derivatives of $\tilde{F}(s)$ at $s = 0$ we get

$$W = \tau + \frac{\lambda s^2}{2(1 - \rho)},$$

where $s^2 = \sigma^2 + \tau^2$ is the second moment of the service time. Using Equation 7.42 we can verify directly that Little's Law $L = \lambda W$ holds for the $M/G/1$ queue under FCFS service discipline.

7.6.2 $G/M/1$ Queue

Now we study a $G/M/1$ queue where customers arrive one at a time and the interarrival times are iid random variables with common cdf $G(\cdot)$, with $G(0) = 0$ and mean $1/\lambda$. The arriving customers form a single queue in an infinite waiting room in front of a single server and demand iid $\exp(\mu)$ service times. Let $X(t)$ be the number of customers in the system at time t. The stochastic process $\{X(t), t \geq 0\}$ is a CTMC if and only if the interarrival times are exponential random variables. Thus in general we cannot use the theory of CTMCs to study the limiting behavior of $X(t)$.

Recall the definitions of X_n, X_n^*, $\hat{X}_n$, π_j, π_j^*, and $\hat{\pi}_j$ from Section 7.1. From Example 7.6 we have

$$\hat{\pi}_j = \pi_j = \pi_j^*, \quad j \geq 0.$$

However, unless the interarrival times are exponential, the arrival process is not a PP, and hence $\hat{\pi}_j \neq p_j$. The next theorem shows that $\{X_n^*, n \geq 0\}$ is a DTMC.

Theorem 7.15   Embedded DTMC in a $G/M/1$ Queue. $\{X_n^*, n \geq 0\}$ *is an irreducible and aperiodic DTMC on* $S = \{0, 1, 2, \cdots\}$ *with one-step transition probability matrix*

$$
P = \begin{bmatrix}
\beta_0 & \alpha_0 & 0 & 0 & 0 & \cdots \\
\beta_1 & \alpha_1 & \alpha_0 & 0 & 0 & \cdots \\
\beta_2 & \alpha_2 & \alpha_1 & \alpha_0 & 0 & \cdots \\
\beta_3 & \alpha_3 & \alpha_2 & \alpha_1 & \alpha_0 & \cdots \\
\vdots & \vdots & \vdots & \vdots & \vdots & \ddots
\end{bmatrix},
\tag{7.44}
$$

where

$$
\alpha_i = \int_0^\infty e^{-\mu t} \frac{(\mu t)^i}{i!} dG(t), \quad i \geq 0,
\tag{7.45}
$$

and

$$
\beta_i = \sum_{j=i+1}^\infty \alpha_j, \quad i \geq 0.
$$

Proof: Let D_n be the number of departures that can occur (assuming there are enough customers in the system) during the nth interarrival time. Since the interarrival times are iid random variables with common distribution $G(\cdot)$, and the service times are iid $\exp(\mu)$, we see that $\{D_n, n \geq 1\}$ is a sequence of iid random variables with common pmf

$$
\begin{aligned}
P(D_n = i) &= P(i \text{ possible departures during an interarrival time}) \\
&= \int_0^\infty e^{-\mu t} \frac{(\mu t)^i}{i!} dG(t) \\
&= \alpha_i.
\end{aligned}
$$

Now, the nth arrival sees X_n^* customers in the system. Hence there are $X_n^* + 1$ customers in the system after the nth customer enters. If $D_{n+1} < X_n^* + 1$, the $(n + 1)$st arrival will see $X_n^* + 1 - D_{n+1}$ customers in the system, else there will be no customers in the system when the next arrival occurs. Hence we get

$$
X_{n+1}^* = \max\{X_n^* + 1 - D_{n+1}, 0\}.
$$

This is identical to Equation 2.12 derived in Example 2.17 on page 21 if we define $Y_n = D_n$. The result then follows from the results in Example 2.17. The DTMC is irreducible and aperiodic since $\alpha_i > 0$ for all $i \geq 0$. ∎

The next theorem gives the result about the limiting distribution of $\{X_n^*, n \geq 0\}$.

Theorem 7.16 $G/M/1$ Queue at Arrival Times. *The DTMC* $\{X_n^*, n \geq 0\}$ *is positive recurrent if and only if*

$$
\rho = \lambda/\mu < 1.
$$

If it is positive recurrent, its limiting distribution is given by

$$
\pi_j^* = \lim_{n \to \infty} P(X_n^* = j) = (1 - \alpha)\alpha^j, \quad j \geq 0
\tag{7.46}
$$

where α is the unique solution in $(0, 1)$ to

$$\alpha = \int_0^\infty e^{-\mu(1-\alpha)t} dG(t) = \tilde{G}(\mu(1-\alpha)). \qquad (7.47)$$

Proof: Since $\{X_n^*, n \geq 0\}$ is the DTMC studied in Example 2.17 on page 21, we can use the results about its limiting distribution from Example 4.26 on page 125. From there we see that the DTMC is positive recurrent if and only if

$$\sum_{k=0}^\infty k\alpha_k > 1.$$

Substituting from Equation 7.45 we get

$$\sum_{k=0}^\infty k\alpha_k = \frac{\mu}{\lambda}.$$

Thus the DTMC is positive recurrent if and only if $\frac{\mu}{\lambda} > 1$, i.e., $\rho < 1$. Let

$$\psi(z) = \sum_{i=0}^\infty z^i \alpha_i.$$

Following the derivation in the proof of Theorem 7.12, we get

$$\psi(z) = \tilde{G}(\mu - \mu z).$$

From Equation 4.46 (using α in place of ρ) we get

$$\pi_j^* = (1 - \alpha)\alpha^j, \quad j \geq 0,$$

where α is the unique solution in $(0, 1)$ to

$$\alpha = \psi(\alpha) = \tilde{G}(\mu - \mu\alpha).$$

This proves the theorem. ∎

Example 7.15   The $M/M/1$ Queue. If the interarrival time distribution is

$$G(x) = 1 - e^{-\lambda x}, \quad x \geq 0$$

the $G/M/1$ queue reduces to $M/M/1$ queue. In this case we get

$$\tilde{G}(s) = \frac{\lambda}{s + \lambda}.$$

Substituting in Equation 7.47 we get

$$\alpha = \frac{\lambda}{\mu(1 - \alpha) + \lambda}.$$

Solving for α we get

$$\alpha = \frac{\lambda}{\mu} = \rho, \text{ or } \alpha = 1.$$

If $\rho < 1$, the $\alpha = \rho$ is the only solution in $(0, 1)$. In this case Equation 7.46 reduces to

$$\pi_j^* = (1 - \rho)\rho^j, \quad j \geq 0.$$

Since the arrival process in this queue is Poisson, we have $p_j = \hat{\pi}_j = \pi_j^*$. Thus we have

$$p_j = (1 - \rho)\rho^j, \quad j \geq 0.$$

This matches with the result in Example 6.36 on page 259, as expected. ∎

The next theorem relates $\{p_j, j \geq 0\}$ and $\{\pi_j^*, j \geq 0\}$.

Theorem 7.17   Limiting Distribution of a $G/M/1$ Queue. *For a $G/M/1$ queue with $\rho = \lambda/\mu < 1$ the limiting distributions $\{p_j, j \geq 0\}$ and $\{\pi_j^*, j \geq 0\}$ are related as follows:*

$$
\begin{align}
p_0 &= 1 - \rho, & (7.48)\\
p_j &= \rho\pi_{j-1}^*, \quad j \geq 1. & (7.49)
\end{align}
$$

Intuition: In steady state, the rate at which customers arrive at the system is λ (inverse of the mean interarrival time). This is called the elementary renewal theorem and will be proved in the next chapter, see Theorem 8.6 on page 376. Let $j \geq 1$ be a fixed number. Each arriving customer sees $j - 1$ customers ahead of him with probability π_{j-1}^*, by definition. Hence, in steady state, the rate of transitions from state $j - 1$ to j in the stochastic process $\{X(t), t \geq 0\}$ is given by $\lambda\pi_{j-1}^*$. Next, the probability that there are j customers in the system in steady state is p_j, by definition. The customer in service always leaves at rate μ, since the service times are exponential. Thus the rate of transitions from state j to $j - 1$ in the $\{X(t), t \geq 0\}$ process in steady state is given by μp_j. In steady state these two rates (one from $j - 1$ to j and the other from j to $j - 1$) must be equal, since the number in the queue moves up and down by one at a time. That is, we must have

$$\mu p_j = \lambda\pi_{j-1}^*.$$

This yields Equation 7.49. Equation 7.48 follows since the p_j's must add up to one. This gives the intuition behind the theorem. The formal proof is postponed to Theorem 9.18 on page 467. ∎

In the next theorem we study the limiting distribution of waiting times (this includes time in service) in a $G/M/1$ queue assuming FCFS service discipline.

Theorem 7.18   Waiting Times in a $G/M/1$ Queue. *The limiting distribution of the waiting time in a stable $G/M/1$ queue with FCFS service discipline is given by*

$$F(x) = \lim_{n \to \infty} P(W_n \leq x) = 1 - e^{-\mu(1-\alpha)x}, \quad x \geq 0. \tag{7.50}$$

Proof: The waiting time of a customer who sees j customers ahead of him is an

Erlang$(j + 1, \mu)$ random variable. Using this we get

$$F(x) = \lim_{n \to \infty} \sum_{j=0}^{\infty} P(W_n \le x | X_n^* = j) P(X_n^* = j)$$

$$= \sum_{j=0}^{\infty} \pi_j^* P(\text{Erl}(j + 1, \mu) \le x).$$

The rest of the proof follows along the same lines as in the case of the $M/M/1$ queue. ∎

From Equation 7.50 we get

$$W = \frac{1}{\mu(1 - \alpha)}.$$

Using Little's Law we get

$$L = \frac{\lambda}{\mu(1 - \alpha)},$$

which can be verified by computing L directly from the limiting distribution of $X(t)$ given Theorem 7.17.

7.7 Retrial Queue

We have already seen the $M/M/1/1$ retrial queue in Example 6.15 on page 213. In this section we generalize it to $M/G/1/1$ queue. We describe the model below.

Customers arrive from outside to a single server according to a PP(λ) and require iid service times with common distribution $G(\cdot)$ and mean τ. There is room only for the customer in service. Thus the capacity is 1, hence the $M/G/1/1$ nomenclature. If an arriving customer finds the server idle, he immediately enters service. Otherwise he joins the "orbit," where he stays for an $\exp(\theta)$ amount of time (called the retrial time) independent of his past and the other customers. At the end of the retrial time he returns to the server, and behaves like a new customer. He persists in conducting retrials until he is served, after which he exits the system. A block diagram of this queueing system is shown in Figure 7.8.

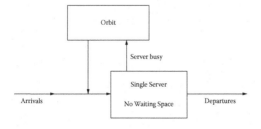

Figure 7.8 *Schematic diagram of a single server retrial queue.*

Let $X(t)$ be the number of customers in the system (those in service + those in orbit) at time t. Note that $\{X(t), t \geq 0\}$ is not a CTMC. It has jumps of size +1 when a new customer arrives, and of size -1 when a customer completes service. Since every arriving customer enters the system (either service or the orbit), and the arrival process is Poisson, we have

$$\hat{\pi}_j = \pi_j^* = \pi_j = p_j, \quad j \geq 0.$$

Thus we can study the limiting behavior of the $\{X(t), t \geq 0\}$ by studying the $\{X_n, t \geq 0\}$ process at departure points, since, as the next theorem shows, it is a DTMC.

Theorem 7.19 Embedded DTMC in an $M/G/1/1$ Retrial Queue. $\{X_n, n \geq 0\}$ *is an irreducible and aperiodic DTMC on $S = \{0, 1, 2, \cdots\}$.*

Proof: Let A_n be the number of arrivals to the queueing system during the nth service time. Since the service times are iid random variables with common distribution $G(\cdot)$, and the arrival process is PP(λ), we see that $\{A_n, n \geq 1\}$ is a sequence of iid random variables. Now, immediately after a service completion, the server is idle. Hence X_n represents the number of customers in the orbit when the nth service completion occurs. Each of these customers will conduct a retrial after iid exp(θ) times. Also, a new arrival will occur after an exp(λ) amount of time. Hence the next service request will occur after an exp($\lambda + \theta X_n$) amount of time. With probability $\theta X_n/(\lambda + \theta X_n)$ this request is from a customer from the orbit, and with probability $\lambda/(\lambda + \theta X_n)$, it is from a new customer. The $(n + 1)$st service starts when this request arrives, during which A_{n+1} new customers arrive and join the orbit. Hence the system dynamics is given by

$$X_{n+1} = \begin{cases} X_n + A_{n+1} & \text{with probability } \lambda/(\lambda + \theta X_n) \\ X_n + A_{n+1} - 1 & \text{with probability } \theta X_n/(\lambda + \theta X_n). \end{cases} \tag{7.51}$$

Since A_{n+1} is independent of history, the above recursion implies that $\{X_n, n \geq 0\}$ is a DTMC. Irreducibility and aperiodicity are obvious. ∎

The next theorem gives the generating function of the limiting distribution of $\{X_n, n \geq 0\}$.

Theorem 7.20 Limiting Distribution of an $M/G/1/1$ Retrial Queue. *The DTMC $\{X_n, n \geq 0\}$ with $\theta > 0$ is positive recurrent if and only if*

$$\rho = \lambda \tau < 1.$$

If it is positive recurrent, its limiting distribution has the generating function given by

$$\phi(z) = \sum_{j=0}^{\infty} \pi_j z^j = (1-\rho) \frac{(1-z)\tilde{G}(\lambda - \lambda z)}{\tilde{G}(\lambda - \lambda z) - z} \cdot \exp\left(-\frac{\lambda}{\theta} \int_z^1 \frac{1 - \tilde{G}(\lambda - \lambda u)}{\tilde{G}(\lambda - \lambda u) - u} du \right),$$
$$\tag{7.52}$$

where

$$\tilde{G}(s) = \int_0^\infty e^{-st} dG(t).$$

Proof: From Equation 7.51 we get

$$
\begin{aligned}
\mathsf{E}(z^{X_{n+1}}) &= \mathsf{E}\left(z^{X_n+A_{n+1}}\frac{\lambda}{\lambda+\theta X_n}\right) + \mathsf{E}\left(z^{X_n+A_{n+1}-1}\frac{\theta X_n}{\lambda+\theta X_n}\right) \\
&= \mathsf{E}(z^{A_{n+1}})\left[\mathsf{E}\left(\frac{\lambda z^{X_n}}{\lambda+\theta X_n}\right) + \mathsf{E}\left(\frac{\theta X_n z^{X_n-1}}{\lambda+\theta X_n}\right)\right]. \quad (7.53)
\end{aligned}
$$

Now let

$$\phi_n(z) = \mathsf{E}(z^{X_n}) \text{ and } \psi_n(z) = \mathsf{E}\left(\frac{\lambda z^{X_n}}{\lambda+\theta X_n}\right).$$

Then

$$
\begin{aligned}
\psi_n(z) + \frac{\theta z}{\lambda}\psi_n'(z) &= \mathsf{E}\left(\frac{\lambda z^{X_n}}{\lambda+\theta X_n}\right) + \frac{\theta z}{\lambda}\mathsf{E}\left(\frac{\lambda X_n z^{X_n-1}}{\lambda+\theta X_n}\right) \\
&= \mathsf{E}\left(\frac{(\lambda+\theta X_n)z^{X_n}}{\lambda+\theta X_n}\right) \quad (7.54) \\
&= \mathsf{E}(z^{X_n}) = \phi_n(z). \quad (7.55)
\end{aligned}
$$

Also, from the proof of Theorem 7.12, we get

$$\mathsf{E}(z^{A_n}) = \tilde{G}(\lambda - \lambda z), \quad n \geq 1.$$

Using Equation 7.55 in the Equation 7.53 we get

$$\psi_{n+1}(z) + \frac{\theta z}{\lambda}\psi_{n+1}'(z) = \tilde{G}(\lambda - \lambda z)(\psi_n(z) + \frac{\theta}{\lambda}\psi_n'(z)). \quad (7.56)$$

Now let

$$\phi(z) = \lim_{n\to\infty} \phi_n(z), \text{ and } \psi(z) = \lim_{n\to\infty} \psi_n(z).$$

Letting $n \to \infty$ in Equation 7.56 and rearranging, we get

$$\frac{\theta}{\lambda}(z - \tilde{G}(\lambda - \lambda z))\psi'(z) = (\tilde{G}(\lambda - \lambda z) - 1)\psi(z)$$

which can be easily integrated to obtain

$$\psi(z) = C' \exp\left(\frac{\lambda}{\theta}\int^z \frac{\tilde{G}(\lambda - \lambda u) - 1}{u - \tilde{G}(\lambda - \lambda u)}du\right),$$

where C' is a constant of integration. By choosing

$$C' = C \exp\left(-\frac{\lambda}{\theta}\int^1 \frac{\tilde{G}(\lambda - \lambda u) - 1}{u - \tilde{G}(\lambda - \lambda u)}du\right),$$

we get

$$\psi(z) = C \exp\left(-\frac{\lambda}{\theta}\int_z^1 \frac{\tilde{G}(\lambda - \lambda u) - 1}{u - \tilde{G}(\lambda - \lambda u)}du\right). \quad (7.57)$$

Letting $n \to \infty$ in Equation 7.55 we get

$$\phi(z) = \psi(z) + \frac{\theta z}{\lambda} \psi'(z).$$

Using Equation 7.56 and 7.57 we get

$$\phi(z) = C \frac{(1-z)\tilde{G}(\lambda - \lambda z)}{\tilde{G}(\lambda - \lambda z) - z} \cdot \exp\left(-\frac{\lambda}{\theta} \int_z^1 \frac{1 - \tilde{G}(\lambda - \lambda u)}{\tilde{G}(\lambda - \lambda u) - u} du\right).$$

The unknown constant C can be evaluated by using $\phi(1) = 1$. This yields (after applying L'Hopital's rule)

$$C = 1 - \rho. \tag{7.58}$$

Hence the theorem follows. ∎

One immediate consequence of Equation 7.52 is that the probability that the system is empty in steady state can be computed as

$$p_0 = \pi_0 = \phi(0) = (1 - \rho) \exp\left(-\frac{\lambda}{\theta} \int_0^1 \frac{1 - \tilde{G}(\lambda - \lambda u)}{\tilde{G}(\lambda - \lambda u) - u} du\right).$$

However, this is not the same as the server being idle, since the server can be idle even if the system is not empty. We can use a Little's Law type argument to show that the server is idle in steady state with probability $1 - \rho$, independent of θ! But this simple fact cannot be deduced by using the embedded DTMC. We shall derive this result formally in Section 9.6.4.

Now, since $p_j = \pi_j$ for all $j \geq 0$, $\phi(z)$ in Equation 7.52 is also the generating function of the limiting distribution of the $\{X(t), t \geq 0\}$ process. Using Equation 7.52 we can compute the expected number of customers in the system in steady state as given in the following theorem.

Theorem 7.21 Expected Number in an $M/G/1/1$ Retrial Queue. *The expected number in steady state in a stable $M/G/1/1$ retrial queue with $\theta > 0$ is given by*

$$L = \rho + \frac{\lambda^2 s^2}{2(1 - \rho)} + \frac{\lambda}{\theta} \frac{\rho}{1 - \rho}, \tag{7.59}$$

where τ and s^2 are the mean and the second moment of the service time.

Proof: We have

$$\begin{aligned} L &= \lim_{t \to \infty} \mathsf{E}(X(t)) \\ &= \sum_{j=0}^{\infty} j p_j = \sum_{j=0}^{\infty} j \pi_j = \left.\frac{d\phi(z)}{dz}\right|_{z=1}. \end{aligned}$$

The theorem follows after evaluating the last expression in straightforward fashion. This involves using L'Hopital's rule twice. ∎

Note that L is a decreasing function of θ. In fact, as $\theta \to \infty$, L of the above theorem converges to the L of a standard $M/G/1$ as given in Theorem 7.13. This makes intuitive sense, since in the limit as $\theta \to \infty$, every customer is always checking to see if the server is idle. Thus as soon as the service is complete a customer from the orbit (if it is not empty) enters service. However, the service discipline is in "random order," rather than in FCFS fashion, although this does not affect the queue length process. As expected, the generating function of the retrial queue reduces to that of the $M/G/1$ queue as $\theta \to \infty$.

7.8 Infinite Server Queue

We have seen the $M/M/s$ queue in Section 7.3.3 and the $M/M/\infty$ queue in Section 7.3.4. Unfortunately, the $M/G/s$ queue proves to be intractable. Surprisingly, $M/G/\infty$ queue can be analyzed very easily. We present that analysis here.

Consider an infinite server queue where customers arrive according to a PP(λ), and request iid service times with common cdf $G(\cdot)$ and mean τ. Let $X(t)$ be the number of customers in such a queue at time t. Suppose $X(0) = 0$. We have analyzed this process in Example 5.17 on page 183. Using the analysis there we see that $X(t)$ is a Poisson random variable with mean $m(t)$, where

$$m(t) = \lambda \int_0^t (1 - G(u))du.$$

We have

$$\lim_{t \to \infty} m(t) = \lambda \tau.$$

Hence the limiting distribution of $X(t)$ is P($\lambda \tau$). Note that this limiting distribution holds even if $X(0) > 0$, since all the initial customers will eventually leave, and do not affect the newly arriving customers.

It is possible to analyze an infinite server queue with non-stationary arrivals. Specifically, suppose the arrival process is an NPP($\lambda(\cdot)$), and the service times are iid with common distribution $G(\cdot)$ and mean τ. We denote such a queue by $M(t)/G/\infty$ queue. This model has found many applications in computing staffing policies in large multiserver queueing systems with non-stationary arrivals. The next theorem gives the main result:

Theorem 7.22 $M(t)/G/\infty$ **Queue.** Let $X(t)$ be the number of customers in an $M(t)/G/\infty$ queue at time t. Suppose $X(0) = 0$. Then $X(t)$ is a P($m(t)$) random variable where

$$m(t) = \int_0^t \lambda(u)(1 - G(t - u))du, \quad t \geq 0. \tag{7.60}$$

Proof: Fix a $t \geq 0$. Consider an arrival occurring at time $u \in [0, t]$. It is in the system at time t (and hence is counted in $X(t)$) if its service time is more than $t - u$, which happens with probability $p(u) = 1 - G(t - u)$. Thus we can think of $X(t)$ as the number of registered events over $(0, t]$ in an NPP($\lambda(\cdot)$) where an event occurring at time u is registered with probability $p(u)$. Then, from Conceptual Exercise 5.17, we see that $X(t)$ is a Poisson process with parameter given in Equation 7.60. ∎

When discussing an $M(t)/G/\infty$ queue, it is often assumed that the system starts at time $-\infty$, that is we specify the rate function $\lambda(t)$ for all $-\infty < t < \infty$. Then we do not need to assume that $X(0) = 0$, and Equation 7.60 can be recast in a more succinct way as follows. First define

$$G_e(t) = \frac{1}{\tau} \int_0^t (1 - G(u))du, \quad 0 \le t < \infty.$$

Note that $G_e(\cdot)$ is a cdf of a continuous non-negative random variable, since it is a differentiable increasing function with $G_e(0) = 0$ and $G_e(\infty) = 1$. It is called the equilibrium distribution associated with G, and we shall learn more about it in Section 8.6. Let S_e be a random variable with cdf G_e. We leave it to the reader (see Computational Exercise 7.70) to show that the number of customers in an $M(t)/G/\infty$ queue starting at $-\infty$ is a Poisson random variable with mean

$$m(t) = \tau \mathsf{E}(\lambda(t - S_e)), \quad -\infty < t < \infty. \tag{7.61}$$

We conclude this chapter with the remark that it is possible to analyze a $G/M/s$ queue with an embedded DTMC chain. The $G/M/\infty$ queue can be analyzed by the methods of renewal processes, to be developed in the next chapter. Note that the $\{X(t), t \ge 0\}$ processes studied in the last three sections are not CTMCs. What kind of processes are these? The search for the answer to this question will lead us into renewal theory, regenerative processes, and Markov regenerative processes. These topics will be covered in the next two chapters.

7.9 Modeling Exercises

7.1 Customers arrive at a taxi stand according to a PP(λ). If a taxi is waiting at the taxi stand, the customer immediately hires it and leaves the taxi stand in the taxi. If there are no taxis available, the customer waits. There is an infinite waiting room for the customers. Independently of the customers, taxis arrive at the taxi stand according to a PP(μ). If a taxi arriving at the taxi stand finds that no customer is waiting, it leaves immediately. Model this system as an $M/M/1$ queue, and specify its parameters.

7.2 A machine produces items one at a time according to a PP(λ). These items are stored in a warehouse of infinite capacity. Demands arise according to a PP(μ). If there is an item in the warehouse when a demand arises, an item is immediately removed to satisfy the demand. Any demands that occur when the warehouse is empty are lost. Let $X(t)$ be the number of items in the warehouse at time t. Model the $\{X(t), t \ge 0\}$ process as a birth and death process.

7.3 Customers arrive at a bank according to a PP(λ). The service times are iid

$\exp(\mu)$. The bank follows the following policy: when there are fewer than four customers in the bank, only one teller is active; for four to nine customers, the bank uses two tellers; and beyond nine customers there are three tellers. Model the number of customers in the bank as a birth and death process.

7.4 Customers arrive according to a PP(λ) at a single-server station and demand iid $\exp(\mu)$ service times. When a customer completes his service, he departs with probability α, or rejoins the queue instantaneously with probability $1 - \alpha$, and behaves like a new customer. The service times are iid $\exp(\mu)$. Model the number of customers in the system as a birth and death process. Is this an $M/M/1$ queue?

7.5 Consider a grocery store checkout queue. When the number of customers in the line is three or fewer, the checkout person does the pricing as well as bagging, taking $\exp(\mu_1)$ time. When there are three or more customers in the line, a bagger comes to help, and the service rate increases to $\mu_2 > \mu_1$, i.e., the reduced service times are now iid $\exp(\mu_2)$. Assume that customers join the checkout line according to a PP(λ). Model the number of customers in the checkout line as a birth and death process.

7.6 Consider a single server queue subject to breakdowns and repairs as follows: the worker stays functional for an $\exp(\theta)$ amount of time and then fails. The repair time is $\exp(\alpha)$. The successive up and down times are iid. However, the server is subject to failures only when it is serving a customer. The service times are iid $\exp(\mu)$. Assume that the failure does not cause any loss of work. Thus if a customer service is interrupted by failure, the service simply resumes after the server is repaired. Let $X(t)$ be the number of customers in this system at time t. Model $\{X(t), t \geq 0\}$ as an $M/G/1$ queue by identifying the correct service time distribution G (or its LST $\tilde{G}(s)$).

7.7 Consider a single server queue that serves customers from k independent sources. Customers from source i arrive according to a PP(λ_i) and need iid $\exp(\mu_i)$ service times. They form a single queue and are served in an FCFS fashion. Let $X(t)$ be the number of customers in the system at time t. Show that $\{X(t), t \geq 0\}$ is the queue-length process in an $M/G/1$ queue. Identify the service distribution G.

7.8 Redo the problem in Modeling Exercise 7.6 assuming that the server can fail even when it is not serving any customers. Is $\{X(t), t \geq 0\}$ the queue length process of an $M/G/1$ queue? Explain. Let X_n be the number of customers in the system after the nth departure. Show that $\{X_n, n \geq 0\}$ is a DTMC and display its transition probability matrix.

7.9 Consider the $\{X(t), t \geq 0\}$ process described in Modeling Exercise 7.2 with the following modification: the machine produces items in a deterministic fashion at a rate of one item per unit time. Model $\{X(t), t \geq 0\}$ as the queue length process in a $G/M/1$ queue.

7.10 Customers arrive according to a PP(λ) at a service station with s distinct servers. Service times at server i are iid $\exp(\mu_i)$, with $\mu_1 > \mu_2 > \cdots > \mu_s$. There is no waiting room. Thus there can be at the most s customers in the system. An incoming customer goes to the fastest available server. If all the servers are busy, he leaves without service. Model this as a CTMC.

7.11 Consider an $M/G/1$ queue where the server goes on vacation if the system is empty upon service completion. If the system is empty upon return from the vacation, the server goes on another vacation; else he starts serving the customers in the system one by one. Successive vacation times are iid. Let X_n be the number of customers in the system after the nth customer departs. Show that $\{X_n, n \geq 0\}$ is a DTMC.

7.12 A service station is staffed with two identical servers. Customers arrive according to a PP(λ). The service times are iid with common distribution $\exp(\mu)$ at either server. Consider the following two routing policies

1. Each customer is randomly assigned to one of the two servers with equal probability.
2. Customers are alternately assigned to the two servers.

Once a customer is assigned to a server he stays in that line until served. Let $X_i(t)$ be the number of customers in line for the ith server. Is $\{X_i(t), t \geq 0\}$ the queue-length process of an $M/M/1$ queue or an $G/M/1$ queue under the two routing schemes? Identify the parameters of the queues.

7.13 Consider the following variation of an $M/G/1$ queue: All customers have iid service times with common cdf G, with mean τ_G and variance σ_G^2. However the customers who enter an empty system have a different service time cdf H with mean τ_H and variance σ_H^2. Let $X(t)$ be the number of customers in the system at time t. Is $\{X(t), t \geq 0\}$ a CTMC? If yes, give its generator matrix. Let X_n be the number of customers in the system after the nth departure. Is $\{X_n, n \geq 0\}$ a DTMC? If yes, give its transition probability matrix.

7.14 Consider a communication node where packets arrive according to a PP(λ). The node is allowed to transmit packets only at times $n = 0, 1, 2 \cdots$, and transmission time of a packet is one unit of time. If a packet arrives at an empty system, it has to wait for the next transmission time to start its transmission. Let $X(t)$ be the number of packets in the system at time t, X_n be the number of packets in the system after the completion of the nth transmission, and $\bar{X}_n$ be the number of packets available for transmission at time n. Is $\{X_n, n \geq 0\}$ a DTMC? If yes, give its transition probabilities. Is $\{\bar{X}_n, n \geq 0\}$ a DTMC? If yes, give its transition probabilities.

7.15 Suppose the customers that cannot enter an $M/M/1/1$ queue (with arrival rate λ and service rate μ) enter service at another single server queue with infinite waiting room. This second queue is called an overflow queue. The service times at the overflow queue are iid $\exp(\theta)$ random variables. Let $X(t)$ be the number of customers at the overflow queue at time t. Model the overflow queue as a $G/M/1$ queue. What is the LST of the interarrival time distribution to the overflow queue?

7.16 Consider two single-server stations in series. Customers arrive to the first station according to a PP(λ). The first queue has no waiting room, and if the server at station one is busy the incoming customers are lost. The customers completing service at the first station enter the second station that has an infinite waiting room. After completing the service at station two, the customers leave the system. The service times at station i are iid exp(μ_i). Let $X_i(t)$ be the number of customers in station i ($i = 1, 2$) at time t. Show that $\{X_2(t), t \geq 0\}$ is a queue length process of a $G/M/1$ queue. Identify the parameters.

7.17 Consider a queueing system with two servers, each with its own queue. Customers arrive according to a PP(λ) to this system, and join the queue in front of server 1 with probability α (these are called type 1 customers) and that in front of server 2 with probability $1 - \alpha$ (these are called type 2 customers). The service times of customers served by server i are iid exp(μ_i), $i = 1, 2$. Each server serves its own queue as long as there are customers in its line, else he goes to help the other server at the other queue. When the two servers cooperate and serve a customer, their combined service rate is $\mu_1 + \mu_2$. Let $X_i(t)$ be the number of type i customers in the system at time t. Is $\{X_i(t), t \geq 0\}$ the queue length process of an $M/M/1$ queue? Why or why not? What about $\{X_1(t) + X_2(t), t \geq 0\}$? What are its parameters?

7.18 Consider a manufacturing facility with n machines in tandem. The ith machine can process a job in an exp(μ_i) amount of time and passes it on to the downstream machine after an instantaneous quality control check. The first machine has an infinite number of jobs waiting for processing, so it is always busy. The job processed by machine i fails the quality control check with probability α_i, and passes it with probability $1 - \alpha_i$. A job coming out of machine i that passes the quality control check is sold and leaves the system. The failed job is reworked by the next machine. If a job processed by machine n fails the inspection it is discarded. Model this as a Jackson network with $n - 1$ nodes and give its parameters.

7.10 Computational Exercises

7.1 Show that the variance of the number of customers in steady state in a stable $M/M/1$ system with arrival rate λ and service rate μ is given by

$$\sigma^2 = \frac{\rho}{(1 - \rho)^2},$$

where $\rho = \lambda/\mu$.

7.2 Let $X^q(t)$ be the number of customers in the queue (not including any in service) at time t in an $M/M/1$ queue with arrival rate λ and service rate μ. Is $\{X^q(t), t \geq 0\}$ a CTMC? Compute the limiting distribution of $X^q(t)$ assuming $\lambda < \mu$. Show that the expected number of customers in the queue (not including the customer in service) is given by

$$L^q = \frac{\rho^2}{1 - \rho}.$$

7.3 Let W_n^q be the time spent in the queue (not including time in service) by the nth arriving customer in an $M/M/1$ queue with arrival rate λ and service rate μ. Compute the limiting distribution of W_n^q assuming $\lambda < \mu$. Compute W^q, the limiting expected value of W_n^q as $n \to \infty$. Using the results of Computational Exercise 7.2 show that $L^q = \lambda W^q$. Thus Little's Law holds when applied to the customers in the queue.

7.4 Let $X(t)$ be the number of customers in the system at time t in an $M/M/1$ queue with arrival rate λ and service rate $\mu > \lambda$. Let

$$T = \inf\{t \geq 0 : X(t) = 0\}.$$

T is called the busy period. Compute $E(T|X(0) = i)$.

7.5 Let T be as in Computational Exercise 7.4. Let N be the total number of customers served during $(0, T]$. Compute $E(N|X(0) = i)$.

7.6 Customers arrive according to $PP(\lambda)$ to a queueing system with two servers. The ith server ($i = 1, 2$) needs $\exp(\mu_i)$ amount of time to serve one customer. Each incoming customer is routed to server 1 with probability p_1 or to server 2 with probability $p_2 = 1 - p_1$, independently. Queue jumping is not allowed. Find the optimum routing probabilities that will minimize the expected total number of customers in the system in steady state.

7.7 Consider a stable $M/M/1$ queue with the following cost structure. A customer who sees i customers ahead of him when he joins the system costs $\$c_i$ to the system. The system charges every customer a fee of $\$f$ upon entry. Show that the long run net revenue is given by

$$\lambda\left(f - \sum_{i=0}^{\infty} c_i \rho^i (1 - \rho)\right).$$

7.8 This is a generalization of Computational Exercise 7.6. A queueing system consists of K servers, each with its own queue. Customers arrive at the system according to a PP(λ). A system controller routes an incoming customer to server k with probability α_k, where $\alpha_1 + \alpha_2 + \cdots + \alpha_K = 1$. Customers assigned to server k receive iid $\exp(\mu_k)$ service times. Assume that $\mu_1 + \mu_2 + \cdots + \mu_K > \lambda$. It costs h_k dollars to hold a customer for one unit of time in queue k (including the time spent in service).

1. What are the feasible values of α_k's so that the resulting system is stable?
2. Compute the expected holding cost per unit time as a function of the routing probabilities α_k ($1 \leq k \leq K$) in the stable region.
3. Compute the optimal routing probabilities α_k that minimize the holding cost per unit time for the entire system.

7.9 Compute the long run fraction of customers who cannot enter the $M/M/1/K$ system described in Section 7.3.2.

7.10 Compute W, the expected time spent in the system by an arriving customer in steady state in an $M/M/1/K$ system, by using Little's Law and Equation 7.20. (If an arriving customer does not enter, his time in the system is zero.) What is the correct value of λ in $L = \lambda W$ as applied to this example?

7.11 Compute W, the expected waiting time of entering customers in steady state in an $M/M/1/K$ system, by using Little's Law and Equation 7.20. What is the correct value of λ in $L = \lambda W$ as applied to this example?

7.12 Suppose there are $0 < i < K$ customers in an $M/M/1/K$ queue at time 0. Compute the expected time when the queue either becomes empty or full.

7.13 Consider the $M/M/1/K$ system of Section 7.3.2 with the following cost structure. Each customer waiting in the system costs $\$c$ per unit time. Each customer entering the system pays $\$a$ as an entry fee to the system. Compute the long run rate of net revenue for this system.

7.14 Consider the system of Modeling Exercise 7.2 with production rate of 10 per hour and demand rate of 8 per hour. Suppose the machine is turned off when the number of items in the warehouse reaches K, and is turned on again when it falls to $K - 1$. Any demands that occur when the warehouse is empty are lost. It costs 5 dollars to produce an item, and 1 dollar to keep an item in the warehouse for one hour. Each item sells for ten dollars.

1. Model this system as an $M/M/1/K$ queue. State the parameters.
2. Compute the long run net income (revenue-production and holding cost) per unit time, as a function of K.
3. Compute numerically the optimal K that maximizes the net income per unit time.

7.15 Consider the $M/M/1$ queue with balking (but no reneging) as described in Section 7.3.6. Suppose the limiting distribution of the number of customers in this queue is $\{p_j, j \geq 0\}$. Using PASTA show that in steady state an arriving customer enters the system with probability $\sum_{j=0}^{\infty} \alpha_j p_j$.

7.16 Consider the $M/M/1$ queue with balking (but no reneging) as described in Section 7.3.6. Suppose the limiting distribution of the number of customers in this queue is $P(\rho)$, where $\rho = \lambda/\mu$. What balking probabilities will produce this limiting distribution?

7.17 Show that the expected number of busy servers in a stable $M/M/s$ queue is λ/μ.

7.18 Derive Equation 7.23. Hence or otherwise compute the expected waiting time of a customer in the $M/M/s$ system in steady state.

7.19 Show that for a stable $M/M/s$ queue of Section 7.3.3

$$L^q = \frac{p_s \rho}{(1-\rho)^2}.$$

Compute W^q explicitly and show that Little's Law $L^q = \lambda W^q$ is satisfied.

7.20 Compute the limiting distribution of the time spent in the queue by a customer in an $M/M/s$ queue. Hence or otherwise compute the limiting distribution of the time spent in the system by a customer in an $M/M/s$ queue.

7.21 Consider two queueing systems. System 1 has s servers, each serving at rate μ. System 2 has a single server, serving at rate $s\mu$. Both systems are subject to $PP(\lambda)$ arrivals. Let L_i be the expected number in system i in steady state, and L_i^q be the expected number of customers in the queue (not including those in service) in System i, $(i = 1, 2)$. Show that

$$L_1 \geq L_2$$

but

$$L_1^q \leq L_2^q.$$

This shows that it is better to have a single efficient server than many inefficient ones if the aim is to minimize the number of customers in the system.

7.22 Consider the finite population queue of Section 7.3.5 with two machines and one repairperson. Suppose every working machine produces revenue at a rate of $\$r$ per unit time. It costs $\$C$ to repair a machine. Compute the long run rate at which the system earns profits (revenue - cost).

7.23 When is the system in Modeling Exercise 7.2 stable? Assuming stability, compute the limiting distribution of the number of items in the warehouse. What fraction of the incoming demands are satisfied in steady state?

7.24 Compute the limiting distribution $\{p_i, 0 \leq i \leq s\}$ of the number of customers in an $M/M/s/s$ queue with arrival rate λ and service rate μ. The quantity p_s is called the blocking probability, and is denoted by $B(s, r)$ where $r = \lambda/\mu$. Show that

$$B(s, r) = \frac{\frac{r^s}{s!}}{\sum_{j=0}^{s} \frac{r^j}{j!}}. \tag{7.62}$$

The above formula is called the Erlang-B formula.

7.25 Let $B(s, r)$ be as in Computational Exercise 7.24. Show that the long run rate at which the customers enter the system is given by $\lambda(1 - B(s, r))$. Also, show that $B(s, r)$ satisfies the recursion

$$B(s, r) = \frac{\rho B(s-1, r)}{s + \rho B(s-1, r)},$$

with initial condition $B(0, r) = 1$.

7.26 When is the system in Modeling Exercise 7.3 stable? Assuming stability, compute the limiting distribution of the number of customers in the bank. What is the steady state probability that three tellers are active?

7.27 When is the system in Modeling Exercise 7.4 stable? Assuming stability, compute the limiting distribution of the number of customers in the system.

7.28 When is the system in Modeling Exercise 7.5 stable? Assuming stability, compute the expected number of customers in the system in steady state.

7.29 Consider the single server queue with N-type control described in Modeling Exercise 6.16. Let $X(t)$ be the number of customers in the system at time t, and $Y(t)$ be 1 if the server is busy and 0 if it is idle at time t. Show that $\{(X(t), Y(t)), t \geq 0\}$ is a CTMC and that it is stable if $\rho = \lambda/\mu < 1$. Assuming it is stable, show that

$$p_{i,j} = \lim_{t \to \infty} P(X(t) = i, Y(t) = j), \quad i \geq 0, \ j = 0, 1,$$

is given by

$$p_{i,0} = \frac{1 - \rho}{N}, \quad 0 \leq i < N,$$

$$p_{i,1} = \frac{\rho}{N}(1 - \rho^i), \quad 1 \leq i < N$$

$$p_{N+n,1} = \frac{\rho}{N}(1 - \rho^N)\rho^n, \quad n \geq 0.$$

7.30 Consider the queueing system of Computational Exercise 7.29. Suppose it costs $\$f$ to turn the server on from the off position, while turning the server off is free of cost. It costs $\$c$ to keep one customer in the system for one unit of time. Compute the long run operating cost per unit of the N-type policy. Show how one can optimally choose N to minimize this cost rate.

7.31 Consider the system of Modeling Exercise 6.31. What is the limiting distribution of the number of customers in the system as seen by an arriving customer of type i? By an entering customer of type i? ($i = 1, 2$)

7.32 Compute the limiting distribution of the CTMC in modeling Exercise 7.10 for the case of $s = 3$. What fraction of the customers are turned away in steady state?

7.33 Consider the Jackson network of single server queues as shown in Figure 7.9. Derive the stability condition. Assuming stability compute

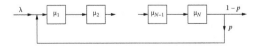

Figure 7.9 *Queueing network for Computational Exercise 7.33.*

1. the expected number of customers in steady state in the network,

2. the fraction of the time the network is completely empty in steady state.

7.34 Do Computational Exercise 7.33 for the network in Figure 7.10.

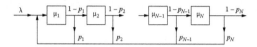

Figure 7.10 *Queueing network for Computational Exercise 7.34.*

7.35 Do Computational Exercise 7.33 for the network in Figure 7.11.

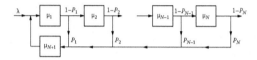

Figure 7.11 *Queueing network for Computational Exercise 7.35.*

7.36 North Carolina State Fair has 35 rides, and it expects to get about 60,000 visitors per day (12 hours) on the average. Each visitor is expected to take 5 rides on the average during his/her visit. Each ride lasts approximately 1 minute and serves an average of 30 riders per batch. Construct an approximate Jackson network model of the rides in the state fair that incorporates all the above data in a judicious fashion. State your assumptions. Is this network stable? Show how to compute the average queue length at a typical ride.

7.37 Consider a network of two nodes in series that operates as follows: customers arrive at the first node from outside according to a PP(λ), and after completing service at node 1 move to node 2, and exit the system after completing service at node 2. The service times at each node are iid $\exp(\mu)$. Node 1 has one server active as long as there are five or fewer customers present at that node, and two servers active otherwise. Node 2 has one server active for up to two customers, two servers for three through ten customers, and three servers for any higher number. If an arriving customer sees a total of i customers at the two nodes, he joins the first node with probability $1/(i + 1)$ and leaves the system without any service with probability $i/(i + 1)$. Compute

1. the condition of stability,

2. the expected number of customers in the network in steady state.

7.38 A 30-mile-long stretch of an interstate highway in Montana has no inlets or exits. This stretch is served by 3 cell towers, stationed at milepost numbers 5, 15, and

25. Each tower serves calls in the ten-mile section around it. Cars enter the highway at milepost zero according to a PP(λ), with $\lambda = 60/hr$. (Ignore the traffic in the reverse direction.) They travel at a constant speed of 100 miles per hour. Each entering car initiates a phone call at rate $\theta = .2$ per minute, i.e., the time until the initiation of a call is an $\exp(\theta)$ random variable. The call duration is exponentially distributed with mean 10 minutes. Once the call is finished the car does not generate any new calls. (Thus each car generates at most one call.) Suppose there is enough channel capacity available that no calls are blocked. When the calling car crosses from the area of one station to the next, the call is seamlessly handed over to the next station. Model this as a Jackson network with five nodes, each having infinite servers. Node 1 is for the first tower, nodes 2 and 3 are for the second tower, and nodes 4 and 5 are for the third tower. Nodes 1, 2, and 4 handle newly initiated calls, while nodes 3 and 5 handle handed-over calls. Tower 1 does not handle any handed-over calls. Note that for infinite server nodes the service time distribution can be general. Let $X_i(t)$ be the number of calls at time t in node $i, 1 \leq i \leq 5$. Compute

1. the service time distribution of the calls in node i,
2. the routing matrix,
3. the expected number of calls handled by the ith station in steady state,
4. the expected number of calls that are handed over from station i to station $i + 1$ per unit time ($i = 1, 2$).

7.39 Consider an open Jackson network with N single-server nodes. Customers arrive from outside the network to the ith node with rate λ_i. A fraction p_i of the customers completing service at node i joins the queue at node $i + 1$ and the rest leave the network permanently, $i = 1, 2, ..., N - 1$. Customers completing service at node N join the queue at node 1 with probability p_N, and the rest leave the network permanently. The service times at node i are $\exp(\mu_i)$ random variables.

1. State the assumptions to model this as a Jackson network.
2. What are the traffic equations for the Jackson network? Solve them.
3. What is the condition of stability?
4. What is the expected number of customers in the network in steady state, assuming the network is stable?

7.40 Show that the probability that a customer in an open Jackson network of Section 7.4 stays in the network forever is zero if $I - R$ is invertible. Also compute the expected number of visits to station j made by a customer who enters station i from outside during his stay in the network.

7.41 For a closed Jackson network of single server queues, show that

1. $\lim_{t\to\infty} P(X_i(t) \geq j) = \rho_i^j \frac{G_N(K)}{G_N(K-j)}, \quad 0 \leq j \leq K.$
2. $L_i = \lim_{t\to\infty} E(X_i(t)) = \sum_{j=1}^{K} \rho_i^j \frac{G_N(K)}{G_N(K-j)}, \quad 1 \leq i \leq N.$

7.42 Generalize the method of computing $G_N(K)$ derived in Example 7.11 to general closed Jackson networks of single-server queues with N nodes and K customers.

7.43 A simple communications network consists of two nodes labeled A and B connected by two one-way communication links: line AB from A to B, and line BA from line from B to A. There are N users at each node. The ith user ($1 \leq i \leq N$) at node A (B) is denoted by A_i (B_i). User A_i has an interactive session set up with user B_i and it operates as follows: User A_i sends a message to user B_i. All the messages generated at node A wait in a buffer at node A for transmission to the appropriate user at node B on line AB in an FCFS fashion. When user B_i receives the message from user A_i, she spends a random amount of time, called think time, to generate a response to it. All the messages generated at node B wait in a buffer at node B for transmission to the appropriate user at node A on line BA in an FCFS fashion. When user A_i receives the message from user B_i, she spends a random amount of time to generate a response to it. This process of messages going back and forth between the pairs of users A_i and B_i continues forever. Suppose all the think times are iid $\exp(\theta)$ random variables, and the message transmission times are iid $\exp(\mu)$ random variables. Model this as a closed Jackson network. What is the expected number of messages in the buffers at nodes A and B in steady state?

7.44 For the closed Jackson network of Section 7.5, define the throughput $TH(i)$ of node i as the rate at which customers leave node i in steady state, i.e.,

$$TH(i) = \sum_{n=0}^{K} \mu_i(n) \lim_{t \to \infty} P(X_i(t) = n).$$

Show that

$$TH(i) = \pi_i \frac{G_N(K)}{G_N(K-1)},$$

where π_i is from Equation 7.33.

7.45 When is the system in Modeling Exercise 7.7 stable? Assuming stability, compute the expected number of customers in the system in steady state.

7.46 When is the system in Modeling Exercise 7.6 stable? Assuming stability, compute the generating function of the limiting distribution of the number of customers in the system.

7.47 Compute the expected number of customers in steady state in an $M/G/1$ system where the arrival rate is one customer per hour and the service time distribution is $PH(\alpha, M)$ where

$$\alpha = [0.5 \ 0.5 \ 0]$$

and

$$M = \begin{bmatrix} -3 & 1 & 1 \\ 0 & -3 & 2 \\ 0 & 0 & -3 \end{bmatrix}.$$

7.48 Compute the expected queue length in an $M/G/1$ queue with the following service time distributions (all with mean $1/\mu$):

1. Exponential with parameter μ,
2. Uniform over $[0, 2/\mu]$,
3. Deterministic with mean $1/\mu$,
4. Erlang with parameters $(k, k\mu)$.

Which distribution produces the largest congestion? Which produces the smallest?

7.49 Consider the $\{X(t), t \geq 0\}$ and the $\{X_n, n \geq 0\}$ processes defined in Modeling Exercise 7.8. Show that the limiting distribution of the two (if they exist) are identical. Let p_n (q_n) be the limiting probability that there are n customers in the system and the server is up (down). Let $p(z)$ and $q(z)$ be the generating functions of $\{p_n, n \geq 0\}$ and $\{q_n, n \geq 0\}$. Show that this system is stable if

$$\frac{\lambda}{\mu} < \frac{\alpha}{\alpha + \theta}.$$

Assuming that the system is stable show that

$$q(z) = \frac{\left(\frac{\mu}{z}\right)\left(\frac{\alpha}{\alpha+\theta} - \frac{\lambda}{\mu}\right)}{\left(\frac{f\mu}{z} - \lambda\right)\left(\frac{\alpha}{\theta} + \frac{\lambda}{\theta}(1-z)\right) - \lambda},$$

and

$$p(z) = \left(\frac{\alpha}{\theta} + \frac{\lambda}{\theta}(1-z)\right) q(z).$$

7.50 Show that the DTMC $\{X_n, n \geq 0\}$ in the Modeling Exercise 7.11 is positive recurrent if $\rho = \lambda\tau < 1$, where λ is the arrival rate and τ is the mean service time. Assuming the DTMC is stable, show that the generating function of the limiting distribution of X_n is given by

$$\phi(z) = \frac{1 - \rho}{m} \cdot \frac{\tilde{G}(\lambda - \lambda z)}{z - \tilde{G}(\lambda - \lambda z)} \cdot (\psi(z) - 1),$$

where $\tilde{G}$ is the LST of the service time, m is the expected number of arrivals during a single vacation, and $\psi(z)$ is the generating function of the number of arrivals during a single vacation.

7.51 Let $X(t)$ be the number of customers at time t in the system described in Modeling Exercise 7.11. Show that $\{X_n, n \geq 0\}$ and $\{X(t), t \geq 0\}$ have the same limiting distribution, assuming it exists. Using the results of Computational Exercise 7.50 show that the expected number of customers in steady state is given by

$$L = \rho + \frac{\lambda^2 s^2}{2(1 - \rho)} + \frac{m^{(2)}}{2m},$$

where s^2 is the second moment of the service time, and $m^{(2)}$ is the second factorial moment of the number of arrivals during a single vacation.

7.52 Let $X(t)$ be the number of customers at time t in an $M/G/1$ queue under N-type control as explained in Modeling Exercise 6.16 for an $M/M/1$ queue. Using the results of Computational Exercises 7.50 and 7.51 establish the condition of stability for this system and compute the generating function of the limiting distribution of $X(t)$ as $t \to \infty$.

7.53 When is the queueing system described in Modeling Exercise 7.12 stable? Assuming stability, compute the expected number of customers in the system in steady state under the two policies. Which policy is better at minimizing the expected number in the system in steady state?

7.54 Analyze the stability of the $\{X(t), t \geq 0\}$ process in Modeling Exercise 7.9. Assuming stability, compute the limiting distribution of the number of items in the warehouse. What fraction of the demands are lost in steady state?

7.55 Show that the DTMC $\{X_n, n \geq 0\}$ in Modeling Exercise 7.13 is positive recurrent if $\rho = \lambda \tau_G < 1$. Assuming the DTMC is stable, show that the generating function of the limiting distribution of X_n is given by

$$\phi(z) = \frac{1 - \lambda \tau_G}{1 - \lambda \tau_G + \lambda \tau_H} \cdot \frac{z\tilde{H}(\lambda - \lambda z) - \tilde{G}(\lambda - \lambda z)}{z - \tilde{G}(\lambda - \lambda z)}.$$

Hint: Use the results of Computational Exercise 4.24.

7.56 Let $X(t)$ be the number of customers at time t in the system described in Modeling Exercise 7.13. Show that $\{X_n, n \geq 0\}$ and $\{X(t), t \geq 0\}$ have the same limiting distribution, assuming it exists. Using the results of Computational Exercise 7.55 show that the expected number of customers in steady state is given by

$$L = \frac{\lambda \tau_H}{1 - \lambda \tau_G + \lambda \tau_H} + \frac{\lambda^2}{2} \cdot \frac{\sigma_H^2 + \tau_H^2 - \sigma_G^2 - \tau_G^2}{1 - \lambda \tau_G + \lambda \tau_H} + \frac{\lambda^2}{2} \cdot \frac{\sigma_G^2 + \tau_G^2}{1 - \lambda \tau_G}.$$

7.57 Show that the DTMC $\{X_n, n \geq 0\}$ in Modeling Exercise 7.14 is positive recurrent if $\lambda < 1$. Assuming the DTMC is stable, compute $\phi(z)$, the generating function of the limiting distribution of X_n as $n \to \infty$.

7.58 Show that the DTMC $\{\bar{X}_n, n \geq 0\}$ in Modeling Exercise 7.14 is positive recurrent if $\lambda < 1$. Assuming the DTMC is stable, compute $\bar{\phi}(z)$, the generating function of the limiting distribution of $\bar{X}_n$ as $n \to \infty$.

7.59 In Modeling Exercise 7.14, is the limiting distribution of $\{X(t), t \geq 0\}$ same as that of $\{X_n, n \geq 0\}$ or $\{\bar{X}_n, n \geq 0\}$? Explain.

7.60 Consider an $M/G/1$ queue where the customers arrive according to a PP(λ) and request iid service times with common mean τ, and variance σ^2. After service completion, a customer leaves with probability p, or returns to the end of the queue with probability $1 - p$, and behaves like a new customer.

1. Compute the mean and variance of the amount of time a customer spends in service during the sojourn time in the system.

2. Compute the condition of stability.

3. Assuming stability, compute the expected number of customers in the system as seen by a departure (from the system) in steady state.

4. Assuming stability, compute the expected number of customers in the system at a service completion (customer may or may not depart at each service completion) in steady state.

7.61 Derive the condition of stability in a $G/M/1$ queue with the interarrival times

$$G(x) = r(1 - e^{-\lambda_1 x}) + (1 - r)(1 - e^{-\lambda_2 x}),$$

where $0 < r < 1$, $\lambda_1 > 0$, $\lambda_2 > 0$, and iid exp(μ) service times. Assume stability and compute the limiting distribution of the number of customers in the system.

7.62 Let $X(t)$ be the number of customers in a $G/M/2$ queue at time t. Let X_n^* be the number of customers as seen by the nth arrival. Show that $\{X_n^*, n \geq 0\}$ is a DTMC, and compute its one-step transition probability matrix. Derive the condition of stability and the limiting distribution of X_n^* as $n \to \infty$.

7.63 Consider the overflow queue of Modeling Exercise 7.15.

1. Compute the condition of stability for the overflow queue.

2. Assuming the overflow queue is stable, compute the pmf of the number of customers in the overflow queue in steady state.

7.64 Consider the following modification to the $M/G/1/1$ retrial queue of Section 7.7. A new customer joins the service immediately if he finds the server free upon his arrival. If the server is busy, the arriving customer leaves immediately with probability c, or joins the orbit with probability $1 - c$, and conducts retrials until he is served. Let X_n and $X(t)$ be as in Section 7.7. Derive the condition of stability and compute the generating function of the limiting distribution of X_n and $X(t)$. Are they the same?

7.65 Consider the retrial queue of Section 7.7 with exp(μ) service times. Show that the results of Section 7.7 are consistent with those of Example 6.38.

7.66 A warehouse stocks Q items. Orders for these items arrive according to a PP(μ). The warehouse follows a $(Q, Q - 1)$ replenishment policy with back orders as follows: If the warehouse is not empty, the incoming demand is satisfied from the existing stock and an order is placed with the supplier for replenishment. If the warehouse is empty, the incoming demand is backlogged and an order is placed with the supplier for replenishment. The lead time, i.e., the amount of time it takes for the order to reach the warehouse from the supplier, is a random variable with distribution $G(\cdot)$. The lead times are iid, and orders may cross, i.e., the orders placed at the supplier may be received out of order. Let $X(t)$ be the number of outstanding orders at time t.

1. Model $\{X(t), t \geq 0\}$ as an $M/G/\infty$ queue.
2. Compute the long run fraction of the time the warehouse is empty.

7.67 QED design of an $M/M/s$ queue. Consider a call center that handles 10,000 calls an hour. Each call lasts approximately 6 minutes. There is infinite waiting room for the calls that find all servers busy. How many servers should we provide if we want to ensure that more than 98% of the customer get to talk to a server with no wait? What is the average queueing time under this system?

7.68 Halfin–Whitt analysis of an $M/M/s/s$ Queue. Let $B(s,r)$ be the Erlang-B formula as given in Computational Exercise 7.24. Let HW be the Halfin–Whitt asymptotic regime as given in Equation 7.25. Show that

$$\lim_{HW} \sqrt{r} B(s,r) = h(-\beta),$$

where h is as given in Equation 7.24. For a given $0 < \alpha < 1$, and $r > 0$, show that there is a unique β so that $\sqrt{r}\alpha = h(-\beta)$. Show that the solution is negative if $r < 2/(\pi\alpha^2)$, otherwise it is positive. Show that the square-root staffing formula $s = r + \beta\sqrt{r}$ produces an $M/M/s/s$ system with blocking probability α for large r.

7.69 Consider the call center of Computational Exercise 7.67, but with no waiting room for blocked calls. Thus a call that finds all servers busy is lost forever. Compute the appropriate staffing level. Resize the system if the arrival rate increases tenfold.

7.70 Derive Equation 7.61.

7.71 Consider an $M(t)/G/\infty$ queue with the following rate function:

$$\lambda(t) = \lambda(1 + \alpha \sin(\beta t)), \quad -\infty < t < \infty.$$

Show that the expected number of customers in this system at time t is

$$\lambda\tau[1 + \alpha(\sin(\beta t)E(\cos(\beta S_e)) - \cos(\beta t)E(\sin(\beta S_e)))].$$

7.72 Consider Computational Exercise 7.71 with $\lambda = \alpha = \beta = 1$. Plot the $\lambda(t)$ and $m(t)$ over $0 \leq t \leq 5$. What does it tell you about when the queue length achieves its maximum relative to where the arrival rate function achieves its maximum?

7.73 Patients arrive at a clinic one by one. The nth patient at a clinic is scheduled to arrive time nd, $n = 0, 1, 2, \cdots$, where $d > 0$ is a fixed number. The nth patient fails to show up for his/her appointment with probability θ, independently of all other patients. There is a single doctor who sees the patients in a first-come first-served fashion. The service times are iid $\exp(\mu)$. Let $X(t)$ be the number of customers in the clinic at time t. Model $\{X(t), t \geq 0\}$ as the queue length process in a $G/M/1$ queue. State the stability condition and compute its limiting distribution assuming stability.

7.74 Customers arrive at an infinite capacity single-server service facility according to $PP(\lambda)$ and request iid service times with mean τ and second moment s^2. Assume $\rho = \lambda\tau < 1$. Let V_i be the value of the service (in dollars) to the i-th customer. Suppose $\{V_i, i \geq 1\}$ are iid $U(0, 1)$ random variables. The manager of the service facility collects a fee p from each customer who opts to join the system. Clearly, an arriving customer will join the system if and only if the service is worth more to him than the service charge p. Otherwise, the customer leaves without service and is permanently lost. All customers who enter the service facility get served in a first-come first-served order. The manager also incurs a cost of h per unit time the customer spends in the system.

1. Let $X(t)$ be the number of customers in the system at time t. Is $\{X(t), t \geq 0\}$ the queue length process of an $M/G/1$ queue? If yes, give its parameters.

2. Compute $L(p)$, the expected number of customers in the system in steady state, as a function of p.

3. Let $G(p)$ be the net profit per unit time earned by the system manager. Show that $G(p)$ is a concave function of p for $p \in [0, 1]$. Numerically find the optimal service charge p that will maximize $G(p)$, assuming $\lambda = 10$, $\rho = .9$, $s^2 = 1$, $h = 1$.

7.75 Consider the system in Modeling Exercise 7.18. A job coming out of machine i that passes the quality control check is sold for d_i. The cost of holding a job in front of machine i is h_i per unit time. Suppose $d_1 < d_2 < \cdots d_n$ and $h_1 = 0 < h_2 < h_3 < \cdots < h_n$. Let L_i be the expected number of jobs in front of machine i in steady state ($2 \leq i \leq n$).

1. Derive the condition of stability and compute L_i, $2 \leq i \leq n$, assuming stability.

2. Compute the long run net profit earned by the system in steady state.

7.76 An arrival process at a service station is either on or off. When it is on, arrivals occur at rate λ, and when it is off, there are no arrivals. The successive on and off times are iid $\exp(\alpha)$ and $\exp(\beta)$, respectively. These arrivals enter a single server queue and request iid $\exp(\mu)$ service times. Let $Z(t)$ be the number of customers in this queueing system at time t.

1. Model this as a $G/M/1$ queue by showing that the interarrival times are iid.

2. What is the common mean and the LST of the interarrival times?

3. What is the condition of stability? What is the distribution of the number of customers in steady state, assuming the queue is stable?

7.77 Consider a single server queue with two classes of customers. Class i customers arrive according to a $PP(\lambda_i)$ and request iid $\exp(\mu)$ service times. (μ is independent of i.) The two classes are independent of each other. Class one customers have pre-emptive priority over class two customers. Thus when a class one customer arrives and the server is serving a class one customer, the arriving customer joins the tail of the queue of all the class one customers. If, on the other hand, the server is serving class two customers (this can happen only if there are no other class one customers in the system), that service is interrupted, and the newly arriving class one customer starts getting served. When there are no more class one customers left in the systems, the server resumes the interrupted service of the class two customer. Thus the server always serves a class one customer if one is present in the system. Within each class customers are served according to FIFO policy. Let $X_i(t)$ be the number of class i customers in the system at time t, and $X(t) = X_1(t) + X_2(t)$ be the total number of customers in the system at time t.

1. Show that $\{X_1(t), t \geq 0\}$ is the queue length process of an $M/M/1$ queueing system. What are its parameters?
2. Show that $\{X(t), t \geq 0\}$ is the queue length process of an $M/M/1$ queueing system. What are its parameters?
3. What is the condition of stability for the two priority system?
4. Let L_i be the expected number of class i customers in the system in steady state, assuming the system is stable. Compute L_1 and L_2.

7.78 Consider the system in Modeling Exercise 7.16. Derive the condition of stability for the system. Assuming stability, compute the expected number of customers in the system in steady state.

7.79 Let $X(t)$ be the number of customers in a queueing system at time t. Suppose $\{X(t), t \geq 0\}$ is a deterministic periodic function of t with period 10. Figure 7.12 shows the sample path over $0 \leq t \leq 10$. Thus there are entries at times 1, 3, 6, 7 and departures at times 5, 8, 9, and 10. In addition there are arrivals at times 2 and 4 that do not enter (hence no jump in $X(t)$ at these two points).

1. Compute the long run fractions $\pi_j^*, \hat{\pi}_j, \alpha$ and α_j for $j = 0, 1, 2, 3$. (See Section 7.1 and Section 7.2.1 for the definitions.) Verify Theorem 7.1.
2. Compute the long run fractions $\pi_j, j = 0, 1, 2, 3$. (See Section 7.1 for the definitions.) Verify Theorem 7.2.
3. Compute the long run fractions $p_j, j = 0, 1, 2, 3$. (See Section 7.1 for the definitions.) Is $p_j = \hat{\pi}_j, \ j \geq 0$? Why or why not?

7.80 Consider the sample path of Computational Exercise 7.79.

1. Compute L, the expected number of customers in the system in the long run.

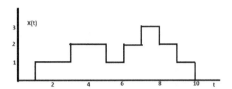

Figure 7.12 *A sample path of the queue length process.*

2. Compute λ, the long run rate of arrivals, and W, the average time spent in the system by all arrivals, assuming FCFS service discipline. Verify that Theorem 7.5 holds.

3. Compute λ_e, the long run rate of entries, and W_e, the average time spent in the system by all entries, assuming FCFS service discipline. Verify that Theorem 7.5 holds.

7.81 Redo Computational Exercise 7.80 assuming LCFS service discipline.

7.82 Consider the tandem queueing network of Example 7.9. Suppose the service rates are under our control, but must satisfy $\mu_1 + \mu_2 + \cdots + \mu_N = \mu$, where $\mu > N\lambda$ is a fixed constant. Find the optimal service rates μ_i^* ($1 \leq i \leq N$) that minimize the long run total holding cost per unit time of all the customers in the network.

CHAPTER 8

Renewal Processes

"Research is seeing what everyone else has seen and thinking what no one else has thought."
—Anonymous

8.1 Introduction

This chapter is devoted to the study of a class of stochastic processes called renewal processes (RPs), and their applications. The RPs play several important roles in the grand scheme of stochastic processes. First, they help remove the stringent distributional assumptions that were needed to build Markov models, namely, the geometric distributions for the DTMCs, and the exponential distributions for the CTMCs. RPs provide us with important tools such as the key renewal theorem (Section 8.5) to deal with general distributions.

Second, RPs provide a unifying theoretical framework for studying the limiting behavior of specialized stochastic processes such as the DTMCs and the CTMCs. Recall that we have seen the discrete renewal theorem in Section 4.5.2 on page 110 and continuous renewal theorem in Section 6.11.2 on page 251. The discrete renewal theorem was used in the study of the limiting behavior of the DTMCs, and the continuous renewal theorem was used in the study of the limiting behavior of the CTMCs. We shall see that RPs appear as embedded processes in the DTMCs and the CTMCs, and the key-renewal theorem provides the unifying tool to obtain convergence results.

Third, RPs lead to important generalizations such as semi-Markov processes (Section 8.9), renewal-reward processes (Section 8.10), and regenerative processes (Section 8.11), to name a few. Renewal-reward processes are very useful in the computation of important performance measures such as long run cost and reward rates. Regenerative processes are the most general class of processes that encompass all the classes of stochastic processes studied in this book. Of course this generality makes them difficult to use in computations. This difficulty is somewhat removed by the Markov-regenerative processes studied in the next chapter. With this overview we begin the study of RPs.

Consider a process of events occurring over time. Let $S_0 = 0$, and S_n be the time of occurrence of the nth event, $n \geq 1$. Assume that

$$0 \leq S_1 \leq S_2 \leq S_3 \leq \cdots,$$

and define

$$X_n = S_n - S_{n-1}, \quad n \geq 1.$$

Thus $\{X_n, n \geq 1\}$ is a sequence of inter-event times. Clearly, $X_n \geq 0$. Since we allow $X_n = 0$, multiple events can occur simultaneously. Next define

$$N(t) = \sup\{n \geq 0 : S_n \leq t\}, \quad t \geq 0. \tag{8.1}$$

Thus $N(t)$ is simply the number of events up to time t. The process $\{N(t), t \geq 0\}$ is called the counting process generated by $\{X_n, n \geq 1\}$. A typical sample path of $\{N(t), t \geq 0\}$ is shown in Figure 8.1. With this notation we are ready to define a renewal process.

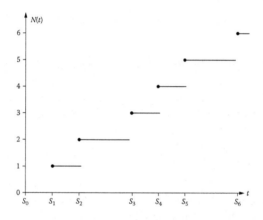

Figure 8.1 *Sample path of a counting process.*

Definition 8.1   Renewal Sequence and Renewal Process. *The sequence $\{S_n, n \geq 1\}$ is called a renewal sequence and the process $\{N(t), t \geq 0\}$ is called a renewal process generated by $\{X_n, n \geq 1\}$ if $\{X_n, n \geq 1\}$ is a sequence of non-negative iid random variables.*

The study of the stochastic process $\{N(t), t \geq 0\}$ is renewal theory. Below we give several examples where RPs are encountered.

Example 8.1   Poisson Process. From the definition of a Poisson process, we see that the RP generated by a sequence of iid $\exp(\lambda)$ random variables is a Poisson process with parameter λ. Thus an RP is a direct generalization of a Poisson process when the inter-event times are iid non-negative random variables that have a general distribution. ■

Example 8.2   Arrival Process in a $G/M/1$ Queue. Let $N(t)$ be the number of arrivals in a $G/M/1$ queue up to time t. By definition (see Section 7.6.2) the inter-arrival times $\{X_n, n \geq 1\}$ in a $G/M/1$ queue are iid random variables. Hence $\{N(t), t \geq 0\}$, the counting process generated by $\{X_n, n \geq 1\}$, is an RP. ∎

Example 8.3   RPs in CTMCs. Let $\{X(t), t \geq 0\}$ be a CTMC on $\{0, 1, 2, \cdots\}$ with $P(X(0) = 0) = 1$. Let S_n be the time of the nth entry into state 0, $n \geq 1$, and let $N(t)$ be the number of entries into state 0 up to t. (Note that we do not count the entry at time 0, since the process was already in state 0 at time 0.) Since $X_n = S_n - S_{n-1}$ is the time between two successive entries into state 0, the Markov property of the CTMC implies that $\{X_n, n \geq 1\}$ is a sequence of iid random variables. Hence $\{N(t), t \geq 0\}$, the counting process generated by $\{X_n, n \geq 1\}$, is an RP. ∎

Example 8.4   Machine Maintenance. Suppose a manufacturing process requires continuous use of a machine. We begin with a new machine at time 0. Suppose we follow the age-replacement policy: replace the machine in use when it fails (called an unplanned replacement) or when it reaches an age of T years (called a planned replacement), where $T > 0$ is a given constant. This policy is useful in reducing the number of unplanned replacements, which tend to be more costly. Let L_i be the lifetime of the ith machine and assume that $\{L_i, i \geq 1\}$ is a sequence of iid non-negative random variables. This process produces several renewal sequences (the reader should convince himself/herself that these are indeed renewal sequences by showing that the inter-event times $\{X_n = S_n - S_{n-1}, n \geq 1\}$ are iid):

1. $S_n = $ time of the nth failure,
2. $S_n = $ time of the nth planned replacement,
3. $S_n = $ time of the nth replacement (planned or otherwise). ∎

Example 8.5   RPs in an $M/G/1$ Queue. Let $X(t)$ be the number of customers at time t in an $M/G/1$ system. Suppose the system is initially empty, i.e., $X(0) = 0$. Let S_n be the nth time a departing customer leaves behind an empty system, i.e., it is the completion time of the nth busy cycle. Since the Poisson process has stationary and independent increments, and the service times are iid, we see that the successive busy cycles are iid, i.e., $\{X_n = S_n - S_{n-1}, n \geq 1\}$ is a sequence of iid random variables. It generates a counting process $\{N(t), t \geq 0\}$ that counts the number of busy cycles completed by time t. It is thus an RP.

If we change the initial state of the system so that an arrival has occurred to an empty system at time 0-, we can come up with a different RP by defining S_n as the nth time an arriving customer enters an empty system. In this case we get an RP where $N(t)$ is the number of busy cycles that start by time t. (Note that we do not count the initial busy cycle.) In general, we can come up with many RPs embedded in a given stochastic process. What about a $G/G/1$ queue? ∎

Example 8.6   RPs in a DTMC. Following Example 8.3 we can identify RPs in a DTMC $\{Z_n, n \geq 0\}$ on $\{0, 1, 2, \cdots\}$ with $P(Z_0 = 0) = 1$. Let S_n be the time of the

nth visit to state 0, $n \geq 1$, and let $N(t)$ be the number of entries into state 0 up to t. (Note that we do not count the visit at time 0, since the process was already in state 0 at time 0.) Then, by the same argument as in Example 8.3, $\{X_n = S_n - S_{n-1}, n \geq 1\}$ is a sequence of positive iid random variables and $\{N(t), t \geq 0\}$ is an RP. Note that in this case renewals can occur only at integer times $n = 1, 2, \cdots$. ∎

The above examples show how pervasive RPs are. Next we characterize an RP.

Theorem 8.1 Characterization of an RP. *An RP generated by a sequence of iid random variables $\{X_n, n \geq 1\}$ with common cdf $G(\cdot)$ is completely characterized by $G(\cdot)$.*

Proof: The renewal sequence $\{S_n, n \geq 1\}$ is clearly completely described by G. Now, for any integer $n \geq 1$, and $0 < t_1 < \cdots < t_n$ and integers $0 \leq k_1 \leq k_2 \leq \cdots \leq k_n$, we have

$$P(N(t_1) = k_1, N(t_2) = k_2, \cdots, N(t_n) = k_n)$$
$$= P(S_{k_1} \leq t_1, S_{k_1+1} > t_1, S_{k_2} \leq t_2, S_{k_2+1} > t_2, \cdots, S_{k_n} \leq t_n, S_{k_n+1} > t_n).$$

Thus all finite dimensional joint distributions of the RP are determined by those of the renewal sequence, which are determined by G. This proves the theorem. ∎

The next theorem shows that the RP "renews" at time $X_1 = S_1$, i.e., it essentially starts anew at time S_1.

Theorem 8.2 *The RP $\{N(t), t \geq 0\}$ generated by $\{X_n, n \geq 1\}$ is stochastically identical to $\{N(t + X_1) - 1, t \geq 0\}$.*

Proof: To show that they are stochastically identical, we have to show that their finite dimensional distributions are identical. Now, for any integer $n \geq 1$, and $0 < t_1 < \cdots < t_n$ and integers $0 \leq k_1 \leq k_2 \leq \cdots \leq k_n$, we have

$$P(N(t_1 + X_1) - 1 = k_1, N(t_2 + X_1) - 1 = k_2, \cdots, N(t_n + X_1) - 1 = k_n)$$
$$= P(N(t_1 + X_1) = k_1 + 1, N(t_2 + X_1) = k_2 + 1, \cdots, N(t_n + X_1) = k_n + 1)$$
$$= P(S_{k_1+1} \leq t_1 + X_1, S_{k_1+2} > t_1 + X_1, S_{k_2+1} \leq t_2 + X_1, S_{k_2+2} > t_2 + X_1,$$
$$\cdots, S_{k_n+1} \leq t_n + X_1, S_{k_n+2} > t_n + X_1)$$
$$= P(S_{k_1+1} - X_1 \leq t_1, S_{k_1+2} - X_1 > t_1, S_{k_2+1} - X_1 \leq t_2, S_{k_2+2} - X_1 > t_2,$$
$$\cdots, S_{k_n+1} - X_1 \leq t_n, S_{k_n+2} - X_1 > t_n)$$
$$= P(S_{k_1} \leq t_1, S_{k_1+1} > t_1, S_{k_2} \leq t_2, S_{k_2+1} > t_2, \cdots, S_{k_n} \leq t_n, S_{k_n+1} > t_n)$$
$$= P(N(t_1) = k_1, N(t_2) = k_2, \cdots, N(t_n) = k_n).$$

Here the second to the last equality follows because $(S_2 - X_1, S_3 - X_1, \cdots, S_{n+1} - X_1)$ has the same distribution as $(S_1, S_2, \cdots, S_n)$. The last equality follows from the proof of Theorem 8.1. ∎

The above theorem provides a justification for calling S_1 the first renewal epoch: the process essentially restarts at that point in time. For the same reason, we call S_n as the nth renewal epoch.

8.2 Properties of $N(t)$

Now that we know how to characterize an RP, we will study it further: its transient behavior, limiting behavior, etc. We shall use the following notation throughout this chapter:

$$G(t) = \mathsf{P}(X_n \leq t), \quad \tau = \mathsf{E}(X_n), \quad s^2 = \mathsf{E}(X_n^2), \quad \sigma^2 = \mathrm{Var}(X_n).$$

We shall assume that

$$G(0-) = 0, \quad G(0+) = G(0) < 1.$$

Thus X_n's are non-negative, but not identically zero with probability 1. This assumption implies that

$$\tau > 0,$$

and is necessary to avoid trivialities.

Theorem 8.3

$$\mathsf{P}(N(t) < \infty) = 1, \quad t \geq 0.$$

Proof: From strong law of large numbers we have

$$\mathsf{P}(\frac{S_n}{n} \to \tau) = 1.$$

Since $\tau > 0$, this implies that

$$\mathsf{P}(S_n \to \infty) = 1.$$

Thus, for $0 \leq t < \infty$,

$$\mathsf{P}(N(t) = \infty) = \mathsf{P}(\lim_{n \to \infty} S_n \leq t)$$
$$= 1 - \mathsf{P}(S_n \to \infty) = 0.$$

Hence the theorem follows. ∎

The next theorem gives the exact distribution of $N(t)$ for a given t.

Theorem 8.4   Marginal Distribution of $N(t)$. *Let*

$$G_k(t) = \mathsf{P}(S_k \leq t), \quad t \geq 0. \tag{8.2}$$

Then

$$p_k(t) = \mathsf{P}(N(t) = k) = G_k(t) - G_{k+1}(t), \quad t \geq 0. \tag{8.3}$$

Proof: Follows along the same lines as the proof of Theorem 5.8 on page 166 by using Equation 8.2. ∎

Example 8.7   Poisson Process. Let $G(t) = 1 - e^{-\lambda t}, t \geq 0$. Then $G_k(t)$ is the cdf of an Erl(k, λ) random variable. Hence

$$G_k(t) = 1 - \sum_{r=0}^{k-1} e^{-\lambda t} \frac{(\lambda t)^r}{r!}, \quad t \geq 0.$$

Hence, from Theorem 8.4, we get

$$p_k(t) = G_k(t) - G_{k+1}(t) = e^{-\lambda t}\frac{(\lambda t)^k}{k!}, \quad t \ge 0.$$

Thus $N(t) \sim P(\lambda t)$. This is to be expected, since $G(t) = 1 - e^{-\lambda t}$ implies the RP $\{N(t), t \ge 0\}$ is a PP(λ). ∎

The function $G_k(\cdot)$ of Equation 8.2 is called the k-fold (Stieltjes-)convolution of G with itself. It can be computed recursively as

$$G_k(t) = \int_0^t G_{k-1}(t-u)dG(u) = \int_0^t G(t-u)dG_{k-1}(u), \quad t \ge 0$$

with initial condition $G_0(t) = 1$ for $t \ge 0$. Let $\tilde{G}(s)$ be the Laplace–Stieltjes transform (LST) of G, defined as

$$\tilde{G}(s) = \int_0^\infty e^{-st}dG(t).$$

Since the LST of a (Stieltjes-)convolution is the product of the LSTs (see Appendix E), it follows that

$$\tilde{G}_k(s) = \tilde{G}(s)^k.$$

Hence

$$\tilde{p}_k(s) = \int_0^\infty e^{-st}dp_k(t) = \tilde{G}_k(s) - \tilde{G}_{k+1}(s) = \tilde{G}(s)^k(1 - \tilde{G}(s)). \qquad (8.4)$$

One of the most important tools of renewal theory is the renewal argument. It is a method of deriving an integral equation for a probabilistic quantity by conditioning on S_1, the time of the first renewal. We explain it by deriving an equation for $p_k(t) = P(N(t) = k)$. Now fix a $k > 0$ and $t \ge 0$. Suppose $S_1 = u$. If $u > t$, we must have $N(t) = 0$. If $u \le t$, we already have one renewal at time u, and the process renews at time u. Hence we can get k renewals by time t if and only if we get $k - 1$ additional renewals in this new renewal process starting at time u up to time t, whose probability is $P(N(t - u) = k - 1)$. Combining these observations we get

$$P(N(t) = k|S_1 = u) = \begin{cases} 0 & \text{if } u > t, \\ P(N(t-u) = k-1) & \text{if } u \le t. \end{cases}$$

Hence,

$$\begin{aligned} p_k(t) &= \int_0^\infty P(N(t) = k|S_1 = u)dG(u) \\ &= \int_0^t P(N(t-u) = k-1)dG(u) \\ &= \int_0^t p_{k-1}(t-u)dG(u). \end{aligned} \qquad (8.5)$$

We also have

$$p_0(t) = \mathsf{P}(N(t) = 0) = \mathsf{P}(S_1 > t) = 1 - G(t). \qquad (8.6)$$

Unfortunately, computing $p_k(t)$ by using Equation 8.5 is no easier than using Equation 8.3. However, when the inter-event times are integer valued random variables, the renewal argument provides a simple recursive method of computing $p_k(t)$. To see this, let

$$\alpha_i = \mathsf{P}(X_n = i), \quad i = 0, 1, 2, \cdots.$$

Since all renewals take place at integer time points, we study $p_k(n)$ for $n = 0, 1, 2, \cdots$. Equation 8.6 reduces to

$$p_0(n) = 1 - \sum_{i=0}^{n} \alpha_i, \quad n = 0, 1, 2, \cdots, \qquad (8.7)$$

and Equation 8.5 reduces to

$$p_k(n) = \sum_{i=0}^{n} \alpha_i p_{k-1}(n - i). \qquad (8.8)$$

The above equation can be used to compute $p_k(n)$ for increasing values of k starting with $p_0(n)$ of Equation 8.7.

This completes our study of the transient behavior of the RP. Next we study its limiting behavior. Unlike in the case of DTMCs and CTMCs, where most of the limiting results were about convergence in distribution, the results in renewal theory are about convergence with probability one (w.p. 1), or almost sure (a.s.) convergence. See Appendix G for relevant definitions. Let $N(\infty)$ be the almost sure limit of $N(t)$ as $t \to \infty$. That is,

$$N(\infty) = \lim_{t \to \infty} N(t), \quad \text{with probability 1.}$$

In other words, $N(\infty)$ is the sample-path wise limit of $N(t)$. It exists since the sample paths of an RP are non-decreasing functions of time. We shall use the following notation:

$$G(\infty) = \lim_{t \to \infty} G(t).$$

Theorem 8.5 **Almost Sure Limit of** $N(t)$**.** *Let* $\{N(t), t \geq 0\}$ *be an RP with common inter-event time cdf* $G(\cdot)$*, with almost sure limit* $N(\infty)$*.*

1. $G(\infty) = 1 \Rightarrow \mathsf{P}(N(\infty) = \infty) = 1$.
2. $G(\infty) < 1 \Rightarrow \mathsf{P}(N(\infty) = k) = G(\infty)^k(1 - G(\infty)), \quad k = 0, 1, 2, \cdots$.

Proof: We have

$$N(\infty) = k < \infty \Leftrightarrow \{X_n < \infty, 1 \leq n \leq k, X_{k+1} = \infty\}.$$

The probability of the event on the right is 0 if $G(\infty) = \mathsf{P}(X_n < \infty) = 1$, and $G(\infty)^k(1 - G(\infty))$ if $G(\infty) < 1$. The theorem follows from this. ∎

Thus if $G(\infty) = 1$ the renewals recur infinitely often, while if $G(\infty) < 1$, the renewals stop occurring after a while. We use this behavior to make the following definition:

Definition 8.2   Recurrent and Transient RP. *A renewal process $\{N(t), t \geq 0\}$ is called*

1. *recurrent if* $P(N(\infty) = \infty) = 1$,
2. *transient if* $P(N(\infty) = \infty) < 1$.

Example 8.8   Transient and Recurrent RPs in a CTMC. Let $\{N(t), t \geq 0\}$ be the renewal process described in Example 8.3. It counts the number of entries into state 0 up to time t. If state 0 is recurrent, the RP $\{N(t), t \geq 0\}$ is recurrent, and if the state 0 is transient, the RP $\{N(t), t \geq 0\}$ is transient. This is because the number of visits to state 0 over $[0, \infty)$ is infinity if state 0 is recurrent and finite if it is transient.
∎

From now on we shall concentrate on recurrent RPs, i.e., we shall assume that $G(\infty) = 1$. In this case $N(t) \to \infty$ with probability 1 as $t \to \infty$. Hence the natural question to ask is the rate at which $N(t)$ approaches ∞ as $t \to \infty$. The next theorem gives us precisely that.

Theorem 8.6   Elementary Renewal Theorem: Almost-Sure Version. *Let $\{N(t), t \geq 0\}$ be a recurrent RP with mean inter-event time $\tau > 0$. Then*

$$\frac{N(t)}{t} \to \frac{1}{\tau}, \quad w.p.1. \tag{8.9}$$

The right-hand side is to be interpreted as 0 if $\tau = \infty$.

Proof: From the definition of the RP and Figure 8.2, we see that $S_{N(t)} \leq t < S_{N(t)+1}$. Hence, when $N(t) > 0$,

Figure 8.2 *Relationship between t, $S_{N(t)}$ and $S_{N(t)+1}$.*

$$\frac{S_{N(t)}}{N(t)} \leq \frac{t}{N(t)} < \frac{S_{N(t)+1}}{N(t)}.$$

Since the RP is recurrent, $N(t) \to \infty$ as $t \to \infty$. Thus, using the strong law of large numbers, we get

$$\lim_{t \to \infty} \frac{S_{N(t)}}{N(t)} = \lim_{n \to \infty} \frac{S_n}{n} = \tau, \quad w.p.1,$$

and

$$\lim_{t \to \infty} \frac{S_{N(t)+1}}{N(t)} = \lim_{n \to \infty} \frac{S_{n+1}}{n} = \lim_{n \to \infty} \frac{S_{n+1}}{n+1} \frac{n+1}{n} = \tau, \quad w.p.1.$$

Hence

$$\limsup_{t\to\infty} \frac{t}{N(t)} \leq \lim_{t\to\infty} \frac{S_{N(t)+1}}{N(t)} = \tau,$$

and

$$\liminf_{t\to\infty} \frac{t}{N(t)} \geq \lim_{t\to\infty} \frac{S_{N(t)}}{N(t)} = \tau.$$

Hence

$$\tau \geq \limsup_{t\to\infty} \frac{t}{N(t)} \geq \liminf_{t\to\infty} \frac{t}{N(t)} \geq \tau.$$

Thus we have

$$\limsup_{t\to\infty} \frac{t}{N(t)} = \liminf_{t\to\infty} \frac{t}{N(t)} = \tau, \quad w.p.1.$$

Thus $t/N(t)$ has a limit given by

$$\lim_{t\to\infty} \frac{t}{N(t)} = \tau, \quad w.p.1.$$

Now, if $\tau < \infty$, we can use the continuity of the function $f(x) = 1/x$ for $x > 0$ to get Equation 8.9. If $\tau = \infty$, we first construct a new renewal process with inter-event times given by $\min(X_n, T)$, for a fixed T, and then let $T \to \infty$. We leave the details to the reader. ∎

Theorem 8.6 makes intuitive sense because the limiting value of $N(t)/t$ is the long run number of renewals per unit time. Thus if the mean inter-event time is 10 minutes, it makes sense that in the long run we should see one renewal every 10 minutes, or $1/10$ renewals every minute.

The following theorem gives a more detailed distributional information about how the RP approaches infinity.

Theorem 8.7 Central Limit Theorem for $N(t)$. *Let $\{N(t), t \geq 0\}$ be a recurrent RP with inter-event times having mean $0 < \tau < \infty$ and variance $\sigma^2 < \infty$. Then*

$$\lim_{t\to\infty} P\left(\frac{N(t) - t/\tau}{\sqrt{\sigma^2 t/\tau^3}} \leq x\right) = \Phi(x) = \int_{-\infty}^{x} \frac{e^{-u^2/2}}{\sqrt{2\pi}} du. \tag{8.10}$$

Proof: We have

$$P(N(t) \geq k) = P(S_k \leq t)$$

and hence

$$P\left(\frac{N(t) - t/\tau}{\sqrt{\sigma^2 t/\tau^3}} \geq \frac{k - t/\tau}{\sqrt{\sigma^2 t/\tau^3}}\right) = P\left(\frac{S_k - k\tau}{\sigma\sqrt{k}} \leq \frac{t - k\tau}{\sigma\sqrt{k}}\right). \tag{8.11}$$

Now let t and k both grow to ∞ so that

$$t/k \to \tau$$

and

$$\frac{t - k\tau}{\sigma\sqrt{k}} \to x,$$

where x is a fixed real number. The above equation implies that

$$\frac{k - t/\tau}{\sqrt{\sigma^2 t/\tau^3}} \to -x.$$

Letting $k, t \to \infty$ appropriately in Equation 8.11, and using the central limit theorem, we get

$$\lim_{t \to \infty} \mathsf{P} \left(\frac{N(t) - t/\tau}{\sqrt{\sigma^2 t/\tau^3}} \geq -x \right) = \lim_{k \to \infty} \mathsf{P} \left(\frac{S_k - k\tau}{\sigma \sqrt{k}} \leq x \right) = \Phi(x). \qquad (8.12)$$

Hence

$$\lim_{t \to \infty} \mathsf{P} \left(\frac{N(t) - t/\tau}{\sqrt{\sigma^2 t/\tau^3}} \leq x \right) = 1 - \Phi(-x) = \Phi(x).$$

This completes the proof. ∎

The above theorem says that for large t, $N(t)$ approaches a normal random variable with mean t/τ and variance $\sigma^2 t/\tau^3$.

Example 8.9 RPs in DTMCs. Let $\{X_n, t \geq 0\}$ be an irreducible, aperiodic, and positive recurrent DTMC with $X_0 = 0$, and consider the RP $\{N(t), t \geq 0\}$ of Example 8.6. We see that the first renewal time S_1 is the same as $\tilde{T}_0$, the time of the first return to state 0, as defined in Equation 4.10. It has mean $\tilde{m}_0$. From Theorems 4.17 and 4.19 on pages 113 and 115, we see that

$$\pi_0 = 1/\tilde{m}_0,$$

where π_0 is the limiting probability that the DTMC is in state 0. Now, from Theorem 8.6, we see that the long run rate of visits to state 0 is given by

$$\lim_{t \to \infty} \frac{N(t)}{t} = \frac{1}{\tilde{m}_0} = \pi_0,$$

with probability 1. ∎

Example 8.10 Machine Maintenance. Suppose a part in a machine is available from two different suppliers, A and B. When the part fails it is replaced by a new one from supplier A with probability .3 and supplier B with probability .7. A part from supplier A lasts for an exponential amount of time with a mean of 8 days, and it takes 1 day to install it. A part from supplier B lasts for an exponential amount of time with a mean of 5 days, and it takes 1/2 a day to install it. Installation times are deterministic. Assume that a failure has occurred at time 0-. Compute the approximate distribution of the number of failures in the first year (not counting the one at time 0-).

Let X_n be the time between the $(n - 1)$st and the nth failures. $\{X_n, n \geq 1\}$ are iid random variables with common distribution given by

$$X_n \sim \begin{cases} 1 + \exp(1/8) & \text{with probability .3,} \\ 0.5 + \exp(1/5) & \text{with probability .7.} \end{cases}$$

Let $N(t)$ be the number of failures up to time t. Then $\{N(t), t \geq 0\}$ is an RP generated by $\{X_n, n \geq 1\}$. Straightforward calculations yield

$$\tau = \mathsf{E}(X_n) = 6.55, \quad \sigma^2 = \text{Var}(X_n) = 39.2725.$$

Hence, from Theorem 8.7, $N(t)$ is approximately normally distributed with mean $t/6.55$ and variance $(39.2725/6.55^3)t$. Thus, in the first year ($t = 365$), the number of failures is approximately normal with mean 55.725 and variance 51.01. ■

8.3 The Renewal Function

We begin with the definition.

Definition 8.3 *Let $\{N(t), t \geq 0\}$ be an RP. The renewal function $M(\cdot)$ is defined as*

$$M(t) = \mathsf{E}(N(t)), \quad t \geq 0. \tag{8.13}$$

The next theorem gives a method of computing the renewal function.

Theorem 8.8 Renewal Function. *Let $\{N(t), t \geq 0\}$ be an RP with common inter-event time cdf G. The renewal function is given by*

$$M(t) = \sum_{k=1}^{\infty} G_k(t), \quad t \geq 0, \tag{8.14}$$

where $G_k(t)$ is as defined in Equation 8.2.

Proof: We have

$$
\begin{aligned}
M(t) &= \sum_{k=0}^{\infty} k \mathsf{P}(N(t) = k) \\
&= \sum_{k=0}^{\infty} k(G_k(t) - G_{k+1}(t)) \quad \text{(Eq. 8.3)} \\
&= \sum_{k=1}^{\infty} G_k(t).
\end{aligned}
$$

This proves the theorem. ■

The next theorem introduces an important equation known as the renewal equation, and gives a simple expression for the LST of the renewal function defined by

$$\tilde{M}(s) = \int_0^{\infty} e^{-st} dM(t), \quad Re(s) > 0.$$

Theorem 8.9 The Renewal Equation. *The renewal function* $\{M(t), t \geq 0\}$ *of a renewal process* $\{N(t), t \geq 0\}$ *with common inter-event time cdf* $G(\cdot)$ *satisfies the renewal equation*

$$M(t) = G(t) + \int_0^t M(t-u)dG(u), \quad t \geq 0. \tag{8.15}$$

The LST of the renewal function is given by

$$\tilde{M}(s) = \frac{\tilde{G}(s)}{1 - \tilde{G}(s)}. \tag{8.16}$$

Proof: We use the renewal argument. Fix a $t \geq 0$, and suppose $S_1 = u$. If $u > t$ the very first renewal is after t, hence $N(t) = 0$, and hence its expected value is zero. On the other hand, if $u \leq t$, then we get one renewal at u, and a new renewal process starts at u, which produces additional $M(t-u)$ expected number of events up to t. Hence we have

$$\mathsf{E}(N(t)|S_1 = u) = \begin{cases} 0 & \text{if } u > t, \\ 1 + M(t-u) & \text{if } u \leq t. \end{cases}$$

Hence we get

$$\begin{aligned} M(t) &= \int_0^\infty \mathsf{E}(N(t)|S_1 = u)dG(u) \\ &= \int_0^t (1 + M(t-u))dG(u) \\ &= G(t) + \int_0^t M(t-u)dG(u). \end{aligned}$$

This gives the renewal equation. Taking LSTs on both sides of Equation 8.15 we get,

$$\tilde{M}(s) = \tilde{G}(s) + \tilde{M}(s)\tilde{G}(s),$$

which yields Equation 8.16. ∎

As in the case of $p_k(t)$, when the inter-event times are integer valued random variables, the renewal argument provides a simple recursive method of computing $M(t)$. To derive this, let

$$\alpha_i = \mathsf{P}(X_n = i), \quad i = 0, 1, 2, \cdots,$$

and

$$\beta_i = \mathsf{P}(X_n \leq i) = \sum_{k=0}^i \alpha_k.$$

Since all renewals take place at integer time points, it suffices to study $M(n)$ for $n = 0, 1, 2, \cdots$. We leave it to the reader to show that Equation 8.15 reduces to

$$M(n) = \beta_n + \sum_{k=0}^n \alpha_k M(n-k), \quad n = 0, 1, 2, \cdots. \tag{8.17}$$

The above equation can be used to compute $M(n)$ recursively for increasing values of n.

The renewal function plays a very important role in the study of renewal processes. The next theorem shows why.

Theorem 8.10   *The renewal function $\{M(t), t \geq 0\}$ completely characterizes the renewal process $\{N(t), t \geq 0\}$.*

Proof: Equation 8.16 can be rewritten as

$$\tilde{G}(s) = \frac{\tilde{M}(s)}{1 + \tilde{M}(s)}. \tag{8.18}$$

Now, the renewal function $\{M(t), t \geq 0\}$ determines its LST $\tilde{M}(s)$, which determines the LST $\tilde{G}(s)$ by the above relation, which determines the cdf $G(\cdot)$, which characterizes the renewal process $\{N(t), t \geq 0\}$, according to Theorem 8.1. Hence the theorem follows. ∎

Example 8.11   Renewal Function for a PP. Consider a renewal process with inter-event time distribution $G(x) = 1 - e^{-\lambda x}$ for $x \geq 0$. We have $\tilde{G}(s) = \lambda/(s + \lambda)$. From Equation 8.16 we get

$$\tilde{M}(s) = \frac{\lambda}{s}.$$

Inverting this we get

$$M(t) = \lambda t, \quad t \geq 0.$$

This is as expected, since the renewal process in this case is a PP(λ). Theorem 8.10 implies that if a renewal process has a linear renewal function, it must be a Poisson process! ∎

Example 8.12   Compute the renewal function for an RP $\{N(t), t \geq 0\}$ generated by $\{X_n, n \geq 0\}$ with common pmf

$$P(X_n = 0) = 1 - \alpha, \quad P(X_n = 1) = \alpha,$$

where $0 < \alpha < 1$.

From Equation 8.17 we get

$$\begin{aligned} M(0) &= 1 - \alpha + (1 - \alpha)M(0), \\ M(n) &= 1 + (1 - \alpha)M(n) + \alpha M(n - 1), \quad n \geq 1. \end{aligned}$$

The above equations can be solved to get

$$M(n) = (n + 1 - \alpha)/\alpha, \quad n \geq 0.$$

Thus

$$M(t) = ([t] + 1 - \alpha)/\alpha, \quad t \geq 0,$$

where $[t]$ is the largest integer not exceeding t. ∎

Next we study the asymptotic behavior of the renewal function.

Theorem 8.11 *Let $M(t)$ be the renewal function of an RP with mean inter-event time $\tau > 0$. Then*

$$M(t) < \infty \text{ for all } t \geq 0.$$

Proof: Let G be the cdf of the inter-event times. Since $\tau > 0$, $G(0) < 1$. Thus there is a $\delta > 0$ such that $G(\delta) < 1$. Define

$$X_n^* = \begin{cases} 0 & \text{if } X_n < \delta, \\ \delta & \text{if } X_n \geq \delta. \end{cases}$$

Let $\{S_n^*, n \geq 1\}$ and $\{N^*(t), t \geq 0\}$ be the renewal sequence and the renewal process generated by $\{X_n^*, n \geq 1\}$. Since $X_n^* \leq X_n$ for all $n \geq 1$, we see that $S_n^* \leq S_n$ for all $n \geq 1$, which implies that $N(t) \leq N^*(t)$ for all $t \geq 0$. Hence

$$M(t) = \mathsf{E}(N(t)) \leq \mathsf{E}(N^*(t)) = M^*(t).$$

We can modify Example 8.12 slightly to get

$$M^*(t) = \frac{[t/\delta] + G(\delta)}{1 - G(\delta)} < \infty.$$

Hence

$$M(t) \leq M^*(t) < \infty,$$

and the result follows. ∎

Theorem 8.12 Elementary Renewal Theorem.

$$\lim_{t \to \infty} \frac{M(t)}{t} = \frac{1}{\tau}, \tag{8.19}$$

where the right-hand side is to be interpreted as zero if $\tau = \infty$.

Before we prove the theorem, it is worth noting that this theorem does not follow from Theorem 8.6, since almost sure convergence does not imply convergence of the expected values. Hence we need to establish this result independently. We need the following result first.

Theorem 8.13

$$\mathsf{E}(S_{N(t)+1}) = \tau(M(t) + 1). \tag{8.20}$$

Proof: We prove this by using renewal argument. Let

$$H(t) = \mathsf{E}(S_{N(t)+1}).$$

We have

$$\mathsf{E}(S_{N(t)+1}|S_1 = u) = \begin{cases} u & \text{if } u > t, \\ u + H(t - u) & \text{if } u \leq t. \end{cases}$$

Hence,

$$
\begin{aligned}
H(t) &= \int_0^\infty \mathsf{E}(S_{N(t)+1}|S_1 = u)dG(u) \\
&= \int_0^t (u + H(t - u))dG(u) + \int_t^\infty udG(u) \\
&= \int_0^\infty udG(u) + \int_0^t H(t - u)dG(u) \\
&= \tau + \int_0^t H(t - u)dG(u).
\end{aligned}
\tag{8.21}
$$

Taking LSTs on both sides of the above equation, we get

$$\tilde{H}(s) = \tau + \tilde{H}(s)\tilde{G}(s).$$

Hence,

$$
\begin{aligned}
\tilde{H}(s) &= \frac{\tau}{1 - \tilde{G}(s)} \\
&= \tau\left[1 + \frac{\tilde{G}(s)}{1 - \tilde{G}(s)}\right] \\
&= \tau(1 + \tilde{M}(s)).
\end{aligned}
$$

Inverting the above equation yields Equation 8.20. ∎

A quick exercise in renewal argument shows that $\mathsf{E}(S_{N(t)}) \neq \tau M(t)$. Thus Equation 8.20 is unusual indeed! With this result we are now ready to prove the elementary renewal theorem.

Proof of Theorem 8.12. Assume $0 < \tau < \infty$. By definition, we have

$$S_{N(t)+1} > t.$$

Hence

$$\mathsf{E}(S_{N(t)+1}) > t.$$

Using Equation 8.20, the above inequality yields

$$\frac{M(t)}{t} > \frac{1}{\tau} - \frac{1}{t}.$$

Hence

$$\liminf_{t \to \infty} \frac{M(t)}{t} \geq \frac{1}{\tau}.
\tag{8.22}$$

Now, fix a $0 < T < \infty$ and define

$$X'_n = \min(X_n, T), \quad n \geq 1.$$

Let $\{N'(t), t \geq 0\}$ be an RP generated by $\{X'_n, n \geq 1\}$, and $M'(t)$ be the corresponding renewal function. Now,

$$S_{N'(t)+1} \leq t + T.$$

Hence, taking expected values on both sides, and using Equation 8.20, we get

$$\tau'(M'(t) + 1) \leq t + T,$$

where $\tau' = E(X'_n)$. Hence we get

$$\frac{M'(t)}{t} \leq \frac{1}{\tau'} + \frac{T - \tau'}{t\tau'}.$$

Since $X'_n \leq X_n$, we see that $M(t) \leq M'(t)$. Hence we get

$$\frac{M(t)}{t} \leq \frac{1}{\tau'} + \frac{T - \tau'}{t\tau'}.$$

This implies

$$\limsup_{t \to \infty} \frac{M(t)}{t} \leq \frac{1}{\tau'}.$$

Now as $T \to \infty$, $\tau' \to \tau$. Hence letting $T \to \infty$ on both sides af the above inequality, we get

$$\limsup_{t \to \infty} \frac{M(t)}{t} \leq \frac{1}{\tau}. \tag{8.23}$$

Combining Equations 8.22 and 8.23 we get Equation 8.19. We leave the case of $\tau = \infty$ to the reader. ∎

This theorem has the same intuitive explanation as Theorem 8.6. We end this section with results about the higher moments of $N(t)$.

Theorem 8.14   Higher Moments of $N(t)$. *Let*

$$M_k(t) = E(N(t)(N(t) - 1) \cdots (N(t) - k + 1)), \quad k = 1, 2, \cdots; \ t \geq 0.$$

Then

$$\tilde{M}_k(s) = k!\tilde{M}(s)^k. \tag{8.24}$$

Proof: From Equation 8.4 we get

$$
\begin{aligned}
\tilde{M}_k(s) &= \sum_{r=k}^{\infty} r(r - 1) \cdots (r - k + 1)\tilde{p}_r(s) \\
&= \sum_{r=k}^{\infty} r(r - 1) \cdots (r - k + 1)\tilde{G}^r(s)(1 - \tilde{G}(s)) \\
&= k! \left(\frac{\tilde{G}(s)}{1 - \tilde{G}(s)}\right)^k \\
&= k!\tilde{M}(s)^k.
\end{aligned}
$$

Thus $M_k(t)/k!$ is a k-fold convolution of $M(t)$ with itself. ∎

8.4 Renewal-Type Equation

In this section we study the type of integral equations that arise from the use of renewal argument. We have already seen two instances of such equations: Equation 8.15 for $M(t)$, and Equation 8.21 for $E(S_{N(t)+1})$. All these equations have the following form:

$$H(t) = D(t) + \int_0^t H(t-u)dG(u), \qquad (8.25)$$

where $G(\cdot)$ is a cdf of a random variable with $G(0-) = 0$ and $G(\infty) = 1$, $D(\cdot)$ is a given function, and H is to be determined. When $D(t) = G(t)$, the above equation is the same as Equation 8.15, and is called the renewal equation. When $D(\cdot)$ is a function other than $G(\cdot)$, Equation 8.25 is called a renewal-type equation. Thus, when $D(t) = \tau$, we get the renewal-type equation for $E(S_{N(t)+1})$, namely, Equation 8.21.

We will need to solve the renewal-type equations arising in applications. The following theorem gives the conditions for the existence and uniqueness of the solution to a renewal-type equation.

Theorem 8.15 Solution of the Renewal-Type Equation. *Suppose*

$$|D(t)| < \infty \quad \text{for all} \quad t \geq 0.$$

Then there exists a unique solution to the renewal-type equation

$$H(t) = D(t) + \int_0^t H(t-u)dG(u),$$

such that

$$|H(t)| < \infty \quad \text{for all} \quad t \geq 0,$$

given by

$$H(t) = D(t) + \int_0^t D(t-u)dM(u), \qquad (8.26)$$

where $M(\cdot)$ is the renewal function associated with $G(\cdot)$.

Proof: Since $M(t) < \infty$ and $G_k(t)$ is a decreasing function of k, Equation 8.14 implies that

$$\lim_{n\to\infty} G_n(t) = 0.$$

We introduce the following convenient notation for (Stieltjes-)convolution:

$$A * B(t) = \int_0^t A(t-u)dB(u).$$

By recursive use of the renewal-type equation, and writing H for $H(t)$ for compactness, we get

$$
\begin{aligned}
H &= D + H * G = D + (D + H * G) * G = D + D * G + H * G_2 = \cdots \\
&= D + D * \sum_{k=1}^{n-1} G_k + H * G_n, \quad n \geq 1.
\end{aligned}
\qquad (8.27)
$$

Letting $n \to \infty$ and using $G_n \to 0$ we get

$$H = D + D * \sum_{k=1}^{\infty} G_k = D + D * M,$$

which is Equation 8.26. Thus we have shown that H as given in Equation 8.26 is a solution to Equation 8.25.

Since D is assumed to be bounded we get

$$c = \sup_{0 \le x \le t} |D(x)| < \infty.$$

Then Equation 8.26 yields

$$|H(t)| \le |D(t)| + |D * M(t)| \le c + cM(t) < \infty.$$

This shows the boundedness of $H(t)$.

To show uniqueness, suppose H_1 and H_2 are two solutions to Equation 8.25. Since both H_1 and H_2 satisfy Equation 8.27, we get

$$H = H_1 - H_2 = (H_1 - H_2) * G_n = H * G_n.$$

Letting $n \to \infty$, we can use the boundedness of H_1 and H_2 to get

$$H = \lim_{n \to \infty} H * G_n = 0.$$

Hence we must have $H_1 = H_2$. This shows uniqueness. ∎

We illustrate with an example below.

Example 8.13   Two-State Machine. We revisit our two-state machine with iid $\exp(\mu)$ up-times. However, when the machine fails after an up-time U, the ensuing down time is cU, for a fixed $c > 0$. Suppose the machine is up at time 0. Let $X(t) = 1$ if the machine is up at time t, and 0 otherwise. Compute

$$H(t) = \mathsf{P}(X(t) = 1).$$

We approach the problem by first deriving a renewal-type equation for $H(t)$ by using the renewal argument.

Let $U_1 \sim \exp(\mu)$ be the first up-time, and $D_1 = cU_1$ be the following down time. The machine becomes as good as new at time $S_1 = U_1 + D_1 = (1 + c)U_1 \sim \exp(\mu/(1 + c))$. Now fix a $t \ge 0$ and suppose $S_1 = u$. If $u > t$, then $X(t) = 1$ if $U_1 > t$. If $u \le t$, the machine regenerates at time u, hence it will be up at time t with probability $H(t - u)$. Combining these observations, we get

$$\mathsf{P}(X(t) = 1 | S_1 = u) = \begin{cases} \mathsf{P}(U_1 > t | S_1 = u) & \text{if } u > t, \\ H(t - u) & \text{if } u \le t. \end{cases}$$

Hence,

$$H(t) = \int_0^{\infty} \mathsf{P}(X(t) = 1 | S_1 = u) dG(u)$$

$$= \int_0^t H(t-u)dG(u) + \int_t^\infty P(U_1 > t|S_1 = u)dG(u), \quad (8.28)$$

where G is the cdf of S_1. Since $P(U_1 > t|S_1 = u) = 0$ if $u \le t$, we get

$$\int_t^\infty P(U_1 > t|S_1 = u)dG(u) = \int_0^\infty P(U_1 > t|S_1 = u)dG(u) = P(U_1 > t) = e^{-\mu t}.$$

Substituting in Equation 8.28, we get

$$H(t) = e^{-\mu t} + \int_0^t H(t-u)dG(u),$$

which is a renewal-type equation with $D(t) = e^{-\mu t}$, and G being the cdf of an $\exp(\mu/(1+c))$ random variable. The RP corresponding to this G is a $PP(\mu/(1+c))$. Hence the renewal function is given by (see Example 8.11)

$$M(t) = \frac{\mu}{1+c}t.$$

Hence the solution given by Equation 8.26 reduces to

$$H(t) = e^{-\mu t} + \int_0^t e^{-\mu(t-u)} \frac{\mu}{1+c} du = \frac{1 + ce^{-\mu t}}{1+c}.$$

Note that we could not have derived the above expression by the method of CTMCs since $\{X(t), t \ge 0\}$ is not a CTMC due the dependence of the up and down times. ∎

The solution in Equation 8.26 is not easy to use in practice since, unlike in the previous example, $M(t)$ is generally not explicitly known. The next theorem gives a transform solution that can be more useful in practice.

Theorem 8.16 *If $D(\cdot)$ has an LST $\tilde{D}(s)$, then $H(\cdot)$ has an LST $\tilde{H}(s)$ given by*

$$\tilde{H}(s) = \frac{\tilde{D}(s)}{1 - \tilde{G}(s)}. \quad (8.29)$$

Proof: Follows by taking LST on both sides of Equation 8.25. ∎

We illustrate with an example. It is an unusual application of renewal-type equations, since generally $D(t)$ is known, and $H(t)$ is unknown in Equation 8.25. However, in the following example, $H(t)$ is known, and $D(t)$ and $G(t)$ are unknown!

Example 8.14 Busy Period in an $M/G/\infty$ Queue. Consider an $M/G/\infty$ queue where customers arrive according to a $PP(\lambda)$ and require iid service times with common cdf $B(\cdot)$. Let $X(t)$ be the number of customers in the system at time t. Now suppose a customer enters an empty system at time 0. Compute the cdf $F(\cdot)$ of the busy period of the system, defined as

$$T = \min\{t \ge 0 : X(t) = 0\}.$$

Define S_n to be the nth time when a customer enters an empty system. Since the arrival process is Poisson, and the service times are iid, we see that $\{S_n, n \geq 1\}$ is a renewal sequence. S_1 is the length of the first busy cycle. Let $H(t) = P(X(t) = 0)$. Using the renewal argument, we get

$$P(X(t) = 0|S_1 = u) = \begin{cases} P(T \leq t|S_1 = u) & \text{if } u > t, \\ H(t - u) & \text{if } u \leq t. \end{cases}$$

Hence, using G as the cdf of S_1, we get

$$\begin{aligned} H(t) &= \int_0^t P(X(t) = 0|S_1 = u)dG(u) \\ &= \int_0^t H(t - u)dG(u) + \int_t^\infty P(T \leq t|S_1 = u)dG(u). \quad (8.30) \end{aligned}$$

Since $P(T \leq t|S_1 = u) = 1$ if $u \leq t$, we get

$$\begin{aligned} \int_t^\infty P(T \leq t|S_1 = u)dG(u) &= \int_0^\infty P(T \leq t|S_1 = u)dG(u) - \int_0^t dG(u) \\ &= P(T \leq t) - G(t). \end{aligned}$$

Substituting in Equation **??**, we get

$$H(t) = P(T \leq t) - G(t) + \int_0^t H(t - u)dG(u). \quad (8.31)$$

We see that $S_1 = T + M$ where $M \sim \exp(\lambda)$ is the time until a customer arrives to the system once it becomes empty at time T, and is independent of T. Since F is the cdf of T and G is cdf of S_1, we see that

$$\tilde{G}(s) = \tilde{F}(s)\frac{\lambda}{s + \lambda}.$$

Also, from the results in Section 7.8, and using the fact that the $M/G/\infty$ queue starts with an arrival to an empty queue at time 0, we see that

$$H(t) = \exp(-\lambda \int_0^t (1 - B(u))du)B(t).$$

Let $\tilde{H}(s)$ be the LST of $H(\cdot)$. Taking the LST of Equation 8.31 we get

$$\tilde{H}(s) = \tilde{F}(s) - \tilde{F}(s)\frac{\lambda}{s + \lambda} + \tilde{H}(s)\tilde{F}(s)\frac{\lambda}{s + \lambda}.$$

Solving for $\tilde{F}(s)$ we get

$$\tilde{F}(s) = \frac{(s + \lambda)\tilde{H}(s)}{s + \lambda\tilde{H}(s)},$$

which is the result we desired. Differentiating $\tilde{F}(s)$ with respect to s and using

$$\lim_{s \to 0} \tilde{H}(s) = H(\infty) = e^{-\lambda\tau}$$

we get

$$E(T) = \frac{e^{\lambda\tau} - 1}{\lambda}. \quad \blacksquare$$

8.5 Key Renewal Theorem

In the previous section we saw several examples of the renewal-type equation, and the methods of solving it. In general explicit solutions to renewal-type equations are hard to come by. Hence in this section we study the asymptotic properties of the solution $H(t)$ as $t \to \infty$. We have done such asymptotic analysis for a discrete renewal-type equation in our study of DTMCs (Equation 4.24 on page 110), and a continuous renewal-type equation in the study of CTMCs (Equation 6.59 on page 251). In this section we shall deal with the general case. We begin with a definition.

Definition 8.4   Periodicity. *A non-negative, non-defective random variable X is called periodic (or arithmetic, or lattice) if there is a $d > 0$ such that*

$$\sum_{k=0}^{\infty} P(X = kd) = 1.$$

The largest such d is called the period (or span) of X. If there is no such d, X is called aperiodic (or non-arithmetic, or non-lattice).

We say that a distribution of X is periodic, (or arithmetic, or lattice) if X is periodic. Similarly we say that an RP is aperiodic if it is generated by aperiodic inter-event times. Note that all continuous random variables on $[0, \infty)$ are aperiodic, and all non-negative integer valued random variables are periodic. However, a discrete random variable is not necessarily periodic. Thus a random variable taking values $\{1, \sqrt{2}\}$ is aperiodic, but one taking values $\{0, \sqrt{2}\}$ is periodic. (What about one that takes values $\{e, \pi\}$?) With the help of this concept, we shall prove the main result in the next theorem.

Theorem 8.17   Key Renewal Theorem. *Let H be a solution to the following renewal-type equation*

$$H(t) = D(t) + \int_0^t H(t - u)dG(u). \tag{8.32}$$

Suppose D is a difference of two non-negative bounded monotone functions and

$$\int_0^{\infty} |D(u)|du < \infty. \tag{8.33}$$

1. *If G is aperiodic with mean $\tau > 0$,*

$$\lim_{t \to \infty} H(t) = \frac{1}{\tau} \int_0^{\infty} D(u)du. \tag{8.34}$$

2. *If G is periodic with period d and mean $\tau > 0$,*

$$\lim_{k \to \infty} H(kd + x) = \frac{d}{\tau} \sum_{k=0}^{\infty} D(kd + x), \quad 0 \le x < d. \tag{8.35}$$

One consequence of the above theorem is that

$$\lim_{t\to\infty} \frac{1}{t} \int_0^t H(u)\,du = \frac{1}{\tau} \int_0^\infty D(u)\,du, \qquad (8.36)$$

whether G is periodic or aperiodic. We can give a simple probabilistic interpretation to the above limit. Let $t > 0$ be fixed, and suppose $U \sim U(0,t)$ is a random variable that is uniformly distributed over $[0,t]$. Then $H(U)$ can be thought of as the value of $H(\cdot)$ at a random time U. We have

$$\mathsf{E}(H(U)) = \frac{1}{t} \int_0^t H(u)\,du.$$

Now, as $t \to \infty$, we can interpret the limit as the expected value of the quantity $H(t)$ observed at a t "picked uniformly over $(0,\infty)$." We sometimes call this limit as $H(t)$ observed at an "arbitrary point in time." The discussion above implies that this limit always exists.

The proof of this theorem is beyond the scope of this book, and we omit it. The hard part is proving that the limit exists. If we assume the limit exists, evaluating the limits is relatively easy, as we have done in the proofs of the discrete renewal theorem (Theorem 4.15 on page 110) and the continuous renewal theorem (Theorem 6.20 on page 251). Feller proves the result under the assumption that D is a "directly Riemann integrable" function. The condition on D assumed above is a sufficient condition for direct Riemann integrability, and is adequate for our purposes. We refer the readers to Feller(1971), Kohlas (1982), or Heyman and Sobel (1982) for proofs. We shall refer to key renewal theorem as KRT from now on.

In the remainder of this section we shall illustrate the usefulness of the KRT by means of several examples. As a first application, we shall prove the following theorem.

Theorem 8.18   Blackwell's Renewal Theorem. *Let $M(\cdot)$ be the renewal function of an RP with mean inter-renewal time $\tau > 0$.*

1. If the RP is aperiodic,

$$\lim_{t\to\infty} [M(t+h) - M(t)] = \frac{h}{\tau}, \quad h \geq 0. \qquad (8.37)$$

2. If the RP is periodic with period d,

$$\lim_{t\to\infty} [M(t+kd) - M(t)] = \frac{kd}{\tau}, \quad k = 0,1,2,\cdots. \qquad (8.38)$$

Proof: We consider the aperiodic case. For a given $h \geq 0$, consider the renewal-type equation 8.25 with the following D function:

$$D(t) = \begin{cases} 1 & \text{if } 0 \leq t \leq h, \\ 0 & \text{if } t > h. \end{cases}$$

The solution is given by Equation 8.26. For $t > h$ this reduces to

$$H(t) = \int_{t-h}^{t} dM(u) = M(t) - M(t-h).$$

From the KRT we get

$$\lim_{t \to \infty} [M(t+h) - M(t)] = \lim_{t \to \infty} H(t+h)$$

$$= \lim_{t \to \infty} H(t) = \frac{1}{\tau} \int_{0}^{\infty} D(u) du = \frac{h}{\tau}.$$

This proves Equation 8.37. Equation 8.38 follows similarly. ∎

It is interesting to note that one can prove that the above theorem is in fact equivalent to the KRT, although the proof is not simple!

Example 8.15 We verify Blackwell's renewal theorem for the two renewal functions we derived in Examples 8.11 and 8.12. In Example 8.11 we have a Poisson process, which is an aperiodic RP with $M(t) = \lambda t$. In this case the mean interrenewal time is $\tau = 1/\lambda$. We get

$$\lim_{t \to \infty} [M(t+h) - M(t)] = \lim_{t \to \infty} \lambda h = \frac{h}{\tau},$$

thus verifying Equation 8.37. In Example 8.12 we have a periodic RP with period 1, with $M(t) = ([t] + 1 - \alpha)/\alpha$. In this case the mean inter-renewal time is $\tau = \alpha$. Thus, if h is a non-negative integer,

$$\lim_{t \to \infty} [M(t+h) - M(t)] = \lim_{t \to \infty} \frac{[t+h] - [t]}{\alpha} = \frac{h}{\tau}.$$

If h is not an integer, the above limit does not exist. This verifies Equation 8.38. ∎

Example 8.16 Asymptotic Behavior of $M(t)$. Let $M(t)$ be the renewal function of an aperiodic RP with inter-event times with finite variance ($\sigma^2 < \infty$). Show that

$$M(t) = \frac{t}{\tau} + \frac{\sigma^2 - \tau^2}{2\tau^2} + o(1), \tag{8.39}$$

where $o(1)$ is a function that approaches 0 as $t \to \infty$.

From Theorem 8.12 we know that

$$\lim_{t \to \infty} \frac{M(t)}{t} = \frac{1}{\tau}.$$

We now derive a renewal-type equation for

$$H(t) = M(t) - t/\tau = \mathsf{E}(N(t) - t/\tau).$$

Using renewal argument we get

$$\mathsf{E}(N(t) - t/\tau | S_1 = u) = \begin{cases} -t/\tau & \text{if } u > t \\ 1 + M(t-u) - t/\tau & \text{if } u \leq t \end{cases}$$

$$= \begin{cases} -t/\tau & \text{if } u > t \\ H(t-u) + 1 - u/\tau & \text{if } u \leq t. \end{cases}$$

Hence we get

$$H(t) = \int_0^\infty \mathsf{E}(N(t) - t/\tau | S_1 = u) dG(u)$$

$$= D(t) + \int_0^t H(t-u) dG(u),$$

where

$$D(t) = \int_0^t (1 - u/\tau) dG(u) - \int_t^\infty (t/\tau) dG(u).$$

Now

$$\frac{1}{\tau} [\int_0^t u \, dG(u) + \int_t^\infty t \, dG(u)]$$

$$= \frac{1}{\tau} \int_0^\infty \min(u, t) dG(u)$$

$$= \frac{1}{\tau} \mathsf{E}(\min(S_1, t))$$

$$= \frac{1}{\tau} \int_0^t (1 - G(u)) du$$

$$= 1 - \frac{1}{\tau} \int_t^\infty (1 - G(u)) du.$$

Using this we can rewrite $D(t)$ in the following form

$$D(t) = \frac{1}{\tau} \int_t^\infty (1 - G(u)) du - (1 - G(t)) = D_1(t) - D_2(t).$$

Thus $D(t)$ is a difference of two monotone functions, and we have

$$\int_0^\infty D_1(t) dt = \int_{t=0}^\infty \frac{1}{\tau} \int_{u=t}^\infty (1 - G(u)) du \, dt$$

$$= \frac{1}{\tau} \int_{u=0}^\infty \int_{t=0}^u (1 - G(u)) dt \, du$$

$$= \frac{1}{\tau} \int_{u=0}^\infty u(1 - G(u)) du$$

$$= \frac{\sigma^2 + \tau^2}{2\tau} < \infty.$$

Also,

$$\int_0^\infty D_2(t) dt = \int_0^\infty (1 - G(t)) dt = \tau < \infty.$$

Hence

$$\int_0^\infty D(t) dt = \frac{\sigma^2 + \tau^2}{2\tau} - \tau = \frac{\sigma^2 - \tau^2}{2\tau}.$$

We can now apply the aperiodic KRT to get

$$\lim_{t\to\infty} \left[M(t) - \frac{t}{\tau} \right] = \frac{\sigma^2 - \tau^2}{2\tau^2}.$$

This implies Equation 8.39. Note that if the coefficient of variation $\sigma^2/\tau^2 < 1$, then t/τ overestimates $M(t)$ for large t, and if $\sigma^2/\tau^2 > 1$, then t/τ underestimates $M(t)$ for large t. For the exponential inter-event times the coefficient of variation is one. In this case t/τ is the exact estimate of $M(t)$, as the RP is a PP in this case. ■

8.6 Recurrence Times

For an aperiodic RP $\{N(t), t \geq 0\}$, let

$$
\begin{aligned}
A(t) &= t - S_{N(t)}, \quad t \geq 0, \\
B(t) &= S_{N(t)+1} - t, \quad t \geq 0, \\
C(t) &= S_{N(t)+1} - S_{N(t)} = X_{N(t)+1} = A(t) + B(t), \quad t \geq 0.
\end{aligned}
$$

Thus $A(t)$ is the total time elapsed since the most recent renewal at or before t; it is called backward recurrence time or current life, or age. $B(t)$ is the time from t until the first renewal after t; it is called the forward recurrence time, or remaining life, or excess life. $C(t)$ is the length of the inter-renewal interval that covers time t; it is called the total life. Figure 8.3 shows the relationships between $A(t)$, $B(t)$, and $C(t)$ graphically.

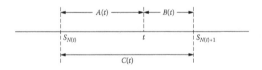

Figure 8.3 *The random variables $A(t)$, $B(t)$, and $C(t)$.*

The stochastic process $\{A(t), t \geq 0\}$ is called the age process, $\{B(t), t \geq 0\}$ is called the remaining life (or excess life) process, and $\{C(t), t \geq 0\}$ is called the total life process. Figures 8.4, 8.5, and 8.6 show the sample paths of the three stochastic processes. The age process increases linearly with rate 1, jumps down to zero at every renewal epoch S_n. The remaining life process starts at X_1, and decreases linearly at rate 1. When it reaches zero at time S_n, it jumps up to X_{n+1}. The total life process has piecewise constant sample paths with upward or downward jumps at S_n. We shall use the renewal argument and the KRT to study these three stochastic processes.

Theorem 8.19 Remaining Life Process. Let $B(t)$ be the remaining life at time t in an aperiodic RP with inter-event time distribution G. Then, for a given $x > 0$,

$$H(t) = P(B(t) > x), \quad t \geq 0,$$

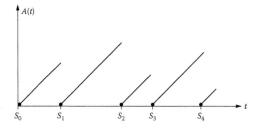

Figure 8.4 *A typical sample path of the age process.*

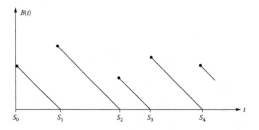

Figure 8.5 *A typical sample path of the excess life process.*

satisfies the renewal-type equation

$$H(t) = 1 - G(x + t) + \int_0^t H(t - u)dG(u), \tag{8.40}$$

and

$$\lim_{t \to \infty} P(B(t) > x) = \frac{1}{\tau} \int_x^\infty (1 - G(u))du. \tag{8.41}$$

Proof: Conditioning on S_1 we get

$$P(B(t) > x | S_1 = u) = \begin{cases} H(t - u) & \text{if } u \le t \\ 0 & \text{if } t < u \le t + x \\ 1 & \text{if } u > t + x. \end{cases}$$

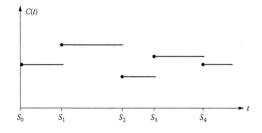

Figure 8.6 *A typical sample path of the total life process.*

Hence we get

$$
\begin{aligned}
H(t) &= \int_0^\infty P(B(t) > x \mid S_1 = u) dG(u) \\
&= 1 - G(x + t) + \int_0^t H(t - u) dG(u),
\end{aligned}
$$

which is Equation 8.40. Now $D(t) = 1 - G(x + t)$ is monotone, and satisfies the condition in Equation 8.33. Since G is assumed to be aperiodic, we can use KRT to get

$$
\lim_{t \to \infty} H(t) = \frac{1}{\tau} \int_0^\infty (1 - G(x + u)) du = \frac{1}{\tau} \int_x^\infty (1 - G(u)) du,
$$

which gives Equation 8.41. ∎

Next we give the limiting distribution of the age process.

Theorem 8.20   Age Process. *Let $A(t)$ be the age at time t in an aperiodic RP with inter-event time distribution G. Its limiting distribution is given by*

$$
\lim_{t \to \infty} P(A(t) \geq x) = \frac{1}{\tau} \int_x^\infty (1 - G(u)) du. \tag{8.42}
$$

Proof: It is possible to prove this theorem by first deriving a renewal-type equation for $H(t) = P(A(t) \geq x)$ and then using the KRT. We leave that to the reader and show an alternate method here. For $t > x$, we have

$$
\{A(t) \geq x\} \Leftrightarrow \{\text{No renewals in } (t - x, t]\} \Leftrightarrow \{B(t - x) > x\}.
$$

Hence

$$
\begin{aligned}
\lim_{t \to \infty} P(A(t) \geq x) &= \lim_{t \to \infty} P(B(t - x) > x) \\
&= \lim_{t \to \infty} P(B(t) > x) \\
&= \frac{1}{\tau} \int_x^\infty (1 - G(u)) du,
\end{aligned}
$$

which is Equation 8.42. ∎

Note that in the limit, $A(t)$ and $B(t)$ are continuous random variables with density $(1 - G(u))/\tau$, even if the inter-event times may be discrete (but aperiodic). Thus the strict or weak inequalities in Equations 8.41 and 8.42 do not make any difference. In the limit the age and excess life are identically distributed! However, they are not independent, as shown in the next theorem.

Theorem 8.21   Joint Distribution of Age and Excess Life. *Suppose the hypotheses of Theorems 8.19 and 8.20 hold. Then*

$$
\lim_{t \to \infty} P(A(t) \geq y, B(t) > x) = \frac{1}{\tau} \int_{x+y}^\infty (1 - G(u)) du. \tag{8.43}
$$

Proof: For $t > y$ we have

$$\{A(t) \geq y, B(t) > x\} \Leftrightarrow \{\text{No renewals in } (t - y, t + x]\} \Leftrightarrow \{B(t - y) > x + y\}.$$

Hence

$$
\begin{aligned}
\lim_{t \to \infty} \mathsf{P}(A(t) \geq y, B(t) > x) &= \lim_{t \to \infty} \mathsf{P}(B(t - y) > x + y) \\
&= \lim_{t \to \infty} \mathsf{P}(B(t) > x + y) \\
&= \frac{1}{\tau} \int_{x+y}^{\infty} (1 - G(u)) du,
\end{aligned}
$$

which is Equation 8.43. ∎

The complementary cdf given in Equation 8.41 occurs frequently in the study of renewal processes. Hence we introduce the following notation:

$$G_e(x) = \frac{1}{\tau} \int_0^x (1 - G(u)) du. \tag{8.44}$$

G_e is a proper distribution of a continuous random variable on $[0, \infty)$, since $\int_0^\infty (1 - G(u)) du = \tau$. It is called the equilibrium distribution corresponding to G. We have seen this distribution in the study of the infinite server queues in Section 7.8.

Let X_e be a random variable with cdf G_e. One can show that

$$\mathsf{E}(X_e) = \int_0^\infty (1 - G_e(u)) du = \frac{\sigma^2 + \tau^2}{2\tau}. \tag{8.45}$$

Unfortunately, since convergence in distribution does not imply convergence of the means, we cannot conclude

$$\lim_{t \to \infty} \mathsf{E}(A(t)) = \lim_{t \to \infty} \mathsf{E}(B(t)) = \frac{\sigma^2 + \tau^2}{2\tau}$$

based on Theorems 8.19 and 8.20. We can, however, show the validity of the above limits directly by using KRT, as shown in the next theorem.

Theorem 8.22 Limiting Mean Excess Life. *Suppose the hypothesis of Theorem 8.19 holds, and $\sigma^2 < \infty$. Then*

$$\lim_{t \to \infty} \mathsf{E}(B(t)) = \frac{\sigma^2 + \tau^2}{2\tau}. \tag{8.46}$$

Proof: Let $H(t) = \mathsf{E}(B(t))$. Conditioning on S_1 we get

$$\mathsf{E}(B(t)|S_1 = u) = \begin{cases} H(t - u) & \text{if } u \leq t, \\ u - t & \text{if } u > t. \end{cases}$$

Hence we get

$$
\begin{aligned}
H(t) &= \int_0^\infty \mathsf{E}(B(t)|S_1 = u) dG(u) \\
&= D(t) + \int_0^t H(t - u) dG(u),
\end{aligned}
$$

where

$$D(t) = \int_t^\infty (u - t)dG(u).$$

Now D is monotone since

$$\frac{d}{dt}D(t) = -(1 - G(t)) \le 0.$$

Also, we can show that

$$\int_0^\infty D(u)du = (\sigma^2 + \tau^2)/2 < \infty.$$

Hence the KRT can be applied to get

$$\lim_{t\to\infty} \mathsf{E}(B(t)) = \frac{1}{\tau}\int_0^\infty D(u)du = \frac{\sigma^2 + \tau^2}{2\tau}.$$

This proves the theorem. ∎

We can derive the next two results in an analogous manner.

Theorem 8.23   Limiting Mean Age. *Suppose the hypothesis of Theorem 8.19 holds, and $\sigma^2 < \infty$. Then*

$$\lim_{t\to\infty} \mathsf{E}(A(t)) = \frac{\sigma^2 + \tau^2}{2\tau}. \tag{8.47}$$

Theorem 8.24   Limiting Mean Total Life. *Suppose the hypothesis of Theorem 8.19 holds, and $\sigma^2 < \infty$. Then*

$$\lim_{t\to\infty} \mathsf{E}(C(t)) = \frac{\sigma^2 + \tau^2}{\tau}. \tag{8.48}$$

Equation 8.48 requires some fine tuning of our intuition. Since $C(t) = X_{N(t)+1}$, the above theorem implies that

$$\lim_{t\to\infty} \mathsf{E}(X_{N(t)+1}) = \frac{\sigma^2 + \tau^2}{\tau} \ge \tau = \mathsf{E}(X_n).$$

Thus, for large t, the inter-renewal time covering t, namely, $X_{N(t)+1}$, is longer in mean than a generic inter-renewal time, say X_n. This counter-intuitive fact is called the inspection paradox or length-biased sampling. One way to rationalize this "paradox" is to think of picking a t uniformly over a very long interval $[0, T]$ over which we have observed an RP. Then it seems plausible the probability that our randomly picked t will lie in a given inter-renewal interval is directly proportional to the length of that interval. Hence such a random t is more likely to fall in a longer interval. This fact is quantified by Equation 8.48.

8.7 Delayed Renewal Processes

In this section we study a seemingly trivial but a useful generalization of an RP. We begin with a definition.

Definition 8.5 Delayed Renewal Process. *A counting process* $\{N(t), t \geq 0\}$ *generated by* $\{X_n, n \geq 1\}$ *as defined by Equation 8.1 is called a delayed RP if* X_1 *has cdf* F, $\{X_n, n \geq 2\}$ *is a sequence of iid random variables with common cdf* G, *and they are independent of* X_1.

To distinguish an RP from a delayed RP we use the notation $\{N^D(t), t \geq 0\}$ to denote a delayed RP. Clearly, if $F = G$, the delayed renewal process reduces to the standard renewal process. Also, the shifted delayed RP $\{N^D(t + X_1) - 1, t \geq 0\}$ is a standard RP. We illustrate with examples where we encounter delayed RPs.

Example 8.17 Let $\{N(t), t \geq 0\}$ be a standard RP. For a fixed $s > 0$ define $N_s(t) = N(s + t) - N(s)$. Then $\{N_s(t), t \geq 0\}$ is a delayed RP. Here F is the distribution of $B(s)$, the excess life at time s in the original RP. ∎

Example 8.18 Delayed RPs in CTMCs. Let $\{X(t), t \geq 0\}$ be an irreducible and recurrent CTMC on state-space $\{0, 1, 2, \cdots\}$. Let $N_j(t)$ be the number of entries into state j up to t. Following the argument in Example 8.3 we see that, if $X(0) = j$, $\{N_j(t), t \geq 0\}$ is an RP. If $X(0) \neq j$, it is a delayed renewal process, since the time of the first entry into state j has a different distribution than the subsequent inter-visit times. ∎

Example 8.19 Delayed RPs in an $M/G/1$ **Queue.** Let $X(t)$ be the number of customers at time t in an $M/G/1$ system. Let $N(t)$ be the number of busy cycles completed by time t. We saw in Example 8.5 that $\{N(t), t \geq 0\}$ is an RP if the system starts empty, that is, if $X(0) = 0$. For any other initial distribution, $\{N(t), t \geq 0\}$ is a delayed RP. ∎

In our study of the delayed renewal process we shall assume that

$$F(0-) = 0, G(0-) = 0, G(0) < 1, \quad F(\infty) = G(\infty) = 1. \qquad (8.49)$$

The delayed renewal process inherits almost all the properties of the standard renewal process. We leave many of them to the reader to prove.

First we study the renewal function for a delayed RP defined as

$$M^D(t) = \mathsf{E}(N^D(t)), \quad t \geq 0.$$

Let $M(t)$ be the renewal function for the standard RP. The next theorem gives the main result about $M^D(t)$.

Theorem 8.25 The Renewal Function for the Delayed RP. *The renewal function* $\{M^D(t), t \geq 0\}$ *of a renewal process* $\{N^D(t), t \geq 0\}$ *satisfies the integral equation*

$$M^D(t) = F(t) + \int_0^t M(t - u)dF(u), \quad t \geq 0. \qquad (8.50)$$

Its LST is given by

$$\tilde{M}^D(s) = \frac{\tilde{F}(s)}{1 - \tilde{G}(s)}. \tag{8.51}$$

Proof: Conditioning on X_1 we get

$$E(N^D(t)|S_1 = u) = \begin{cases} 0 & \text{if } u > t, \\ 1 + M(t-u) & \text{if } u \le t. \end{cases}$$

Hence we get

$$\begin{aligned} M^D(t) &= \int_0^\infty E(N^D(t)|S_1 = u)dF(u) \\ &= \int_0^t (1 + M(t-u))dF(u) \\ &= F(t) + \int_0^t M(t-u)dF(u). \end{aligned}$$

This gives Equation 8.50. Taking LSTs on both sides of it we get,

$$\tilde{M}^D(s) = \tilde{F}(s) + \tilde{M}(s)\tilde{F}(s).$$

Substituting for $\tilde{M}(s)$ from Equation 8.16, we get Equation 8.51. ∎

The renewal function $\{M^D(t), t \ge 0\}$ also inherits almost all the properties of the renewal function of a standard RP, except that it does not uniquely characterize the delayed RP.

When we use renewal argument in a delayed RP, we generally get an equation of the following type:

$$H^D(t) = C(t) + \int_0^t H(t-x)dF(x), \quad t \ge 0. \tag{8.52}$$

This is not a renewal-type equation, since H^D appears on the left, but not on the right. The next theorem gives a result about the limiting behavior of the function H^D satisfying the above equation.

Theorem 8.26 *Let $C(t)$ and $H(t)$ be bounded functions with finite limits as $t \to \infty$. Let H^D be given as in Equation 8.52, where F is a cdf of a non-negative non-defective random variable. Then*

$$\lim_{t\to\infty} H^D(t) = \lim_{t\to\infty} C(t) + \lim_{t\to\infty} H(t). \tag{8.53}$$

Proof: Since $H(\cdot)$ is a bounded function with a finite limit h (say), we see that the functions $H_t(\cdot)$ defined by

$$H_t(x) = \begin{cases} H(t-x) & \text{if } 0 \le x \le t \\ 0 & \text{if } x > t \end{cases}$$

form a sequence of bounded functions with

$$\lim_{t \to \infty} H_t(x) = h, \quad x \geq 0.$$

Hence by bounded convergence theorem

$$\lim_{t \to \infty} \int_0^t H(t-x)dF(x) = \lim_{t \to \infty} \int_0^\infty H_t(x)dF(x) = \int_0^\infty h\,dF(x) = h.$$

The theorem follows from this. ∎

As an application of the above theorem we study the asymptotic behavior of $M^D(t)$ in the following example.

Example 8.20 Asymptotic Behavior of $M^D(t)$. Let $M^D(t)$ be the renewal function of a delayed RP. Let

$$\tau_F = \mathsf{E}(X_1), \quad \sigma_F^2 = \mathrm{Var}(X_1) < \infty,$$

$$\tau = \mathsf{E}(X_n), \quad \sigma^2 = \mathrm{Var}(X_n) < \infty, \quad n \geq 2.$$

Show that

$$M^D(t) = \frac{t}{\tau} + \frac{\sigma^2 + \tau^2 - 2\tau\tau_F}{2\tau^2} + o(1), \tag{8.54}$$

where $o(1)$ is a function that approaches 0 as $t \to \infty$.

We first derive an integral equation for

$$H^D(t) = M^D(t) - t/\tau.$$

Using renewal argument we get

$$\mathsf{E}(N^D(t) - t/\tau | S_1 = u) = \begin{cases} -t/\tau & \text{if } u > t, \\ 1 + M(t-u) - t/\tau & \text{if } u \leq t. \end{cases}$$

Hence, after rearrangements, we get

$$H^D(t) = C(t) + \int_0^t H(t-u)dF(u),$$

where

$$H(t) = M(t) - t/\tau,$$

and

$$C(t) = \int_0^t (1 - u/\tau)dF(u) - \int_t^\infty (t/\tau)dF(u).$$

Now, from Example 8.16 we have

$$\lim_{t \to \infty} H(t) = \frac{\sigma^2 - \tau^2}{2\tau^2},$$

and

$$\lim_{t \to \infty} C(t) = 1 - \tau_F/\tau.$$

We can now apply Theorem 8.26 to get

$$\lim_{t \to \infty} H^D(t) = \frac{\sigma^2 - \tau^2}{2\tau^2} + 1 - \frac{\tau_F}{\tau}.$$

This implies Equation 8.54. ∎

Next we study a special case of a delayed RP called the equilibrium renewal process as defined below.

Definition 8.6 Equilibrium Renewal Process. *A delayed RP is called an equilibrium RP if the cdf F is related to G as follows:*

$$F(x) = G_e(x) = \frac{1}{\tau} \int_0^x (1 - G(u))du, \quad x \geq 0.$$

An equilibrium RP is denoted by $\{N^e(t), t \geq 0\}$. Recall that G_e is the limiting distribution of the excess life in a standard renewal process; see Theorem 8.19. Thus $\{N^e(t), t \geq 0\}$ can be thought of as the limit of the shifted RP $\{N_s(t), t \geq 0\}$ of Example 8.17 as $s \to \infty$. We prove two important properties of the equilibrium RP below.

Theorem 8.27 Renewal Function of an Equilibrium RP. *Let $\{N^e(t), t \geq 0\}$ be an equilibrium RP as in Definition 8.6. Then*

$$M^e(t) = \mathsf{E}(N^e(t)) = \frac{t}{\tau}, \quad t \geq 0. \tag{8.55}$$

Proof: From the definition of an equilibrium RP, we get

$$\tilde{F}(s) = \tilde{G}_e(s) = \frac{1 - \tilde{G}(s)}{\tau s}.$$

Substituting in Equation 8.51, we get

$$\tilde{M}^e(s) = \frac{\tilde{F}(s)}{1 - \tilde{G}(s)} = \frac{1}{\tau s} \frac{1 - \tilde{G}(s)}{1 - \tilde{G}(s)} = \frac{1}{\tau s}.$$

Inverting this we get Equation 8.55. ∎

Note that $M^e(t)$ is a linear function of t, but $\{N^e(t), t \geq 0\}$ is not a PP. Does this contradict our statement in Example 8.11 that linear renewal function implies that the RP is a PP? Not at all, since $\{N^e(t), t \geq 0\}$ is a delayed RP, and not an RP. If it was an RP, i.e., if $G_e = G$, then it would be a PP. This indeed is the case, since it is possible to show that $G = G_e$ if and only if G is exponential!

Theorem 8.28 Excess Life in an Equilibrium RP. *Let $B^e(t)$ be the excess life at time t in an equilibrium RP as in Definition 8.6. Then,*

$$\mathsf{P}(B^e(t) \leq x) = G_e(x), \quad t \geq 0, \ x \geq 0.$$

Proof: Let $H^e(t) = P(B^e(t) > x)$ and $H(t) = P(B(t) > x)$ where $B(t)$ is the excess life in an RP with common inter-event time cdf G. Using G_e as the cdf of S_1 and following the proof of Theorem 8.19 we get

$$H^e(t) = 1 - G_e(x + t) + \int_0^t H(t - u)dG_e(u).$$

Taking the LST on both sides, we get

$$\tilde{H}^e(s) = 1 - c(s) + \tilde{H}(s)\tilde{G}_e(s), \tag{8.56}$$

where

$$c(s) = \int_0^\infty e^{-st}dG_e(x + t).$$

On the other hand, from Equation 8.40, we have

$$H(t) = 1 - G(x + t) + \int_0^t H(t - u)dG(u).$$

Taking LST of the above equation we get

$$\tilde{H}(s) = \frac{1 - b(s)}{1 - \tilde{G}(s)}, \tag{8.57}$$

where

$$b(s) = \int_0^\infty e^{-st}dG(x + t).$$

Taking into account the jump of size $G_e(x)$ at $t = 0$ in the function $G_e(x + t)$, we can show that

$$c(s) = G_e(x) + \frac{1 - b(s)}{s\tau}. \tag{8.58}$$

Substituting Equation 8.58 and 8.57 in 8.56, and simplifying, we get

$$\tilde{H}^e(s) = 1 - G_e(x).$$

Hence we must have

$$H^e(t) = 1 - G_e(x),$$

for all $t \geq 0$. This proves the theorem. ∎

Thus the distribution of the excess life in an equilibrium RP does not change with time. This is another manifestation of the "equilibrium"! We leave it to the reader to derive the next result from this.

Theorem 8.29 $\{N^e(t), t \geq 0\}$ *has stationary increments.*

An interesting application of an equilibrium RP to $G/M/\infty$ queue is described in Computational Exercise 8.52.

8.8 Alternating Renewal Processes

Consider a stochastic process $\{X(t), t \geq 0\}$ with state-space $\{0, 1\}$. Suppose the process starts in state 1 (also called the "up" state). It stays in that state for U_1 amount of time and then jumps to state 0 (also called the "down" state). It stays in state 0 for D_1 amount of time and then goes back to state 1. This process repeats forever, with U_n being the nth up-time, and D_n the nth down-time. The nth up-time followed by the nth down-time is called the nth cycle. A sample path of such a process is shown in Figure 8.7.

Figure 8.7 *A typical sample path of an alternating renewal process.*

Definition 8.7   Alternating Renewal Process. *The stochastic process $\{X(t), t \geq 0\}$ as described above is called an alternating renewal process if $\{(U_n, D_n), n \geq 1\}$ is a sequence of iid non-negative bivariate random variables.*

We abbreviate "alternating renewal process" as ARP. Note that an ARP does not count any events even though it has the term "renewal process" in its name. The next theorem gives the main result about the ARPs.

Theorem 8.30   Distribution of ARP. *Suppose the ARP $\{X(t), t \geq 0\}$ starts in state 1 at time zero. Then $H(t) = P(X(t) = 1)$ satisfies the renewal-type equation*

$$H(t) = P(U_1 > t) + \int_0^t H(t - u)dG(u), \qquad (8.59)$$

where $G(x) = P(U_1 + D_1 \leq x)$. If G is aperiodic and $E(U_1 + D_1) < \infty$,

$$\lim_{t \to \infty} P(X(t) = 1) = \frac{E(U_1)}{E(U_1) + E(D_1)}. \qquad (8.60)$$

If G is periodic with period d, then the above limit holds if $t = nd$ and $n \to \infty$.

Proof: Let S_n be the nth time the ARP enters state 1, i.e.,

$$S_n = \sum_{i=1}^n (U_i + D_i), \quad n \geq 1.$$

Then $\{S_n, n \geq 1\}$ is a renewal sequence. Since the ARP renews at time $S_1 = U_1 + D_1$, we get

$$P(X(t) = 1 | S_1 = u) = \begin{cases} H(t-u) & \text{if } u \leq t, \\ P(U_1 > t | S_1 = u) & \text{if } u > t. \end{cases}$$

Using $G(u) = P(S_1 \leq u)$, we get

$$\begin{aligned} H(t) &= \int_0^\infty P(X(t) = 1 | S_1 = u) dG(u) \\ &= \int_0^t H(t-u) dG(u) + \int_t^\infty P(U_1 > t | S_1 = u) dG(u). \quad (8.61) \end{aligned}$$

We use the same trick as in Example 8.13: since $P(U_1 > t | S_1 = u) = 0$ if $u \leq t$, we get

$$\int_t^\infty P(U_1 > t | S_1 = u) dG(u) = \int_0^\infty P(U_1 > t | S_1 = u) dG(u) = P(U_1 > t).$$

Substituting in Equation 8.61 we get Equation 8.59. Now, $P(U_1 > t)$ is bounded and monotone, and

$$\int_0^\infty P(U_1 > u) du = E(U_1) < \infty.$$

Hence we can use KRT to get, assuming S_1 is aperiodic,

$$\lim_{t \to \infty} H(t) = \lim_{t \to \infty} P(X(t) = 1) = \frac{E(U_1)}{E(S_1)} = \frac{E(U_1)}{E(U_1) + E(D_1)}.$$

This proves the theorem. The periodic case follows similarly. ∎

The above theorem is intuitively obvious if one interprets the limiting probability that $X(t) = 1$ as the fraction of the time the ARP is up. Thus, if the successive up-times are 30 minutes long on the average, and the down-times are 10 minutes long on the average, then it is reasonable to expect that the ARP will be up 75% of the time in the long run. What is non-intuitive about the theorem is that it is valid even if the U_n and D_n are dependent random variables. All we need to assume is that successive cycles of the ARP are independent. This fact makes the ARP a powerful tool. We illustrate its power with several examples.

Example 8.21   Two-State Machine. The CTMC $\{X(t), t \geq 0\}$ developed in Example 6.4 on page 207 to describe a two-state machine is an ARP with $U_n \sim \exp(\mu)$ and $D_n \sim \exp(\lambda)$. In this example U_n and D_n are also independent. Since $S_1 = U_1 + D_1$ is aperiodic, we can use Equation 8.60 to get

$$\lim_{t \to \infty} P(X(t) = 1) = \frac{E(U_1)}{E(U_1) + E(D_1)} = \frac{\lambda}{\lambda + \mu},$$

and hence

$$\lim_{t \to \infty} P(X(t) = 0) = \frac{E(D_1)}{E(U_1) + E(D_1)} = \frac{\mu}{\lambda + \mu}.$$

This matches with the limiting distribution of $\{X(t), t \geq 0\}$ computed in Example 6.34 on page 258. ∎

Example 8.22   Excess Life. In this example we use ARPs to derive the limiting distribution of the excess life in a standard RP given in Equation 8.41.

For a fixed $x > 0$, define

$$X(t) = \begin{cases} 1 & \text{if } B(t) > x, \\ 0 & \text{if } B(t) \leq x. \end{cases}$$

Then $\{X(t), t \geq 0\}$ is a stochastic process with state-space $\{0, 1\}$ (see Figure 8.8) with

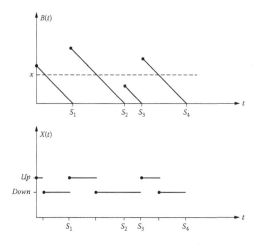

Figure 8.8 *The excess life process and the induced ARP.*

$$\begin{aligned} U_n &= \max(X_n - x, 0), \\ D_n &= \min(X_n, x), \\ U_n + D_n &= X_n. \end{aligned}$$

Note that U_n and D_n are dependent, but $\{(U_n, D_n), n \geq 0\}$ is a sequence of iid bivariate random variables. Hence $\{X(t), t \geq 0\}$ is an ARP. Then, assuming X_n is aperiodic, we can use Theorem 8.30 to get

$$\begin{aligned} \lim_{t \to \infty} P(B(t) > x) &= \lim_{t \to \infty} P(X(t) = 1) \\ &= \frac{E(U_1)}{E(U_1) + E(D_1)} \\ &= \frac{E(\max(X_1 - x, 0))}{E(X_1)} \\ &= \frac{1}{\tau} \int_x^\infty (1 - G(u)) du \end{aligned}$$

which is Equation 8.41, as desired. ■

Example 8.23 $M/G/1/1$ **Queue.** Consider an $M/G/1/1$ queue with arrival rate λ and iid service times with mean τ. Let $X(t)$ be the number of customers in the system at time t. Compute the limiting distribution of $X(t)$ as $t \to \infty$.

The stochastic process $\{X(t), t \geq 0\}$ has state-space $\{0, 1\}$, where the downtimes $\{D_n, n \geq 1\}$ are iid $\exp(\lambda)$ and the up-times $\{U_n, n \geq 1\}$ are iid service times. Hence $\{X(t), t \geq 0\}$ is an aperiodic ARP (i.e., $S_1 = U_1 + D_1$ is aperiodic). Hence

$$
\begin{aligned}
p_1 &= \lim_{t\to\infty} \mathsf{P}(X(t) = 1) \\
&= \frac{\mathsf{E}(U_1)}{\mathsf{E}(U_1) + \mathsf{E}(D_1)} \\
&= \frac{\tau}{\tau + 1/\lambda} \\
&= \frac{\rho}{1 + \rho},
\end{aligned}
$$

where $\rho = \lambda\tau$. Hence

$$
p_0 = \lim_{t\to\infty} \mathsf{P}(X(t) = 0) = \frac{1}{1 + \rho}. \quad \blacksquare
$$

Example 8.24 $G/M/1/1$ **Queue.** Consider a $G/M/1/1$ queue with iid inter-arrival times with common cdf G and mean $1/\lambda$, and iid $\exp(\mu)$ service times. Let $X(t)$ be the number of customers in the system at time t. Compute the limiting distribution of $X(t)$ as $t \to \infty$.

The stochastic process $\{X(t), t \geq 0\}$ has state-space $\{0, 1\}$. The up-times $\{U_n, n \geq 1\}$ are iid $\exp(\mu)$ and the cycle lengths $\{U_n + D_n, n \geq 1\}$ form a sequence of iid random variables, although U_n and D_n are dependent. Hence $\{X(t), t \geq 0\}$ is an ARP. Now, let $N(t)$ be the number of arrivals (who may or may not enter) up to time t and A_n be the time of the nth arrival. Then

$$
U_1 + D_1 = A_{N(U_1)+1}.
$$

Hence, from Theorem 8.13, we get

$$
\mathsf{E}(U_1 + D_1 | U_1 = t) = \frac{1}{\lambda}(1 + M(t)),
$$

where $1/\lambda$ is the mean inter-arrival time, and $M(t) = \mathsf{E}(N(t))$. Hence

$$
\begin{aligned}
\mathsf{E}(U_1 + D_1) &= \int_0^\infty \frac{1}{\lambda}(1 + M(t))\mu e^{-\mu t}\, dt \\
&= \frac{1}{\lambda}\left(1 + \int_0^\infty M(t)\mu e^{-\mu t}\, dt\right) \\
&= \frac{1}{\lambda}\left(1 + \int_0^\infty e^{-\mu t}\, dM(t)\right)
\end{aligned}
$$

$$= \frac{1}{\lambda}\left(1 + \tilde{M}(\mu)\right)$$

$$= \frac{1}{\lambda}\left(1 + \frac{\tilde{G}(\mu)}{1 - \tilde{G}(\mu)}\right)$$

$$= \frac{1}{\lambda(1 - \tilde{G}(\mu))}.$$

Assuming G is aperiodic, we get

$$p_1 = \lim_{t \to \infty} \mathsf{P}(X(t) = 1) = \frac{\mathsf{E}(U_1)}{\mathsf{E}(U_1 + D_1)} = \rho(1 - \tilde{G}(\mu)),$$

where $\rho = \lambda/\mu$. Hence

$$p_0 = \lim_{t \to \infty} \mathsf{P}(X(t) = 0) = 1 - \rho(1 - \tilde{G}(\mu)). \quad \blacksquare$$

Next we define a delayed ARP, along the same lines as in the definition of the delayed RP.

Definition 8.8 Delayed Alternating Renewal Process. *The stochastic process* $\{X(t), t \geq 0\}$ *with state-space* $\{0, 1\}$ *with* U_n *being the nth up-time, and* D_n *the nth down-time, is called a delayed alternating renewal process if* $\{(U_n, D_n), n \geq 2\}$ *is a sequence of iid non-negative bivariate random variables, and is independent of* (U_1, D_1).

Thus a delayed ARP behaves like a standard ARP from the second cycle onward. The main result about the delayed ARP is given in the next theorem.

Theorem 8.31 Distribution of Delayed ARP. *Suppose* $\{X(t), t \geq 0\}$ *is a delayed ARP that enters state 1 at time* $S_1 = U_1 + D_1$. *If* $\mathsf{P}(S_1 < \infty) = 1$ *and* $U_2 + D_2$ *is aperiodic with* $\mathsf{E}(U_2 + D_2) < \infty$,

$$\lim_{t \to \infty} \mathsf{P}(X(t) = 1) = \frac{\mathsf{E}(U_2)}{\mathsf{E}(U_2) + \mathsf{E}(D_2)}. \quad (8.62)$$

If $U_1 + D_1$ *and* $U_2 + D_2$ *both have a common period d, then the above limit holds if* $t = nd$ *and* $n \to \infty$.

Proof: Let F be the cdf of S_1. Conditioning on S_1 we get

$$\mathsf{P}(X(t) = 1) = \int_0^t H(t - u)dF(u) + \mathsf{P}(U_1 > t),$$

where $H(t)$ is as defined in Theorem 8.30 for a standard ARP with sojourn times $\{(U_n, D_n), n \geq 2\}$. Then from the proof of Theorem 8.26, we get

$$\lim_{t \to \infty} \mathsf{P}(X(t) = 1) = \lim_{t \to \infty} H(t) + \lim_{t \to \infty} \mathsf{P}(U_1 > t) = \frac{\mathsf{E}(U_2)}{\mathsf{E}(U_2) + \mathsf{E}(D_2)} + \lim_{t \to \infty} \mathsf{P}(U_1 > t).$$

Here we used the facts that $F(\infty) = 1$, $H(t) \leq 1$, and $U_2 + D_2$ is aperiodic. $\quad \blacksquare$

This theorem says that the limiting behavior of a delayed ARP is not affected by its behavior over the first cycle, as long as it terminates with probability 1. Finally, one can prove the validity of Theorems 8.30 and 8.31 even if the sojourn in states 1 and 0 during one cycle are not contiguous intervals.

8.9 Semi-Markov Processes

In this section we study a class of stochastic processes obtained by relaxing the exponential sojourn time assumption in the CTMCs. Specifically, we consider a stochastic process $\{X(t), t \geq 0\}$ of Section 6.1 with a countable state-space S. It starts in the initial state X_0 at time $t = 0$. It stays there for a sojourn time Y_1 and then jumps to state X_1. In general it stays in state X_n for a duration given by Y_{n+1} and then jumps to state X_{n+1}, $n \geq 0$. Let $N(t)$ be the number of jumps up to time t. We assume that the condition in Equation 6.1 on page 201 holds. Then $\{N(t), t \geq 0\}$ is well defined, and the sequence $\{X_0, (X_n, Y_n), n \geq 1\}$ can be used to define the process $\{X(t), t \geq 0\}$ by

$$X(t) = X_{N(t)}, \quad t \geq 0. \tag{8.63}$$

In this section we study the case where $\{X(t), t \geq 0\}$ belongs to a particular class of stochastic processes called the semi-Markov processes (SMP), as defined below.

Definition 8.9   Semi-Markov Process. *The stochastic process $\{X(t), t \geq 0\}$ as defined by Equation 8.63 is called an SMP if it has a countable state-space S, and the sequence $\{X_0, (X_n, Y_n), n \geq 1\}$ satisfies*

$$P(X_{n+1} = j, Y_{n+1} \leq y | X_n = i, Y_n, X_{n-1}, Y_{n-1}, \cdots, X_1, Y_1, X_0)$$
$$= P(X_1 = j, Y_1 \leq y | X_0 = i) = G_{i,j}(y), \quad i, j \in S, \ n \geq 0. \tag{8.64}$$

Thus the semi-Markov process has Markov property at every jump epoch, hence the name "semi"-Markov. In comparison, a CTMC has Markov property at every time t. Note that unlike in the CTMCs, we allow $P(X_{n+1} = i | X_n = i) = G_{i,i}(\infty) > 0$. Also, unlike in the CTMCs, the variables X_{n+1} and Y_{n+1} in an SMP are allowed to depend on each other.

Define the kernel of an SMP as the matrix

$$G(y) = [G_{i,j}(y)]_{i,j \in S}, \quad y \geq 0.$$

An SMP is completely described by giving its kernel and the initial distribution

$$a_i = P(X(0) = i), \quad i \in S,$$

if we further assume that the SMP enters state i with probability a_i at time 0.

Clearly $\{X_n, n \geq 0\}$ is a DTMC (called the embedded DTMC in the SMP) with transition probabilities

$$p_{i,j} = G_{i,j}(\infty) = P(X_{n+1} = j | X_n = i), \quad i, j \in S.$$

If there is a state $i \in S$ for which $G_{i,j}(y) = 0$ for all $j \in S$ and all $y \geq 0$, then state i must be absorbing. Hence we set $p_{i,i} = 1$ in such a case. With this convention, we see that

$$\sum_{j \in S} p_{i,j} = 1, \quad i \in S.$$

Next, let

$$G_i(y) = \sum_{j \in S} G_{i,j}(y) = \mathsf{P}(S_1 \leq y | X_0 = i), \quad i \in S, y \geq 0.$$

Thus G_i is the cdf of the sojourn time in state i. We illustrate with several examples.

Example 8.25   Alternating Renewal Process. An ARP $\{X(t), t \geq 0\}$ is an SMP if the sojourn time U_n is independent of D_n. In this case the kernel is given by

$$G(y) = \begin{bmatrix} 0 & \mathsf{P}(D_n \leq y) \\ \mathsf{P}(U_n \leq y) & 0 \end{bmatrix}. \quad \blacksquare$$

Example 8.26   CTMC as an SMP. A CTMC $\{X(t), t \geq 0\}$ with generator matrix Q is an SMP with kernel $G(y) = [G_{i,j}(y)]$ with $G_{i,i}(y) = 0$ and

$$G_{i,j}(y) = (1 - e^{-q_i y})p_{i,j}, \quad i \neq j,$$

where $q_i = -q_{i,i}$ and $p_{i,j} = q_{i,j}/q_i$ for $i \neq j$. If $q_i = 0$ we define $p_{i,i} = 1$ and $G_{i,j}(y) = 0$, $y \geq 0$. Thus in a CTMC the sojourn time in the current state and the state after the jump are independent. $\blacksquare$

Example 8.27   Series System. Consider a system consisting of N components in series, i.e., it needs all components in working order in order to function properly. The life time of the ith component is $\exp(\lambda_i)$ random variable. At time 0 all the components are up. As soon as any of the components fails, the system fails, and the repair starts immediately. The repair time of the ith component is a non-negative random variable with cdf $H_i(\cdot)$. Once repaired, a component is as good as new. When the system is down, no more failures occur. Let $X(t) = 0$ if the system is functioning at time t, and $X(t) = i$ if component i is down (and hence under repair) at time t. We shall show that $\{X(t), t \geq 0\}$ is an SMP and compute its kernel.

The state-space is $\{0, 1, 2, \cdots, N\}$. Suppose the system is in state 0 at time t. Then all components are up and the system state changes to $i \in \{1, 2, \cdots, N\}$ if the first item to fail is item i. Thus the sojourn time in state 0 is $\exp(\lambda)$ where

$$\lambda = \sum_{i=1}^{N} \lambda_i,$$

and the next state is i with probability λ_i/λ. Once the system enters state $i \in \{1, 2, \cdots, N\}$, repair starts on component i. Thus the sojourn time in this state has cdf H_i, at the end of which the system jumps to state 0. Combining all these observations we see that $\{X(t), t \geq 0\}$ is an SMP with kernel

$$
G(y) = \begin{bmatrix}
0 & \frac{\lambda_1}{\lambda}(1 - e^{-\lambda y}) & \frac{\lambda_2}{\lambda}(1 - e^{-\lambda y}) & \cdots & \frac{\lambda_N}{\lambda}(1 - e^{-\lambda y}) \\
H_1(y) & 0 & 0 & \cdots & 0 \\
H_2(y) & 0 & 0 & \cdots & 0 \\
\vdots & \vdots & \vdots & \ddots & \vdots \\
H_N(y) & 0 & 0 & \cdots & 0
\end{bmatrix}.
$$

We had seen this system in Modeling Exercise 6.9 on page 276 where the repair-times were assumed to be exponential random variables. Thus the CTMC there is a special case of the SMP developed here. ∎

Armed with these examples of SMPs we now study the limiting behavior of the SMPs. We need the following preliminaries. Define

$$T_j = \min\{t \geq Y_1 : X(t) = j\}, \quad j \in S,$$

where Y_1 is the first sojourn time. Thus if the SMP starts in state j, T_j is the first time it returns to state j (after leaving it at time Y_1.) If it starts in a state $i \neq j$, then T_j is the first time it enters state j. Now let

$$\tau_i = E(Y_1 | X(0) = i), \quad i \in S,$$

and

$$\tau_{i,j} = E(T_j | X(0) = i), \quad i, j \in S.$$

The next theorem shows how to compute the $\tau_{i,j}$'s. It also extends the concept of first-step analysis to SMPs.

Theorem 8.32   First Passage Times in SMPs. *The mean first passage times $\{\tau_{i,j}\}$ satisfy the following equations:*

$$\tau_{i,j} = \tau_i + \sum_{k \neq j} p_{i,k} \tau_{k,j}. \tag{8.65}$$

Proof: Follows along the same line as the derivation of Equation 6.55 on page 248, using τ_i in place of $1/q_i$ as $E(Y_1 | X(0) = i)$. ∎

Theorem 8.33   *Suppose the embedded DTMC $\{X_n, n \geq 0\}$ in an SMP $\{X(t), t \geq 0\}$ has transition probability matrix $P = [p_{i,j}]$ that is irreducible and recurrent. Let π be a positive solution to*

$$\pi = \pi P.$$

Then the mean return time to state j is given by

$$\tau_{j,j} = \frac{\sum_{i \in S} \pi_i \tau_i}{\pi_j}, \quad j \in S. \tag{8.66}$$

Proof: Follows along the same line as the derivation of Equation 6.56 on page 249, and using τ_i in place of $1/q_i$ as $\mathsf{E}(Y_1|X(0) = i)$. ∎

Next we introduce the concept of irreducibility and periodicity for SMPs.

Definition 8.10 *An SMP $\{X(t), t \geq 0\}$ is said to be irreducible (and recurrent) if its embedded DTMC is irreducible (and recurrent).*

It follows from Theorem 8.33 that if the mean return time to any state j in an irreducible recurrent SMP is finite, it is finite for all states in the SMP. Hence we can make the following definition.

Definition 8.11 *An irreducible SMP $\{X(t), t \geq 0\}$ is said to be positive recurrent if the mean return time to state j is finite for any $j \in S$.*

It follows from Theorem 8.33 that a necessary and sufficient condition for positive recurrence of an irreducible recurrent SMP is

$$\sum_{i \in S} \pi_i \tau_i < \infty,$$

where π_i and τ_i are as in Theorem 8.33. Note that if the first return time is aperiodic (periodic with period d) for any one state, it is aperiodic (periodic with period d) for all states in an irreducible and recurrent SMP. This motivates the next definition.

Definition 8.12 *An irreducible and recurrent SMP is called aperiodic if the first passage time T_i, starting with $X_0 = i$, is an aperiodic random variable for any state i. If it is periodic with period d, the SMP is said to be periodic with period d.*

With these preliminaries we can now state the main result about the limiting behavior of SMPs in the next theorem. Intuitively, we consider a (possibly delayed) ARP associated with the SMP such that the ARP is in state 1 whenever the SMP is in state j, and zero otherwise. The length of a cycle in this ARP is the time between two consecutive visits by the SMP to state j, hence the expected length of the cycle is $\tau_{j,j}$. During this cycle the SMP spends an expected time τ_j in state j. Hence the long run fraction of the time spent in state j is given by $\tau_j/\tau_{j,j}$, which gives the limiting distribution of the SMP. Since the behavior in the first cycle does not matter (as long as it is finite with probability 1), the same result holds no matter what state the SMP starts in, as long as the second and the subsequent cycles are defined to start with an entry into state j.

Theorem 8.34 Limiting Behavior of SMPs. *Let $\{X(t), t \geq 0\}$ be an irreducible, positive recurrent and aperiodic SMP with kernel G. Let π be a positive solution to*

$$\pi = \pi G(\infty).$$

Then $\{X(t), t \geq 0\}$ has a limiting distribution $[p_j, j \in S]$, and it is given by

$$p_j = \lim_{t \to \infty} \mathsf{P}(X(t) = j|X(0) = i) = \frac{\pi_j \tau_j}{\sum_{k \in S} \pi_k \tau_k}, \quad j \in S. \qquad (8.67)$$

If the SMP is periodic with period d, then the above limit holds if $t = nd$ and $n \to \infty$.

Proof: Fix a $j \in S$ and define

$$Z(t) = \begin{cases} 1 & \text{if } X(t) = j, \\ 0 & \text{if } X(t) \neq j. \end{cases}$$

Let S_n be the nth entry into state j. The probabilistic structure of the SMP $\{X(t), t \geq 0\}$ implies that $\{Z(t), t \geq 0\}$ is an ARP if $X(0) = j$, else it is a delayed ARP. In any case consider the cycle $[S_1, S_2)$. We have $X(S_1) = j$. The sojourn time U_2 of the delayed ARP in state 1 has mean τ_j. Let D_2 be the time spent by the delayed ARP in state 0 in the second cycle. The second cycle ends as soon as the SMP re-enters state j. Hence we have

$$E(U_2 + D_2) = E(S_2 - S_1) = E(T_j | X(0) = j) = \tau_{j,j}.$$

Suppose $i \in S$ is such that T_i is aperiodic. Since the embedded chain is irreducible, the SMP has a positive probability of visiting every state $k \in S$ during $(0, T_j)$ starting from state j. This implies that every T_j must be aperiodic. Also, since $G_i(\infty) = 1$ for all i, and the DTMC is recurrent, $T_j < \infty$ with probability 1 starting from any state $X(0) = i$. Thus the hypothesis of Theorem 8.31 is satisfied. Hence

$$\lim_{t \to \infty} P(X(t) = j | X(0) = i) = \lim_{t \to \infty} P(Z(t) = 1)$$

$$= \frac{E(U_2)}{E(U_2) + E(D_2)} = \frac{\tau_j}{\tau_{j,j}} = \frac{\pi_j \tau_j}{\sum_{k \in S} \pi_k \tau_k},$$

where the last equality follows from Equation 8.66. This proves the theorem. ∎

Equation 8.67 implies that the limiting distribution of the SMP depends on the sojourn time distributions only through their means! This insensitivity of the limiting distribution is very interesting and useful. It has generated a lot of literature investigating similar insensitivity results in other contexts.

We illustrate with examples.

Example 8.28   Alternating Renewal Process. Let $\{X(t), t \geq 0\}$ be the ARP of Example 8.25 with independent U_n and D_n. The embedded DTMC has the P matrix given by

$$P = G(\infty) = \begin{bmatrix} 0 & 1 \\ 1 & 0 \end{bmatrix}.$$

This is an irreducible recurrent matrix, and $\pi = \begin{bmatrix} 1 & 1 \end{bmatrix}$ satisfies $\pi = \pi P$. Hence Equation 8.67 yields

$$p_0 = \frac{E(D_1)}{E(U_1) + E(D_1)}, \quad p_1 = \frac{E(U_1)}{E(U_1) + E(D_1)}.$$

This matches with Equation 8.60, although that equation was derived without assuming independence of U_n and D_n. ∎

Example 8.29   CTMC as SMP. Let $\{X(t), t \geq 0\}$ be an irreducible positive recurrent CTMC on state-space S with generator matrix Q. Then it has a limiting distribution $p = [p_j, j \in S]$ that is given by the unique solution to

$$pQ = 0, \quad \sum_{i \in S} p_i = 1.$$

We saw in Example 8.26 that $\{X(t), t \geq 0\}$ is an SMP with an embedded DTMC with transition probability matrix $P = [p_{i,j}]$ where $p_{i,j} = q_{i,j}/q_i$ if $i \neq j$, $p_{i,i} = 0$, and $q_i = -q_{i,i}$. Let π be the solution to $\pi = \pi P$. From Theorem 6.26 on page 257 we have

$$p_j = \frac{\pi_j/q_j}{\sum_{i \in S} \pi_i/q_i}, \quad j \in S.$$

This is the same as Equation 8.67 since the expected sojourn time in state i is given by $\tau_i = 1/q_i$. Thus the theory of SMP produces consistent results when applied to CTMCs. ∎

Example 8.30   Series System. Compute the limiting distribution of the series system of Example 8.27.

The embedded DTMC has transition probability matrix given by

$$P = G(\infty) = \begin{bmatrix} 0 & \frac{\lambda_1}{\lambda} & \frac{\lambda_2}{\lambda} & \cdots & \frac{\lambda_N}{\lambda} \\ 1 & 0 & 0 & \cdots & 0 \\ 1 & 0 & 0 & \cdots & 0 \\ \vdots & \vdots & \vdots & \ddots & \vdots \\ 1 & 0 & 0 & \cdots & 0 \end{bmatrix}.$$

A solution to $\pi = \pi P$ is given by

$$\pi_0 = \lambda, \quad \pi_i = \lambda_i, \quad 1 \leq i \leq N.$$

Let r_i be the mean repair time of component i. Then we have

$$\tau_0 = 1/\lambda, \quad \tau_i = r_i, \quad 1 \leq i \leq N.$$

Using Equation 8.67 we get

$$p_0 = \frac{\pi_0 \tau_0}{\sum_{i=0}^{N} \pi_i \tau_i} = \frac{1}{1 + \sum_{i=1}^{N} \lambda_i r_i}$$

and

$$p_j = \frac{\pi_j \tau_j}{\sum_{i=0}^{N} \pi_i \tau_i} = \frac{\lambda_j r_j}{1 + \sum_{i=1}^{N} \lambda_i r_i}, \quad 1 \leq j \leq N. \quad ∎$$

Unlike in the study of CTMCs, the study of the limiting distribution of the SMPs cannot stop at the limiting distribution of $X(t)$ as $t \to \infty$, since knowing the value of $X(t)$ at time t is, in general, not enough to determine the future of an SMP. We also need to know the distribution of the remaining sojourn time at that time. We will have to wait for the theory of Markov regenerative processes in the next chapter to settle this question completely.

8.10 Renewal Processes with Costs/Rewards

We studied cost/reward models for the DTMCs in Section 4.8 and for the CTMCs in Section 6.13. Following that tradition we now study cost/reward models associated with RPs.

Let $\{N(t), t \geq 0\}$ be a standard RP generated by an iid sequence $\{X_n, n \geq 1\}$. We think of X_n as the length of the nth cycle between renewals. Suppose a reward R_n is earned at the end of the nth cycle, i.e., at time $S_n = X_1 + \cdots + X_n$. Define $Z(t)$ to be the total reward earned up to time t, i.e.,

$$Z(t) = \sum_{n=1}^{N(t)} R_n, \quad t \geq 0. \tag{8.68}$$

Note that $R(t) = 0$ if $N(t) = 0$. With this notation we are ready to define a renewal reward process.

Definition 8.13 Renewal Reward Process. *The stochastic process $\{Z(t), t \geq 0\}$ defined by Equation 8.68 is called a renewal reward process if $\{(X_n, R_n), n \geq 1\}$ is a sequence of iid bivariate random variables.*

We say that the process $\{Z(t), t \geq 0\}$ is generated by $\{(X_n, R_n), n \geq 1\}$. Note that X_n is a non-negative random variable, representing the length of the nth cycle, but R_n can be positive or negative. Thus the sample paths of a renewal reward process may go up or down, as shown in Figure 8.9. We illustrate with several examples.

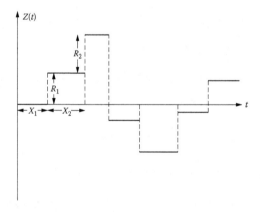

Figure 8.9 *A typical sample path of a renewal reward process.*

Example 8.31 RP as a Renewal Reward Process. Suppose $R_n = 1$ for $n \geq 1$. Then $Z(t) = N(t)$. Thus an RP is a special case of a renewal reward process. ∎

Example 8.32 CPP as a Renewal Reward Process. Let $\{Z(t), t \geq 0\}$ be a CPP with batch arrival rate λ and $P(\text{Batch Size} = k) = a_k$, $k \geq 0$. Then $\{Z(t), t \geq 0\}$ is a renewal reward process where $\{R_n, n \geq 1\}$ is a sequence of iid random variables with common pmf $[a_k, k \geq 0]$. It is independent of $\{X_n, n \geq 1\}$, which is a sequence of iid exp(λ) random variables. ∎

Example 8.33 Machine Maintenance. Consider the age replacement policy described in Example 8.4, where we replace a machine upon failure or upon reaching age T. Recall that L_i is the lifetime of the ith machine and $\{L_i, i \geq 1\}$ is a sequence of iid non-negative random variables. Replacing a machine by a new one costs $\$C_r$, and failure costs $\$C_f$. Let $Z(t)$ be the total cost incurred up to time t. Show that $\{Z(t), t \geq 0\}$ is a renewal reward process.

Let S_n be the time when nth replacement occurs ($S_0 = 0$), and $N(t)$ be the number of replacements up to time t. Then

$$S_n - S_{n-1} = X_n = \min(L_n, T).$$

Thus $\{X_n, n \geq 1\}$ is a sequence of iid random variables, and $\{N(t), t \geq 0\}$ is an RP generated by it. The cost R_n, incurred at time S_n, is given by

$$R_n = \begin{cases} C_r & \text{if } L_n > T, \\ C_f + C_r & \text{if } L_n \leq T. \end{cases}$$

Here we have implicitly assumed that if $L_n = T$, then we pay for the failure and then replace the machine. With the above expression for the cost R_n, we see that $\{(X_n, R_n), n \geq 1\}$ is a sequence of iid bivariate random variables. It is clear that $\{Z(t), t \geq 0\}$ is generated by $\{(X_n, R_n), n \geq 1\}$, and hence it is a renewal reward process. ∎

Computing the distribution of $Z(t)$ is rather difficult. Hence we study its asymptotic properties as $t \to \infty$.

Theorem 8.35 Almost-Sure ERT for Renewal Reward Processes. *Let* $\{Z(t), t \geq 0\}$ *be a renewal reward process generated by* $\{(X_n, R_n), n \geq 1\}$, *and suppose*

$$r = \mathsf{E}(R_n) < \infty, \quad \tau = \mathsf{E}(X_n) < \infty.$$

Then

$$\lim_{t \to \infty} \frac{Z(t)}{t} = \frac{r}{\tau}, \quad w.p.1. \tag{8.69}$$

Proof: Using Equation 8.68, we get

$$\frac{Z(t)}{t} = \frac{\sum_{n=1}^{N(t)} R_n}{N(t)} \cdot \frac{N(t)}{t}.$$

Since $N(t) \to \infty$ with probability 1, we can use the strong law of large numbers to get

$$\lim_{t \to \infty} \frac{\sum_{n=1}^{N(t)} R_n}{N(t)} = r, \quad w.p.1.$$

Furthermore, Theorem 8.6 yields

$$\frac{N(t)}{t} \to \frac{1}{\tau}, \quad w.p.1.$$

Combining these two assertions we get Equation 8.69. ∎

Next we derive the expected value version of the above result, i.e., $\lim E(Z(t))/t = r/\tau$. As in the case of the elementary renewal theorem, this conclusion does not follow from Theorem 8.35, and has to be established independently. We need two results before we can prove this.

Theorem 8.36 *Suppose the hypothesis of Theorem 8.35 holds. Then*

$$\mathsf{E}\left(\sum_{n=1}^{N(t)+1} R_n\right) = r(M(t) + 1), \tag{8.70}$$

where $M(t) = \mathsf{E}(N(t))$.

Proof: Let

$$H(t) = \mathsf{E}\left(\sum_{n=1}^{N(t)+1} R_n\right).$$

Using the renewal argument we get

$$\mathsf{E}\left(\sum_{n=1}^{N(t)+1} R_n \,\Big|\, X_1 = u\right) = \begin{cases} \mathsf{E}(R_1 | X_1 = u) & \text{if } u > t, \\ \mathsf{E}(R_1 | X_1 = u) + H(t-u) & \text{if } u \le t. \end{cases}$$

Hence, using $G(u) = \mathsf{P}(X_n \le u)$, we get

$$\begin{aligned} H(t) &= \int_0^\infty \mathsf{E}(R_1 | X_1 = u) dG(u) + \int_0^t H(t-u) dG(u) \\ &= r + \int_0^t H(t-u) dG(u). \end{aligned}$$

Solving this renewal-type equation by using Equation 8.26, we get

$$H(t) = r + \int_0^t r \, dM(u) = r(1 + M(t)).$$

This proves the theorem. ∎

Theorem 8.37 *Suppose the hypothesis of Theorem 8.35 holds. Then*

$$\lim_{t \to \infty} \frac{\mathsf{E}(R_{N(t)+1})}{t} = 0. \tag{8.71}$$

Proof: Note that $R_{N(t)+1}$ depends upon $X_{N(t)+1}$, which has a different distribution than X_1 as we had seen in our study of total life process in Section 8.6. Thus

$E(R_{N(t)+1}) \neq E(R_1)$, and hence this theorem does not follow trivially. We prove it by deriving a renewal equation for

$$H(t) = E(R_{N(t)+1}).$$

Using the renewal argument we get

$$E(R_{N(t)+1}|X_1 = u) = \begin{cases} E(R_1|X_1 = u) & \text{if } u > t, \\ H(t-u) & \text{if } u \leq t. \end{cases}$$

Hence, using $G(u) = P(X_n \leq u)$, we get

$$H(t) = D(t) + \int_0^t H(t-u)dG(u),$$

where

$$D(t) = \int_t^\infty E(R_1|X_1 = u)dG(u).$$

Now

$$\begin{aligned} |D(t)| &= |\int_t^\infty E(R_1|X_1 = u)dG(u)| \leq \int_t^\infty |E(R_1|X_1 = u)|dG(u) \\ &\leq \int_0^\infty |E(R_1|X_1 = u)|dG(u) \leq E(|R_1|) < \infty. \end{aligned} \tag{8.72}$$

Hence Theorem 8.15 yields

$$H(t) = D(t) + \int_0^\infty D(t-u)dM(u).$$

Also, Equation 8.72 implies that $|D(t)| \to 0$ as $t \to \infty$. Hence, for a given $\epsilon > 0$, there exists a $T < \infty$ such that $|D(t)| < \epsilon$ for $t \geq T$. Then, for all $t \geq T$, we have

$$\begin{aligned} \frac{H(t)}{t} &= \frac{D(t)}{t} + \int_0^{t-T} \frac{D(t-u)}{t}dM(u) + \int_{t-T}^t \frac{D(t-u)}{t}dM(u) \\ &\leq \frac{\epsilon}{t} + \epsilon\frac{M(t-T)}{t} + E(|R_1|)\frac{M(t) - M(t-T)}{t}. \end{aligned}$$

Now, as $t \to \infty$, $M(t) - M(t-T) \to T/\tau$. Hence we get

$$\lim_{t\to\infty} \frac{H(t)}{t} \leq \lim_{t\to\infty} \frac{\epsilon}{t} + \epsilon \lim_{t\to\infty} \frac{M(t-T)}{t} + \lim_{t\to\infty} E(|R_1|)\frac{M(t) - M(t-T)}{t} = \frac{\epsilon}{\tau},$$

since the first and the third limit on the right-hand side are zero, and the second limit is $1/\tau$ from the elementary renewal theorem. Since $\epsilon > 0$ was arbitrary, we get Equation 8.71. ∎

Now we are ready to prove the next theorem.

Theorem 8.38 ERT for Renewal Reward Processes. *Suppose the hypothesis of Theorem 8.35 holds. Then*

$$\lim_{t\to\infty} \frac{E(Z(t))}{t} = \frac{r}{\tau}. \tag{8.73}$$

Proof: Write

$$Z(t) = \sum_{n=1}^{N(t)+1} R_n - R_{N(t)+1}.$$

Hence

$$\frac{\mathsf{E}(Z(t))}{t} = \frac{\mathsf{E}\left(\sum_{n=1}^{N(t)+1} R_n\right)}{t} - \frac{\mathsf{E}(R_{N(t)+1})}{t}$$

$$= r\frac{1 + M(t)}{t} - \frac{\mathsf{E}(R_{N(t)+1})}{t},$$

where we have used Theorem 8.36. Now let $t \to \infty$. Theorem 8.37 and the elementary renewal theorem yield

$$\lim_{t \to \infty} \frac{\mathsf{E}(Z(t))}{t} = \frac{r}{\tau}$$

as desired. ∎

The above theorem is very intuitive and useful: it says that the long run expected rate of reward is simply the ratio of the expected reward in one cycle and the expected length of that cycle. What is surprising is that the reward does not have to be independent of the cycle length. This is what makes the theorem so useful in applications. We end this section with two examples.

Example 8.34   Machine Maintenance. Compute the long run expected cost per unit time of the age replacement policy described in the machine maintenance model of Example 8.33.

Suppose the lifetimes $\{L_i, i \geq 1\}$ are iid random variables with common cdf $F(\cdot)$. Then we have

$$\tau = \mathsf{E}(X_n) = \mathsf{E}(\min(L_n, T)) = \int_0^T (1 - F(u))du.$$

$$r = \mathsf{E}(R_n) = C_r + C_f F(T).$$

Hence the long run cost rate is given by

$$\lim_{t \to \infty} \frac{\mathsf{E}(Z(t))}{t} = \frac{r}{\tau} = \frac{C_r + C_f F(T)}{\int_0^T (1 - F(u))du}.$$

Clearly, as T increases, the cost rate of the planned replacements decreases, but the cost rate of the failures increases. Hence one would expect that there is an optimal T which minimizes the total cost rate. The actual optimal value of T depends upon C_r, C_f and F. For example, if $L_n \sim \exp(\lambda)$, we get

$$\lim_{t \to \infty} \frac{\mathsf{E}(Z(t))}{t} = \lambda C_f + \frac{\lambda C_r}{1 - e^{-\lambda T}}.$$

This is a monotonically decreasing function of T, implying that the optimal T is infinity, i.e., the machine should be replaced only upon failure. In retrospect, this is to be expected, since a machine with $\exp(\lambda)$ life time is always as good as new! ∎

In our study of renewal reward processes, we have assumed that the reward R_n is earned at the end of the nth cycle. This is not necessary. The results of this section remain valid no matter how the reward is earned over the cycle as long as the total reward earned over the nth cycle is R_n and $\{(X_n, R_n), n \geq 1\}$ is a sequence of iid bivariate random variables. We use this fact in the next example.

Example 8.35 Total Up-Time. Let $\{X(t), t \geq 0\}$ be the ARP as defined in Section 8.8. Let $W(t)$ be the total time spent in state 1 by the ARP up to time t. Show that

$$\lim_{t \to \infty} \frac{W(t)}{t} = \frac{\mathsf{E}(U_1)}{\mathsf{E}(U_1) + \mathsf{E}(D_1)}, \quad w.p.1.$$

$\{W(t), t \geq 0\}$ can be seen to be a renewal reward process with $U_n + D_n$ as the nth cycle duration and U_n as the reward earned over the nth cycle. Note that the reward is earned continuously at rate $X(t)$ during the cycle. Thus,

$$\tau = \mathsf{E}(U_1 + D_1), \quad r = \mathsf{E}(U_1).$$

Hence the result follows from Theorem 8.35. ∎

The results of this section remain valid for "delayed" renewal reward processes, i.e., when $\{(X_n, R_n), n \geq 2\}$ is a sequence of iid bivariate random variables, and is independent of (X_1, R_1). We shall omit the proofs.

Theorems 8.35 and 8.38 deal with what we had earlier called the "average cost case." It is possible to study the "discounted cost" case as well. The results are left as Computational Exercise 8.49.

8.11 Regenerative Processes

We study a class of stochastic processes called "regenerative processes" in this section. Intuitively, a regenerative process $\{X(t), t \geq 0\}$ has "regeneration points" $\{S_n, n \geq 1\}$, so that the probabilistic behavior of the stochastic process during the nth cycle, i.e., $\{X(t), S_{n-1} \leq t < S_n\}$, is the same for each $n \geq 1$, and independent from cycle to cycle. This idea is made precise in the following definition.

Definition 8.14 Regenerative Process. *A stochastic process $\{X(t), t \geq 0\}$ is called a regenerative process (RGP) if there exists a non-negative random variable S_1 so that*

1. $\mathsf{P}(S_1 = 0) < 1$, $\mathsf{P}(S_1 < \infty) = 1$,

2. $\{X(t), t \geq 0\}$ and $\{X(t + S_1), t \geq 0\}$ are stochastically identical, and

3. $\{X(t + S_1), t \geq 0\}$ is independent of $\{X(t), 0 \leq t < S_1\}$.

The above definition has several important implications. Existence of S_1 implies the existence of an infinite sequence of increasing random variables $\{S_n, n \geq 1\}$

so that $\{X(t), t \geq 0\}$ and $\{X(t + S_n), t \geq 0\}$ are stochastically identical, and $\{X(t + S_n), t \geq 0\}$ is independent of $\{X(t), 0 \leq t < S_n\}$. We say that S_n is the nth regeneration epoch, since the stochastic process loses all its memory and starts afresh at times S_n. The interval $[S_{n-1}, S_n)$ is called the nth regenerative cycle $(S_0 = 0)$. The probabilistic behavior of the RGP over consecutive regenerative cycles is independent and identical. In particular, $\{X(S_n), n \geq 0\}$ are iid random variables.

Definition 8.15 Delayed Regenerative Process. *A stochastic process* $\{X(t), t \geq 0\}$ *is called a delayed RGP if there exists a non-negative random variable* S_1 *so that*

1. $P(S_1 < \infty) = 1$,

2. $\{X(t + S_1), t \geq 0\}$ *is independent of* $\{X(t), 0 \leq t < S_1\}$, *and*

3. $\{X(t + S_1), t \geq 0\}$ *is a regenerative process.*

Thus for a delayed RGP we get a sequence of regenerative epochs $\{S_n, n \geq 0\}$ $(S_0 = 0)$. The behavior of the process over different regenerative cycles is independent. However, the behavior over the first cycle may be different than the behavior over the later cycles. We now give several examples of RGPs.

Example 8.36 Alternating Renewal Process. Consider an ARP $\{X(t), t \geq 0\}$ generated by the iid bivariate sequence $\{(U_n, D_n), n \geq 1\}$ with $U_1 + D_1 < \infty$ with probability 1. It is an RGP with regenerative epochs

$$S_n = \sum_{i=1}^{n}(U_i + D_i), \quad n \geq 1,$$

since every time the process enters state 1, it regenerates. If $\{X(t), t \geq 0\}$ is a delayed ARP, it is a delayed RGP. ∎

Example 8.37 Recurrence Times. The renewal process $\{N(t), t \geq 0\}$ with renewal sequence $\{S_n, n \geq 1\}$ is not an RGP. However the age process $\{A(t), t \geq 0\}$ and the excess life process $\{B(t), t \geq 0\}$ are RGPs with regenerative epochs $\{S_n, n \geq 1\}$. If $\{N(t), t \geq 0\}$ is a delayed RP, then $\{A(t), t \geq 0\}$ and $\{B(t), t \geq 0\}$ are delayed RGPs. ∎

Example 8.38 CTMCs as RGPs. Let $\{X(t), t \geq 0\}$ be an irreducible recurrent CTMC on $\{0, 1, 2, \cdots\}$ with $X(0) = 0$. Let S_n be the nth entry into state 0. The Markov property of the CTMC implies that it regenerates at S_n, for $n \geq 1$. Hence it is an RGP with regeneration epochs $\{S_n, n \geq 1\}$. If $X(0) \neq 0$, then it is a delayed RGP with this sequence of regenerative epochs. Now suppose S_n is the time of nth entry into a set of states $\{0, 1, \cdots, k\}$. Are $\{S_n, n \geq 1\}$ the regenerative epochs of $\{X(t), t \geq 0\}$? ∎

Example 8.39 SMPs as RGPs. Let $\{X(t), t \geq 0\}$ be an SMP as defined in Section 8.9. Suppose it has entered state 0 at time 0. Let S_n be the nth entry into state 0. The Markov property of the SMP at all points of transition implies that it is an RGP that regenerates at S_n, for $n \geq 1$. If $X(0) \neq 0$, then it is a delayed RGP with this sequence of regenerative epochs. ∎

Example 8.40 $G/G/1$ **Queue.** Let $X(t)$ be the number of customers in a $G/G/1$ queue. Let S_n be the nth time when a customer enters an empty system. From the independence of the inter-arrival times and the service times, it follows that the system loses dependence on the history at times $S_1, S_2, S_3, \cdots$. A sufficient condition for $P(S_1 < \infty) = 1$ is that the mean service time be less than the mean inter-arrival time. If the process starts with a customer entering an empty system at time 0, $\{X(t), t \geq 0\}$ is an RGP, otherwise it is a delayed RGP. ∎

Next we study the limiting behavior of the RGPs. The main result is given in the next theorem.

Theorem 8.39 **Limiting Distribution of an RGP.** *Let* $\{X(t), t \geq 0\}$ *be an RGP with state-space* $(-\infty, \infty)$ *with right continuous sample paths with left limits. Let* S_1 *be the first regeneration epoch and* $U_1(x)$ *be the time that the process spends in the interval* $(-\infty, x]$ *during* $[0, S_1)$. *If* S_1 *is aperiodic with* $\mathsf{E}(S_1) < \infty$,

$$F(x) = \lim_{t \to \infty} \mathsf{P}(X(t) \leq x) = \frac{\mathsf{E}(U_1(x))}{\mathsf{E}(S_1)}. \tag{8.74}$$

If S_1 *is periodic with period d, the above limit holds if* $t = nd$ *and* $n \to \infty$.

Proof: Fix an $x \in (-\infty, \infty)$, and define $H(t) = \mathsf{P}(X(t) \leq x)$. Since the process regenerates at time S_1, we have

$$\mathsf{P}(X(t) \leq x | S_1 = u) = \begin{cases} H(t - u) & \text{if } u \leq t, \\ \mathsf{P}(X(t) \leq x | S_1 = u) & \text{if } u > t. \end{cases}$$

Using $G(u) = \mathsf{P}(S_1 \leq u)$, we get

$$\begin{aligned} H(t) &= \int_0^\infty \mathsf{P}(X(t) \leq x | S_1 = u) dG(u) \\ &= D(t) + \int_0^t H(t - u) dG(u), \end{aligned} \tag{8.75}$$

where

$$D(t) = \mathsf{P}(X(t) \leq x, S_1 > t).$$

It can be shown that the assumptions about the sample paths ensure that $D(\cdot)$ satisfies the conditions in KRT (Theorem 8.17). Now define

$$Z(t) = \begin{cases} 1 & \text{if } X(t) \leq x, \\ 0 & \text{if } X(t) > x. \end{cases}$$

Then we have

$$\mathsf{E}(U_1(x)) = \mathsf{E}\left(\int_0^{S_1} Z(t) dt\right)$$

$$= \int_0^\infty E\left(\int_0^u Z(t)dt \,\Big|\, S_1 = u\right) dG(u)$$

$$= \int_0^\infty \int_0^u E(Z(t)|S_1 = u) dt \, dG(u)$$

$$= \int_0^\infty \int_0^u P(X(t) \le x | S_1 = u) dt \, dG(u)$$

$$= \int_0^\infty \int_t^\infty P(X(t) \le x | S_1 = u) dG(u) dt$$

$$= \int_0^\infty P(X(t) \le x, S_1 > t) dt.$$

Using the KRT we get

$$\lim_{t\to\infty} P(X(t) \le x) = \lim_{t\to\infty} H(t) = \frac{1}{E(S_1)} \int_0^\infty D(t) dt$$

$$= \frac{1}{E(S_1)} \int_0^\infty P(X(t) \le x, S_1 > t) dt = \frac{E(U_1(x))}{E(S_1)}.$$

This proves the theorem. ∎

Several observations about the above theorem are in order. Let $U_1(\infty) = \lim_{x\to\infty} U_1(x)$. Since the state-space of the RGP is $(-\infty, \infty)$ and $S_1 < \infty$ with probability 1, we have $U_1(\infty) = S_1$ and $E(U_1(\infty)) = E(S_1)$. This implies

$$F(\infty) = \lim_{t\to\infty} P(X(t) < \infty) = \frac{E(U_1(\infty))}{E(S_1)} = 1.$$

Thus the RGP satisfying the hypothesis of the above theorem has a proper limiting distribution.

If the state-space of the RGP is discrete, say $\{0, 1, 2 \cdots\}$, we can define $U_{1,j}$ as the time the RGP spends in state j over the first regenerative cycle. In that case the above theorem implies that the RGP has a proper limiting pmf given by

$$p_j = \lim_{t\to\infty} P(X(t) = j) = \frac{E(U_{1,j})}{E(S_1)}. \tag{8.76}$$

Finally, Theorem 8.39 continues to hold if $\{X(t), t \ge 0\}$ is a delayed RGP, except we now use the second cycle to do the computation. Thus, for a delayed RGP satisfying the hypothesis of Theorem 8.39, we have

$$\lim_{t\to\infty} P(X(t) \le x) = \frac{E(U_2(x))}{E(S_2 - S_1)}. \tag{8.77}$$

We illustrate with an example.

Example 8.41 Suppose a coin is tossed repeatedly and independently with $p = P(H)$, and $q = 1 - p = P(T)$ on any one toss. What is the expected number of tosses needed to observe the sequence HHTT?

Let Y_n be the outcome of the nth toss. Let $X_0 = 1$ and for $n \geq 1$ define

$$X_n = \begin{cases} 1 & \text{if } Y_n = H, Y_{n+1} = H, Y_{n+2} = T, Y_{n+3} = T, \\ 0 & \text{otherwise.} \end{cases}$$

Let $S_0 = 0$ and define $S_{k+1} = \min\{n > S_k : X_n = 1\}$, for $k \geq 0$. Note that since the coin tosses are iid, $\{S_{k+1} - S_k, k \geq 1\}$ are iid and have the same distribution as $S_1 + 3$. Let $X(t) = X_{[t]}$, where $[t]$ is the largest integer less than or equal to t. Thus $\{X(t), t \geq 0\}$ is a periodic delayed RGP with period 1, and $\{S_k, k \geq 1\}$ are the regeneration epochs. During each regenerative cycle it spends 1 unit of time in state 1. Hence

$$\lim_{n \to \infty} P(X(n) = 1) = \lim_{n \to \infty} P(X_n = 1) = \frac{1}{E(S_2 - S_1)}.$$

However we know that

$$P(X_n = 1) = P(Y_n = H, Y_{n+1} = H, Y_{n+2} = T, Y_{n+3} = T) = p^2 q^2.$$

Hence

$$E(S_2 - S_1) = E(S_1 + 3) = \frac{1}{p^2 q^2}.$$

Since the number of tosses needed to observe HHTT is $S_1 + 3$, we see that the expected number of tosses needed to observe HHTT is $1/p^2 q^2$. Will this method work for the sequence HTTH? ∎

8.11.1 RGPs with Costs/Rewards

We next consider the cost/reward models in RGPs. Let $X(t)$ be the state of a system at time t, and assume that $\{X(t), t \geq 0\}$ is a regenerative process. Suppose the system earns rewards at rate $r(x)$ whenever it is in state x. Then the total reward earned by the system up to time t is given by

$$\int_0^t r(X(u)) du,$$

and the reward rate up to time t is given by

$$\frac{1}{t} \int_0^t r(X(u)) du.$$

The next theorem gives the limiting behavior of the reward rate as $t \to \infty$.

Theorem 8.40 Reward Rates in RGPs. Let $\{X(t), t \geq 0\}$ be a regenerative process on $S = (-\infty, \infty)$ with proper limiting distribution $F(\cdot)$. Let $r : S \to (-\infty, \infty)$ be bounded from either above or below. Then

1. $\lim_{t \to \infty} \frac{1}{t} \int_0^t r(X(u)) du = \int_{-\infty}^{\infty} r(u) dF(u)$, w.p. 1,

2. $\lim_{t \to \infty} \frac{1}{t} E\left(\int_0^t r(X(u)) du \right) = \int_{-\infty}^{\infty} r(u) dF(u)$.

Proof: Let $\{S_n, n \geq 1\}$ (with $S_0 = 0$) be a sequence of regenerative epochs in $\{X(t), t \geq 0\}$. Now define

$$Z(t) = \int_0^t r(X(u))du$$

and

$$R_n = \int_{S_{n-1}}^{S_n} r(X(u))du, \quad n \geq 1.$$

Since $\{(S_n - S_{n-1}, R_n), n \geq 1\}$ is a sequence of iid bivariate random variables, we see that $\{Z(t), t \geq 0\}$ is a renewal reward process. From Theorems 8.35 and 8.38 we get

$$\lim_{t \to \infty} \frac{1}{t} \int_0^t r(X(u))du = \frac{E(R_1)}{E(S_1)}, \quad \text{w.p. 1,}$$

and

$$\lim_{t \to \infty} \frac{1}{t} E\left(\int_0^t r(X(u))du\right) = \frac{E(R_1)}{E(S_1)}.$$

Now, from Theorem 8.39, we see that the time spent in the infinitesimal interval $(x, x + dx]$ by the system over the first regenerative cycle is given by $E(S_1)dF(x)$. Then

$$E(R_1) = E\left(\int_0^{S_1} r(X(u))du\right) = \int_{-\infty}^{\infty} r(x)E(S_1)dF(x).$$

Hence

$$\frac{E(R_1)}{E(S_1)} = \int_{-\infty}^{\infty} r(x)dF(x).$$

Thus the theorem follows. ∎

If the state-space of the RGP is discrete, say $\{0, 1, 2 \cdots\}$, and the limiting pmf of the RGP is given by $\{p_j, j \geq 0\}$ as in Equation 8.76, the long run reward rate in Theorem 8.40 reduces to

$$\sum_{j=0}^{\infty} r(j)p_j. \tag{8.78}$$

We illustrate the above theorem with an example below.

Example 8.42 Optimal Clearing. Suppose a manufacturing facility produces finished goods one at a time according to a renewal process with mean production time $\tau < \infty$ per item and stores them in a warehouse. As soon as there are k items in the warehouse we clear it by shipping them to the retailers. The clearing is instantaneous. It costs $\$h$ to hold an item in the warehouse per unit time, and it costs $\$c$ to clear the warehouse. What is optimal value of k that minimizes the long run holding plus clearing cost per unit time?

Let $X(t)$ be the number of items in the warehouse at time t. Suppose $X(0) = 0$ and let S_n be the time of the nth clearing time of the warehouse. It is obvious that

$\{X(t), t \geq 0\}$ is a regenerative process on state-space $\{0, 1, \cdots, k-1\}$ with regeneration epochs $\{S_n, n \geq 1\}$. A typical sample path of the $\{X(t), t \geq 0\}$ process (with $k = 5$) is shown in Figure 8.10. We see that

$$P(S_1 < \infty) = 1, \quad E(S_1) = k\tau.$$

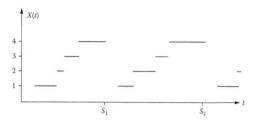

Figure 8.10 *A typical sample path of a regenerative process.*

Hence, from Equation 8.76 we get

$$p_j = \lim_{t \to \infty} P(X(t) = j) = \frac{E(U_{1,j})}{E(S_1)} = \frac{\tau}{k\tau} = \frac{1}{k}, \quad 0 \leq j \leq k - 1.$$

We compute C_h, the long run expected holding cost per unit time first. The system incurs waiting cost at a rate jh per unit time if there are j items in the warehouse. We can use Equation 8.78 to get

$$C_h = \sum_{j=0}^{k-1} jhp_j = \frac{h}{k} \sum_{j=0}^{k-1} j = \frac{1}{2} h(k - 1).$$

To compute C_c, the long run clearing cost per unit time, we use Theorem 8.38 for the renewal reward processes to get

$$C_c = \frac{c}{k\tau}.$$

Thus the $C(k)$, the long run total cost per unit time is given by

$$C(k) = \frac{1}{2} h(k - 1) + \frac{c}{k\tau}.$$

This is a convex function of k and is minimized at

$$k^* = \sqrt{\frac{2c}{h\tau}}.$$

Since k^* must be an integer we check the two integers near the above solution to see which is optimal. ∎

8.11.2 Little's Law

We gave a sample path proof of Little's Law in Theorem 7.5 on page 313. Here we give an alternate proof under the assumption that the queueing system is described by an RGP $\{X(t), t \geq 0\}$, where $X(t)$ is the number of customers in the system at time t.

Theorem 8.41 **Little's Law for RGPs.** *Suppose $\{X(t), t \geq 0\}$ is an RGP on $\{0, 1, 2, \cdots\}$ with regeneration epochs $\{S_n, n \geq 1\}$ where S_n is the nth time it jumps from state 0 to 1 (i.e., a customer enters an empty system). Suppose $\mathsf{E}(S_1) < \infty$. Then the limits L, λ, and W as defined by Equations 7.12, 7.13, and 7.14 on page 313, exist and are related by*

$$L = \lambda W.$$

Proof: Let N be the number of customers who enter the system over $[0, S_1)$. Thus it includes the arrival at time 0, but not the one at time S_1. By the definition of S_1, the number of departures up to time S_1 is also N. Then using the renewal reward theorem (Theorem 8.38) we see that the limits L, λ, and W exist, and are given by

$$L = \mathsf{E}\left(\int_0^{S_1} X(u)du\right) / \mathsf{E}(S_1), \tag{8.79}$$

$$\lambda = \mathsf{E}(N)/\mathsf{E}(S_1), \tag{8.80}$$

$$W = \mathsf{E}\left(\sum_{n=1}^{N} W_n\right) / \mathsf{E}(N). \tag{8.81}$$

Now define

$$I_n(t) = \begin{cases} 1 & \text{if the nth customer is in the system at time } t, \\ 0 & \text{otherwise.} \end{cases}$$

Then for $1 \leq n \leq N$ we have

$$W_n = \int_0^{S_1} I_n(t)dt,$$

since these customers depart by time S_1. Also, for $0 \leq t < S_1$, we have

$$X(t) = \sum_{n=1}^{N} I_n(t),$$

since the right-hand side simply counts the number of customers in the system at time t. Combining these equations we get

$$\sum_{n=1}^{N} W_n = \sum_{n=1}^{N} \int_0^{S_1} I_n(t)dt = \int_0^{S_1} \sum_{n=1}^{N} I_n(t)dt = \int_0^{S_1} X(t)dt.$$

Substituting in Equation 8.79 we get

$$L = \mathsf{E}\left(\int_0^{S_1} X(u)du\right) / \mathsf{E}(S_1)$$

$$= \mathsf{E}\left(\sum_{n=1}^{N} W_n\right) / \mathsf{E}(S_1)$$

$$= \frac{E(N)}{E(S_1)} \cdot \frac{\left(\sum_{n=1}^{N} W_n\right)}{E(S_1)}$$

$$= \lambda W$$

as desired. ∎

We end this chapter with the observation that the difficulty in using Theorem 8.39 as a computational tool is in the computation of $E(U_1(x))$ and $E(S_1)$. This is generally the result of the fact that the sample paths of the RGP over $[0, S_1)$ can be quite complicated with multiple visits to the interval $(-\infty, x]$. In the next chapter we shall study the Markov RGPs, which alleviate this problem by using "smaller" S_1, but in the process giving up the assumption that the system loses dependence on the history completely at time S_1. This leads to a richer structure and a more powerful computational tool. So we march on to Markov RGPs!

8.12 Computational Exercises

8.1 Consider an $M/M/1/K$ queue that is full at time 0. Let $S_0 = 0$, and S_n be the arrival time of the nth customer who sees the system full upon arrival. Show that $\{S_n, n \geq 1\}$ is a renewal sequence. What does the corresponding RP count? Do the same analysis for the $M/G/1/K$ and the $G/M/1/K$ queue.

8.2 Consider a birth and death process on $S = \{\cdots, -2, -1, 0, 1, 2, \cdots\}$ with $\lambda_n = \lambda$ and $\mu_n = \mu$ for all $n \in S$. Let S_n be the time of the nth downward jump. Show that $\{S_n, n \geq 1\}$ is a renewal sequence. What is the corresponding RP?

8.3 Consider the RP of Computational Exercise 8.2. Is this renewal process transient or recurrent?

8.4 Let $X(t)$ be the number of customers at time t in a $G/G/1$ queue with iid inter-arrival times and iid service times. Suppose at time 0 a customer enters an empty system and starts service. Find an embedded renewal sequence in $\{X(t), t \geq 0\}$.

8.5 Let $\{N(t), t \geq 0\}$ be an RP with iid inter-renewal times with common pdf

$$g(t) = \lambda^2 t e^{-\lambda t}, \quad t \geq 0.$$

Compute $P(N(t) = k)$ for $k = 0, 1, 2, \cdots$.

8.6 Let $\{N(t), t \geq 0\}$ be an RP with integer valued inter-renewal times with common pmf

$$P(X_n = 0) = 1 - \alpha, \ P(X_n = 1) = \alpha, \quad n \geq 1,$$

where $0 < \alpha < 1$. Compute $P(N(t) = k)$ for $k = 0, 1, 2, \cdots$. (First do this for non-negative integer-valued t.)

8.7 Let $\{N(t), t \geq 0\}$ be an RP with common inter-renewal time pmf

$$P(X_n = i) = \alpha^{i-1}(1 - \alpha), \quad i \geq 1,$$

where $0 < \alpha < 1$. Compute $P(N(t) = k)$ for $k = 0, 1, 2, \cdots$. (First do this for non-negative integer-valued t.)

8.8 Let $\{N(t), t \geq 0\}$ be an RP with common inter-renewal time pmf

$$P(X_n = 1) = .8, \ P(X_n = 2) = .2, \quad n \geq 1.$$

Compute $P(N(t) = k)$ for $k = 0, 1, 2, \cdots$. (First do this for non-negative integer-valued t.)

8.9 Let $\{N(t), t \geq 0\}$ be an RP with common inter-renewal time pmf

$$P(X_n = 0) = .2, \ P(X_n = 1) = .3, \ P(X_n = 2) = .5, \quad n \geq 1.$$

Compute $P(N(t) = k)$ for $k = 0, 1, 2, \cdots$. (First do this for non-negative integer-valued t.)

8.10 Let $\{N(t), t \geq 0\}$ be an RP with common inter-renewal time pdf

$$g(t) = r\lambda_1 e^{-\lambda_1 t} + (1 - r)\lambda_2 e^{-\lambda_2 t}, \quad t \geq 0.$$

Compute the LST $\tilde{p}_k(s)$ of $p_k(t) = P(N(t) = k)$ for $k = 0, 1, 2, \cdots$. Compute $p_0(t)$ and $p_1(t)$.

8.11 Consider the machine maintenance problem of Example 8.4. Suppose the machine lifetimes (in years) are iid $U(2, 5)$, and they are replaced upon failure or upon reaching age 3. Compute

1. the long run rate of replacements,
2. the long run rate of planned replacements,
3. the long run rate of failures.

8.12 In Computational Exercise 8.11 compute the asymptotic distribution of the number of replacements.

8.13 Let $\{Y_n, n \geq 0\}$ be a DTMC on $\{0, 1\}$ with the following transition probability matrix

$$\begin{bmatrix} \alpha & 1 - \alpha \\ 1 - \beta & \beta \end{bmatrix}.$$

Analyze the asymptotic behavior of $N(n)$ = number of visits to state 0 during $\{1, 2, \cdots, n\}$, assuming that $Y_0 = 0$.

8.14 Let $\{Y(t), t \geq 0\}$ be a CTMC on $\{0, 1\}$ with the following generator matrix

$$\begin{bmatrix} -\lambda & \lambda \\ \mu & -\mu \end{bmatrix}.$$

Analyze the asymptotic behavior of $N(t)$ = number of visits to state 0 during $(0, t]$, assuming that $Y(0) = 0$.

8.15 Compute the renewal function for the RP in Computational Exercise 8.10.

8.16 Compute the renewal function for the RP in Computational Exercise 8.8.

8.17 Compute the renewal function for the RP in Computational Exercise 8.14.

8.18 Derive a renewal-type equation for $E(S_{N(t)+k})$, $k \geq 1$, and solve it.

8.19 Compute $M^*(t) = E(N^*(t))$ in terms of $M(t) = E(N(t))$, where $N^*(t)$ and $N(t)$ are as defined in Conceptual Exercise 8.2.

8.20 Let $M_i(t) = E(N_i(t))$ ($i = 1, 2$) where $N_i(t)$ is as defined in Conceptual Exercise 8.3. Compute $\tilde{M}_i(s)$ in terms of $\tilde{G}(s)$ for $i = 1, 2$.

8.21 Derive a renewal-type equation for $P(A(t) \leq x)$, where $A(t)$ is the age at time t in an RP. Show that the KRT is applicable and compute the limiting distribution of $A(t)$ as $t \to \infty$, assuming that the RP is aperiodic.

8.22 Derive a renewal-type equation for $E(A(t)B(t))$, where $A(t)$ $(B(t))$ is the age (excess-life) at time t in an RP. Show that the KRT is applicable and compute the limiting value of $E(A(t)B(t))$ as $t \to \infty$, assuming that the RP is aperiodic. Using this compute the limiting covariance of $A(t)$ and $B(t)$ as $t \to \infty$.

8.23 Derive an integral equation for $p(t) = P(N(t)$ is odd) for an RP $\{N(t), t \geq 0\}$ by conditioning on the first renewal time. Is this a renewal-type equation? Solve it explicitly when the RP is PP(λ).

8.24 Consider the two-state CTMC of Computational Exercise 8.14. Let $W(t)$ be the time spent by this process in state 0 during $(0, t]$. Derive a renewal-type equation for $E(W(t))$, and solve it using the transform method.

8.25 Let $\{N(t), t \geq 0\}$ be an RP with inter-renewal time distribution G. Using the renewal argument derive the following renewal-type equation for $M_k(t) = E(N(t) (N(t) - 1) \cdots (N(t) - k + 1))$, $(k \geq 1)$, with $M_0(t) = 1$, and $M_1(t) = M(t)$:

$$M_k(t) = k \int_0^t M_{k-1}(t - u)dG(u) + \int_0^t M_k(t - u)dG(u).$$

Hence derive an expression for $\tilde{M}_k(s)$ in terms of $\tilde{G}(s)$.

8.26 Derive a renewal-type equation for $E(A(t)^k)$, where $A(t)$ is the age at time t in an RP. Assume that the inter-renewal times are aperiodic with finite $(k+1)$st moment. Show that the KRT is applicable and compute the limiting value of $E(A(t)^k)$ as $t \to \infty$.

8.27 Derive a renewal-type equation for $E(C(t)^k)$, where $C(t)$ is the total life at time t in an RP. Assume that the inter-renewal times are aperiodic with finite $(k + 1)$st moment. Show that the KRT is applicable and compute the limiting value of $E(C(t)^k)$ as $t \to \infty$.

8.28 Derive a renewal-type equation for $E(A(t)^k C(t)^m)$, where $A(t)$ is the age and $C(t)$ is the total life at time t in an RP. Assume that the inter-renewal times are aperiodic with finite $(k + m + 1)$st moment. Show that the KRT is applicable and compute the limiting value of $E(A(t)^k C(t)^m)$ as $t \to \infty$.

8.29 Derive an integral equation for $P(B^D(t) > x)$, where $B^D(t)$ is the excess life at time t in a delayed RP. Compute its limiting value.

8.30 Consider a $G/G/1/1$ queue with inter-arrival time cdf G and service time cdf F. Using ARPs, compute the limiting probability that the server is busy.

8.31 Compute the expected length of a busy period started by a single customer in a stable $M/G/1$ queue of Section 7.6.1, by constructing an appropriate ARP. (Use Equation 7.41 on page 339.)

8.32 Consider an $M/M/1/K$ queue of Section 7.3.2 with limiting distribution given by Equation 7.19. Construct an appropriate ARP to compute the expected time when the system becomes full (i.e., it enters state K from $K - 1$) for the first time after it becomes non-full (i.e., after it enters state $K - 1$ from K).

8.33 A particle moves on n sites arranged in a circle as follows: it stays at the ith site for a random amount of time with cdf F_i and mean μ_i and then moves to the adjacent site in the clockwise direction. Let H be the cdf of the time it takes to complete the circle, and assume that it is aperiodic with mean μ. Furthermore, assume that the successive sojourn times are independent. Construct an appropriate ARP to compute the limiting probability that the particle is on site i.

8.34 For the series system in Example 8.27 define $Y(t) = 0$ if the system is up at time t and 1 if it is down at time t. Assume that the system is up at time 0, and show that $\{Y(t), t \geq 0\}$ is an ARP, and compute the long run probability that the system is up. Verify the result with the one obtained in Example 8.30.

8.35 Do Computational Exercise 8.34 by constructing an appropriate two-state SMP.

8.36 Let $\{Y(t), t \geq 0\}$ be as in Computational Exercise 8.14. Suppose $Y(0) = 0$. Show that $\{Y(t), t \geq 0\}$ is an ARP. Compute $P(Y(t) = 0)$ by solving Equation 8.59 by using the transform methods.

8.37 In the machine maintenance model of Example 8.34, suppose the machine lifetimes are iid $U(0, a)$ random variables. Compute the optimal age replacement parameter T that minimizes the long run expected total cost per unit time.

8.38 Demands occur according to a $PP(\lambda)$ at a warehouse that has S items in it initially. As soon as the number of items in the warehouse decreases to $s(< S)$, the inventory is instantaneously replenished to S. Let $X(t)$ be the number of items in the warehouse at time t. Is $\{X(t), t \geq 0\}$ a (1) a CTMC, (2) an SMP, (3) a RGP? Compute its limiting distribution.

8.39 Let $X(t)$ be the number of customers in a queueing system. In which of the following systems is $\{X(t), t \geq 0\}$ an SMP? Why or why not?

1. An $M/G/1/1$ system,
2. A $G/M/1/1$ system,
3. A $G/G/1/1$ system.

8.40 Do Computational Exercise 8.38 if the demands arrive according to an RP with common inter-demand time cdf G.

8.41 A machine is subject to shocks that arrive according to $PP(\lambda)$. Each shock causes damage that can be represented by an integer-valued random variable with pmf $\{\alpha_j, j \geq 1\}$. The damages are additive, and when the total damage exceeds a threshold K, the machine breaks down. The repair time has cdf $A(\cdot)$, and successive repair times are iid. Shocks have no effect during repair, and the machine is as good as new once the repair completes. Model this system by an appropriate SMP. Show its kernel.

8.42 Let $\{W(t), t \geq 0\}$ be as defined in Example 8.35 and let U and D be the generic up and down times. Define $H(t) = \mathsf{E}(W(t)) - \frac{\mathsf{E}(U)}{\mathsf{E}(U+D)}t$. Show that H satisfies the renewal-type equation

$$H(t) = \mathsf{E}(\min(U, t)) - \frac{\mathsf{E}(U)\mathsf{E}(\min(U + D, t))}{\mathsf{E}(U + D)} + \int_0^t H(t - u)dG(u),$$

where G is the cdf of $U + D$. Assuming that G is aperiodic, show that

$$\lim_{t \to \infty} H(t) = \frac{1}{2} \cdot \frac{\mathsf{E}(U)\mathsf{E}((U + D)^2) - \mathsf{E}(U + D)\mathsf{E}(U^2)}{(\mathsf{E}(U + D))^2}.$$

8.43 Let X_n be the amount of inventory in a warehouse at the beginning of day n. Suppose $X_0 = S > 0$, a given number. Let Y_n be the size of the demand on day n and assume that $\{Y_n, n \geq 1\}$ is a sequence of iid random variables with common cdf G. As soon as the inventory goes below $s < S$, it is instantaneously replenished to S. Is $\{X_n, n \geq 0\}$ a (1) DTMC, (2) SMP, (3) RGP? Explain why or why not. Show that

$$\lim_{n \to \infty} \mathsf{P}(X_n \geq x) = \frac{1 + M(S - x)}{1 + M(S - s)}, \quad s \leq x \leq S,$$

where $M(\cdot)$ is the renewal function corresponding to the cdf G.

8.44 Use the renewal equation derived in the proof of Theorem 8.37 to compute the limit of $\mathsf{E}(R_{N(t)+1})$ as $t \to \infty$.

8.45 What is the long run fraction of customers who are turned away in an $M/G/1/1$ queue? In a $G/M/1/1$ queue?

8.46 The patients in a hospital are classified as belonging to the following units: (1) coronary care unit, (2) intensive care unit, (3) ambulatory unit, (4) extended care unit, and (5) home or dead. As soon as a patient goes home or dies, a new patient is admitted to the coronary unit. The successive units the patient visits form a DTMC with transition probability matrix given below:

$$\begin{bmatrix} 0 & 1 & 0 & 0 & 0 \\ 0 & 0 & 1 & 0 & 0 \\ 0.1 & 0 & 0 & 0.9 & 0 \\ 0.1 & 0.1 & 0.1 & 0.5 & 0.2 \\ 0 & 0 & 0 & 0 & 1 \end{bmatrix}.$$

The patient spends on the average 1.7 days in the coronary care unit, 2.2 days in the intensive care unit, 8 days in the ambulatory unit, and 16 days in the extended care unit. Let $X(t)$ be the state of the patient in the hospital (assume it has exactly one patient at all times; extension to more than one patient is easy if the patient movements are iid) at time t. Model it as an SMP with four states and compute the long run probability that patient is in the ambulatory unit.

8.47 A company classifies its employees into four grades, labeled 1, 2, 3, and 4. An employee's stay in grade i is determined by two random variables: the promotion time A_i, and the tolerance time B_i. The employee stays in grade i for $\min(A_i, B_i)$ amount of time. If $A_i \leq B_i$, he moves to grade $i + 1$. If $A_i > B_i$, he quits, and is instantaneously replaced by a new employee in grade 1. Since there is no promotion from grade 4, we set $A_4 = \infty$. Let $X(t)$ be the grade of the employee working at time t.

1. Model $\{X(t), t \geq 0\}$ as an SMP. Explicitly state any assumptions needed to do this, and display the kernel.
2. Assume $A_i \sim \exp(\lambda_i)$ and $B_i \sim \text{Erl}(2, \mu_i)$, and that they are independent. Compute the limiting distribution of $X(t)$.

8.48 Sixteen underground bunkers are connected by tunnels as shown in Figure 8.11. This complex serves as a residence of a military despot, who spends a random amount of time in a bunker and then moves to any of the adjacent ones with equal probability. Suppose the mean time spent in bunker i during one stay is τ_i. Compute the long run probability that the despot is in bunker i. (You may use symmetry to analyze the embedded DTMC.)

8.49 Consider the renewal reward process of Section 8.10. Suppose the rewards are discounted with continuous discount factor $\alpha > 0$. Let $D(t)$ be the total discounted reward earned up to time t, i.e.,

$$D(t) = \sum_{n=1}^{N(t)} e^{-\alpha S_n} R_n.$$

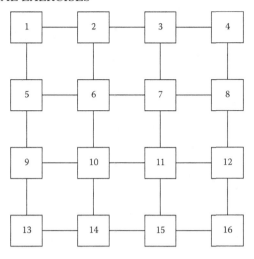

Figure 8.11 *The underground maze of 16 bunkers.*

Show that

$$\lim_{t\to\infty} \mathsf{E}(D(t)) = \frac{\mathsf{E}(R_1 e^{-\alpha S_1})}{1 - \mathsf{E}(e^{-\alpha S_1})}.$$

8.50 Suppose demands arise according to a PP(λ) at a warehouse that initially has S items. A demand is satisfied instantly if there are items in the warehouse, else the demand is lost. When the warehouse becomes empty, it places an order for S items from the supplier. The order is fulfilled after a random amount of time (called the lead-time) with mean L days. Suppose it costs $\$h$ to store an item in the warehouse for one day. The warehouse manager pays $\$c$ to buy an item and sells it for $\$p$. The order processing cost is $\$d$, regardless of the size of the order. Using an appropriate renewal reward process, compute the long run cost rate for this policy. Find the value of S that will minimize this cost rate for the following parameters: $\lambda = 2$ per day, $h = 1$, $c = 70$, $p = 80$, $d = 50$, $L = 3$.

8.51 Consider the parking lot of Modeling Exercise 6.24, and let $N_0(t)$ be the number of customers who arrived there up to time t. Suppose $\{N_0(t), t \geq 0\}$ is an RP with common inter-arrival time cdf $A(\cdot)$. Let $N_k(t)$ be the number of customers who arrived up to time t and found spaces 1 through k occupied ($1 \leq k \leq K$). $\{N_k(t), t \geq 0\}$ is called the overflow process from space k.

1. Show that $\{N_k(t), t \geq 0\}$ is an RP and that the LST of its inter-renewal times is given by

$$\phi_k(s) = \frac{\phi_{k-1}(s+\mu)}{1 - \phi_{k-1}(s) + \phi_{k-1}(s+\mu)},$$

where $\phi_0(s)$ is the LST of $A(\cdot)$.

2. Let $X_k(t)$ be 1 if space k is occupied, and zero otherwise. Show that

$\{X_k(t), t \geq 0\}$ is the queue-length process in a $G/M/1/1$ queue with arrival process$\{N_{k-1}(t), t \geq 0\}$ and iid $\exp(\mu)$ service times. Compute the limiting distribution of $X_k(t)$ as $t \to \infty$, and show that long run fraction of the time the kth space is occupied (called its utilization) is given by

$$\frac{1 - \phi_{k-1}(\mu)}{\mu \tau_{k-1}},$$

where τ_k is the mean inter-renewal time in $\{N_k(t), t \geq 0\}$. Show that

$$\tau_k = \tau_{k-1}/\phi_{k-1}(\mu),$$

where τ_0 is the mean of $A(\cdot)$.

3. Compute the space utilizations for each space if there are six parking spaces and customers arrive every 2 minutes in a deterministic fashion and stay in the lot for an average of 15 minutes.

8.52 Consider a $G/M/\infty$ queue with infinite number of servers, common inter-arrival time cdf G with mean $1/\lambda$ and $\exp(\mu)$ service times. Suppose that at time 0 the system is in steady state. In particular the arrival process is assumed to have started at $-\infty$ so that at time 0 it is in equilibrium. Let $\{T_n, n \geq 0\}$ be the arrival times of customers who arrived before time 0, indexed in reverse, so that $0 > T_1 > T_2 > \cdots$. Now, let X_i be 1 if the customer who arrived at time T_i is still in the system at time 0, and 0 otherwise. Thus $X = \sum_{i=1}^{\infty} X_i$ is the total number of customers in the system at time 0 (i.e., in steady state.)

1. Show that cdf of $-T_1$ is G_e, the equilibrium cdf associated with G, and $\{T_i - T_{i+1}, i \geq 1\}$ is a sequence of iid random variables with common cdf G, and is independent of $-T_1$.

2. Show that

$$E(X_i) = \frac{\lambda}{\mu} \tilde{G}(\mu)^{i-1}(1 - \tilde{G}(\mu)), \quad i \geq 1.$$

3. Using the above, show that the expected number of customers in steady state in a $G/M/\infty$ queue is given by

$$E(X) = \frac{\lambda}{\mu}.$$

Show this directly by using Little's Law.

4. Show that for $i > j$,

$$E(X_i X_j) = \frac{\lambda}{2\mu} \tilde{G}(\mu)^{i-j} \tilde{G}(2\mu)^{j-1}(1 - \tilde{G}(2\mu)).$$

5. Using the above, show that

$$E(X(X - 1)) = \frac{\lambda}{\mu} \cdot \frac{\tilde{G}(\mu)}{1 - \tilde{G}(\mu)}.$$

8.53 Functionally identical machines are available from N different vendors. The lifetimes of the machines from vendor i are iid $\exp(\lambda_i)$ random variables. The machines from different vendors are independent of each other. We use a "cyclic" replacement policy parameterized by a fixed positive number T as follows: suppose we are currently using a machine from vendor i. If it is less than T time units old upon failure, it is replaced by a machine from vendor $i + 1$ if $i < N$ and vendor 1 if $i = N$. If the machine is at least T time units old upon failure, the replacement is from vendor i. Replacements are instantaneous. Let $X(t) = i$ if the machine in use at time t is from vendor i.

1. Is $\{X(t), t \geq 0\}$ an SMP? If it is, give its kernel G.
2. What is the long run fraction of the time that a machine from vendor i is in use?
3. Suppose a machine from vendor i costs $\$c_i$. What is the long run cost per unit time of operating this policy?

8.54 April One Computers provides the following warranty on all its hard drives for its customers that sign a long-term contract with it: it will replace a malfunctioning drive with a new one for free any number of times for up to one year after the initial purchase. If the hard drive fails after one year, the customer must purchase a new one with a new one-year free replacement warranty. Suppose the lifetimes of these drives are iid exponential variables with common mean $1/\lambda$. The customer pays c for each new hard drive that is not replaced for free. It costs the company d to make the drive, ($d < c$). All failed hard drives are discarded. Suppose a customer has signed a long-term contract with April One Computers for his hard drive. Let $Z(t)$ be the total cost (excluding the initial purchase cost) to the customer over the interval $(0, t]$, and $Y(t)$ be the total profits to the April One Computers from this contract over the period $(0, t]$. Assume that replacement is instantaneous.

1. Show that $\{Z(t), t \geq 0\}$ and $\{Y(t), t \geq 0\}$ are renewal reward processes.
2. Compute the expected cost per year to the customer and the expected profit per year to the producer.

8.55 A machine is subject to shocks that arrive one at a time according to a renewal process with inter-renewal time cdf F. If the machine is working when a shock occurs it fails with probability $\alpha \in (0, 1)$, independent of history. When the machine fails, it undergoes repairs; the repairtimes being iid $\exp(\lambda)$ random variables. The shocks have no effect on the machine under repair. The repaired machine is as good as new. Let $X(t)$ be the number of working machines at time t. Assume that a shock has occurred at time zero and $X(0+) = 0$.

1. Is $\{X(t), t \geq 0\}$ a CTMC for all distributions F? Why or why not?
2. Is $\{X(t), t \geq 0\}$ an SMP? Why or why not?
3. Is $\{X(t), t \geq 0\}$ an ARP? Why or why not?
4. Is $\{X(t), t \geq 0\}$ a regenerative process? Why or why not?

8.56 Suppose the lifetime of a machine is an $\exp(\mu)$ random variable. A repairperson visits a machine at random times as follows. If the machine is working when the repairperson visits, he simply goes away and returns after d amount of time, where d is a fixed positive constant. If the machine is down when the repairperson visits, he repairs the machine in r amount of time, where r is a fixed positive constant. Then he goes away and returns after d amount of time. The machine is as good as new after the repair is complete. Let $N(t)$ be the number of repair completions during $(0, t]$. Furthermore, let $X(t)$ be 0 if the machine is working at time t, 1 if it is down at time t, and 2 if it is under repair at time t. Suppose a repair has just completed at time 0.

1. Is $\{N(t), t \geq 0\}$ a renewal process? Why or why not?
2. Compute

$$\lim_{t \to \infty} \frac{N(t)}{t}.$$

3. Is $\{X(t), t \geq 0\}$ a semi-Markov process? Why or why not? Describe its kernel if it is an SMP.
4. Is $\{X(t), t \geq 0\}$ a regenerative process? Why or why not?

8.57 A single server works on an infinite supply of jobs. The amount of time it takes the server to work on the jobs are iid exponential with rate μ. If the service of a job takes less than b units of time, the server immediately starts on a new job. Otherwise, it takes a break for $a > 0$ units of time and then starts a new job immediately afterwards. Suppose that at time zero, the server starts a new job. Let $A(t) = 1$ if the server is working at time t, and $A(t) = 0$ otherwise.

1. Show that $\{A(t), t \geq 0\}$ is an ARP.
2. Compute $\lim_{t \to \infty} P(A(t) = 1)$.

8.58 Suppose that $\{A(t), t \geq 0\}$ and $\{B(t), t \geq 0\}$ are respectively the age and remaining life processes associated with the renewal process $\{N(t), t \geq 0\}$.

1. Derive a renewal-type equation for $P(A(t) > B(t))$.
2. Find $\lim_{t \to \infty} P(A(t) > B(t))$.
3. Suppose now that times between renewal points are exponentially distributed with mean $1/\lambda$. Determine $P(A(t) > B(t))$ for any $t > 0$.

8.59 Tour boats that make trips on a river leave every T minutes. Potential customers arrive according to a Poisson process with rate λ per minute. Customers are impatient so that an arriving customer may leave before the boat departs. The amount of times the customers are willing to wait are assumed to be iid non-negative random variables with with common cdf $G(\cdot)$ and the corresponding pdf $g(\cdot)$. A fixed cost of $K > 0$ is incurred for each trip of the boat and a reward of $r > 0$ is earned for each customer joining a trip. Assume that the boat's capacity is unlimited and that the boat makes the trip even if there is nobody on the boat.

1. Compute the long run net revenue per unit time.

2. Suppose that $G(\cdot)$ is uniform$(0, a)$. Give an expression for the optimal value of T that maximizes the long run net revenue per unit time.

8.60 The lifetime of an oilpump (in days) is an exp(λ) random variable and the repairtime (in days) is an exp(μ) random variable. The successive up-times and repairtimes are independent. When the pump is working, it produces g gallons of oil per day. The oil is consumed at a rate of $c < g$ gallons a day. The excess oil is stored in a tank of unlimited capacity. Thus the oil reserves increase by $g - c$ gallons a day when the pump is working. When the pump fails, the oil reserves decrease by c gallons a day. When the pump is repaired, it is turned on as soon as the tank is empty. Suppose it costs h cents per gallon per day to store the oil. Using the renewal reward theory, compute the storage cost per day in the long run.

8.61 A machine is used to manufacture an item that requires 45 minutes of machining. The machine is subject to failures and repairs as follows: The machine works for an exponential amount of time with mean 24 hours before it fails. Repairs take one hour to complete on the average. The machine is as good as new after the repairs are complete. Consider two cases:

Case 1: When the machine fails, the item it is working on is damaged and has to be discarded. The machine starts with a new piece when it is repaired.

Case 2: When the machine fails, the item it is working on is undamaged and the machine continues working on it when it is repaired.

Suppose the machine is up at time zero and starts working on a new item. Let $N(t)$ be the number of items manufactured up to time t. For each case,

1. show that $\{N(t), t \geq 0\}$ is an RP,

2. compute the long run rate of production.

8.62 A machine shop consists of K independent and identical machines. Each machine works for an exp(μ) amount of time before it fails. During its lifetime each machine produces revenue at a rate of $\$R$ per unit time. When a machine fails it needs to be replaced by a new and identical machine at a cost of $\$C_m$. Any number of machines can be replaced simultaneously. The replacement operation is instantaneous. The repairperson charges an additional $\$C_v$ per visit, regardless of how many machines are replaced on a given visit. Consider the following policy: Wait until the number of working machines falls below k, (where $0 < k \leq K$ is a fixed integer) and then replace all the failed machines simultaneously.

1. Compute $g(k)$, the long run cost rate of following this policy.

2. Compute the optimal value of the parameter k that minimizes this cost rate for the following data: Mean lifetime of the machine is 10 days, the revenue rate is $100 per day, the replacement cost is $750 per machine, and the visit charge is $250.

8.63 The inventory of gasoline at a gas pump is managed as follows: The underground gas tank holds 40,000 gallons. Demand rate for gas is continuous and deterministic: 4,000 gallons a day. When the level of gas in the tank falls to 10,000 gallons, an order is placed for a new tankerful of gas. Unfortunately, the delivery time is a continuous random variable distributed uniformly over $(2,3)$ days. When a gas tanker arrives, it fills up the underground tank (instantaneously) and leaves. Demand occurring when the gas tank is empty is lost.

1. Compute the long run fraction of the time the gas station is out of gas.

2. Suppose the gas station has to pay \$500 each time the gas tanker visits the gas station, regardless of how much gas is delivered. The manager sells the gas for 10 cents a gallon more than the purchase price. Compute the long run rate of net income at the gas station.

8.64 A machine produces items continuously at a rate r per unit time as long as it is functioning. The demand for these items occurs at a constant rate d per unit time. We assume that $d < r$ so that the stock of the items increases at rate $r - d$ when the machine is up. The lifetime of the machine is $\exp(\mu)$. The production stops when either the machine fails or the stock level reaches a pre-specified level S. When the production is halted the stock reduces at rate d. If the production stops due to machine failure, repair starts immediately and lasts an $\exp(\lambda)$ amount of time, and the machine is as good as new when the repair is complete. The production starts again when the machine is up and the stock level reaches zero. The machine can fail only if it is producing items.

1. Let $X(t)$ be the stock level at time t. Assume that $X(0) = 0$ and the machine is up at time 0. Show that $X(t)$ is a regenerative process.

2. Compute the long run fraction of the time that the machine is up (regardless of whether it is producing items).

3. Compute the limit of $E(X(t))$ as $t \to \infty$.

8.65 Let $\{N(t), t \geq 0\}$ be a PP(λ). By conditioning on an appropriate event time in the PP, derive a renewal type equation for $H(t) = P(N(t) \text{ is even})$. Compute the LST of $H(t)$ and invert it to get an explicit expression for $H(t)$.

8.66 Let $\{N(t), t \geq 0\}$ be a renewal process generated by $\{X_n, n \geq 1\}$ with common cdf $G(\cdot)$, mean τ and second moment $s^2 < \infty$. Assume that X_n is aperiodic. Let $A(t)$ be the age at time t and $B(t)$ be the remaining life at time t in this process at time t, $t \geq 0$. Let $Z(t) = \min(A(t), B(t))$.

1. Derive a renewal-type equation for $E(Z(t)), t \geq 0$.

2. Using the key renewal theorem compute $\lim_{t \to \infty} E(Z(t))$.

3. Show that $\{Z(t), t \geq 0\}$ a regenerative process.

4. Let $x \geq 0$ be a fixed real number. Using the theory of RGPs, compute the $\lim_{t \to \infty} P(Z(t) > x)$.

8.67 Here is a very simple model of how an e-retailer tracks the customers visiting its website. It categorizes a customer as satisfied or unsatisfied. A customer may have zero or one outstanding order with the e-retailer. When a satisfied (unsatisfied) customer with zero outstanding orders visits its website, he/she places a new order with probability .3 (.1), or leaves the website without placing an order with probability .7 (.9). If a customer does not place an order, then his satisfaction status does not change. If the customer places an order, he/she awaits its delivery, during which time he/she does not return to the website. The delivery times (in days) are iid $U(2, 7)$ continuous random variables. When the delivery is received, the customer state becomes unsatisfied if the delivery time exceeds 5 days, otherwise he/she becomes satisfied. (Assume that the product delivered is always satisfactory). Assume that a customer starts in the satisfied state and continues this way forever. Suppose a satisfied (unsatisfied) customer returns (after the last visit with no purchase, or after the last delivery) to the website after an exponential amount of time with mean 10 days (20 days).

1. State the assumptions needed to model the state of the customer by a four-state semi-Markov process $\{X(t), t \geq 0\}$. Give the kernel $G(\cdot)$ of the SMP. Is the SMP irreducible? Is it periodic or aperiodic? Is it recurrent or transient?

2. Compute the long run fraction of the time a customer spends in satisfied state.

3. Suppose the customer gets a satisfactory delivery at time 0. Compute the expected time when he/she becomes unsatisfied again.

4. Suppose the average size of a purchase by a satisfied customer is 100 dollars and that by an unsatisfied customer is 30 dollars. Compute the long run amount of revenue per year the e-retailer gets from a single customer.

5. Suppose the e-retailer gives every unsatisfied customer (every time they have an unsatisfactory experience with the delivery) a coupon that gives them 50% off their next purchase. As a result of this offer, an unsatisfied customer instantly becomes a satisfied customer. What is the long run revenue per year from a single customer under this policy? Should the e-retailer follow this policy?

8.13 Conceptual Exercises

8.1 Suppose $\{X_n, n \geq 0\}$ is a DTMC on $\{0, 1, 2, \cdots\}$. Suppose we define $S_0 = 0$ and $S_n = \min\{k > S_{n-1} : X_k = X_{S_{k-1}}\}$, $n \geq 1$. Is $\{S_n, n \geq 1\}$ a renewal sequence? (Hint: Consider if X_0 is a constant or not.)

8.2 Let $\{N(t), t \geq 0\}$ be an RP. Suppose each event is registered with probability p, independent of everything else. Let $N^*(t)$ be the number of registered events up to t. Is $\{N^*(t), t \geq 0\}$ an RP?

8.3 Let $\{N(t), t \geq 0\}$ be an RP. Define

$$N_1(t) = \left[\frac{N(t)}{2}\right],$$

$$N_2(t) = \left[\frac{N(t) + 1}{2}\right],$$

where $[x]$ is the largest integer less than or equal to x. Is $\{N_1(t), t \geq 0\}$ an RP? What about $\{N_2(t), t \geq 0\}$?

8.4 Complete the proof of Theorem 8.6 when $\tau = \infty$, by showing that the limsup and liminf of $N(t)/t$ are both zero.

8.5 Derive Equation 8.17 by directly using the renewal argument.

8.6 Complete the proof of Theorem 8.12 when $\tau = \infty$, by showing that the limsup and liminf of $M(t)/t$ are both zero.

8.7 Show that a random variable with the following pmf is aperiodic:
$$P(X = e) = .5, \quad P(X = \pi) = .5.$$

8.8 Let $\{N^D(t), t \geq 0\}$ be a delayed RP satisfying Equation 8.49. Show that
$$P(N^D(t) < \infty) = 1, \quad \text{for all} \quad t \geq 0.$$

8.9 Let $\{N^D(t), t \geq 0\}$ be a delayed RP satisfying Equation 8.49. Show that
$$
\begin{aligned}
P(N^D(t) = 0) &= 1 - F(t), \\
P(N^D(t) = k) &= F * (G_{k-1} - G_k)(t), \quad k \geq 1.
\end{aligned}
$$

8.10 Let $\{N^D(t), t \geq 0\}$ be a delayed RP satisfying Equation 8.49. Show that
$$P(\lim_{t \to \infty} N^D(t) = \infty) = 1.$$

8.11 Let $\{N^D(t), t \geq 0\}$ be a delayed RP satisfying Equation 8.49. Show that
$$\lim_{t \to \infty} \frac{N^D(t)}{t} = \frac{1}{\tau}, \quad w.p.1,$$
where $\tau = E(X_2)$.

8.12 Let $\{M^D(t), t \geq 0\}$ be the renewal function of a delayed RP satisfying Equation 8.49. Show that
$$\lim_{t \to \infty} \frac{M^D(t)}{t} = \frac{1}{\tau}.$$
where $\tau = E(X_2)$.

8.13 Let $\{X(t), t \geq 0\}$ be an ARP as given in Definition 8.7. Let $N_i(t)$ be the number of entries into state i ($i = 0, 1$) over $(0, t]$. Is $\{N_i(t), t \geq 0\}$ an RP?

8.14 Let $N_j(t)$ be as in Example 8.18. Show that the limiting values of $N_j(t)/t$ and $E(N_j(t))/t$ are independent of the initial state of the CTMC.

8.15 Let $\{N^e(t), t \geq 0\}$ be an equilibrium renewal process. Show that the distribution of $N^e(s + t) - N^e(s)$ is independent of s, i.e., $\{N^e(t), t \geq 0\}$ has stationary increments. Are the increments independent?

8.16 Let $\{X_n, n \geq 0\}$ be an irreducible, aperiodic, positive recurrent DTMC with limiting distribution $\{\pi_j, j \in S\}$. Use regenerative processes to show that the mean inter-visit time to state j is given by $1/\pi_j$.

8.17 Let $\{(X_n, B_n, C_n), n \geq 1\}$ be a sequence of iid tri-variate random variables. Suppose X_n's are non-negative, representing the length of the nth cycle, B_n the benefit accrued at the end of the nth cycle, and C_n the cost incurred at the end of the nth cycle. Let $Z_b(t)$ $(Z_c(t))$ be the total benefit (cost) up to time t. Show that the long run cost to benefit ratio is given by

$$\lim_{t \to \infty} \frac{Z_c(t)}{Z_b(t)} = \lim_{t \to \infty} \frac{\mathsf{E}(Z_c(t))}{\mathsf{E}(Z_b(t))} = \frac{\mathsf{E}(C_1)}{\mathsf{E}(R_1)}.$$

8.18 Derive the following analog of Theroem 8.41:

$$L_q = \lambda W_q,$$

where W_q is the expected time in the queue (time in the system - the time in service), and L_q is the expected number in the queue (number in the system - number in service).

8.19 Let $\{X(t), t \geq 0\}$ be a positive recurrent and aperiodic SMP with state-space S, kernel G, and limiting distribution $\{p_i, i \in S\}$. Suppose we incur cost at rate c_i whenever the SMP is in state i. Show that the long run rate at which we incur cost is given by $\sum_{i \in S} p_i c_i$.

8.20 Consider the system in Conceptual Exercise 8.19. Suppose the costs are discounted continuously with a discount factor $\alpha > 0$. Let $\phi(i)$ be the total expected discounted cost incurred over the infinite horizon given that the SMP enters state i at time 0. Let

$$\gamma_i = \frac{c_i}{\alpha}(1 - \tilde{G}_i(\alpha)),$$

where $\tilde{G}_i(\cdot)$ is the LST of the sojourn time in state i. Show that the $\phi(i)$'s satisfy the following equation:

$$\phi(i) = \gamma_i + \sum_{j \in S} \tilde{G}_{i,j}(\alpha)\phi(j).$$

CHAPTER 9

Markov Regenerative Processes

Once Sherlock Holmes and Dr. Watson went camping. They set up their tent under the trees in a campground, had their outdoor meals, and went to sleep. In the middle of the night Sherlock woke up Dr. Watson and said,
"My dear Dr. Watson, tell me what you see."
Dr. Watson opened his eyes and said, "I see a beautiful night sky through the trees."
"And, my dear Watson, what does that tell you?"
"It tell me that it is a clear night, and the world is peacefully sleeping."
"What else does it tell you?"
"I don't know what else you want me say. Why, what does it tell you, Sherlock?"
Sherlock gravely replied, "Elementary, my dear Watson. It tells me that someone has stolen our tent."

9.1 Definitions and Examples

Markov renewal theory is a natural generalization of renewal theory, and as the name suggests, it combines the concepts from Markov chains and renewal processes. We begin with a definition.

Definition 9.1 Markov Renewal Sequence. *A sequence of bivariate random variables* $\{(X_n, S_n), n \geq 0\}$ *is called a Markov renewal sequence (MRS) if*
1. $S_0 = 0$, $S_{n+1} \geq S_n$, $X_n \in I$, *where* I *is a discrete set,*
2. For all $n \geq 0$,

$$P(X_{n+1} = j, S_{n+1} - S_n \leq x | X_n = i, S_n, X_{n-1}, S_{n-1}, \cdots, X_0, S_0)$$
$$= P(X_{n+1} = j, S_{n+1} - S_n \leq x | X_n = i)$$
$$= P(X_1 = j, S_1 \leq x | X_0 = i).$$

Now define $Y_n = S_n - S_{n-1}$ for $n \geq 1$. There is a one to one correspondence between the sequence $\{(X_n, S_n), n \geq 0\}$ and the sequence $\{X_0, (X_n, Y_n), n \geq 1\}$. In fact we have used the sequence $\{X_0, (X_n, Y_n), n \geq 1\}$ earlier to define a CTMC in Definition 6.1 on page 202 and an SMP in Definition 8.9 on page 408. In a similar vein we use Markov renewal sequences to build a class of processes called the Markov regenerative processes (MRGPs) that encompasses CTMCs and

SMPs. We allow the state-spaces of such a process to be discrete, say $\{0, 1, 2, \cdots\}$, or continuous, say $[0, \infty)$.

Definition 9.2 Markov Regenerative Process. *A stochastic process $\{Z(t), t \geq 0\}$ with state-space S is called a Markov regenerative process if there is a Markov renewal sequence $\{(X_n, S_n), n \geq 0\}$ such that $\{Z(t + S_n), t \geq 0\}$ given $\{Z(u) : 0 \leq u < S_n, X_n = i\}$ is stochastically identical to $\{Z(t), t \geq 0\}$ given $X_0 = i$ and is conditionally independent of $\{Z(u) : 0 \leq u < S_n, X_n = i\}$ given $X_n = i$.*

Several comments on the implications of the above definition are in order. In most applications we find that $X_n = Z(S_n+)$ or $X_n = Z(S_n-)$. One can see that if $\{(X_n, S_n), n \geq 0\}$ is Markov renewal sequence then $\{X_n, n \geq 0\}$ is a DTMC. Thus in most applications $\{Z(S_n+), n \geq 0\}$ or $\{Z(S_n-), n \geq 0\}$ is a DTMC. Hence sometimes an MRGP is also called a process with an embedded Markov chain. We also see that

$$\mathsf{P}(Z(t + S_n) \leq x | Z(u) : 0 \leq u < S_n, X_n = i) = \mathsf{P}(Z(t) \leq x | X_0 = i).$$

A typical sample path of an MRGP is shown in Figure 9.1, where we have assumed that the state-space of the MRGP is discrete, and $X_n = Z(S_n-)$.

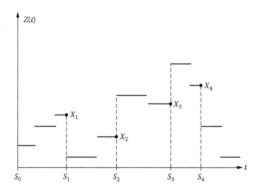

Figure 9.1 *A typical sample path of a Markov regenerative process.*

In the examples below we verify that a given process is an MRGP by identifying the embedded MRS. To do this we need to know how to describe an MRS. The next theorem answers this question. We need the following notation. Let

$$a_i = \mathsf{P}(X_0 = i), \quad i \in I.$$

The vector $a = [a_i]_{i \in I}$ is called the initial distribution. Also, let

$$G_{i,j}(y) = \mathsf{P}(X_1 = j, S_1 \leq y | X_0 = i), \quad i, j \in I. \tag{9.1}$$

Let $G(y) = [G_{i,j}(y)]_{i,j \in I}$. We call $G = G(\cdot)$ the kernel of the MRS.

Theorem 9.1 Characterization of an MRS. *An MRS $\{(X_n, S_n), n \geq 0\}$ is completely described by the initial distribution a and the kernel G.*

Proof: Describing the sequence $\{(X_n, S_n), n \geq 0\}$ is equivalent to describing the sequence $\{X_0, (X_n, Y_n), n \geq 1\}$. Thus it suffices to show that

$$P(X_0 = i_0, X_1 = i_1, Y_1 \leq y_1, \cdots, X_n = i_n, Y_n \leq y_n)$$
$$= a_{i_0} G_{i_0, i_1}(y_1) \cdots G_{i_{n-1}, i_n}(y_n), \tag{9.2}$$

for $n \geq 1$, $i_k \in I$ and $y_k \geq 0$ for $1 \leq k \leq n$. We leave it to the reader to show this by induction. ∎

We get the following as a corollary to the above theorem. We leave the proof to the reader.

Theorem 9.2 *Let $\{(X_n, S_n), n \geq 0\}$ be an MRS with initial distribution a and kernel G. Then $\{X_n, n \geq 0\}$ is a DTMC with the same initial distribution a and transition probability matrix*

$$P = G(\infty),$$

with the convention that $p_{i,i} = 1$ if $\sum_{j \in I} G_{i,j}(\infty) = 0$.

We illustrate all the concepts introduced so far by examples.

Example 9.1   MRGPs vs. RGPs. It is easy to get confused between an MRGP and an RGP (regenerative process). In this example we make the distinction between the two clearer. The future of the MRGP from $t = S_n$ depends on its past up to time S_n only through X_n. The future of an RGP from $t = S_n$ onwards is completely independent of its past up to time S_n. In particular, if $\{X_n, n \geq 0\}$ is a sequence of iid random variables, then an MRGP is an RGP. Thus an RGP is a special case of an MRGP. If $\{X_n, n \geq 0\}$ is an irreducible recurrent DTMC, then we see that an MRGP is also an RGP, since every time X_n visits a given state, say zero, the process starts afresh, and this will happen infinitely often. However, when $\{X_n, n \geq 0\}$ is transient, an MRGP is not a regenerative process.

For example, let $\{N(t), t \geq 0\}$ be a recurrent RP with common inter-renewal time cdf A. Let S_n be the nth renewal time, with $S_0 = 0$. Define $X_n = N(S_n+) = n$. Then $\{(X_n, S_n), n \geq 0\}$ is an MRS with kernel $G = [G_{i,j}]$ where the non-zero elements of G are as given below:

$$G_{i,i+1}(x) = A(x), \quad i \geq 0.$$

Thus $\{N(t), t \geq 0\}$ is an MRGP with $\{(X_n, S_n), n \geq 0\}$ as an embedded MRS. However, it is not a regenerative process. This shows that the set of MRGPs is a strict superset of the set of RGPs. ∎

Example 9.2   CTMCs as MRGPs. Let $\{Z(t), t \geq 0\}$ be a CTMC with state-space $I = \{0, 1, 2, \cdots\}$ and generator matrix Q. Show that $\{Z(t), t \geq 0\}$ is an MRGP.

Let $S_0 = 0$ and S_n be the nth jump epoch in the CTMC. Define $X_n = Z(S_n+)$. Then, from Definition 6.1 on page 202, it follows that $\{(X_n, S_n), n \geq 0\}$ is an MRS with kernel

$$G_{i,j}(y) = \frac{q_{i,j}}{q_i}(1 - e^{-q_i y}), \quad i, j \in I, \ i \neq j, \ q_i > 0, \ y \geq 0.$$

If $q_i = 0$, we define $G_{i,j}(y) = 0$ for all $j \in I$. We also define $G_{i,i}(y) = 0$ for all $i \in I$. The Markov property of the CTMC implies that $\{Z(t + S_n), t \geq 0\}$ depends on $\{Z(t), 0 \leq t < S_n, X_n\}$ only through X_n. Thus $\{Z(t), t \geq 0\}$ is an MRGP. ∎

Example 9.3 SMPs as MRGPs. Let $\{Z(t), t \geq 0\}$ be an SMP with state-space $\{0, 1, 2, \cdots\}$ and kernel G. Show that $\{Z(t), t \geq 0\}$ is an MRGP.

We assume that the SMP has entered the initial state at time 0. Let $S_0 = 0$ and S_n be the nth jump epoch in the SMP. Define $X_n = Z(S_n+)$. Then, from Definition 8.9 on page 408, it follows that $\{(X_n, S_n), n \geq 0\}$ is an MRS with kernel G. The Markov property of the SMP at jump epochs implies that $\{Z(t + S_n), t \geq 0\}$ depends on $\{Z(t), 0 \leq t < S_n, X_n\}$ only through X_n. Thus $\{Z(t), t \geq 0\}$ is an MRGP. ∎

Example 9.4 $M/G/1$ **Queue.** Let $Z(t)$ be the number of customers in an $M/G/1$ queue with PP(λ) arrivals and common service time (s.t.) cdf $B(\cdot)$. Show that $\{Z(t), t \geq 0\}$ is an MRGP.

Assume that either the system is empty initially, or a service has just started at time 0. Let S_n be the time of the nth departure, and define $Y_n = S_n - S_{n-1}$, and $X_n = Z(S_n+)$. Note that if $X_n > 0$, Y_n is the nth service time. If $X_n = 0$, then Y_n is the sum of an exp(λ) random variable and the nth service time. Then, due to the memoryless property of the PP and the iid service times, we get, for $i > 0$ and $j \geq i - 1$,

$$
\begin{aligned}
G_{i,j}(y) &= \mathsf{P}(X_{n+1} = j, S_{n+1} - S_n \leq x | X_n = i, S_n, X_{n-1}, S_{n-1}, \cdots, X_0, S_0) \\
&= \mathsf{P}(X_{n+1} = j, Y_{n+1} \leq y | X_n = i) \\
&= \mathsf{P}(j - i + 1 \text{ arrivals during a s.t. and s.t.} \leq y) \\
&= \int_0^y e^{-\lambda t} \frac{(\lambda t)^{j-i+1}}{(j - i + 1)!} dB(t).
\end{aligned}
$$

For $i = 0$ and $j \geq 0$ we get

$$
\begin{aligned}
G_{0,j}(y) &= \mathsf{P}(X_{n+1} = j, S_{n+1} - S_n \leq x | X_n = 0, S_n, X_{n-1}, S_{n-1}, \cdots, X_0, S_0) \\
&= \mathsf{P}(X_{n+1} = j, Y_{n+1} \leq y | X_n = 0) \\
&= \mathsf{P}(j \text{ arrivals during a s.t. and s.t.} + \text{idle time} \leq y) \\
&= \int_0^y (1 - e^{-\lambda(y-t)}) e^{-\lambda t} \frac{(\lambda t)^j}{j!} dB(t).
\end{aligned}
$$

Thus $\{(X_n, S_n), n \geq 0\}$ is an MRS with kernel G. The queue length process from time S_n onwards depends on the history up to time S_n only via X_n. Hence $\{(Z(t), t \geq 0\}$ is an MRGP. Note that we had seen in Section 7.6.1 that $\{X_n, n \geq 0\}$ is an embedded DTMC with transition probability matrix $P = G(\infty)$. One can show that this is consistent with Equation 7.36 on page 337. ∎

Example 9.5 $G/M/1$ **Queue.** Let $Z(t)$ be the number of customers in a $G/M/1$ queue with common inter-arrival time (i.a.t.) cdf $A(\cdot)$, and iid $\exp(\mu)$ service times. Show that $\{Z(t), t \geq 0\}$ is an MRGP.

Assume that an arrival has occurred at time 0. Let S_n be the time of the nth arrival, and define $X_n = Z(S_n-)$, and $Y_n = S_n - S_{n-1}$, the nth inter-arrival time. Due to the memoryless property of the PP and the iid service times, we get, for $0 < j \leq i+1$,

$$G_{i,j}(y) = \mathsf{P}(i+1-j \text{ departures during an i.a.t. and i.a.t.} \leq y)$$

$$= \int_0^y e^{-\mu t} \frac{(\mu t)^{i+1-j}}{(i+1-j)!} dA(t),$$

and

$$G_{i,0}(y) = A(y) - \sum_{j=1}^{i+1} G_{i,j}(y).$$

Thus $\{(X_n, S_n), n \geq 0\}$ is an MRS. The queue length process from time S_n onwards depends on the history up to time S_n only via X_n. Hence $\{(Z(t), t \geq 0\}$ is an MRGP. Note that we had seen in Section 7.6.2 that $\{X_n, n \geq 0\}$ is an embedded DTMC with transition probability matrix $P = G(\infty)$. One can show that this is consistent with Equation 7.44 on page 342. ∎

Example 9.6 $M/G/1/1$ **Retrial Queue.** Consider an $M/G/1/1$ retrial queue as described in Section 7.7 of Chapter 7. Let $R(t)$ be the number of customers in orbit and $I(t)$ be the number of customers in service at time t. Let $X(t) = R(t) + I(t)$ be the number of customers in the system at time t. The service times are iid with common cdf $B(\cdot)$. The arrival process is PP(λ), and retrial times are iid $\exp(\theta)$. Show that $\{(R(t), I(t)), t \geq 0\}$ and $\{X(t), t \geq 0\}$ are MRGPs.

Assume that the server is idle initially. Let S_n be the time of the nth service completion, and define $X_n = X(S_n+)$, and $Y_n = S_n - S_{n-1}$, the nth idle time plus the following service time. Since the server is idle after a service completion, X_n is also the number of customers in the orbit at the nth service completion. Thus, if $X_n = i$, the $(n+1)$st idle time is $\exp(\lambda_i)$, where $\lambda_i = \lambda+i\theta$. Due to the memoryless property of the PP and the iid service times, we get, for $0 \leq i \leq j+1$,

$$G_{i,j}(y) = \frac{\lambda}{\lambda_i} \mathsf{P}(j - i \text{ arrivals during a s.t. and s.t. + idle time} \leq y)$$

$$+ \frac{i\theta}{\lambda_i} \mathsf{P}(j - i + 1 \text{ arrivals during a s.t. and s.t. + idle time} \leq y)$$

$$= \frac{\lambda}{\lambda_i} \int_0^y (1 - e^{-\lambda_i(y-t)}) e^{-\lambda t} \frac{(\lambda t)^{j-i}}{(j-i)!} dB(t)$$

$$+ \frac{i\theta}{\lambda_i} \int_0^y (1 - e^{-\lambda_i(y-t)}) e^{-\lambda t} \frac{(\lambda t)^{j-i+1}}{(j-i+1)!} dB(t).$$

Thus $\{(X_n, S_n), n \geq 0\}$ is an MRS with kernel G. The $\{(R(t), I(t)), t \geq 0\}$ and $\{X(t), t \geq 0\}$ processes from time S_n onwards depend on the history up to time S_n

only via X_n. Hence both are MRGPs with the same embedded MRS. Note that we had seen in Theorem 7.19 on page 346 that $\{X_n, n \geq 0\}$ is an embedded DTMC in the $\{X(t), t \geq 0\}$ process. It is an embedded DTMC in the $\{(R(t), I(t)), t \geq 0\}$ process as well. ■

9.2 Markov Renewal Process and Markov Renewal Function

Let $\{(X_n, S_n), n \geq 0\}$ be a given Markov renewal sequence with kernel G. Define the counting process

$$N(t) = \sup\{n \geq 0 : S_n \leq t\}. \tag{9.3}$$

We say that the process $\{N(t), t \geq 0\}$ is regular if $N(t) < \infty$ with probability 1 for all $0 \leq t < \infty$. We derive a sufficient condition for regularity next. First we need the following notation:

$$G_i(y) = \mathsf{P}(Y_1 \leq y | X_0 = i) = \sum_{j \in I} G_{i,j}(y). \tag{9.4}$$

Theorem 9.3 Sufficient Condition for Regularity. *The process* $\{N(t), t \geq 0\}$ *defined by Equation 9.3 is regular if there exists an* $\epsilon > 0$ *and a* $\delta > 0$ *such that*

$$G_i(\delta) < 1 - \epsilon, \quad i \in I. \tag{9.5}$$

Proof: Let $\{Y_n^*, n \geq 1\}$ be a sequence of iid random variables with common pmf

$$\mathsf{P}(Y_n^* = 0) = 1 - \epsilon, \quad \mathsf{P}(Y_n^* = \delta) = \epsilon. \tag{9.6}$$

Let $\{S_n^*, n \geq 1\}$ be the renewal sequence and $\{N^*(t), t \geq 0\}$ be the RP generated by $\{Y_n^*, n \geq 1\}$. Equations 9.5 and 9.6 imply that

$$\mathsf{P}(Y_n \leq t) \leq \mathsf{P}(Y_n^* \leq t), \quad t \geq 0, \ n \geq 1.$$

Hence

$$\mathsf{P}(S_n \leq t) \leq \mathsf{P}(S_n^* \leq t), \quad t \geq 0, \ n \geq 1,$$

which implies

$$\mathsf{P}(N(t) \geq k) \leq \mathsf{P}(N^*(t) \geq k), \quad k \geq 1.$$

However $N^*(t) < \infty$ with probability 1, from Theorem 8.3 on page 373. Hence $N(t) < \infty$ with probability 1. ■

The condition in Equation 9.5 is not necessary, and can be relaxed. We refer the reader to Cinlar (1975) for weaker conditions. For a regular process $\{N(t), t \geq 0\}$, define

$$X(t) = X_{N(t)}. \tag{9.7}$$

It can be seen that $\{X(t), t \geq 0\}$ is an SMP with kernel G. We say that $\{X(t), t \geq 0\}$ is the SMP generated by the MRS $\{(X_n, S_n), n \geq 0\}$. Define $N_j(t)$ to be the number of entries into state j by the SMP over $(0, t]$, and

$$M_{i,j}(t) = \mathsf{E}(N_j(t) | X_0 = i), \quad i, j \in I, \ t \geq 0. \tag{9.8}$$

Thus $M_{i,j}(t)$ is the expected number of transitions into state j over $(0, t]$ starting from $X(0) = i$.

Definition 9.3 Markov Renewal Process. *The vector valued process $\{\mathbf{N}(t) = [N_j(t)]_{j \in I}, t \geq 0\}$ is called the Markov renewal process generated by the Markov renewal sequence $\{(X_n, S_n), n \geq 0\}$.*

Next, following the development in the renewal theory, we define the Markov renewal function.

Definition 9.4 Markov Renewal Function. *The matrix $M(t) = [M_{i,j}(t)]_{i,j \in I}$ is called the Markov renewal function generated by the Markov renewal sequence $\{(X_n, S_n), n \geq 0\}$.*

Theorem 9.4 *Suppose the condition in Equation 9.5 holds. Then*

$$\sum_{j \in I} M_{i,j}(t) \leq \frac{1}{\epsilon} \left(\frac{t}{\delta} + 1 \right) < \infty, \quad t \geq 0.$$

Proof: Consider the process $\{N^*(t), t \geq 0\}$ introduced in the proof of Theorem 9.3. Since $N(t) = \sum_{j \in I} N_j(t)$, we have

$$\begin{aligned}
\sum_{j \in I} M_{i,j}(t) &= \sum_{j \in I} \mathsf{E}(N_j(t)|X_0 = i) \\
&= \mathsf{E}(N(t)|X_0 = i) \\
&\leq \mathsf{E}(N^*(t)) \leq \frac{1}{\epsilon} \left(\frac{t}{\delta} + 1 \right) < \infty.
\end{aligned}$$

This yields the theorem. ∎

We define the concept of matrix convolution next. We shall find it very useful in the study of Markov renewal processes. Let $A(t) = [A_{i,j}(t)]$ and $B(t) = [B_{i,j}(t)]$ be two matrices of functions for which the product $A(t)B(t)$ is defined. A matrix $C(t) = [C_{i,j}(t)]$ is called the convolution of A and B, written $C(t) = A * B(t)$, if

$$C_{i,j}(t) = \sum_k \int_0^t A_{i,k}(t-u)dB_{k,j}(u) = \sum_k \int_0^t dA_{i,k}(u)B_{k,j}(t-u).$$

The proof of the next theorem introduces the concept of Markov renewal argument. We urge the reader to become adept at using it.

Theorem 9.5 Markov Renewal Equation. *Let $M(t)$ be the Markov renewal function generated by the Markov renewal sequence $\{(X_n, S_n), n \geq 0\}$ with kernel G. $M(t)$ satisfies the following Markov renewal equation:*

$$M(t) = G(t) + G * M(t). \tag{9.9}$$

Proof: Condition on X_1 and S_1 to get

$$E(N_j(t)|X_0 = i, X_1 = k, S_1 = u) = \begin{cases} \delta_{k,j} + M_{k,j}(t-u) & \text{if } u \le t, \\ 0 & \text{if } u > t. \end{cases}$$

Here the Kronecker delta function is defined as $\delta_{i,j} = 1$ if $i = j$, and 0 otherwise. Hence we get

$$
\begin{aligned}
M_{i,j}(t) &= \int_0^\infty \sum_{k \in I} E(N_j(t)|X_0 = i, X_1 = k, S_1 = u) dG_{i,k}(u) \\
&= \int_0^t \sum_{k \in I} (\delta_{k,j} + M_{k,j}(t-u)) dG_{i,k}(u) \\
&= \int_0^t dG_{i,j}(u) + \int_0^t \sum_{k \in I} M_{k,j}(t-u) dG_{i,k}(u) \\
&= G_{i,j}(t) + [G * M(t)]_{i,j}.
\end{aligned}
$$

The above equation, in matrix form, is Equation 9.9. ∎

A k-fold convolution of matrix A with itself is denoted by A^{*k}. Using this notation we get the following analog of Theorem 8.8 on page 379.

Theorem 9.6 *Suppose the condition in Equation 9.5 holds. Then*

$$M(t) = \sum_{k=1}^\infty G^{*k}(t), \quad t \ge 0.$$

Proof: Iterating Equation 9.9, and writing M for $M(t)$ etc, we get

$$
\begin{aligned}
M &= G + G * M \\
&= G + G * (G + G * M) = G + G^{*2} + G^{*2} * M \\
&= \sum_{k=1}^n G^{*k} + G^{*n} * M, \quad n \ge 1.
\end{aligned}
\tag{9.10}
$$

Note that one consequence of the regularity condition of Equation 9.5 is that

$$\lim_{n \to \infty} G^{*n} e = 0,
\tag{9.11}$$

where e is a vector of all ones. Then Theorem 9.4 implies that

$$\lim_{n \to \infty} G^{*n} * M = 0.$$

Hence the theorem follows by letting $n \to \infty$ in Equation 9.10. ∎

Define the LST of a matrix as follows

$$\tilde{A}(s) = [\tilde{A}_{i,j}(s)] = \left[\int_0^\infty e^{-st} dA_{i,j}(t) \right].$$

Let $C(t) = A * B(t)$ for $t \geq 0$. Then one can show that

$$\tilde{C}(s) = \tilde{A}(s)\tilde{B}(s). \tag{9.12}$$

Using this notation we get the following analog of Theorem 8.9 on page 380.

Theorem 9.7 *Suppose the condition in Equation 9.5 holds. The LST of the Markov renewal function is given by*

$$\tilde{M}(s) = [I - \tilde{G}(s)]^{-1}\tilde{G}(s). \tag{9.13}$$

Proof: Taking the LST on both sides of Equation 9.9, we get

$$\tilde{M}(s) = \tilde{G}(s) + \tilde{G}(s)\tilde{M}(s).$$

Thus

$$(I - \tilde{G}(s))\tilde{M}(s) = \tilde{G}(s).$$

The regularity condition is one sufficient condition under which the inverse of $(I - \tilde{G}(s))$ exists. Hence the theorem follows. ∎

Note that if the SMP is recurrent and $X_0 = j$ then $\{N_j(t), t \geq 0\}$ is a standard renewal process, and $M_{j,j}(t)$ is its renewal function; while $\{N_i(t), t \geq 0\}$ $(i \neq j)$ is a delayed renewal process, and $M_{i,j}(t)$ is the delayed renewal function. Thus the limiting behavior of the Markov renewal processes and functions can be derived from the corresponding theorems from Chapter 8.

9.3 Key Renewal Theorem for MRPs

Equation 9.9 is called the Markov renewal equation and plays the same role as the renewal equation in renewal theory. The Markov renewal argument, namely the technique of conditioning on X_1 and S_1, generally yields an integral equation of the following form

$$H(t) = D(t) + G * H(t), \tag{9.14}$$

where G is the kernel of the Markov renewal sequence. Such an equation is called a Markov renewal-type equation. We study such an equation in this section. The next theorem gives the solution to the Markov renewal-type equation.

Theorem 9.8   Solution of Markov Renewal-Type Equation. *Suppose that*

$$\sup_{i \in I} |D_{i,j}(t)| \leq d_j(t) < \infty, \quad j \in I, \ t \geq 0, \tag{9.15}$$

and G satisfies the condition in Theorem 9.3. Then there exists a unique solution to the Markov renewal-type Equation 9.14 such that

$$\sup_{i \in I} |H_{i,j}(t)| \leq h_j(t) < \infty, \quad j \in I, \ t \geq 0, \tag{9.16}$$

and is given by

$$H(t) = D(t) + M * D(t), \tag{9.17}$$

where $M(t)$ is the Markov renewal function associated with G.

Proof: First we verify that the solution in Equation 9.17 satisfies Equation 9.14. Dropping (t) for ease of notation we get

$$H = D + M * D = D + (G + G * M) * D = D + G * (D + M * D) = D + G * H.$$

Here the first equality follows from Equation 9.17, the second from Equation 9.9, the third from associativity of matrix convolutions, and the last from Equation 9.14. Thus any H satisfying Equation 9.17 satisfies Equation 9.14.

Next we establish Equation 9.16. Suppose Equation 9.15 holds and let

$$c_j(t) = \sup_{0 \le u \le t} d_j(u).$$

From Equation 9.17 we get

$$
\begin{aligned}
|H_{i,j}(t)| &= \left| D_{i,j}(t) + \int_0^t \sum_{k \in I} dM_{i,k}(u) D_{k,j}(t - u) \right| \\
&\le d_j(t) + c_j(t) \sum_{k \in I} \int_0^t dM_{i,k}(u) \\
&\le c_j(t) \left(1 + \sum_{k \in I} M_{i,k}(t)\right) \\
&\le c_j(t) \left(1 + \frac{1}{\epsilon} \left(\frac{t}{\delta} + 1\right)\right),
\end{aligned}
$$

where the last inequality follows from Theorem 9.4. Thus Equation 9.16 follows if we define the right-hand side above as $h_j(t)$.

To show uniqueness, suppose H_1 and H_2 are two solutions to Equation 9.14 satisfying Equation 9.16. Then $H = H_1 - H_2$ satisfies Equation 9.16 and $H = G * H$. Iterating this n times we get

$$H = G^{*n} * H.$$

Using

$$\hat{h}_j(t) = \sup_{0 \le u \le t} h_j(u)$$

we get

$$
\begin{aligned}
|H_{i,j}(t)| &\le \left| \int_0^t \sum_{k \in I} dG_{i,k}^{*n}(u) H_{k,j}(t - u) \right| \\
&\le \hat{h}_j(t) \sum_{k \in I} G_{i,k}^{*n}(t).
\end{aligned}
$$

Letting $n \to \infty$ and using Equation 9.11 we see that the right-hand side goes to 0. Hence $H = 0$, or, $H_1 = H_2$. This shows uniqueness. ∎

In general solving the generalized Markov renewal equation is a formidable task, and we rarely attempt it. Hence we turn our attention to its limiting behavior as $t \to \infty$. The next theorem is the key renewal theorem for Markov renewal processes. We shall use the following notation from Section 8.9:

$$T_j = \min\{t \ge S_1 : X(t) = j\}, \quad j \in I,$$

$$\tau_i = \mathsf{E}(S_1|X(0) = i), \quad i \in I,$$
$$\tau_{i,j} = \mathsf{E}(T_j|X(0) = i), \quad i, j \in I.$$

Theorem 9.9 Key Markov Renewal Theorem. *Let H satisfy the following Markov renewal-type equation*

$$H(t) = D(t) + G * H(t). \tag{9.18}$$

1. *Suppose the condition of Theorem 9.3 is satisfied.*
2. *The SMP generated by the Markov renewal sequence is irreducible, aperiodic, positive recurrent, and has the limiting distribution $\{p_j, j \in I\}$.*
3. *Suppose $D(t) = [D_{i,j}(t)]$ satisfies*

   3a. $|D_{i,j}(t)| \leq d_j(t) < \infty, \quad i \in I, \ t \geq 0$,
   3b. $D_{i,j}(t) \ (i, j \in I)$ *is a difference of two non-negative bounded monotone functions*,
   3c. $\int_0^\infty \sum_{k \in I} \frac{p_k}{\tau_k} |D_{k,j}(u)| du < \infty, \quad j \in I.$

Then

$$\lim_{t \to \infty} H_{i,j}(t) = \int_0^\infty \sum_{k \in I} \frac{p_k}{\tau_k} D_{k,j}(u) du. \tag{9.19}$$

Proof: Since the condition of Theorem 9.3 is satisfied, and the condition [3a] holds, we see that the unique solution to Equation 9.18 is given by Equation 9.17. In scalar form we get

$$\begin{aligned}
H_{i,j}(t) &= D_{i,j}(t) + \sum_{k \in I} \int_0^t dM_{i,k}(u) D_{k,j}(t - u) \\
&= D_{i,j}(t) + \int_0^t dM_{i,i}(u) D_{i,j}(t - u) \\
&\quad + \sum_{k \neq i} \int_0^t dM_{i,k}(u) D_{k,j}(t - u).
\end{aligned} \tag{9.20}$$

Conditions [3a], [3b], and [3c], and the fact that $M_{i,i}(t)$ is a standard renewal function of a renewal process with mean inter-renewal time $\tau_{i,i}$, implies that we can use the key renewal theorem (Theorem 8.17 on page 389) to show that

$$\lim_{t \to \infty} [D_{i,j}(t) + \int_0^t dM_{i,i}(u) D_{i,j}(t - u)] = \frac{1}{\tau_{i,i}} \int_0^\infty D_{i,j}(u) du. \tag{9.21}$$

Since $M_{i,k}(t)$ is a delayed renewal function with common mean inter-renewal time $\tau_{k,k}$, we can use an argument similar to that in the proof of Theorem 8.26 on page 399, to get

$$\lim_{t \to \infty} \int_0^t dM_{i,k}(u) D_{k,j}(t - u) = \frac{1}{\tau_{k,k}} \int_0^\infty D_{k,j}(u) du. \tag{9.22}$$

Letting $t \to \infty$ in Equation 9.20, and using Equations 9.21 and 9.22, we get

$$
\begin{aligned}
\lim_{t \to \infty} H_{i,j}(t) &= \lim_{t \to \infty} D_{i,j}(t) + \lim_{t \to \infty} \sum_{k \in I} \int_0^t dM_{i,k}(u) D_{k,j}(t - u) \\
&= \sum_{k \in I} \frac{1}{\tau_{k,k}} \int_0^\infty D_{k,j}(u) du \\
&= \sum_{k \in I} \frac{p_k}{\tau_k} \int_0^\infty D_{k,j}(u) du,
\end{aligned}
$$

where the last equality follows from Equation 8.67 on page 411. ∎

This theorem appears as Theorem 4.17 of Chapter 10 of Cinlar (1975), and later in Kohlas (1982) and Heyman and Sobel (1982). We do not consider the periodic case here, since it is more involved in terms of notation. We refer the reader to Cinlar (1975) for further details.

9.4 Semi-Markov Processes: Further Results

Let $\{(X_n, S_n), n \geq 0\}$ be a Markov renewal sequence with kernel G that satisfies the regularity condition of Theorem 9.3. Let $\{X(t), t \geq 0\}$ be the semi-Markov process generated by this Markov renewal sequence. We have studied the limiting behavior of this SMP in Section 8.9 by means of an alternating renewal process. In this section we derive the results of Theorem 8.34 by using the key Markov renewal theorem (Theorem 9.9).

Define

$$
p_{i,j}(t) = P(X(t) = j | X(0) = i), \quad i, j \in I, \ t \geq 0.
$$

Note that by $X(0) = i$ we implicitly mean that the SMP has just entered state i at time 0. We use the Markov renewal argument to derive a Markov renewal-type equation for

$$
P(t) = [p_{i,j}(t)]
$$

in the theorem below.

Theorem 9.10 Markov Renewal-Type Equation for $P(t)$. *The matrix $P(t)$ satisfies the following Markov renewal-type equation:*

$$
P(t) = D(t) + G * P(t), \tag{9.23}
$$

where $D(t)$ is a diagonal matrix with

$$
D_{i,i}(t) = 1 - G_i(t) = 1 - \sum_{j \in I} G_{i,j}(t), \quad i \in I. \tag{9.24}
$$

Proof: Condition on X_1 and S_1 to get

$$
P(X(t) = j | X_0 = i, X_1 = k, S_1 = u) = \begin{cases} p_{k,j}(t - u) & \text{if } u \leq t, \\ \delta_{i,j} & \text{if } u > t. \end{cases}
$$

Hence we get

$$
\begin{aligned}
p_{i,j}(t) &= \int_0^\infty \sum_{k \in I} \mathsf{P}(X(t) = j | X_0 = i, X_1 = k, S_1 = u) dG_{i,k}(u) \\
&= \int_0^t \sum_{k \in I} p_{k,j}(t - u) dG_{i,k}(u) + \int_t^\infty \sum_{k \in I} \delta_{i,j} dG_{i,k}(u) \\
&= [G * P(t)]_{i,j} + \delta_{i,j}(1 - G_i(t)),
\end{aligned}
$$

which, in matrix form, yields the desired Markov renewal-type equation. ∎

Next we use the key Markov renewal theorem 9.9 to study the limiting behavior of $P(t)$. This may seem circular, since the key Markov renewal theorem uses $\{p_j, j \in I\}$, the limiting distribution of the SMP, in the limit! We assume that this limiting distribution is obtained by using Theorem 8.34 on page 411. Then we show that the limiting distribution produced by applying the key Markov renewal theorem produces the same limiting distribution.

Theorem 9.11   Limiting Distribution of an SMP. *Let* $\{X(t), t \geq 0\}$ *be an irreducible, positive recurrent, and aperiodic SMP with kernel* G. *Let* π *be a positive solution to*

$$
\pi = \pi G(\infty).
$$

Then $\{X(t), t \geq 0\}$ *has a limiting distribution* $[p_j, j \in I]$, *and it is given by*

$$
p_j = \lim_{t \to \infty} \mathsf{P}(X(t) = j | X(0) = i) = \frac{\pi_j \tau_j}{\sum_{k \in S} \pi_k \tau_k}, \quad j \in I. \tag{9.25}
$$

Proof: Theorem 9.10 shows that $P(t)$ satisfies the Markov renewal-type equation 9.23 with the $D(t)$ matrix as given in Equation 9.24. It is easy to verify that all the conditions of Theorem 9.9 are satisfied. The $\{p_j, j \in I\}$ in the condition 2 is given by Equation 8.67 on page 411. Note that

$$
\int_0^\infty D_{i,i}(t) dt = \int_0^\infty (1 - G_i(t)) dt = \tau_i.
$$

Hence Equation 9.19 reduces to

$$
\begin{aligned}
\lim_{t \to \infty} p_{i,j}(t) &= \int_0^\infty \sum_{k \in I} \frac{p_k}{\tau_k} D_{k,j}(t) dt \\
&= \frac{p_j}{\tau_j} \int_0^\infty D_{j,j}(u) du \\
&= \frac{p_j}{\tau_j} \tau_j = p_j.
\end{aligned}
$$

This proves the theorem. ∎

We had remarked in Section 8.9 that the study of the limiting behavior of the SMPs cannot stop at the limiting distribution of $X(t)$ as $t \to \infty$, since knowing the value of $X(t)$ at time t is, in general, not enough to determine the future of an SMP. We also need to know the distribution of the remaining sojourn time at that time. We proceed to do that here. Toward this end, define

$$B(t) = S_{N(t)+1} - t, \quad Z(t) = X_{N(t)+1}. \tag{9.26}$$

Thus $B(t)$ is the time until the next transition after t (or the remaining sojourn time in the current state at time t), and $Z(t)$ is the state after the next transition after t. To complete the study of the limiting distribution, we study

$$H_{i,j}(t) = P(X(t) = j, B(t) > x, Z(t) = k | X(0) = i), \tag{9.27}$$

where $x \geq 0$ and $k \in I$ are fixed. The next theorem gives the limit of the above quantity as $t \to \infty$.

Theorem 9.12 Limiting Behavior of an SMP. *Let $\{X(t), t \geq 0\}$ be an irreducible, positive recurrent, and aperiodic SMP with kernel G with limiting distribution $[p_j, j \in I]$. Then*

$$\lim_{t \to \infty} H_{i,j}(t) = \frac{p_j}{\tau_j} \int_x^\infty (G_{j,k}(\infty) - G_{j,k}(u)) du, \tag{9.28}$$

where $H_{i,j}(t)$ is as defined in Equation 9.27.

Proof: Condition on X_1 and S_1 to get

$$P(X(t) = j, B(t) > x, Z(t) = k | X_0 = i, X_1 = r, S_1 = u)$$

$$= \begin{cases} H_{r,j}(t - u) & \text{if } u \leq t, \\ 0 & \text{if } t < u \leq t + x, \\ \delta_{i,j}\delta_{r,k} & \text{if } u > t + x. \end{cases}$$

Hence we get

$H_{i,j}(t)$

$$= \int_0^\infty \sum_{r \in I} P(X(t) = j, B(t) > x, Z(t) = k | X_0 = i, X_1 = r, S_1 = u) dG_{i,r}(u)$$

$$= \int_0^t \sum_{r \in I} H_{r,j}(t - u) dG_{i,r}(u) + \int_{t+x}^\infty \sum_{r \in I} \delta_{i,j}\delta_{r,k} dG_{i,r}(u)$$

$$= \delta_{i,j} \int_{t+x}^\infty dG_{i,k}(u) + \int_0^t \sum_{r \in I} H_{r,j}(t - u) dG_{i,r}(u)$$

$$= \delta_{i,j}(G_{i,k}(\infty) - G_{i,k}(t + x)) + [G * H(t)]_{i,j},$$

which, in matrix form, yields

$$H(t) = D(t) + G * H(t),$$

where $D(t)$ is a diagonal matrix with

$$D_{i,i}(t) = G_{i,k}(\infty) - G_{i,k}(t + x).$$

It is easy to see that all the conditions of Theorem 9.9 are satisfied. Hence Equation 9.19 yields

$$
\begin{aligned}
\lim_{t \to \infty} H_{i,j}(t) &= \int_0^\infty \sum_{r \in I} \frac{p_r}{\tau_r} D_{r,j}(u) du \\
&= \frac{p_j}{\tau_j} \int_0^\infty (G_{j,k}(\infty) - G_{j,k}(u + x)) du,
\end{aligned}
$$

which reduces to Equation 9.28 as desired. ∎

The next theorem follows immediately from the above theorem.

Theorem 9.13 *Under the conditions of Theorem 9.12*

$$
\lim_{t \to \infty} \mathsf{P}(X(t) = j, B(t) > x | X(0) = i) = p_j \frac{1}{\tau_j} \int_x^\infty (1 - G_j(u)) du, \qquad (9.29)
$$

and

$$
\lim_{t \to \infty} \mathsf{P}(B(t) > x | X(t) = j) = \frac{1}{\tau_j} \int_x^\infty (1 - G_j(u)) du, \qquad (9.30)
$$

Proof: We have

$$
\mathsf{P}(X(t) = j, B(t) > x | X(0) = i) = \sum_{k \in I} \mathsf{P}(X(t) = j, B(t) > x, Z(t) = k | X(0) = i).
$$

Hence, letting $t \to \infty$ and using Equation 9.28 we get

$$
\lim_{t \to \infty} \mathsf{P}(X(t) = j, B(t) > x | X(0) = i) = \sum_{k \in I} \frac{p_j}{\tau_j} \int_x^\infty (G_{j,k}(\infty) - G_{j,k}(u)) du.
$$

$$(9.31)$$

Equation 9.29 follows by using

$$
\sum_{k \in I} (G_{j,k}(\infty) - G_{j,k}(t)) = 1 - G_j(t).
$$

to simplify the right-hand side of Equation 9.31. To derive Equation 9.30, write

$$
\begin{aligned}
&\mathsf{P}(X(t) = j, B(t) > x | X(0) = i) \\
&= \mathsf{P}(X(t) = j | X(0) = i) \mathsf{P}(B(t) > x | X(0) = i, X(t) = j)
\end{aligned}
$$

Taking limits as $t \to \infty$ on both sides and using Equation 9.29, we see that

$$
p_j \frac{1}{\tau_j} \int_x^\infty (1 - G_j(u)) du = p_j \lim_{t \to \infty} \mathsf{P}(B(t) > x | X(0) = i, X(t) = j).
$$

Hence we get

$$
\lim_{t \to \infty} \mathsf{P}(B(t) > x | X(0) = i, X(t) = j) = \frac{1}{\tau_j} \int_x^\infty (1 - G_j(u)) du.
$$

Since the right-hand side is independent of i we get the desired result in the theorem. ∎

The above theorem has an interesting interpretation: in steady state, the state of the SMP is j with probability p_j, and given that the state is j, the remaining sojourn time in that state has the same distribution as the equilibrium distribution associated with G_j. In hindsight, this is to be expected. We close this section with an example.

Example 9.7   Remaining Service Time in an $M/G/1$ Queue. Let $Z(t)$ be the number of customers at time t in an $M/G/1$ queue with PP(λ) arrivals and iid service times with common cdf $F(\cdot)$ and common mean τ. Define $U(t)$ to be the remaining service time of the customer in service at time t, if $Z(t) > 0$. If $Z(t) = 0$, define $U(t) = 0$. Show that

$$\lim_{t \to \infty} \mathsf{P}(U(t) > x | Z(t) > 0) = \frac{1}{\tau} \int_x^\infty (1 - F(u)) du. \qquad (9.32)$$

Let $S_0 = 0$. If $Z(S_n+) = 0$ define S_{n+1} to be the time of arrival of the next customer, and if $Z(S_n+) > 0$ define S_{n+1} to be the time of the next departure. Let $X_n = Z(S_n+)$. Then it can be seen that $\{(X_n, S_n), n \geq 0\}$ is a Markov renewal sequence and

$$G_j(t) = \begin{cases} F(t) & \text{if } j > 0, \\ 1 - e^{-\lambda t} & \text{if } j = 0. \end{cases}$$

Let $\{X(t), t \geq 0\}$ be the SMP generated by the Markov renewal sequence $\{(X_n, S_n), n \geq 0\}$. Let $B(t)$ be as defined in Equation 9.26. Then it is clear that if $X(t) > 0$, then $B(t) = U(t)$. From Equation 9.30 we see that, for a stable queue,

$$\lim_{t \to \infty} \mathsf{P}(B(t) > x | X(t) = j) = \frac{1}{\tau} \int_x^\infty (1 - F(u)) du, \quad j > 0. \qquad (9.33)$$

This yields the desired result in Equation 9.32, since $Z(t) > 0$ is equivalent to $X(t) > 0$. Thus, in steady state when the server is busy, the remaining service time distribution is given by the equilibrium distribution associated with the service time distribution. Thus the expected remaining service time in steady state, given that the server is busy, is given by $\mathsf{E}(S^2)/2\mathsf{E}(S)$, where S is a generic random variable representing the services time. This fact is very important in many waiting time calculations. Another curious fact to note is that Equation 9.33 fails if $X(t) = j$ on the left-hand side is replaced by $Z(t) = j$! ∎

9.5 Markov Regenerative Processes

We began this chapter with the definition of an MRGP in Definition 9.2. Markov regenerative processes (MRGPs) are to the Markov renewal sequences what regenerative processes are to renewal sequences. Let $\{Z(t), t \geq 0\}$ be an MRGP with an embedded MRS $\{(X_n, S_n), n \geq 0\}$. We begin by assuming that the $Z(t)$ and X_n both take values in a discrete set I. We have seen several examples of MRGPs in Section 9.1.

As with the SMPs we study the transient behavior of the MRGPs with countable state-space I by concentrating on

$$H_{i,j}(t) = \mathsf{P}(Z(t) = j | X_0 = i), \quad i, j \in I.$$

The next theorem gives the Markov renewal-type equation satisfied by $H(t) = [H_{i,j}(t)]$.

Theorem 9.14 Transient Distribution of an MRGP. *Let $\{Z(t), t \geq 0\}$ be an MRGP with embedded MRS $\{(X_n, S_n), n \geq 0\}$ with kernel G. Assume that $Z(t) \in I$ for all $t \geq 0$ and $X_n \in I$ for all $n \geq 0$, where I is a discrete set. Let*

$$D_{i,j}(t) = \mathsf{P}(Z(t) = j, S_1 > t | X_0 = i), \quad i, j \in I,$$

and $D(t) = [D_{i,j}(t)]$. Then $H(t)$ satisfies the following Markov renewal-type equation:

$$H(t) = D(t) + G * H(t).$$

Proof: Condition on X_1 and S_1 to get

$$\mathsf{P}(Z(t) = j | X_0 = i, X_1 = k, S_1 = u)$$
$$= \begin{cases} H_{k,j}(t - u) & \text{if } u \leq t, \\ \mathsf{P}(Z(t) = j | X_0 = i, X_1 = k, S_1 = u) & \text{if } u > t. \end{cases}$$

Hence we get

$$\begin{aligned} H_{i,j}(t) &= \int_t^\infty \sum_{k \in I} \mathsf{P}(Z(t) = j | X_0 = i, X_1 = k, S_1 = u) dG_{i,k}(u) \\ &\quad + \sum_{k \in I} \int_0^t H_{k,j}(t - u) dG_{i,k}(u) \\ &= \mathsf{P}(Z(t) = j, S_1 > t | X_0 = i) + [G * H(t)]_{i,j}, \end{aligned}$$

which yields the Markov renewal-type equation in the theorem. ∎

Note that $D(t)$ contains the information about the behavior of the MRGP over the first cycle $(0, S_1)$. Thus the above theorem relates the behavior of the process at time t to its behavior over the first cycle. In practice computing $D(t)$ is not easy, and solving the Markov renewal-type equation to obtain $H(t)$ is even harder. Hence, following the now well-trodden path, we study its limiting behavior in the next theorem.

Theorem 9.15 Limiting Behavior of MRGPs. *Let $\{Z(t), t \geq 0\}$ be an MRGP with embedded MRS $\{(X_n, S_n), n \geq 0\}$ with kernel G satisfying conditions of Theorem 9.3. Assume that $Z(t) \in I$ for all $t \geq 0$ and $X_n \in I$ for all $n \geq 0$, where I is a discrete set. Let $\alpha_{k,j}$ be the expected time spent by the MRGP in state j during $(0, S_1)$ starting with $X_0 = k$. Furthermore, suppose that the sample paths of the MRGP are right continuous with left limits, the SMP $\{X(t), t \geq 0\}$ generated by the MRS $\{(X_n, S_n), n \geq 0\}$ is irreducible, aperiodic, positive recurrent, with limiting distribution $[p_j, j \in I]$. Then*

$$\lim_{t \to \infty} \mathsf{P}(Z(t) = j | X_0 = i) = \sum_{k \in I} p_k \frac{\alpha_{k,j}}{\tau_k}, \quad i, j \in I. \tag{9.34}$$

Proof: Consider the Markov renewal-type equation derived in Theorem 9.14. It is straightforward to verify that the conditions of Theorem 9.9 are satisfied. Thus Equation 9.19 can be used to compute the limit of $H_{i,j}(t) = P(Z(t) = j | X_0 = i)$ as follows:

$$\lim_{t \to \infty} H_{i,j}(t) = \int_0^\infty \sum_{k \in I} \frac{p_k}{\tau_k} D_{k,j}(u) du. \tag{9.35}$$

Now,

$$
\begin{aligned}
\alpha_{k,j} &= \text{E(time spent by the MRGP in state } j \text{ during } (0, S_1) | X_0 = k) \\
&= \text{E} \left(\int_0^{S_1} 1_{\{Z(t)=j\}} dt \,\middle|\, X_0 = k \right) \\
&= \int_{u=0}^\infty \text{E} \left(\int_{t=0}^{S_1} 1_{\{Z(t)=j\}} dt \,\middle|\, X_0 = k, S_1 = u \right) dG_k(u) \\
&= \int_{u=0}^\infty \int_{t=0}^u P(Z(t) = j | X_0 = k, S_1 = u) dt\, dG_k(u) \\
&= \int_{t=0}^\infty \int_{u=t}^\infty P(Z(t) = j | X_0 = k, S_1 = u) dG_k(u) dt \\
&= \int_{t=0}^\infty P(Z(t) = j, S_1 > t | X_0 = k) dt \\
&= \int_{t=0}^\infty D_{k,j}(t) dt.
\end{aligned}
$$

Substituting in Equation 9.35 we get Equation 9.34. ∎

Note that the limiting distribution of the MRGP given in Equation 9.34 is independent of the initial distribution of the MRS, or, initial value of X_0. The distribution itself can be intuitively explained as follows: Since every time the SMP $\{X(t), t \geq 0\}$ visits state k it spends τ_k amount of time there, we can interpret $\alpha_{k,j}/\tau_k$ as the time time spent by the MRGP in state j per unit time spent by the SMP in state k. Since p_k is the fraction of the time spent in state k by the SMP in the long run, we can compute

$$
\begin{aligned}
\lim_{t \to \infty} &P(Z(t) = j | X_0 = i) \\
&= \text{Long run fraction of the time spent by the MRGP in state } j \\
&= \sum_{k \in I} [\text{Long run fraction of the time spent by the SMP in state } k] \times \\
&\quad [\text{Long run time spent by the MRGP in state } j \text{ per unit time spent by} \\
&\quad \text{the SMP in state } k] \\
&= \sum_{k \in I} p_k \frac{\alpha_{k,j}}{\tau_k}.
\end{aligned}
$$

Now let us relax the assumption that the $Z(t)$ and X_n take values in the same discrete set I. Suppose $Z(t) \in S$ for all $t \geq 0$ and $X_n \in I$ for all $n \geq 0$. We can see that 9.34 remains valid even in this case if S is also discrete. If S is continuous, say $(-\infty, \infty)$, we can proceed as follows: fix an $x \in S$, and define

$$Y(t) = 1_{\{Z(t) \leq x\}}, \quad t \geq 0.$$

Then it is clear that $\{Y(t), t \geq 0\}$ is also an MRGP with the same embedded MRS $\{(X_n, S_n), n \geq 0\}$. We do need to assume that the sample paths of $\{Z(t), t \geq 0\}$ are sufficiently nice so that the sample paths of $\{Y(t), t \geq 0\}$ are right continuous with left limits. Next define $\alpha_k(x)$ as the expected time spent by the $\{Z(t), t \geq 0\}$ process in the set $(-\infty, x]$ during $(0, S_1)$ starting with $X_0 = k$. Then we can show that

$$\lim_{t \to \infty} \mathsf{P}(Z(t) \leq x | X_0 = i) = \sum_{k \in I} p_k \frac{\alpha_k(x)}{\tau_k}, \quad i \in I, \ x \in S.$$

We illustrate with two examples.

Example 9.8 SMPs as MRGPs. Let $\{X(t), t \geq 0\}$ be an SMP with kernel G. We saw in Example 9.3 that an SMP is a special case of an MRGP. Since

$$X(t) = X_0, \quad 0 \leq t < S_1,$$

we have

$$\alpha_{k,j} = \begin{cases} \tau_j & \text{if } k = j, \\ 0 & \text{if } k \neq j. \end{cases}$$

Substituting in Equation 9.34 we get

$$\lim_{t \to \infty} \mathsf{P}(X(t) = j | X_0 = i) = \sum_{k \in I} p_k \frac{\alpha_{k,j}}{\tau_k} = p_j,$$

as expected. ∎

Example 9.9 Machine Maintenance. Consider a machine consisting of two identical components in series. The lifetimes of the components are iid $\exp(\mu)$ random variables. A single repairperson repairs these components, the repair times being iid random variables with common cdf $A(\cdot)$ and mean τ. The components are as good as new after repairs. Assume that the functioning component can fail even if the machine fails. Compute the long run fraction of the time that the machine is working.

Let $Z(t)$ be the number of functioning components at time t. Let $S_0 = 0$. If $Z(S_n+) = 2$, define S_{n+1} as the time of first failure after S_n. If $Z(S_n+) = 0$ or 1, define S_{n+1} as the time of completion of repair of the item under repair at time S_n+. Define $X_n = Z(S_n+)$. Note that $X_n \in \{1, 2\}$ even if $Z(t) \in \{0, 1, 2\}$. Then $\{(X_n, S_n), n \geq 0\}$ is a Markov renewal sequence with kernel

$$G(x) = \begin{bmatrix} G_{1,1}(x) & G_{1,2}(x) \\ G_{2,1}(x) & G_{2,2}(x) \end{bmatrix},$$

where

$$G_{1,1}(x) = \mathsf{P}(\text{Repair time} \leq x, \text{one failure during repair} | X_0 = Z(0) = 1)$$

$$= \int_0^x (1 - e^{-\mu t}) dA(t),$$

$$G_{1,2}(x) = P(\text{Repair time} \le x, \text{no failure during repair} | X_0 = Z(0) = 1)$$

$$= \int_0^x e^{-\mu t} dA(t),$$

$$G_{2,1}(x) = P(\text{Failure time} \le x | X_0 = Z(0) = 2)$$

$$= 1 - e^{-2\mu x},$$

$$G_{2,2}(x) = 0.$$

Thus

$$G_1(x) = G_{1,1}(x) + G_{1,2}(x) = A(x), \quad \tau_1 = \tau$$

and

$$G_2(x) = G_{2,1}(x) + G_{2,2}(x) = 1 - e^{-2\mu x}, \quad \tau_2 = 1/2\mu.$$

The transition probability matrix of $\{X_n, n \ge 0\}$ is

$$P = G(\infty) = \begin{bmatrix} 1 - \tilde{A}(\mu) & \tilde{A}(\mu) \\ 1 & 0 \end{bmatrix},$$

where $\tilde{A}(\mu) = \int_0^\infty e^{-\mu t} dA(t)$. To use Equation 9.34 we need to compute the limiting distribution of the SMP $\{X(t), t \ge 0\}$ generated by the MRS $\{(X_n, S_n), n \ge 0\}$ described above. We do that using Theorem 8.34 on page 8.34. We see that

$$\pi = [\pi_1 \ \pi_2] = [1 \ \tilde{A}(\mu)]$$

satisfies $\pi = \pi P$. Substituting in Equation 8.67 we get

$$p_1 = \lim_{t \to \infty} P(X(t) = 1) = \frac{\pi_1 \tau_1}{\pi_1 \tau_1 + \pi_2 \tau_2} = \frac{2\mu\tau}{2\mu\tau + \tilde{A}(\mu)},$$

$$p_2 = \lim_{t \to \infty} P(X(t) = 2) = \frac{\pi_2 \tau_2}{\pi_1 \tau_1 + \pi_2 \tau_2} = \frac{\tilde{A}(\mu)}{2\mu\tau + \tilde{A}(\mu)}.$$

From Equation 9.35, we see that the long run fraction of the time the machine is up is given by

$$\lim_{t \to \infty} P(Z(t) = 2) = p_1 \frac{\alpha_{1,2}}{\tau_1} + p_2 \frac{\alpha_{2,2}}{\tau_2}. \tag{9.36}$$

Thus we need to compute $\alpha_{1,2}$ and $\alpha_{2,2}$. We have

$$\alpha_{1,2} = 0, \quad \alpha_{2,2} = \frac{1}{2\mu}.$$

Substituting in the previous equation and simplifying, we see that the long run fraction of the time the machine is up is given by

$$\frac{\tilde{A}(\mu)}{2\mu\tau + \tilde{A}(\mu)}.$$

Note that this is the same as p_2! We leave it to the reader (see Computational Exercise 9.3) to show that

$$\lim_{t \to \infty} P(Z(t) = 1) = \frac{2(1 - \tilde{A}(\mu))}{2\mu\tau + \tilde{A}(\mu)}. \tag{9.37}$$

This is different than p_1. The limiting probability that $Z(t)$ is 0 can now be easily determined. ∎

9.6 Applications to Queues

In this section we apply the theory of MRGPs to four queueing systems: the birth and death queues, the $M/G/1$ queue, the $G/M/1$ queue, and the $M/G/1/1$ retrial queue.

9.6.1 The Birth and Death Queues

Let $Z(t)$ be the number of customers in a queueing system at time t. Assume that $\{Z(t), t \geq 0\}$ is a birth and death process on $\{0, 1, 2, \cdots\}$ with birth parameters $\lambda_i > 0$ for $i \geq 0$, and death parameter $\mu_0 = 0$, $\mu_i > 0$ for $i \geq 1$. We studied the limiting behavior of this process in Example 6.35 on page 258. There we saw that such a queueing system is stable if

$$\sum_{i=0}^{\infty} \rho_i < \infty,$$

where $\rho_0 = 1$, and

$$\rho_i = \frac{\lambda_0 \lambda_1 \cdots \lambda_{i-1}}{\mu_1 \mu_2 \cdots \mu_i}, \quad i \geq 1,$$

and its limiting distribution is given by

$$p_j = \frac{\rho_j}{\sum_{i=0}^{\infty} \rho_i}, \quad j \geq 0.$$

Thus p_j is the steady state probability that there are j customers in the system. From Section 7.1 recall that π_j^* is the steady state probability that an entering customer sees j customers ahead of him at the time of entry. We saw in Theorem 7.3 that if $\lambda_i = \lambda$ for all $i \geq 0$, we can apply PASTA (Theorem 7.3 on page 305, and noting that all arriving customers enter) to see that $\pi_j^* = p_j$. We use the theory of MRGPs to derive a relationship between π_j^* and p_j in the general case in the next theorem.

Theorem 9.16 The Birth and Death Queue. *For a stable birth and death queue,*

$$\pi_j^* = \frac{\lambda_j p_j}{\sum_{k=0}^{\infty} \lambda_k p_k}, \quad j = 0, 1, 2, \cdots. \tag{9.38}$$

Proof: Let $S_0 = 0$, and S_n be the time of the nth upward jump in the $\{Z(t), t \geq 0\}$ process. Thus the nth entry to the queueing system takes place at time S_n. Let $X_n =$

$Z(S_n-)$, the number of customers as seen by the nth entering customer. (This was denoted by X_n^* in Section 7.1.) Note that $X_n = k$ implies that $Z(S_n+) = k + 1$. Since $\{Z(t), t \geq 0\}$ is a CTMC, $\{(X_n, S_n), n \geq 0\}$ is a Markov renewal sequence. Note that $\{Z(t), t \geq 0\}$ stays the same or decreases over $[0, S_1)$, and $\{X_n, n \geq 0\}$ is a DTMC with transition probabilities

$$p_{k,j} = m_{k,j} \frac{\lambda_j}{\lambda_j + \mu_j}, \quad 0 \leq j \leq k + 1, \tag{9.39}$$

where

$$m_{k,k+1} = 1, \quad m_{k,j} = \prod_{r=j+1}^{k+1} \frac{\mu_r}{\lambda_r + \mu_r}, \quad 0 \leq j \leq k.$$

We can interpret $m_{k,j}$ as the probability that the $\{Z(t), t \geq 0\}$ process visits state j over $[0, S_1)$ starting with $X_0 = k$. Since

$$\pi_j^* = \lim_{n \to \infty} P(X_n = j), \quad j \geq 0, \tag{9.40}$$

we see that $\{\pi_j^*, j \geq 0\}$ satisfy

$$\pi_j^* = \sum_{k=\max(j-1,0)}^{\infty} \pi_k^* p_{k,j} = \sum_{k=\max(j-1,0)}^{\infty} \pi_k^* m_{k,j} \frac{\lambda_j}{\lambda_j + \mu_j}, \quad j \geq 0. \tag{9.41}$$

We can treat $\{Z(t), t \geq 0\}$ as an MRGP with the embedded MRS $\{(X_n, S_n), n \geq 0\}$. Let $\{X(t), t \geq 0\}$ be the SMP generated by the MRS $\{(X_n, S_n), n \geq 0\}$. We change the notation slightly and use

$$p_j = \lim_{t \to \infty} P(Z(t) = j), \quad j \geq 0, \tag{9.42}$$

and

$$\bar{p}_j = \lim_{t \to \infty} P(X(t) = j), \quad j \geq 0. \tag{9.43}$$

We use Theorem 8.34 on page 411 to get

$$\bar{p}_j = \frac{\pi_j^* \tau_j}{\sum_{k=0}^{\infty} \pi_k^* \tau_k}, \quad j \geq 0.$$

Now, let $\alpha_{k,j}$ be as defined in Theorem 9.15. Note that the MRGP can visit state j at most once during $[0, S_1)$, $m_{k,j}$ is the probability that the MRGP visits state j over $(0, S_1)$ starting with $X_0 = k$, and $1/(\lambda_j + \mu_j)$ is the average time it spends in state j once its reaches state j. These observations can be combined to yield

$$\alpha_{k,j} = m_{k,j} \frac{1}{\lambda_j + \mu_j}, \quad 0 \leq j \leq k + 1.$$

Using the above results in Theorem 9.15 we get

$$p_j = \sum_{k=j-1}^{\infty} \bar{p}_k \frac{\alpha_{k,j}}{\tau_k} = \frac{\sum_{k=j-1}^{\infty} \pi_k^* \alpha_{k,j}}{\sum_{k=0}^{\infty} \pi_k^* \tau_k}, \quad j \geq 0.$$

Now we have

$$
\begin{aligned}
\sum_{k=j-1}^{\infty} \pi_k^* \alpha_{k,j} &= \sum_{k=j-1}^{\infty} \pi_k^* m_{k,j} \frac{1}{\lambda_j + \mu_j} \\
&= \frac{1}{\lambda_j} \sum_{k=j-1}^{\infty} \pi_k^* m_{k,j} \frac{\lambda_j}{\lambda_j + \mu_j} \\
&= \frac{1}{\lambda_j} \sum_{k=j-1}^{\infty} \pi_k^* p_{k,j} \quad \text{(from Eq. 9.39)} \\
&= \frac{\pi_j^*}{\lambda_j} \quad \text{(from Eq. 9.41)}.
\end{aligned}
$$

Hence $p_j \propto \pi_j^*/\lambda_j$, or $\pi_j^* \propto \lambda_j p_j$. Since the π_j^* must add up to 1, we get Equation 9.38 as desired. ∎

The relation in Equation 9.38 has a simple intuitive explanation: $\lambda_j p_j$ is the rate at which state $j+1$ is entered from state j in steady state. Hence it must be proportional to the probability that an entering customer sees j customers ahead of him.

Now let π_j be the limiting probability that a departing customer leaves behind j customers in the system. One can define an appropriate Markov renewal sequence and follow the steps of the proof of Theorem 9.16 to show that

$$
\pi_j = \frac{\mu_{j+1} p_{j+1}}{\sum_{k=1}^{\infty} \mu_k p_k}, \quad j = 0, 1, 2, \cdots. \tag{9.44}
$$

However, for a positive recurrent birth and death process we have

$$
\lambda_j p_j = \mu_{j+1} p_{j+1}, \quad j \geq 0.
$$

Hence we get

$$
\pi_j = \frac{\lambda_j p_j}{\sum_{k=0}^{\infty} \lambda_k p_k} = \pi_j^*, \quad j = 0, 1, 2, \cdots. \tag{9.45}
$$

Thus, in steady state, the distribution of the number of customers as seen by an arrival is the same as seen by a departure. This is a probabilistic proof of the general Theorem 7.2 on page 303.

9.6.2 The $M/G/1$ Queue

Let $Z(t)$ be the number of customers in a stable $M/G/1$ queue at time t with arrival rate λ and service time cdf $B(\cdot)$ with mean τ. We showed in Example 9.4 that $\{Z(t), t \geq 0\}$ is an MRGP with the embedded Markov renewal sequence $\{(X_n, S_n), n \geq 0\}$ as defined there. Let $\{X(t), t \geq 0\}$ be the SMP generated by this MRS. Let p_j and $\bar{p}_j$ be as defined in Equations 9.42 and 9.43. Also define

$$
\pi_j = \lim_{n \to \infty} \mathsf{P}(X_n = j), \quad j \geq 0.
$$

Thus π_j is the probability that a departing customer leaves behind j customers in steady state. We had used PASTA and other properties of general queueing systems to show that $\pi_j = p_j$ in Example 7.6. Here we derive this result by using the theory of MRGPs.

Theorem 9.17 $M/G/1$ **Queue.** *For a stable $M/G/1$ queue*

$$p_j = \pi_j, \quad j \geq 0.$$

Proof: We have shown in Theorem 7.11 that $\{X_n, n \geq 0\}$ is a DTMC with transition probability matrix P as given there. We need the following quantities to apply Theorem 8.34 on page 411:

$$\tau_0 = \frac{1}{\lambda} + \tau, \quad \tau_j = \tau, \quad j \geq 1.$$

Using Equation 8.67 we get

$$\bar{p}_j = \frac{\pi_j \tau_j}{\sum_{k=0}^{\infty} \pi_k \tau_k}, \quad j \geq 0.$$

From Equation 7.41 on page 339 we get

$$\pi_0 = 1 - \lambda\tau.$$

Hence we have

$$\sum_{k=0}^{\infty} \pi_k \tau_k = \tau + \pi_0/\lambda = 1/\lambda.$$

Thus we get

$$\bar{p}_j = \lambda\pi_j\tau_j. \tag{9.46}$$

In order to use Theorem 9.15 we need to compute $\alpha_{k,j}$, the expected time spent by the MRGP in state j over $[0, S_1)$ starting with $X_0 = k$. Note that the cdf of the sojourn time of the SMP in state $k > 0$ is given by $G_k(x) = B(x)$. For $j \geq k > 0$ we have

$$
\begin{aligned}
\alpha_{k,j} &= \mathsf{E}\left(\int_{t=0}^{S_1} 1_{\{Z(t)=j\}} dt \,\middle|\, X_0 = k\right) \\
&= \int_{u=0}^{\infty} \mathsf{E}\left(\int_{t=0}^{S_1} 1_{\{Z(t)=j\}} dt \,\middle|\, X_0 = k, S_1 = u\right) dG_k(u) \\
&= \int_{u=0}^{\infty} \int_{t=0}^{u} \mathsf{P}(Z(t) = j \,|\, X_0 = k, S_1 = u) dt \, dB(u) \\
&= \int_{u=0}^{\infty} \int_{t=0}^{u} \mathsf{P}(j - k \text{ arrivals in } [0, t)) dt \, dB(u) \\
&= \int_{u=0}^{\infty} \int_{t=0}^{u} e^{-\lambda t} \frac{(\lambda t)^{j-k}}{(j-k)!} dt \, dB(u) \\
&= \frac{1}{\lambda}\left[1 - \sum_{r=0}^{j-k} \alpha_r\right]
\end{aligned}
$$

where

$$\alpha_i = \int_0^\infty e^{-\lambda t} \frac{(\lambda t)^i}{i!} dB(t), \quad i \geq 0.$$

In a similar way we can show that

$$\alpha_{0,0} = 1/\lambda, \quad \alpha_{0,j} = \alpha_{1,j}, \quad j \geq 1.$$

All other $\alpha_{k,j}$'s are zero. Substituting in Equation 9.34 we get

$$\begin{aligned}
p_j &= \sum_{k=0}^\infty \bar{p}_k \frac{\alpha_{k,j}}{\tau_k} \\
&= \lambda \sum_{k=0}^j \pi_k \alpha_{k,j} \quad \text{(Use Eq. 9.46)} \\
&= \lambda \left(\pi_0 \alpha_{0,j} + \sum_{k=1}^j \pi_k \alpha_{k,j} \right) \\
&= \pi_j,
\end{aligned}$$

where the last equality follows by using the balance equation $\pi = \pi P$ repeatedly to simplify the right-hand side. ∎

We cannot use the theory of MRGPs to prove $\pi_j^* = p_j$ in a similar way. However, we have seen in Example 7.6 that $\pi_j^* = \pi_j$, and hence we have a direct proof of PASTA for the $M/G/1$ queue.

9.6.3 The $G/M/1$ Queue

Let $Z(t)$ be the number of customers in a stable $G/M/1$ queue at time t with common inter-arrival time cdf $A(\cdot)$ with mean $1/\lambda$ and iid $\exp(\mu)$ service times. We showed in Example 9.5 that $\{Z(t), t \geq 0\}$ is an MRGP with the embedded Markov renewal sequence $\{(X_n, S_n), n \geq 0\}$ as defined there. Let $\{X(t), t \geq 0\}$ be the SMP generated by this MRS. Let p_j, $\bar{p}_j$ and π_j^* be as defined in Equations 9.42, 9.43, and 9.40. Here we use the theory of MRGPs to give a computational proof of Theorem 7.17.

Theorem 9.18 Limiting Distribution of a $G/M/1$ Queue. *For a $G/M/1$ queue with $\rho = \lambda/\mu < 1$ the limiting distributions $\{p_j, j \geq 0\}$ and $\{\pi_j^*, j \geq 0\}$ are related as follows:*

$$\begin{aligned}
p_0 &= 1 - \rho, \\
p_j &= \rho \pi_{j-1}^*, \quad j \geq 1.
\end{aligned}$$

Proof: We have shown in Theorem 7.15 on page 342 that $\{X_n, n \geq 0\}$ (it was denoted by $\{X_n^*, n \geq 0\}$ there) is a DTMC with transition probability matrix P as given in Equation 7.44. We need the following quantities to apply Theorem 8.34 on page 411:

$$\tau_j = 1/\lambda, \quad j \geq 0.$$

Substituting in Equation 8.67 we get

$$\bar{p}_j = \pi_j^*, \quad j \geq 0. \tag{9.47}$$

In order to use Theorem 9.15 we need to compute $\alpha_{k,j}$. Note that the cdf of the sojourn time of the SMP in state $k \geq 0$ is given by $G_k(x) = A(x)$. Going through the same calculations as in the proof of Theorem 9.17 we get, for $1 \leq j \leq k+1$

$$\alpha_{k,j} = \frac{1}{\mu}\left[1 - \sum_{r=0}^{k+1-j} \alpha_r\right],$$

where

$$\alpha_i = \int_0^\infty e^{-\mu t}\frac{(\mu t)^i}{i!}dA(t), \quad i \geq 0.$$

All other $\alpha_{k,j}$'s are zero. Substituting in Equation 9.35 we get, for $j \geq 1$,

$$\begin{aligned}
p_j &= \sum_{k=0}^\infty \bar{p}_k \frac{\alpha_{k,j}}{\tau_k} \\
&= \lambda\sum_{k=0}^\infty \pi_k^* \alpha_{k,j} \quad \text{(Use Eq. 9.47)} \\
&= \frac{\lambda}{\mu}\sum_{k=j-1}^\infty \pi_k^*\left[1 - \sum_{r=0}^{k+1-j}\alpha_r\right] \\
&= \rho\pi_{j-1}^*,
\end{aligned}$$

where the last equality follows by using the balance equation $\pi^* = \pi^* P$ repeatedly to simplify the right-hand side. Finally, we have

$$p_0 = 1 - \sum_{j=1}^\infty p_j = 1 - \rho\sum_{j=1}^\infty \pi_{j-1}^* = 1 - \rho.$$

This completes the proof. ■

9.6.4 The $M/G/1/1$ Retrial Queue

Consider the $M/G/1/1$ retrial queue as described in Example 9.6. Using the notation there we see that $\{Z(t) = (R(t), I(t)), t \geq 0\}$ is an MRGP with the MRS $\{(X_n, S_n), n \geq 0\}$ defined there. Let $\{X(t), t \geq 0\}$ be the SMP generated by this MRS. Note that this is different than the $X(t)$ defined in Example 9.6. Let $\bar{p}_j$ be as defined in Equations 9.42. We have derived the limiting distribution of the embedded DTMC $\{X_n, n \geq 0\}$ in Theorem 7.20. We use those results to derive the limiting distribution of $Z(t) = (R(t), I(t))$ as $t \to \infty$.

Let

$$p_{(j,i)} = \lim_{t\to\infty} P((R(t), I(t)) = (j,i)), \quad j \geq 0, \ i = 0, 1, \tag{9.48}$$

and define the generating functions

$$\phi_i(z) = \sum_{j=0}^{\infty} z^j p_{(j,i)}, \quad i = 0, 1.$$

Theorem 9.19 $M/G/1/1$ **Retrial Queue.** *Suppose* $\rho = \lambda\tau < 1$. *Then*

$$\phi_0(z) = (1 - \rho) \exp\left(-\frac{\lambda}{\theta} \int_z^1 \frac{1 - \tilde{G}(\lambda - \lambda u)}{\tilde{G}(\lambda - \lambda u) - u} du\right), \tag{9.49}$$

$$\phi_1(z) = (1 - \rho) \frac{\tilde{G}(\lambda - \lambda z) - 1}{z - \tilde{G}(\lambda - \lambda z)} \cdot \exp\left(-\frac{\lambda}{\theta} \int_z^1 \frac{1 - \tilde{G}(\lambda - \lambda u)}{\tilde{G}(\lambda - \lambda u) - u} du\right). \tag{9.50}$$

Proof: We have shown in Theorem 7.19 on page 346 that $\{X_n, n \geq 0\}$ is a DTMC. Let

$$\pi_k = \lim_{n \to \infty} \mathsf{P}(X_n = k), \quad k \geq 0.$$

In Theorem 7.20 we computed the generating function

$$\phi(z) = \sum_{k=0}^{\infty} z^k \pi_k.$$

We also derived a peculiar, but useful, generating function

$$\psi(z) = \lambda \sum_{k=0}^{\infty} z^k \frac{\pi_k}{\lambda + k\theta}$$

in Equation 7.57 on page 347. From there we see that

$$\psi(1) = 1 - \rho.$$

Next we compute the relevant quantities.

$$\tau_k = \frac{1}{\lambda + k\theta} + \tau, \quad k \geq 0,$$

$$\alpha_{k,(k,0)} = \frac{1}{\lambda + k\theta}, \quad k \geq 0,$$

$$\alpha_{k,(j,0)} = 0, \quad \text{if } k \neq j.$$

The quantities $\alpha_{k,(j,1)}$ are more complicated; however, we do not need them, as we shall see. We have

$$\sum_{k=0}^{\infty} \pi_k \tau_k = \tau + \sum_{k=0}^{\infty} \frac{\pi_k}{\lambda + k\theta}$$

$$= \tau + \frac{\psi(1)}{\lambda} = \frac{1}{\lambda}.$$

This is to be expected, since the left-hand side is the expected time between two successive departures in steady state, and hence must equal $1/\lambda$, the expected time between two consecutive arrivals. Substituting in Equation 8.67 we get

$$\bar{p}_j = \lambda \pi_j \tau_j. \tag{9.51}$$

Substituting in Equation 9.35 we get

$$
\begin{aligned}
p_{(j,0)} &= \sum_{k=0}^{\infty} \bar{p}_k \frac{\alpha_{k,(j,0)}}{\tau_k} \\
&= \lambda \sum_{k=0}^{\infty} \pi_k \alpha_{k,(j,0)} \quad \text{(Use Eq. 9.51)} \\
&= \frac{\lambda}{\lambda + j\theta} \pi_j.
\end{aligned}
$$

Thus

$$\phi_0(z) = \sum_{j=0}^{\infty} z^j \frac{\lambda}{\lambda + j\theta} \pi_j = \psi(z).$$

The Equation 9.49 now follows from Equations 7.57 and 7.58.

Now let $\phi(z)$ be the limiting generating function of X_n as derived in Equation 7.52 on page 346. We saw in Section 7.7 that it is also the limiting generating function of $R(t) + I(t)$ as $t \to \infty$. Hence we get

$$\phi(z) = \phi_0(z) + z\phi_1(z).$$

Substituting for $\phi(z)$ and $\phi_0(z)$ we get the expression for $\phi_1(z)$ in Equation 9.50. ∎

Note that the limiting probability that the server is idle is given by

$$\phi_0(1) = 1 - \rho,$$

as expected. It is a bit surprising that this is independent of the retrial rate θ as long as it is positive! We could not derive this result in our earlier analysis of the retrial queue in Section 7.7.

9.7 Modeling Exercises

9.1 Let $\{Z(t), t \geq 0\}$ be a birth and death process on $S = \{0, 1, 2, \cdots\}$ with birth parameters $\lambda_i > 0$ for $i \geq 0$, and death parameters $\mu_i > 0$ for $i \geq 1$. Let S_n be the nth downward jump epoch and define $X_n = X(S_n+)$. Assume that $S_0 = 0$. Show that $\{Z(t), t \geq 0\}$ is an MRGP with $\{(X_n, S_n), n \geq 0\}$ as the embedded MRS. Compute its kernel.

9.2 Let $Z(t)$ be the number of customers in an $M/G/1/K$ queue with PP(λ) arrivals and common service time cdf $B(\cdot)$. Assume that either the system is empty initially, or a service has just started at time 0. Let S_n be the time of the nth departure, and define $X_n = Z(S_n+)$. Show that $\{Z(t), t \geq 0\}$ is an MRGP with $\{(X_n, S_n), n \geq 0\}$ as the embedded MRS. Compute its kernel.

9.3 Let $Z(t)$ be the number of customers in a $G/M/1/K$ queue with common inter-arrival time cdf $A(\cdot)$, and iid $\exp(\mu)$ service times. Assume that an arrival has occurred at time 0. Let S_n be the time of the nth arrival (who may or may not enter), and define $X_n = Z(S_n-)$. Show that $\{Z(t), t \geq 0\}$ is an MRGP with $\{(X_n, S_n), n \geq 0\}$ as the embedded MRS. Compute its kernel.

9.4 Consider a closed queueing network with two single-server nodes and N customers. After completing service at node 1 (2) a customer joins node 2 (1). The service times at node i are iid $\exp(\mu_i)$, $(i = 1, 2)$. Let $Z(t)$ be the number of customers at node 1 at time t, and S_n be the time of the nth service completion at node 2, which is also an arrival instant at node 1. Assume $S_0 = 0$, i.e., a service completion at node 2 has occurred at time 0. Let $X_n = Z(S_n-)$. Show that $\{Z(t), t \geq 0\}$ is an MRGP with $\{(X_n, S_n), n \geq 0\}$ as the embedded MRS. Compute its kernel.

9.5 Consider a workshop with N machines and a single repairperson. Lifetimes of the machines are iid $\exp(\mu)$, and the repair times are iid with common cdf $R(\cdot)$. Let $Z(t)$ be the number of down machines at time t, S_n the nth repair completion time, and $X_n = Z(S_n+)$. Assume that a repair has just been completed at time 0. Show that $\{Z(t), t \geq 0\}$ is an MRGP with $\{(X_n, S_n), n \geq 0\}$ as the embedded MRS. Compute its kernel.

9.6 Consider the queueing network of Modeling Exercise 9.4. Suppose one of the N customers is tagged, and let S_n be the time of his nth return to node 1. Assume $S_0 = 0$, i.e., the tagged customer joins node 1 at time 0. Let $X_n = Z(S_n-)$. Show that $\{Z(t), t \geq 0\}$ is an MRGP with $\{(X_n, S_n), n \geq 0\}$ as the embedded MRS. Compute its kernel.

9.7 A machine can exist in one of $N + 1$ states labeled $\{0, 1, 2, \cdots, N\}$, with state 0 representing a new machine, and state N representing a failed machine, and the in-between states indicating increasing levels of deterioration. Left to itself, the machine changes states according to a DTMC with transition probability matrix $H = [h_{i,j}]$. The maintenance policy calls for repairing the machine whenever it reaches a state k or more, where $0 < k \leq N$ is a given integer. Suppose the repair takes one unit of time and transforms the machine from state $i(\geq k)$ to state $j(< k)$ with probability $a_{i,j}$. Let S_n be the nth repair completion time and X_n be the state of the machine immediately after the nth repair completion. Assume that $S_0 = 0$. Let $Z(t)$ be the state of the machine at time t. Show that $\{Z(t), t \geq 0\}$ is an MRGP with $\{(X_n, S_n), n \geq 0\}$ as the embedded MRS. Compute its kernel.

9.8 A machine is maintained by a robot. The lifetime of a new machine is an $\exp(\mu)$ random variable. When the machine fails, it is repaired by a robot in a fixed amount of time r. After repair the machine is as good as new. After repair, the robot needs a down time of a fixed amount d. If the machine fails when the robot is down, it has to wait for the robot to become available again. Let $Z(t)$ be 0 if the machine is down and waiting for repair, 1 if the machine is under repair, and 2 if the machine is up. Show that $\{Z(t), t \geq 0\}$ is an MRGP.

9.9 A high-speed network transmits cells (constant length packets of data) over communication channels. At its input ports it exercises access control by dropping incoming cells if the input rate gets too high. One such control mechanism is described here. The controller generates r tokens at times $n = 0, 1, 2, \cdots$ into a token pool of size M. Tokens that exceed the capacity are lost. The cells arrive according to a PP(λ). If there is a token available when a cell arrives, it grabs one token and enters the network immediately, else the cell is prohibited from entering the network and is permanently lost. Let $X(t)$ be the number of tokens in the token pool at time t. Model $\{X(t), t \geq 0\}$ as an MRGP.

9.10 Consider following modification to the access control mechanism of Modeling Exercise 9.9. The incoming cells are queued up in a buffer of size L. When an incoming cell finds no tokens, it waits in the buffer and enters the network as soon as the next token becomes available. Any cells arriving when the buffer is full are lost. Thus the token pool and the buffer cannot be simultaneously non-empty. Let $Y(t)$ be the number of cells in the buffer at time t, and define $Z(t) = X(t) - Y(t)$. Model $\{Z(t), t \geq 0\}$ as an MRGP.

9.11 Customers arrive at a service station according to a Poisson process with rate λ. Servers arrive at this station according to an independent renewal process with iid inter-arrival times with mean τ and second moment s^2. Each incoming server removes each of the waiting customers with probability $\alpha > 0$ in an independent fashion, and departs immediately. Let $X(t)$ be the number of customers in the system at time t. Model $\{X(t), t \geq 0\}$ as an MRGP by identifying an appropriate MRS embedded in it.

9.8 Computational Exercises

9.1 Consider a system that cycles through N states labeled $1, 2, \cdots, N$ starting in state 1. It stays in state i for a random amount of time with cdf H_i and then jumps to state $i + 1$ if $i < N$ and state 1 if $i = N$. Let S_n be the time of the nth transition, with $S_0 = 0$, and X_n be the state of the system after the nth jump. Compute the LST $\tilde{M}(s)$ of the Markov renewal function $M(t)$ for the MRS $\{(X_n, S_n), n \geq 0\}$.

9.2 Let $\{Z(t), t \geq 0\}$ be a CTMC with generator matrix Q. For the MRS described in Example 9.2, compute the LST $\tilde{M}(s)$ of the Markov renewal function $M(t)$.

9.3 Derive Equation 9.37, and then compute the limiting probability that 0 components are functional in the machine in Example 9.9.

9.4 Consider a two state CTMC $\{X(t), t \geq 0\}$ on $\{0, 1\}$ with generator matrix

$$Q = \begin{bmatrix} -\lambda & \lambda \\ \mu & -\mu \end{bmatrix}.$$

Let $H_{i,j}(t)$ be the expected time spent in state j by the CTMC up to time t starting in state i at time 0. Derive a Markov renewal-type equation for $H(t)$. Solve it by inverting its LST $\tilde{H}(s)$.

9.5 Consider an $M/M/1$ queue with balking as follows: an incoming customer joins the system with probability $1/(j+1)$ if he sees j customers ahead of him. Compute the steady state probability that

1. an entering customer sees j customers ahead of him when he joins,
2. a potential arrival joins the system.

9.6 Let p_j be the limiting probability that there are j customers in an $M/G/1/K$ queue, and π_j^* be the limiting probability that an entering customer sees j customers ahead of him when he enters. Using the theory of MRGPs, establish the relationship between the p_j's and the π_j^*'s.

9.7 Compute the long run behavior of the access control scheme described in Modeling Exercise 9.9 for the special case when $M = r$.

9.8 Consider the following variation of the $M/G/1/1$ retrial queue. Suppose that after a service completion a customer departs with probability $1 - p$, or rejoins the orbit with probability p and behaves like a new customer. Study the limiting distribution of this system using appropriate MRGPs.

9.9 Consider yet another variation of the $M/G/1/1$ retrial queue. Suppose that an arriving customer joins service immediately if he finds the server free upon arrival. Else he departs with probability $1 - p$, or joins the orbit with probability p and conducts retrials until getting served. Study the limiting distribution of this system using appropriate MRGPs.

9.10 Using the MRGP developed in Modeling Exercise 9.8, compute the long run fraction of the time that the machine is up.

9.11 Consider the MRGP developed in Modeling Exercise 9.11. Let X_n be the number of customers left behind after the nth server departs. Compute the limiting value of $E(X_n)$ as $n \to \infty$. Compute the limiting value of $E(X(t))$ as $t \to \infty$.

9.12 A workshop has $k \geq 1$ machines that are maintained by a single repairperson. The lifetimes of the machines are iid $\exp(\mu)$ random variables and repairtimes are constant, denoted by r. The repairperson repairs the machines one at a time in the order of failure, and after repair the machines are as good as new. Let $X(t)$ be the number of working machines at time t. Suppose we observe the system every time a repair begins or ends or both. Let S_n be the time of n-th such event and define $X_n = X(S_n+)$. What is the state space S of $\{X_n, n \geq 0\}$? Show that $\{(X_n, S_n), n \geq 0\}$ is a Markov renewal sequence. Compute its kernel $G(x) = [G_{i,j}(x)]$. Compute

$$p_j = \lim_{t \to \infty} P(X(t) = j),$$

in terms of the limiting distribution of $\{X_n, n \geq 0\}$.

9.13 Let $\{Y(t), t \geq 0\}$ be a CTMC with state-space $\{0, 1\}$ and transition rates $q_{0,1} = \alpha$ and $q_{1,0} = \beta$. When the CTMC is in state i, customers arrive at a facility according to $PP(\theta_i)$, $(i = 0, 1)$. Customers arriving while the CTMC is in state i are called type i customers. Let $A_i(t)$ be the number of type i customers that arrive at the facility over $(0, t]$. Let

$$H_{i,j}(t) = E(A_j(t)|Y(0) = i), \quad i, j = 0, 1.$$

Derive a Markov renewal-type equation for $H(t) = [H_{i,j}(t)]$ and solve for $H(t)$ using LSTs.

9.14 A machine is subject to shocks that arrive one at a time according to a renewal process with inter-renewal time cdf F. If the machine is working when a shock occurs it fails with probability $\alpha \in (0, 1)$, independent of history. When the machine fails, it undergoes repairs; the repairtimes being iid $\exp(\lambda)$ random variables. The shocks have no effect on the machine under repair. The repaired machine is as good as new. Let $X(t)$ be the number of working machines (0 or 1) at time t. Assume that a shock has occurred at time zero and $X(0+) = 0$. Let S_n be the time of the nth shock. Let $X_n = X(S_n+)$, the state of the machine just after the nth shock. Show that $\{X(t), t \geq 0\}$ is an MRGP with the embedded Markov renewal sequence $\{(X_n, S_n), n \geq 0\}$. Compute the limiting distribution of $\{X(t), t \geq 0\}$.

9.15 Let $X(t)$ be the number of customers at time t in a $G/M/2/2$ system, with common continuous inter-arrival time cdf $A(\cdot)$, and iid $\exp(\mu)$ service times. Let S_n be the time of the n-th arrival (not necessarily entry). Assume $S_0 = 0$. Let $X_n = X(S_n-)$. Using this, show that $\{X(t), t \geq 0\}$ is an MRGP. Compute the limiting distribution of $X(t)$ as $t \to \infty$.

9.16 Suppose the lifetime of a machine is an $\exp(\mu)$ random variable. A repairperson visits the machine at random times as follows. If the machine is working when the repairperson visits, he simply goes away and returns after d amount of time, where d is a fixed positive constant. If the machine is down when the repairperson visits, he repairs the machine in r amount of time, where r is a fixed positive constant. Then he goes away and returns after d amount of time. The machine is as good as new after the repair is complete. Let $X(t)$ be 0 if the machine is working at time t, 1 if it is down but not under repair at time t, and 2 if it is under repair at time t. Suppose a repair has just been completed at time 0. Show that $\{X(t), t \geq 0\}$ is an MRGP. Compute its limiting distribution.

9.17 Consider Computational Exercise 7.73. Show that $\{X(t), t \geq 0\}$, as defined there, is an MRGP. Does $P(X(t) = j)$ have a limit as $t \to \infty$? Compute the long run fraction of the time that there are j patients in the system. Suppose the waiting cost of each patient is w per unit time, and the idle time of the doctor costs h per unit time. Compute the long run cost per unit time as a function of d.

9.9 Conceptual Exercises

9.1 Complete the proof of Theorem 9.1.

9.2 Prove Theorem 9.2.

9.3 Consider the Markov renewal sequence embedded in a CTMC as described in Example 9.2. Show that the regularity condition in Equation 9.5 is satisfied if and only if the CTMC is uniformizable.

9.4 Derive Equation 9.12.

9.5 Let $\{(X_n, S_n), n \geq 0\}$ be a given Markov renewal sequence with kernel G. Show that the process $\{X(t), t \geq 0\}$ defined by Equation 9.7 is an SMP with kernel G.

CHAPTER 10

Diffusion Processes

Suppose you're on a game show, and you're given the choice of three doors. Behind one door is a car, behind the others, goats. You pick a door, say #1, and the host, who knows what's behind the doors, opens another door, say #3, which has a goat. He says to you, "Do you want to pick door #2?" Is it to your advantage to switch your choice of doors?

A question asked of Marilyn vos Savant by a reader. Marilyn said yes, while countless others said, "It does not matter." This created a heated controversy, and produced several papers, with one reader telling Marilyn: "You are the goat!"

10.1 Brownian Motion

In this chapter we study a class of stochastic processes called the diffusion processes. Intuitively speaking these processes are continuous-time, continuous state-space processes and their sample paths are everywhere continuous but nowhere differentiable. The history of diffusion processes begins with the botanist Brown, who in 1827 observed that grains of pollen suspended in a liquid display a kind of erratic motion. This motion came to be known as Brownian motion. Einstein later used physical principles to do a mathematical analysis of this motion. Wiener later provided a rigorous probabilistic foundation for Brownian motion, and hence sometimes it is also called the Wiener process. Diffusion processes are built upon the simpler process of Brownian motion.

We begin with a formal definition. It uses the concept of stationary and independent increments introduced in Definition 5.5 on page 168. We also use the notation $\mathcal{R}$ to denote the real line $(-\infty, \infty)$, and $N(\mu, \sigma^2)$ to denote a normal random variable (or its distribution) with mean μ and variance σ^2.

Definition 10.1 Brownian Motion (BM). *A stochastic process $\{X(t), t \geq 0\}$ with state-space $\mathcal{R}$ is called a Brownian motion (BM) with drift parameter $\mu \subset \mathcal{R}$ and variance parameter $\sigma^2 > 0$ if*

1. for each t ≥ 0, $X(t) \sim N(\mu t, \sigma^2 t)$,

2. it has stationary and independent increments.

We shall denote a BM with parameters μ and σ by $BM(\mu, \sigma)$.

Definition 10.2   Standard Brownian Motion (SBM). *A BM(0,1) is called a standard Brownian motion (SBM).*

We reserve the notation $\{B(t), t \geq 0\}$ for the SBM. Some important properties of a Brownian motion are given in the following theorem.

Theorem 10.1   Basic Properties of a BM. *Let $\{B(t), t \geq 0\}$ be an SBM, and define $X(t) = \mu t + \sigma B(t)$.*

1. $\{X(t), t \geq 0\}$ is a BM(μ, σ).

2. $X(0) = B(0) = 0$ with probability 1.

3. $\{X(t), t \geq 0\}$ is a Markov process.

4. $\{-X(t), t \geq 0\}$ is a BM($-\mu, \sigma$).

Proof: Parts 1 and 2 follow from the definition of a BM. To see part 3, we have

$$
\begin{aligned}
&P(X(t+s) \leq x | X(s) = y, X(u) : 0 \leq u \leq s) \\
&= P(X(t+s) - X(s) \leq x - y | X(s) = y, X(u) : 0 \leq u \leq s) \\
&= P(X(t+s) - X(s) \leq x - y) \quad \text{(Independent Increments)} \\
&= P(X(t) - X(0) \leq x - y) \quad \text{(Stationary Increments)} \\
&= \Phi\left(\frac{x - y - \mu t}{\sigma \sqrt{t}}\right),
\end{aligned}
$$

where $\Phi(\cdot)$ is the cdf of a standard normal random variable, and the last equality follows from part 1 of the definition of a BM. Part 4 follows from the definition of a BM(μ, σ). ■

The last property mentioned in the above theorem implies that if $\{B(t), t \geq 0\}$ is an SBM, then so is $\{-B(t), t \geq 0\}$. We say that an SBM is symmetric. Let $\{X(t), t \geq 0\}$ be a BM(μ, σ), x be a given real number and define

$$
Y(t) = x + X(t), \quad t \geq 0.
$$

We call $\{Y(t), t \geq 0\}$ a BM(μ, σ) starting at x.

We next compute the finite dimensional joint pdf of an SBM. By definition, $B(t)$ is a $N(0, t)$ random variable, and has the density $\phi(t, x)$ given by

$$
\phi(t, x) = \frac{1}{\sqrt{2\pi t}} e^{-\frac{x^2}{2t}}, \quad t > 0, \; x \in \mathcal{R}. \tag{10.1}
$$

Since Brownian motion has such a nice structure we can give much more detailed results about its finite dimensional joint pdfs.

Theorem 10.2   Joint Distributions of an SBM. *Let* $\{B(t), t \geq 0\}$ *be an SBM, and* $0 < t_1 < t_2 < \cdots < t_n$. *Then* $[B(t_1), B(t_2), \cdots, B(t_n)]$ *is a multivariate* $N(\mu, \Sigma)$ *random variable with the mean vector* $\mu = [\mu_i]$, $1 \leq i \leq n$, *and variance-covariance matrix* $\Sigma = [\sigma_{ij}]$, $1 \leq i, j \leq n$, *given by*

$$\mu_i = 0, \quad 1 \leq i \leq n,$$

and

$$\sigma_{ij} = \min\{t_i, t_j\}, \quad 1 \leq i, j \leq n.$$

Proof: Let $t_0 = 0$ and define $Y_i = B(t_i) - B(t_{i-1})$, $1 \leq i \leq n$. Then from the definition of the SBM we see that $Y = [Y_1, Y_2, \cdots, Y_n]^\top$ is a multivariate normal with mean vector $\mu_Y = 0$ and variance-covariance matrix $\Sigma_Y = \text{diag}(t_1 - t_0, t_2 - t_1, \cdots, t_n - t_{n-1})$. We see that $B = [B(t_1), B(t_2), \cdots, B(t_n)]^\top = AY$ where $A = [a_{ij}]$ is given by

$$a_{i,j} = \begin{cases} 1 & \text{if } i \geq j, \\ 0 & \text{if } i < j. \end{cases}$$

Hence B is a multivariate normal random vector. Its mean vector and variance-covariance matrix can be computed by

$$\mu = A\mu_Y = 0, \quad \Sigma = A\Sigma_Y A^\top.$$

The result follows from this. ∎

Since all the finite dimensional joint pdfs in an SBM are multivariate normal, explicit expressions can be computed for many probabilistic quantities. For example, the joint density $f(x, y)$ of $B(s)$ and $B(t)$ $(0 < s < t)$ is

$$f(x, y) = \phi(s, x)\phi(t - s, y - x) = \frac{1}{2\pi\sqrt{s(t - s)}} e^{-\frac{x^2}{2s} - \frac{(y-x)^2}{2(t-s)}}. \tag{10.2}$$

One can show that, given $B(t) = x$, $B(s)$ $(0 < s < t)$ is a normal random variable with mean xs/t and variance $s(t - s)/t$. A special case of this arises when we consider the SBM conditioned on $B(1) = 0$. Such a process is called the Brownian bridge, since it is anchored at 0 at times 0 and 1. We can see that $B(t)$ $(0 \leq t \leq 1)$ in a Brownian bridge is a normal random variable with mean 0 and variance $t(1 - t)$.

We next state two more basic properties of the BM without proof. First we need two definitions.

Definition 10.3 Stopping Times. *A random variable T defined on the same probability space as a stochastic process* $\{X(t), t \geq 0\}$ *is called a stopping time for the* $\{X(t), t \geq 0\}$ *if the event* $\{T \leq t\}$ *is completely determined by* $\{X(u) : 0 \leq u \leq t\}$.

For example,

$$T = \min\{t \geq 0 : X(t) \geq 2\}$$

is a stopping time, but

$$T = \min\{t \geq 0 : X(t) \geq X(1)\}$$

is not. In the latter case we cannot determine if $T \leq 1/2$ without observing $X(1)$, which is not part of $\{X(u) : 0 \leq u \leq 1/2\}$.

Definition 10.4 Strong Markov Property. *A Markov process* $\{X(t), t \geq 0\}$ *is said to be a strong Markov process (or to have strong Markov property) if, for any stopping time* T, *the finite dimensional distributions of* $\{X(t), t \geq T\}$, *given* $\{X(u) : 0 \leq u \leq T\}$, *depend only on* $X(T)$.

The above definition is mathematically imprecise as it stands, but its meaning is clear: The future of the process from any stopping time onwards depends on the history up to that stopping time only via the value of the process at that stopping time. We will need an enormous mathematical apparatus from measure theory to make it precise, and we shall not do so here. For the same reason, it is difficult to come up with an example of a Markov process that is not strong Markov. The DTMCs and the CTMCs that we have studied so far have strong Markov property. We state the following theorem without proof.

Theorem 10.3 *A* $BM(\mu, \sigma)$ *is a strong Markov process.*

10.2 Sample Path Properties of BM

In this section we shall study several important properties of the sample paths of a BM. We state the first property in the next theorem.

Theorem 10.4 *The sample paths of* $\{X(t), t \geq 0\}$ *are everywhere continuous and nowhere differentiable with probability 1.*

This is one of the deep results about Brownian motion. It is much stronger than asserting that the sample paths as a function of t are continuous for almost all values of $t \geq 0$. This property is valid even for the sample paths of a Poisson process since they have a countable number of jumps with probability one. But none of the sample paths of a Poisson process is continuous everywhere. What the above theorem asserts is that, with probability one, a randomly occurring sample path of a Brownian motion is a continuous function of t for all $t \geq 0$. Even more surprisingly, it asserts that, with probability one, a randomly occurring sample path of a Brownian motion is nowhere differentiable. Thus the sample paths of a Brownian motion are indeed very crazy functions of t.

We will not give formal proof of the above theorem. Instead we shall provide a way to understand such a bizarre behavior by proposing an explicit method of constructing the sample paths of an SBM as a limit of a sequence of simple symmetric random walks. Recall from Example 2.10 on page 16 that $\{X_n, n \geq 0\}$ is called a simple symmetric random walk if it is a DTMC on all integers with transition probabilities

$$p_{i,i+1} = 1/2, \quad p_{i,i-1} = 1/2, \quad i = 0, \pm 1, \pm 2, \cdots.$$

Using the notation $[x]$ to denote the integer part of x, define a sequence of continuous time stochastic processes $\{X^k(t), t \geq 0\}$, indexed by $k \geq 1$, as follows:

$$X^k(t) = \frac{X_{[kt]}}{\sqrt{k}}, \quad t \geq 0. \tag{10.3}$$

Figure 10.1 shows the typical sample paths of the $\{X^k(t), 0 \leq t < 1\}$ processes

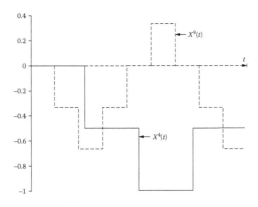

Figure 10.1 *Typical sample paths of the* $\{X^k(t), 0 \leq t < 1\}$ *processes for* $k = 4$ *and* 9.

for $k = 4$ and $k = 9$, corresponding to the sample path $[X_0, X_1 \cdots, X_9] = [0, -1, -2, -1, 0, 1, 0, -1, -2, -1]$. Now define

$$X^*(t) = \lim_{k \to \infty} X^k(t), \quad t \geq 0, \tag{10.4}$$

where the limit is in the almost sure sense. (See Appendix H for the relevant definitions.) Thus a sample path of $\{X_n, n \geq 0\}$ produces a unique (in the almost sure sense) sample path of $\{X^*(t), t \geq 0\}$. We have the following result:

Theorem 10.5 Random Walk to Brownian Motion. *Suppose* $|X_0| < \infty$ *with probability 1. Then the process* $\{X^*(t), t \geq 0\}$ *defined by Equation 10.4 exists, and is an SBM.*

Proof: We will not show the existence. Let $\{Y_n, n \geq 0\}$ be a sequence of iid random variables with common pmf

$$P(Y_n = 1) = .5, \quad P(Y_n = -1) = .5, \quad n \geq 1.$$

We have

$$E(Y_n) = 0, \quad \text{Var}(Y_n) = 1, \quad n \geq 1.$$

Since $\{X_n, n \geq 0\}$ is a simple random walk, we see that

$$X_n = X_0 + \sum_{i=1}^{n} Y_i. \tag{10.5}$$

Then

$$
\begin{aligned}
X^*(t) &= \lim_{k \to \infty} \frac{X_0 + \sum_{i=1}^{[kt]} Y_i}{\sqrt{k}} \\
&= \lim_{k \to \infty} \frac{\sqrt{[kt]}}{\sqrt{k}} \frac{X_0 + \sum_{i=1}^{[kt]} Y_i}{\sqrt{[kt]}} \\
&= \sqrt{t} N(0,1) \quad \text{(in distribution)} \\
&= N(0,t) \quad \text{(in distribution)}.
\end{aligned}
$$

This shows part 1 of the Definition 10.1 with $\mu = 0$ and $\sigma = 1$.

Next we show that $\{X^*(t), t \geq 0\}$ has stationary and independent increments. From Equation 10.5 we see that $\{X^k(t), t \geq 0\}$ has stationary and independent increments as long as the endpoints of the intervals over which the increments are computed are integer multiples of $1/k$. Hence in the limit, as $k \to \infty$, $\{X^*(t), t \geq 0\}$ has stationary increments as long as the intervals have rational start and endpoints, and hence for all intervals. Thus $\{X^*(t), t \geq 0\}$ satisfies both parts of the Definition 10.1. ∎

This explicit construction of an SBM is useful in getting an intuitive understanding of the behavior of its sample paths. For example, the sample paths of $\{X^k(t), t \geq 0\}$ are piecewise constant functions of time t with jumps of $\pm 1/\sqrt{k}$ at all times t that are integer multiples of $1/k$. Hence as $k \to \infty$ the sample paths become continuous everywhere. Furthermore, for any t the finite derivative of the $\{X^k(t), t \geq 0\}$ process is given by

$$
\frac{X^k(t + 1/k) - X^k(t)}{1/k} = \pm\sqrt{k}.
$$

Thus in the limit the finite derivative has limsup of $+\infty$ and liminf of $-\infty$. Thus in the limit the sample paths of an SBM are nowhere differentiable.

10.3 Kolmogorov Equations for Standard Brownian Motion

Let $p(t, x)$ be the density of $B(t)$. Since the SBM is a Markov process, we can expect to derive differential equations for $p(t, x)$, much along the lines of the Chapman–Kolmogorov equations we derived in Theorem 6.4 on page 222. However, since the state-space of the SBM is continuous, we need a different machinery to derive these equations. The formal derivation of these equations is rather involved, and we refer the reader to advanced books on the subject, such as Chung (1967). Here we present an "engineering" derivation, which glosses over some of the technicalities. The following theorem gives the main differential equation, which is equivalent to the forward equations derived in Theorem 6.4. We shall use the following notation for the partial derivatives:

$$
p_t(t, x) = \frac{\partial p(t, x)}{\partial t},
$$

$$p_x(t,x) = \frac{\partial p(t,x)}{\partial x},$$

$$p_{xx}(t,x) = \frac{\partial^2 p(t,x)}{\partial x^2}.$$

We also use the notation $f_Z(\cdot)$ to denote the density of a random variable Z. If $Z = c$ with probability one, then f_Z is interpreted to be the Dirac delta function concentrated at c.

Theorem 10.6   Kolmogorov Equation for SBM. *Let $\{B(t), t \geq 0\}$ be an SBM. The density $p(t,x)$ of $B(t)$ satisfies the following partial differential equation:*

$$p_t(t,x) = \frac{1}{2} p_{xx}(t,x), \quad x \in \mathcal{R}, \ t \geq 0, \tag{10.6}$$

with the boundary condition

$$p(0,x) = f_{B(0)}(x),$$

where $f_{B(0)}$ is the density of $B(0)$.

Proof: We assume that $p(t,x)$ is continuously differentiable in t and twice continuously differentiable in x. Write

$$B(t) = (B(t) - B(t-h)) + B(t-h).$$

Note that $B(t-h)$ is independent of $B(t) - B(t-h)$ due to the independence of the increments. Furthermore, stationarity of increments implies that $B(t) - B(t-h)$ is identical to $B(h) - B(0) = B(h) \sim N(0,h)$ in distribution. Conditioning on $B(t-h)$ we get

$$p(t,x) = \int_{\mathcal{R}} p(t-h,y) f_{B(h)}(x-y) dy, \tag{10.7}$$

where $f_{B(h)}(\cdot)$ is the density of $B(h)$. Using Taylor series expansion for $p(t-h,y)$ around (t,x), we get

$$p(t-h,y) = p(t,x) - h p_t(t,x) + (y-x) p_x(t,x) + \frac{(y-x)^2}{2} p_{xx}(t,x) + o((y-x)^2).$$

Substituting in Equation 10.7 we get

$$
\begin{aligned}
p(t,x) &= \int_{\mathcal{R}} [p(t,x) - h p_t(t,x) + (y-x) p_x(t,x) \\
&\quad + \frac{1}{2}(y-x)^2 p_{xx}(t,x) + o((y-x)^2)] f_{B(h)}(x-y) dy \\
&= p(t,x) - h p_t(t,x) - \mathsf{E}(B(h)) p_x(t,x) + \frac{1}{2}\mathsf{E}(B(h)^2) p_{xx}(t,x) + o(h) \\
&= p(t,x) - h p_t(t,x) + \frac{h}{2} p_{xx}(t,x) + o(h).
\end{aligned}
$$

Here we have used the fact that all odd moments of $B(h)$ are zero, and

$$\mathsf{E}(B(h)^{2k}) = \frac{(2k)!}{2^k k!} h^k,$$

to obtain the $o(h)$ term. Dividing by h we get

$$p_t(t, x) = \frac{1}{2} p_{xx}(t, x) + \frac{o(h)}{h}.$$

Letting $h \to 0$ we get Equation 10.6. ∎

Equation 10.6 is the forward differential equation. It is also possible to derive the backward differential equations, but we do not present it here. Under the initial condition $B(0) = 0$, the above equation has a unique solution given by

$$p(t, x) = \phi(t, x), \quad t > 0, \ x \in \mathcal{R}, \tag{10.8}$$

which is the density of a $N(0, t)$ random variable. We shall find this method of deriving partial differential equations quite useful when dealing with Brownian motion and related processes. The following theorem gives an extension of the above theorem to a $BM(\mu, \sigma)$.

Theorem 10.7 Kolmogorov Equation for BM. *Let $\{X(t), t \geq 0\}$ be a $BM(\mu, \sigma)$ and let $p(t, x)$ be the density of $X(t)$. It satisfies the following partial differential equation:*

$$p_t(t, x) = -\mu p_x(t, x) + \frac{\sigma^2}{2} p_{xx}(t, x), \quad x \in \mathcal{R}, \ t \geq 0, \tag{10.9}$$

with the boundary condition

$$p(0, x) = f_{X(0)}(x).$$

Proof: See Computational Exercise 10.8. ∎

10.4 First Passage Times

Following our plan of studying any class of stochastic processes, we now study the first passage times in an SBM. Define

$$T_a = \min\{t \geq 0 : B(t) = a\} \tag{10.10}$$

as the first passage time to the state a. Note that this is a well-defined random variable since the sample paths of an SBM are everywhere continuous with probability 1. It is clear that T_a is a stopping time, since one can tell if the SBM has visited state a by time t by looking at the sample path of the SBM over $[0, t]$. The next theorem gives the pdf of T_a. It uses a clever argument called the "reflection principle," which uses the symmetry of the SBM and its strong Markov property.

Theorem 10.8 Pdf of T_a. *Let T_a be as defined in Equation 10.10. If $a \neq 0$, the pdf of T_a is given by*

$$f_{T_a}(t) = \frac{|a|}{\sqrt{2\pi t^3}} e^{-\frac{a^2}{2t}}, \quad t > 0. \tag{10.11}$$

If $a = 0$, $T_a = 0$ with probability 1.

Proof: Suppose $a > 0$. We have

$$P(T_a \leq t) = P(T_a \leq t, B(t) \geq a) + P(T_a \leq t, B(t) < a).$$

Since the paths of an SBM are continuous, if a sample path is above a at time t, it must have crossed a at or before t. Thus $\{B(t) \geq a\} \Rightarrow \{T_a \leq t\}$. Hence

$$P(T_a \leq t, B(t) \geq a) = P(B(t) \geq a) = P(B(t) > a).$$

Next, if $T_a \leq t$, then due to strong Markov property, the $\{B(t + T_a) - B(T_a), t \geq 0\}$ is an SBM and is independent of $\{B(u) : 0 \leq u \leq T_a\}$. Thus

$$
\begin{aligned}
& P(T_a \leq t, B(t) < a) \\
= \ & P(T_a \leq t, B(T_a + t - T_a) - B(T_a) < 0) \\
= \ & P(T_a \leq t, B(T_a + t - T_a) - B(T_a) > 0) \ \text{(SBM is symmetric)} \\
= \ & P(T_a \leq t, B(t) > a) = P(B(t) > a).
\end{aligned}
$$

Combining the above observations we get

$$P(T_a \leq t) = 2P(B(t) > a) = 2 \int_a^\infty \phi(t, u) du. \tag{10.12}$$

Taking derivatives on both sides with respect to t (this is a bit tedious), we get the pdf given in Equation 10.11. The case of $a < 0$ follows in a symmetric fashion. Finally, consider the case $a = 0$. Then $B(0) = 0$ implies that $T_0 = 0$. ∎

Direct computations show that

$$E(T_a) = \int_0^\infty t f_{T_a}(t) dt = \infty, \quad (a \neq 0).$$

Thus the SBM reaches any state a eventually with probability 1, but the expected time to do so is infinity, except when $a = 0$. This should remind the reader of the definitions of null recurrence of DTMCs and CTMCs. In some sense, an SBM is a null recurrent Markov process, although we have not formally defined this concept for continuous state-space Markov processes. In retrospect, this is not surprising, since an SBM is a limit of a simple symmetric random walk which is a null recurrent DTMC.

From the above theorem we immediately get the distribution of the maximum and the minimum of an SBM.

Theorem 10.9   The Max and Min of an SBM. *The densities of*

$$U(t) = \max\{B(u) : 0 \leq u \leq t\}, \quad L(t) = \min\{B(u) : 0 \leq u \leq t\},$$

are given by

$$f_{U(t)}(a) = f_{L(t)}(-a) = \sqrt{\frac{2}{\pi t}} e^{-\frac{a^2}{2t}}, \quad a \geq 0.$$

Proof: We see that, for $a \geq 0$,

$$\{U(t) \geq a\} \Leftrightarrow \{T_a \leq t\}.$$

From Equation 10.12 we get

$$P(U(t) \geq a) = P(T_a \leq t) = 2 \int_a^\infty \phi(t, u) du.$$

The result follows from taking derivatives with respect to a. The result about the minimum follows by using the fact that $\{-B(t), t \geq 0\}$ is also an SBM. ∎

Extending this analysis to a BM(μ, σ) with $\mu \neq 0$ is not straightforward, since we do not have symmetry property when $\mu \neq 0$. We follow a different, and more general, approach to analyze this problem. To be precise, let $\{X(t), t \geq 0\}$ be a BM(μ, σ), and, for $a > 0$, define

$$T_a = \min\{t \geq 0 : X(t) = a\}, \tag{10.13}$$

and let

$$\psi(s) = E(e^{-sT_a}),$$

be its LST. The main result is given in the next theorem. It uses the infinitesimal version of the first-step analysis that we have used before in the CTMCs and the DTMCs. In this analysis we will need to consider a BM starting from any state $x \in \mathcal{R}$.

Theorem 10.10 LST of T_a. *Let $a > 0$. Then*

$$\psi(s) = \exp(a(\mu - \sqrt{\mu^2 + 2\sigma^2 s})/\sigma^2). \tag{10.14}$$

$$P(T_a < \infty) = \begin{cases} 1 & \text{if } \mu \geq 0, \\ e^{2\mu a/\sigma^2} & \text{if } \mu < 0, \end{cases} \tag{10.15}$$

$$E(T_a) = \begin{cases} a/\mu & \text{if } \mu > 0, \\ \infty & \text{if } \mu \leq 0. \end{cases} \tag{10.16}$$

Proof: Let $x \leq a$ be a fixed real number and define

$$Y(t, x) = x + X(t), \quad t \geq 0. \tag{10.17}$$

We have seen that $\{Y(t, x), t \geq 0\}$ is a BM starting in state x. Define

$$T_a(x) = \min\{t \geq 0 : Y(t, x) \geq a\}.$$

Note that $T_a(0)$ as defined above is the same as T_a as defined in Equation 10.13, since $a > 0$. We shall do an infinitesimal first-step analysis and derive a differential equation for the LST

$$\psi(s, x) = E(e^{-sT_a(x)}).$$

We know that $Y(h, x) - Y(0, x) \sim N(\mu h, \sigma^2 h)$. Let $Y(h, x) - Y(0, x) = y$. If $x + y > a$, $T_a(x) \approx h$. Else, due to the Markov property of the BM, $T_a(x)$ has the same distribution as $h + T_a(x + y)$. Using this argument and $f_h(\cdot)$ as the pdf of $Y(h, x) - Y(0, x)$ we get, for $x < a$,

$$\begin{aligned} \psi(s, x) &= E(e^{-sT_a(x)}) \\ &= \int_{y \in \mathcal{R}} E(e^{-sT_a(x)} | Y(h, x) - Y(0, x) = y) f_h(y) dy \end{aligned}$$

$$= \int_{y \in \mathcal{R}} \mathsf{E}(e^{-s(T_a(x+y)+h)}) f_h(y) dy$$

$$= \int_{y \in \mathcal{R}} e^{-sh} \psi(s, x + y) f_h(y) dy.$$

Now, for a fixed s, the Taylor expansion of $\psi(s, x + y)$ around x yields

$$\psi(s, x + y) = \psi(s, x) + y\psi'(s, x) + \frac{1}{2} y^2 \psi''(s, x) + o(y^2),$$

where the superscripts $'$ and $''$ denote the first and second derivative with respect to x. Substituting in the previous equation we get

$$\begin{aligned}
\psi(s, x) &= \int_{y \in \mathcal{R}} e^{-sh} [\psi(s, x) + y\psi'(s, x) + \frac{1}{2} y^2 \psi''(s, x) + o(y^2)] f_h(y) dy \\
&= e^{-sh} [\psi(s, x) + \mathsf{E}(Y(h, x) - Y(0, x)) \psi'(s, x) + \\
&\quad \frac{1}{2} \mathsf{E}((Y(h, x) - Y(0, x))^2) \psi''(s, x) + o(h)] \\
&= e^{-sh} [\psi(s, x) + \mu h \psi'(s, x) + \frac{1}{2} \sigma^2 h \psi''(s, x)] + o(h).
\end{aligned}$$

Using $e^{-sh} = 1 - sh + o(h)$ and collecting the $o(h)$ terms we get

$$\psi(s, x) = (1 - sh) \psi(s, x) + \mu h \psi'(s, x) + \frac{1}{2} \sigma^2 h \psi''(s, x) + o(h).$$

Dividing by h and letting $h \to 0$, we get

$$s\psi(s, x) - \mu \psi'(s, x) - \frac{1}{2} \sigma^2 \psi''(s, x) = 0. \tag{10.18}$$

This is a second-order differential equation with constant coefficients. Hence it has a solution of the form

$$\psi(s, x) = A e^{\theta x}, \quad -\infty < x < a,$$

where A is an arbitrary constant. Substituting in Equation 10.18, we get

$$s - \mu \theta - \frac{1}{2} \sigma^2 \theta^2 = 0.$$

This has two solutions:

$$\theta_1 = \frac{-\mu + \sqrt{\mu^2 + 2\sigma^2 s}}{\sigma^2}, \quad \theta_2 = \frac{-\mu - \sqrt{\mu^2 + 2\sigma^2 s}}{\sigma^2}.$$

Now, one expects that as $x \to -\infty$, $T_a(x) \to \infty$, and hence $\psi(s, x) \to 0$. Thus we should use the solution $\theta = \theta_1$. Thus we get

$$\psi(s, x) = A \exp(\theta_1 x).$$

Finally, we have the boundary condition

$$\psi(s, a) = 1,$$

since $T_a(a) = 0$. Using this we get

$$\psi(s, x) = \exp(\theta_1(x - a)).$$

The required LST is given by

$$\psi(s) = \psi(s, 0).$$

This yields Equation 10.14. Equations 10.15 and 10.16 can be derived by using the properties of the LST. ∎

The case of $a < 0$ can be handled by simply studying the $-a$ case in a BM$(-\mu, \sigma)$.

Next let $a < 0$ and $b > 0$ and define

$$T_{a,b} = \min\{t \geq 0 : X(t) \in \{a, b\}\}. \tag{10.19}$$

Thus $T_{a,b}$ is the first time the BM(μ, σ) reaches either a or b. Following Theorem 10.10 we give the main result in the following theorem.

Theorem 10.11   Mean of $T_{a,b}$. *Let*

$$\theta = -\frac{2\mu}{\sigma^2}. \tag{10.20}$$

Then

$$P(X(T_{a,b}) = b) = \frac{\exp(\theta a) - 1}{\exp(\theta a) - \exp(\theta b)}, \tag{10.21}$$

$$E(T_{a,b}) = \frac{b(\exp(\theta a) - 1) - a(\exp(\theta b) - 1)}{\mu(\exp(\theta a) - \exp(\theta b))}. \tag{10.22}$$

In particular, when $\mu = 0$,

$$P(X(T_{a,b}) = b) = \frac{|a|}{|a| + b}, \tag{10.23}$$

$$E(T_{a,b}) = |a|b/\sigma^2. \tag{10.24}$$

Proof: Let $x \in [a, b]$ be a given real number and $Y(t, x)$ be as in Equation 10.17. Define

$$T_{a,b}(x) = \min\{t \geq 0 : Y(t, x) \geq b \text{ or } Y(t, x) \leq a\}. \tag{10.25}$$

Then $T_{a,b}(0)$ is the same as $T_{a,b}$ of Equation 10.19. Now, let

$$v(x) = P(Y(T_{a,b}(x), x) = b).$$

Using the argument and notation from the proof of Theorem 10.10, we get, for $a < x < b$,

$$
\begin{aligned}
v(x) &= \int_{y \in R} v(x + y) f_h(y) dy \\
&= \int_{y \in R} [v(x) + yv'(x) + \frac{1}{2} y^2 v''(x) + o(y^2)] f_h(y) dy \\
&= v(x) + \mu h v'(x) + \frac{\sigma^2 h}{2} v''(x) + o(h).
\end{aligned}
$$

Dividing by h and letting $h \to 0$, we get

$$\frac{1}{2}\sigma^2 v''(x) + \mu v'(x) = 0.$$

The solution to the above equation is

$$v'(x) = Ce^{\theta x},$$

where

$$\theta = -\frac{2\mu}{\sigma^2}.$$

This yields

$$v(x) = Ae^{\theta x} + B,$$

where A and B are arbitrary constants. Using the boundary conditions

$$v(a) = 0, \quad v(b) = 1,$$

we get the complete solution as

$$v(x) = \frac{\exp(\theta a) - \exp(\theta x)}{\exp(\theta a) - \exp(\theta b)}. \tag{10.26}$$

Substituting $x = 0$ gives Equation 10.21.

To derive Equation 10.22, let

$$m(x) = \mathsf{E}(T_{a,b}(x)).$$

Then, for $a < x < b$, we have

$$
\begin{aligned}
m(x) &= \int_{y \in \mathcal{R}} [h + m(x + y)] f_h(y) dy \\
&= h + \int_{y \in \mathcal{R}} [m(x) + ym'(x) + \frac{1}{2} y^2 m''(x) + o(y^2)] f_h(y) dy \\
&= h + m(x) + \mu h m'(x) + \frac{1}{2} \sigma^2 h m''(x) + o(h).
\end{aligned}
$$

Simplifying and dividing by h and letting $h \to 0$, we get

$$\frac{1}{2}\sigma^2 m''(x) + \mu m'(x) = -1.$$

The solution to the above equation is

$$m'(x) = Ce^{\theta x} - 1/\mu,$$

which yields

$$m(x) = Ae^{\theta x} + B - x/\mu,$$

where A and B are arbitrary constants. Using the boundary conditions

$$m(a) = m(b) = 0,$$

we get the complete solution as

$$m(x) = \frac{(a - b)\exp(\theta x) + b\exp(\theta a) - a\exp(\theta b)}{\mu(\exp(\theta a) - \exp(\theta b))} - \frac{x}{\mu}.$$

The required solution is given by $m(0)$ and is given in Equation 10.22. This proves the theorem. The results for the case $\mu = 0$ are obtained by taking the limits of Equations 10.21 and 10.22 as $\mu \to 0$. ∎

We shall show an alternate method of deriving the results of the above theorem by using the concept of Martingales in Section 10.7.

10.5 Reflected SBM

We begin with the definition of a reflected SBM.

Definition 10.5   Reflected SBM. *Let $\{B(t), t \geq 0\}$ be an SBM. The process $\{Y(t), t \geq 0\}$ defined by*

$$Y(t) = |B(t)|, \quad t \geq 0, \tag{10.27}$$

is called an SBM reflected at the origin, or simply a reflected SBM.

The nomenclature makes intuitive sense because one can obtain a sample path of $\{Y(t), t \geq 0\}$ by reflecting the parts of the sample path of an SBM that lie below zero around the horizontal axis $x = 0$ in the (t, x) plane, as shown in Figure 10.2.

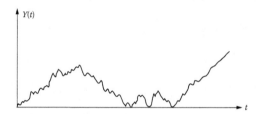

Figure 10.2 *A typical sample path of an SBM reflected at 0.*

We also say that $x = 0$ is a reflecting boundary. The following theorem lists the important properties of a reflected SBM.

Theorem 10.12   Properties of Reflected SBM. *Let $\{Y(t), t \geq 0\}$ be a reflected Brownian motion. Then*

1. The state-space of $\{Y(t), t \geq 0\}$ is $[0, \infty)$.

2. $\{Y(t), t \geq 0\}$ is a Markov process.

3. The increments of $\{Y(t), t \geq 0\}$ are stationary, but not independent.

4. For a fixed $t \geq 0$, $Y(t)$ has density given by

$$f_{Y(t)}(y) = 2\phi(t, y), \quad t \geq 0, \ y \geq 0.$$

5. $\mathsf{E}(Y(t)) = \sqrt{\frac{2t}{\pi}}, \quad \mathrm{Var}(Y(t)) = \left(1 - \frac{2}{\pi}\right)t.$

Proof: Part 1 is obvious from the definition. To show part 2, we exploit the following consequence of the symmetry of the SBM

$$\mathsf{P}(-y \le B(t+s) \le y | B(t) = x) = \mathsf{P}(-y \le B(t+s) \le y | B(t) = -x).$$

Now let $0 \le t_1 < t_2 < \cdots < t_n < t_n + s$. We have

$$
\begin{aligned}
\mathsf{P}(Y(t_n + s) &\le y | Y(t_i) = y_i, 1 \le i \le n) \\
&= \mathsf{P}(-y \le B(t_n + s) \le y | B(t_i) = \pm y_i, 1 \le i \le n) \\
&= \mathsf{P}(-y \le B(t_n + s) \le y | B(t_i) = y_i, 1 \le i \le n) \quad \text{(by symmetry)} \\
&= \mathsf{P}(-y \le B(t_n + s) \le y | B(t_n) = y_n) \quad \text{(SBM is Markov)} \\
&= \mathsf{P}(-y - y_n \le B(t_n + s) - B(t_n) \le y - y_n | B(t_n) = y_n) \\
&= \int_{-y}^{y} \phi(s, x - y_n) dx \quad \text{(SBM has ind. inc.).}
\end{aligned}
$$

The last equality follows because $B(t_n + s) - B(t_n)$ is independent of $B(t_n)$ and is normally distributed with mean 0 and variance s. This proves the Markov property. It also implies that given $Y(t) = x$, the density of the increment $Y(t+s) - Y(t)$ is $\phi(s, y) + \phi(s, -y - 2x) = \phi(s, y) + \phi(s, y + 2x)$. This shows the stationarity of increments. Since the density depends on x, the increments are clearly not independent. That proves part 3. Part 4 follows by taking the derivatives with respect to y on both sides of

$$\mathsf{P}(Y(t) \le y) = \mathsf{P}(-y \le B(t) < y) = \int_{-y}^{y} \phi(t, u) du,$$

and using the symmetry of the ϕ density. Part 5 follows by direct integrations. ∎

The symmetry of the SBM can be used to study its reflection across any horizontal line $x = a$, not just across the t−axis $x = 0$. In case $a > 0$, we consider the process that is reflected downward, so that the reflected process $\{Y_a(t), t \ge 0\}$ has state-space $(-\infty, a]$ and is defined by

$$Y_a(t) = a - |a - B(t)|, \quad t \ge 0.$$

Similarly, in case $a < 0$, we consider the process that reflected upward, so that the reflected process $\{Y_a(t), t \ge 0\}$ has state-space $[a, \infty)$ and is defined by

$$Y_a(t) = |B(t) - a| + a, \quad t \ge 0.$$

For $a = 0$, we could use either of the above two definitions. The Definition 10.6 corresponds to using the upward reflection. A typical sample path of such a process is shown in Figure 10.3.

Theorem 10.13 *Let $S_a = [a, \infty)$ if $a \le 0$, and $(-\infty, a]$ if $a > 0$. The density $p(t, y)$ of $Y_a(t)$ is given by*

$$p(t, y) = \phi(t, y) + \phi(t, 2a - y), \quad y \in S_a. \tag{10.28}$$

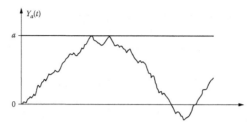

Figure 10.3 *A typical sample path of a BM reflected downward at $a > 0$.*

Proof: Consider the case $a > 0$. Using symmetry, we get for $x \geq 0$,

$$P(a - Y_a(t) \leq x) = P(a - x \leq B(t) \leq a + x) = \int_{a-x}^{a+x} \phi(t, u) du.$$

Making the change of variables $y = a - x$, the above equation yields

$$P(Y_a(t) \geq y) = \int_{y}^{2a-y} \phi(t, u) du, \quad y \in S_a.$$

Taking the derivatives with respect to y we can derive the density given in Equation 10.28. The case $a \leq 0$ is similar. ∎

How does the partial differential equation of Theorem 10.6 account for the reflecting boundaries? We state the result in the following theorem.

Theorem 10.14   Kolmogorov Equation for the Reflected SBM. *Let $p(t, y)$ be the density $Y_a(t)$. It satisfies the partial differential equation and the boundary conditions of Theorem 10.6, and it satisfies an additional boundary condition:*

$$p_y(t, a) = 0, \quad t \geq 0.$$

Proof: The proof of the partial differential equation is as in Theorem 10.6. The proof of the boundary condition is technical, and beyond the scope of this book. ∎

Note that the partial derivative in the boundary condition of the above theorem is to be interpreted as the right derivative if $a \leq 0$ and the left derivative if $a > 0$. The reader should verify that the density in Equation 10.28 satisfies the partial differential equation and the boundary condition.

10.6 Reflected BM and Limiting Distributions

Analysis of a reflected BM(μ, σ) is much harder when $\mu \neq 0$ due to the lack of symmetry. In fact a reflected BM cannot be defined as in Equation 10.27, and we need a new definition given below. Another name for Brownian motion that is restricted to lie in a given region ($[0, \infty)$, for the reflected Brownian motion, for example) is regulated Brownian motion.

Definition 10.6 Reflected BM. *Let* $\{X(t), t \geq 0\}$ *be a* BM(μ, σ). *The process* $\{Y(t), t \geq 0\}$ *defined by*

$$Y(t) = X(t) - \inf_{0 \leq u \leq t} X(u), \quad t \geq 0, \tag{10.29}$$

is called a BM *reflected at the origin, or simply a reflected* BM.

Note that $|X(t)|$ is a very different process than the reflected BM defined above. The two are identical (in distribution) if $\mu = 0$. Since we do not have symmetry, we derive the transient distribution of a reflected BM by deriving the Kolmogorov equation satisfied by its density in the next theorem.

Theorem 10.15 Kolmogorov Equation for a Reflected BM. *Let* $\{Y(t), t \geq 0\}$ *be a* BM(μ, σ) *reflected at the origin, and let* $p(t, y)$ *be the density of* $Y(t)$. *It satisfies the following partial differential equation:*

$$p_t(t, y) = -\mu p_y(t, y) + \frac{1}{2}\sigma^2 p_{yy}(t, y), \quad y > 0, \ t \geq 0, \tag{10.30}$$

with boundary conditions

$$p(0, y) = f_{Y(0)}(y),$$

$$\mu p(t, 0) = \frac{1}{2}\sigma^2 p_y(t, 0), \quad t > 0.$$

Proof: The proof of the partial differential equation is as in Theorem 10.7. The first boundary condition is as in Theorem 10.7. The proof of the second boundary condition is technical, and beyond the scope of this book. ∎

One can similarly study a Brownian motion constrained to lie in the interval $[a, b]$ with $X(0) \in [a, b]$. A typical sample path of such a BM is shown in Figure 10.4. Thus

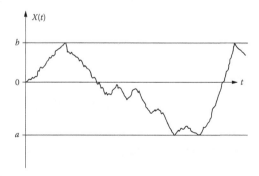

Figure 10.4 *A typical sample path of a BM constrained to lie in* $[a, b]$.

it is reflected up at a and down at b. Giving the functional form of such a BM along the same lines as Equation 10.29 is not possible. However, we give the Kolmogorov equation satisfied by its density in the next theorem. The proof is omitted.

Theorem 10.16   Kolmogorov Equation for a BM on $[a, b]$**.** *Let* $a < 0 < b$*, and* $\{Y(t), t \geq 0\}$ *be a* $\mathrm{BM}(\mu, \sigma)$ *reflected up at* a *and down at* b*. Let* $p(t, y)$ *be the density of* $Y(t)$*. It satisfies the following partial differential equation:*

$$p_t(t, y) = -\mu p_y(t, y) + \frac{1}{2}\sigma^2 p_{yy}(t, y), \quad t \geq 0, \ y \in [a, b], \tag{10.31}$$

with boundary conditions

$$\mu p(t, a) = \frac{1}{2}\sigma^2 p_y(t, a), \quad t > 0$$

$$\mu p(t, b) = \frac{1}{2}\sigma^2 p_y(t, b), \quad t > 0.$$

Note that $p_y(t, a)$ is to be interpreted as the right derivative, and $p_y(t, b)$ as the left derivative. Although, one can solve the above equation analytically, the result is complicated, and hence we do not give it here. Below we study the solution when $t \to \infty$.

Theorem 10.17   Limiting Distribution. *Let* $p(t, y)$ *be as in Theorem 10.16. The limiting distribution*

$$p(y) = \lim_{t \to \infty} p(t, y), \quad y \in [a, b],$$

is given by

$$p(y) = \frac{\theta \exp(-\theta y)}{\exp(-\theta a) - \exp(-\theta b)}, \quad y \in [a, b], \tag{10.32}$$

where

$$\theta = -\frac{2\mu}{\sigma^2}.$$

Proof: We shall assume that the limiting distribution $[p(y), a \leq y \leq b]$ exists. Hence we expect

$$\lim_{t \to \infty} p_t(t, y) = 0, \quad y \in [a, b].$$

Substituting in Equation 10.31 we get

$$p''(y) + \theta p'(y) = 0,$$

with $\theta = -2\mu/\sigma^2$. The solution to the above equation is given by

$$p(y) = A \exp(-\theta y) + B, \quad y \in [a, b].$$

The boundary conditions in Theorem 10.16 reduce to

$$\theta p(a) = -p'(a),$$
$$\theta p(b) = -p'(b).$$

Both these conditions yield only one equation: $B = 0$. The remaining constant A can be evaluated by using

$$\int_a^b p(y)dy = 1.$$

This yields the solution given in Equation 10.32. ∎

10.7 BM and Martingales

We begin with the definition of a Martingale.

Definition 10.7 **Martingales.** *A discrete time real valued stochastic process* $\{X_n, n \geq 0\}$ *is called a Martingale if*

$$E(|X_n|) < \infty, \quad n \geq 0$$

and

$$\mathsf{E}(X_{n+m}|X_n, X_{n-1}, \cdots, X_0) = X_n, \quad n, m \geq 0. \tag{10.33}$$

A continuous time real valued stochastic process $\{X(t), t \geq 0\}$ *is called a Martingale if*

$$E(|X(t)|) < \infty, \quad t \geq 0$$

and

$$\mathsf{E}(X(t+s)|X(u) : 0 \leq u \leq s) = X(s), \quad s, t \geq 0. \tag{10.34}$$

Example 10.1 **Random Walk Martingale.** Let $\{Y_n, n \geq 1\}$ be a sequence of iid random variables with $\mathsf{E}(Y_n) = 0$. Define

$$X_0 = 0, \quad X_n = Y_1 + Y_2 + \cdots + Y_n, \quad n \geq 1.$$

Then $E(X_n) = 0$ and

$$\mathsf{E}(X_{n+m}|X_n, X_{n-1}, \cdots, X_0) = X_n + \mathsf{E}(\sum_{k=1}^{m} Y_{n+k}) = X_n, \quad n \geq 0.$$

Hence $\{X_n, n \geq 0\}$ is a Martingale. ∎

Example 10.2 **Brownian Motion Martingale.** Let $\{B(t), t \geq 0\}$ be an SBM. Then $E(B(t)) = 0$. Using the stationarity and independence of increments of the SBM we get

$$
\begin{aligned}
\mathsf{E}&(B(t+s)|B(u) : 0 \leq u \leq s) \\
&= \mathsf{E}(B(t+s) - B(s) + B(s)|B(u) : 0 \leq u \leq s) \\
&= \mathsf{E}(B(t+s) - B(s)|B(u) : 0 \leq u \leq s) \\
&\quad + \mathsf{E}(B(s)|B(u) : 0 \leq u \leq s) \\
&= B(s).
\end{aligned}
$$

Thus an SBM is a Martingale. ∎

We next define a useful generalization of the concept of a Martingale.

Definition 10.8   Generalized Martingales. *A discrete time real valued stochastic process $\{X_n, n \geq 0\}$ is called a Martingale with respect to another discrete time stochastic process $\{Y_n, n \geq 0\}$ if*

$$E(|X_n|) < \infty, \quad n \geq 0$$

and

$$\mathsf{E}(X_{n+m}|Y_n, Y_{n-1}, \cdots, Y_0) = X_n, \quad n, m \geq 0. \tag{10.35}$$

A continuous time real valued stochastic process $\{X(t), t \geq 0\}$ is called a Martingale with respect to another continuous time stochastic process $\{Y(t), t \geq 0\}$ if

$$E(|X(t)|) < \infty, \quad t \geq 0$$

and

$$\mathsf{E}(X(t+s)|Y(u) : 0 \leq u \leq s) = X(s), \quad s, t \geq 0. \tag{10.36}$$

From Definitions 10.7 and 10.8 it follows that when $\{X_n, n \geq 0\}$ is a Martingale with respect to $\{X_n, n \geq 0\}$, we simply say that $\{X_n, n \geq 0\}$ is a Martingale. We illustrate with several examples.

Example 10.3   Linear Martingale. Let $\{X(t), t \geq 0\}$ be a BM(μ, σ), and define

$$Y(t) = X(t) - \mu t, \quad t \geq 0.$$

Show that $\{Y(t), t \geq 0\}$ is a Martingale with respect to $\{X(t), t \geq 0\}$.

First, we have $E(Y(t)) = 0$. Using the stationarity and independence of increments of the BM we get

$$
\begin{aligned}
&\mathsf{E}(Y(t+s)|X(u) : 0 \leq u \leq s)\\
&= \mathsf{E}(X(t+s) - \mu(t+s)|X(u) : 0 \leq u \leq s)\\
&= \mathsf{E}(X(t+s) - X(s) - \mu t + X(s) - \mu s|X(u) : 0 \leq u \leq s)\\
&= \mathsf{E}(X(t+s) - X(s) - \mu t|X(u) : 0 \leq u \leq s)\\
&\quad +\mathsf{E}(X(s) - \mu s|X(u) : 0 \leq u \leq s)\\
&= X(s) - \mu s = Y(s).
\end{aligned}
$$

Thus $\{Y(t), t \geq 0\}$ is a Martingale with respect to $\{X(t), t \geq 0\}$. ∎

Example 10.4   Quadratic Martingale. Let $\{B(t), t \geq 0\}$ be the SBM, and define

$$Y(t) = B^2(t) - t, \quad t \geq 0.$$

Show that $\{Y(t), t \geq 0\}$ is a Martingale with respect to $\{B(t), t \geq 0\}$.

We have $E(Y(t)) = 0$. Using the stationarity and independence of increments of the SBM, we get

$$\mathsf{E}(Y(t+s)|B(u) : 0 \leq u \leq s)$$

$$= \quad E(B^2(t+s) - (t+s)|B(u) : 0 \le u \le s)$$
$$= \quad E((B(t+s) - B(s) + B(s))^2 - (t+s)|B(u) : 0 \le u \le s)$$
$$= \quad E((B(t+s) - B(s))^2|B(u) : 0 \le u \le s)$$
$$\quad + E(2(B(t+s) - B(s))B(s)|B(u) : 0 \le u \le s)$$
$$\quad + E(B(s)^2|B(u) : 0 \le u \le s) - (t+s)$$
$$= \quad t + B^2(s) - (t+s)$$
$$= \quad B^2(s) - s = Y(s).$$

Thus $\{B^2(t) - t, t \ge 0\}$ is a Martingale with respect to $\{B(t), t \ge 0\}$. ∎

Example 10.5   Exponential Martingale. Let $\{X(t), t \ge 0\}$ be a BM(μ, σ). For a $\theta \in \mathcal{R}$ define

$$Y(t) = \exp\left(\theta X(t) - (\theta\mu + \frac{1}{2}\theta^2\sigma^2)t\right), \quad t \ge 0.$$

Show that $\{Y(t), t \ge 0\}$ is a Martingale with respect to $\{X(t), t \ge 0\}$.

We leave it to the reader (see Computational Exercise 10.29) to show that

$$E(Y(t)) = 1, \quad t \ge 0.$$

Using the stationarity and independence of increments of the BM, we get

$$E(Y(t+s)|X(u) : 0 \le u \le s)$$
$$= \quad E(\exp(\theta X(t+s) - (\theta\mu + \frac{1}{2}\theta^2\sigma^2)(t+s))|X(u) : 0 \le u \le s)$$
$$= \quad E(\exp(\theta(X(t+s) - X(s) + X(s)) -$$
$$\quad (\theta\mu + \frac{1}{2}\theta^2\sigma^2)(t+s))|X(u) : 0 \le u \le s)$$
$$= \quad \exp(-(\theta\mu + \frac{1}{2}\theta^2\sigma^2)(t+s)) \cdot$$
$$\quad E(\exp(\theta(X(t+s) - X(s)))\exp(\theta X(s))|X(u) : 0 \le u \le s)$$
$$= \quad \exp(\theta X(s) - (\theta\mu + \frac{1}{2}\theta^2\sigma^2)(t+s)) \cdot$$
$$\quad E(\exp(\theta(X(t+s) - X(s)))|X(u) : 0 \le u \le s)$$
$$= \quad \exp(\theta X(s) - (\theta\mu + \frac{1}{2}\theta^2\sigma^2)(t+s))E(\exp(\theta X(t)))$$
$$= \quad \exp(\theta X(s) - (\theta\mu + \frac{1}{2}\theta^2\sigma^2)s) = Y(s).$$

The penultimate equality follows from the expression for the LST of a N$(\mu t, \sigma^2 t)$ random variable. Thus $\{Y(t), t \ge 0\}$ is a Martingale with respect to $\{X(t), t \ge 0\}$. As a special case we see that $\exp(\theta B(t) - \theta^2 t/2)$ is a Martingale with respect to $\{B(t), t \ge 0\}$. ∎

One of the main results about Martingales is the optional sampling theorem (also called the stopping theorem), stated below for the continuous time case.

Theorem 10.18 Optional Sampling Theorem. *Let* $\{X(t), t \geq 0\}$ *be a Martingale and* T *be a finite stopping time, that is, there is a* $t < \infty$ *such that* $\mathsf{P}(T \leq t) = 1$. *Then*

$$\mathsf{E}(X(T)) = \mathsf{E}(X(0)).$$

We do not include the proof of this theorem here. Similar theorem holds for the discrete time Martingales as well. The condition given in the above theorem is only a sufficient condition, and there are many other weaker sufficient conditions under which the same conclusion holds. We find that the condition given here is the easiest to check in practice. We illustrate the theorem by using it to derive the results of Theorem 10.11.

Example 10.6 Let $\{X(t), t \geq 0\}$ be a BM(μ, σ), with $\mu \neq 0$. Let $a < 0$ and $b > 0$ be given, and $T_{a,b}$ be the first time the BM visits a or b. From Example 10.5 we conclude that $\exp(\theta X(t))$ is a Martingale if we choose

$$\theta = -\frac{2\mu}{\sigma^2}.$$

Since $X(t) \sim N(\mu t, \sigma^2 t)$, we see that $X(t)$ will eventually go below a or above b with probability 1. Hence $\mathsf{P}(T_{a,b} < \infty) = 1$. However, $T_{a,b}$ is not a finite stopping time. Fix a $t < \infty$ and Define $\tilde{T} = \min(T, t)$. Then $\tilde{T}$ is a bounded stopping time. Hence, using Theorem 10.18 we get

$$\mathsf{E}(\exp(\theta X(\tilde{T}))) = \mathsf{E}(\exp(\theta X(0))) = 1.$$

Now, we have

$$
\begin{aligned}
1 &= \mathsf{E}(\exp(\theta X(\tilde{T}))) \\
&= \exp(\theta a)\mathsf{P}(X(T_{a,b}) = a; T_{a,b} \leq t) + \exp(\theta b)\mathsf{P}(X(T_{a,b}) = b; T_{a,b} \leq t) \\
&\quad + \mathsf{E}(\exp(\theta X(t)); T_{a,b} > t). \tag{10.37}
\end{aligned}
$$

Now

$$\mathsf{E}(\exp(\theta X(t)); T_{a,b} > t) \leq \exp(\max(\theta a, \theta b))\mathsf{P}(T_{a,b} > t).$$

The right hand side decreases to zero since $\mathsf{P}(T_{a,b} > t)$ goes to zero as $t \to \infty$. Letting $t \to \infty$ in Equation 10.37 we get

$$1 = \exp(\theta a)\mathsf{P}(X(T_{a,b}) = a) + \exp(\theta b)\mathsf{P}(X(T_{a,b}) = b).$$

Since $\mathsf{P}(X(T_{a,b}) = a) + \mathsf{P}(X(T_{a,b}) = b) = 1$, we can solve the above equation to get Equation 10.21.

To derive $\mathsf{E}(T_{a,b})$ we use the linear Martingale $X(t) - \mu t$ of Example 10.3. Using Theorem 10.18 we get

$$\mathsf{E}(X(\tilde{T}) - \mu\tilde{T}) = \mathsf{E}(X(0) - \mu \cdot 0) = 0.$$

Thus we have

$$\mathsf{E}(\tilde{T}) = \mathsf{E}(X(\tilde{T}))/\mu. \tag{10.38}$$

Now

$$\mathsf{E}(X(\tilde{T})) = a\mathsf{P}(X(T_{a,b}) = a; T_{a,b} \le t) + b\mathsf{P}(X(T_{a,b}) = b; T_{a,b} \le t) + \mathsf{E}(X(t); T_{a,b} > t),$$

and

$$\mathsf{E}(\tilde{T}) = \mathsf{E}(T_{a,b}; T_{a,b} \le t) + t\mathsf{P}(T_{a,b} > t).$$

Now $T_{a,b} > t$ implies $a < X(t) < b$. Hence one can show that

$$t\mathsf{P}(T_{a,b} > t) \le t\mathsf{P}(a < X(t) < b) \to 0 \text{ as } t \to \infty,$$

and

$$\mathsf{E}(X(t); T_{a,b} > t) \le \max(|a|, b)\mathsf{P}(T_{a,b} > t) \to 0 \text{ as } t \to \infty.$$

Hence, we have

$$\lim_{t \to \infty} \mathsf{E}(X(\tilde{T})) = a\mathsf{P}(X(T_{a,b}) = a) + b\mathsf{P}(X(T_{a,b}) = b).$$

and

$$\lim_{t \to \infty} \mathsf{E}(\tilde{T}) = \mathsf{E}(T_{a,b}).$$

Hence, letting $t \to \infty$ in Equation 10.38, we get

$$\mathsf{E}(T_{a,b}) = (a\mathsf{P}(X(T_{a,b}) = a) + b\mathsf{P}(X(T_{a,b}) = b))/\mu,$$

Simplifying this we get Equation 10.22. The results for the case $\mu = 0$ can be obtained by letting $\mu \to 0$ in the above results. ∎

10.8 Cost/Reward Models

Let $X(t)$ be the state of a system at time t, and suppose $\{X(t), t \ge 0\}$ is a BM(μ, σ). Suppose the system incurs costs at rate $f(u, x)$ whenever it is in state x at time u. Using $\alpha \ge 0$ as a continuous discount factor, we see that the total discounted cost over $[0, t]$ is given by

$$\int_0^t e^{-\alpha u} f(u, X(u)) du, \quad t \ge 0.$$

Note that $\{X(u), 0 \le u \le t\}$ is a continuous function with probability 1. Hence, if f is sufficiently nice (for example, continuous), the integral above is well defined as a Riemann integral with probability one. Since the integral is a random variable itself, one can, in theory, compute its distribution. Here we focus our attention on computing its expected value, called the expected total discounted cost (ETDC), as defined below.

$$c(t) = \mathsf{E}\left(\int_0^t e^{-\alpha u} f(u, X(u)) du\right), \quad t \ge 0.$$

Interchanging the integral and the expectation we get

$$c(t) = \int_0^t e^{-\alpha u} \mathsf{E}(f(u, X(u))) du, \quad t \ge 0.$$

Since we know the distribution of $X(u)$, we can compute $\mathsf{E}(f(u, X(u)))$ as a function of u, and then evaluate the above integral as a standard Riemann integral (assuming f is reasonably nice). We illustrate with an example.

Example 10.7 Quadratic Cost Model. Suppose $f(u, x) = \beta x^2$, where $\beta > 0$ is a fixed constant. Thus the cost rate at time u does not depend on u. Compute $c(t)$. We have

$$
\begin{aligned}
c(t) &= \mathsf{E}\left(\int_0^t e^{-\alpha u} \beta X^2(u) du \right) \\
&= \int_0^t e^{-\alpha u} \beta \mathsf{E}(X^2(u)) du \\
&= \int_0^t e^{-\alpha u} \beta (\sigma^2 u + \mu^2 u^2) du \\
&= \beta \sigma^2 (1 - e^{-\alpha t}(1 + \alpha t))/\alpha^2 + 2\beta\mu^2(1 - e^{-\alpha t}(1 + \alpha t + \tfrac{1}{2}\alpha^2 t^2))/\alpha^3.
\end{aligned}
$$

If we let $t \to \infty$, we get the ETDC over the infinite horizon as

$$
c(\infty) = \lim_{t \to \infty} c(t) = 2\beta(\mu^2 + \frac{\alpha}{2}\sigma^2)/\alpha^3.
$$

If we let $\alpha \to 0$ in the expression for $c(t)$ we get the expected total cost over $[0, t]$ as

$$
\beta\left(\frac{1}{2}\sigma^2 t^2 + \frac{1}{3}\mu^2 t^3 \right).
$$

This implies that the average cost per unit time $c(t)/t$ goes to infinity as $t \to \infty$. ∎

The $c(t)$ represents the ETDC over $[0, t]$, where t is a fixed constant. In many applications we are interested in the ETDC over $[0, T]$, where T is a random variable, typically a stopping time for the underlying process. Here we consider the case where $T = T_{a,b}$, where $T_{a,b}$ is the first passage time to the set $\{a, b\}$, as defined in Equation 10.19. We further assume that the cost rate in state x is $f(x)$. That is, we are interested in

$$
c = \mathsf{E}\left(\int_0^{T_{a,b}} e^{-\alpha u} f(X(u)) du \right).
$$

Using $g(\cdot)$ as the pdf of $T_{a,b}$, we can compute this integral as follows:

$$
\begin{aligned}
c &= \mathsf{E}\left(\int_0^{T_{a,b}} e^{-\alpha u} f(X(u)) du \right) \\
&= \int_0^\infty \mathsf{E}\left(\int_0^{T_{a,b}} e^{-\alpha u} f(X(u)) du \Big| T_{a,b} = t \right) g(t) dt \\
&= \int_0^\infty \left(\int_0^t e^{-\alpha u} \mathsf{E}(f(X(u))|T_{a,b} = t) du \right) g(t) dt.
\end{aligned}
$$

This is generally too hard to compute for two reasons: first, the density g is not an easy function to deal with, and second, the conditional distribution of $X(u)$ given

$T_{a,b} = t$, needed in computing the expectation inside the integral, is equally hard to compute. Hence we develop an alternate method below. We first compute a more general quantity

$$c(x) = \mathsf{E}\left(\int_0^{T_{a,b}(x)} e^{-\alpha u} f(Y(u,x)) du \,\Big|\, Y(0,x) = x\right), \quad a \le x \le b,$$

where $\{Y(t,x) = x + X(t), t \ge 0\}$ is a Brownian motion starting at x, and $T_{a,b}(x)$ is the first time $Y(t,x)$ visits a or b, as defined in Equation 10.25. Then we have

$$c = c(0).$$

The next theorem describes the differential equation satisfied by $c(x)$.

Theorem 10.19   ETDC over $[0, T_{a,b}(x)]$. *The function $c(x)$ satisfies the following differential equation:*

$$\frac{1}{2}\sigma^2 \frac{d^2 c(x)}{dx^2} + \mu \frac{dc(x)}{dx} - \alpha c(x) = -f(x). \tag{10.39}$$

The boundary condition is $c(a) = c(b) = 0$.

Proof: We do an infinitesimal first-step analysis and derive a differential equation for $c(x)$. We know that $Y(h,x) - Y(0,x) = X(h) - X(0) \sim N(\mu h, \sigma^2 h)$. Using $f_h(\cdot)$ as the pdf of $Y(h,x) - Y(0,x)$ we get

$$c(x) = \mathsf{E}(\int_0^{T_{a,b}(x)} e^{-\alpha u} f(Y(u,x)) du \,\Big|\, Y(0,x) = x)$$

$$= \int \mathsf{E}(\int_0^{T_{a,b}(x)} e^{-\alpha u} f(Y(u,x)) du \,\Big|\, Y(h,x) - Y(0,x) = y, Y(0,x) = x) f_h(y) dy$$

$$= \int [f(x)h + \mathsf{E}(\int_h^{T_{a,b}(x)} e^{-\alpha u} f(Y(u,x)) du \,\Big|\, Y(h,x) = x + y) + o(h)] f_h(y) dy$$

$$= \int [f(x)h + \mathsf{E}(e^{-\alpha h} \int_0^{T_{a,b}(x+y)} e^{-\alpha u} f(Y(u,x+y)) du \,\Big|\, Y(0,x+y) = x + y)$$

$$\quad + o(h)] f_h(y) dy$$

$$= \int [f(x)h + e^{-\alpha h} c(x+y) + o(h)] f_h(y) dy.$$

Here we have used the fact that, given $Y(h,x) = x + y$, $\{Y(t,x), t \ge h\}$ is stochastically identical to $\{Y(t, x+y), t \ge 0\}$, and the integrals with respect to y are over the entire real line. Now, for a fixed h, the Taylor expansion of $c(x+y)$ around x yields

$$c(x+y) = c(x) + yc'(x) + \frac{1}{2}y^2 c''(x) + \cdots.$$

Substituting in the previous equation we get

$$c(x) = \int_{y \in \mathcal{R}} [f(x)h + e^{-\alpha h}(c(x) + yc'(x) + \frac{1}{2}y^2 c''(x) + \cdots)] f_h(y) dy$$

$$= \quad f(x)h + e^{-\alpha h}[c(x) + \mathsf{E}(X(h) - X(0))c'(x)$$
$$+ \frac{1}{2}\mathsf{E}((X(h) - X(0))^2)c''(x) + \cdots]$$
$$= \quad f(x)h + e^{-\alpha h}[c(x) + \mu h c'(x) + \frac{1}{2}\sigma^2 h c''(x)] + o(h).$$

Using $e^{-\alpha h} = 1 - \alpha h + o(h)$ and collecting the $o(h)$ terms we get

$$c(x) = f(x)h + (1 - \alpha h)c(x) + \mu h c'(x) + \frac{\sigma^2 h}{2}c''(x) + o(h).$$

Dividing by h and letting $h \to 0$, we get

$$f(x) - \alpha c(x) + \mu c'(x) + \frac{1}{2}\sigma^2 c''(x) = 0. \tag{10.40}$$

Rearranging this yields Equation 10.39. The boundary conditions follow since $T_{a,b}(a) = T_{a,b}(b) = 0$. ■

Next we give a solution to Equation 10.39 in the special case when $\alpha = 0$. First define:

$$\theta = -2\mu/\sigma^2,$$
$$s(x) = \exp(\theta x),$$
$$v(x) = \frac{s(a) - s(x)}{s(a) - s(b)}.$$

One can verify by direct substitution that, when $\alpha = 0$ and $\mu \neq 0$, the solution to Equation 10.39 is given by

$$c(x) = \frac{1}{\mu}\left(v(x)\int_x^b (1 - s(b-u))f(u)du + (1 - v(x))\int_a^x (s(a-u) - 1)f(u)du\right),$$
$$\tag{10.41}$$

for $a \leq x \leq b$. When $\mu = 0$, the above equation further reduces to

$$c(x) = \frac{2}{\sigma^2}\left(\frac{x-a}{b-a}\int_x^b (b-u)f(u)du + \frac{b-x}{b-a}\int_a^x (u-a)f(u)du\right), \quad a \leq x \leq b.$$
$$\tag{10.42}$$

We apply the above results to a simple control policy in the example below.

Example 10.8   Control Policy for an SBM. Suppose the state of a system evolves according to an SBM. When the system is in state x, the system incurs holding cost at rate cx^2. We have an option of instantaneously changing the state of the system to zero at any time by paying a fee of K. Once the system state is reset, it evolves as an SBM until the next intervention. Since the stochastic evolution is symmetric, we use the following control policy parameterized by a single parameter $b > 0$: reset the system state back to state 0 when it reaches state b or $-b$. Compute the value of b that minimizes the long run cost per unit time of operating the system.

Let $Y(t)$ be the state of the system at time t, and let

$$T = \min\{t \geq 0 : Y(t) = b \text{ or } Y(t) = -b\}.$$

Then $Y(T+) = Y(0) = 0$, and the system regenerates at time T. Let $c(0)$ be the expected holding cost over $(0, T]$ starting from state 0. From Equation 10.42 we get

$$c(0) = c \left(\int_0^b (b - u)u^2 du + \int_{-b}^0 (u + b)u^2 du \right) = cb^4/6. \tag{10.43}$$

Thus the total cost over a cycle $(0, T]$ is $K + cb^4/6$. The expected length of the cycle is given by Equation 10.24 as $E(T) = b^2$. From the results on renewal reward processes we get the long run cost per unit time as

$$\frac{K + cb^4/6}{b^2}.$$

This is a convex function over $b > 0$ and is minimized at

$$b^* = \left(\frac{6K}{c} \right)^{1/4}.$$

Note that b^* increases as c decreases or K increases, as expected. ∎

10.9 Stochastic Integration

In the previous section we saw integrals of the form

$$\int_0^t f(u, X(u)) du$$

where $\{X(t), t \geq 0\}$ is a BM. If f is a continuous function, we see that $f(u, X(u))$ is a continuous function everywhere on $[0, t]$ with probability 1. Hence, the above integral can be defined as the standard Riemann integral for almost every sample path. In this section we define integrals of the form

$$\int_0^t f(u, X(u)) dX(u). \tag{10.44}$$

If the sample paths of $\{X(t), t \geq 0\}$ were differentiable functions, with derivative $X'(t)$, we could define the above integral as

$$\int_0^t f(u, X(u))X'(u) du.$$

However, we have seen in Section 10.2 that, with probability one, the sample paths of a BM are nowhere differentiable. Hence we cannot use this definition. If the sample paths of the BM had bounded variation (i.e., finite total variation), we could use the Stieltjes integral to define the integral. However, we shall show below that the total variation of an SBM (and hence of any BM) is infinite. We begin with the definition of the total variation and the quadratic variation of a standard Brownian motion.

Definition 10.9   Total and Quadratic Variation. *For an SBM $\{B(t), t \geq 0\}$, and $p > 0$, define the p-th variation process of B by*

$$V_B^p(t) = \lim_{k \to \infty} \sum_{n=1}^{k} \left| B\left(t\frac{n}{k}\right) - B\left(t\frac{n-1}{k}\right) \right|^p \tag{10.45}$$

if the limit exists. The cases $p = 1$ and $p = 2$ are referred to as the total variation and quadratic variation of the SBM, respectively.

Although we have defined the concept of p-variation for an SBM, the same definition can be used to define the p-variation of any continuous time stochastic process.

Theorem 10.20 *With probability one*

$$\begin{aligned}
V_B^1(t) &= \infty, & t > 0, \\
V_B^2(t) &= t, & t \geq 0, \\
V_B^p(t) &= 0, & t \geq 0, p > 2.
\end{aligned}$$

Proof: Let $X^k(t)$ be as defined in Equation 10.3. We begin by computing the total variation of the $\{X^k(t), t \geq 0\}$ process over $[0, t]$. Since the sample paths of $\{X^k(t), t \geq 0\}$ are piecewise constant functions of time t, we see that the total variation of X^k over $[0, t]$ is just the sum of the absolute values of all the jumps over $[0, t]$. Since the jumps in the sample paths of $\{X^k(t), t \geq 0\}$ are of size $\pm 1/\sqrt{k}$ at all integer multiples of $1/k$, we get

$$\begin{aligned}
V_{X^k}^1(t) &= \sum_{n=1}^{[kt]} \left| X^k\left(t\frac{n}{k}\right) - X^k\left(t\frac{n-1}{k}\right) \right| \\
&= \sum_{n=1}^{[kt]} \frac{1}{\sqrt{k}} \\
&= \frac{[kt]}{\sqrt{k}}.
\end{aligned}$$

Similarly, the quadratic variation of X^k is given by

$$\begin{aligned}
V_{X^k}^2(t) &= \sum_{n=1}^{[kt]} \left| X^k\left(t\frac{n}{k}\right) - X^k\left(t\frac{n-1}{k}\right) \right|^2 \\
&= \sum_{n=1}^{[kt]} \left(\frac{1}{\sqrt{k}}\right)^2 \\
&= \frac{[kt]}{k}.
\end{aligned}$$

Now, we know that $\{X^k(t), t \geq 0\}$ converges to $\{B(t), t \geq 0\}$ as $k \to \infty$. Hence total and quadratic variations of $\{X^k(t), t \geq 0\}$ converge to that of $\{B(t), t \geq 0\}$ as

$k \to \infty$. (This needs to be proved, but we omit this step.) Hence we get

$$V_B^1(t) = \lim_{k \to \infty} V_{X^k}^1(t) = \lim_{k \to \infty} \frac{[kt]}{\sqrt{k}} = \infty,$$

and

$$V_B^2(t) = \lim_{k \to \infty} V_{X^k}^2(t) = \lim_{k \to \infty} \frac{[kt]}{k} = t.$$

The result about higher-order variation follows similarly. This "proves" the theorem. ∎

Thus we cannot use Stieltjes integrals to define the integral in Equation 10.44. What we are facing is a new kind of integral, and it was properly defined first by Ito. We shall use a severely simplified version of that definition here. (See Appendix G for the definition of convergence in mean square.)

Definition 10.10   Ito Integral. *Let $f(t, x)$ be a continuous real-valued function in t and x. Suppose*

$$\int_0^t E(f^2(u, B(u))) du < \infty.$$

Then the Ito integral of $f(t, B(t))$ with respect to $B(t)$ is defined as

$$\int_0^t f(u, B(u)) dB(u)$$

$$= \lim_{k \to \infty} \sum_{n=1}^k f\left(t\frac{n-1}{k}, B\left(t\frac{n-1}{k}\right)\right)\left(B\left(t\frac{n}{k}\right) - B\left(t\frac{n-1}{k}\right)\right) \quad (10.46)$$

Here the limit is defined in the mean-squared sense.

Note that the above definition is very similar to the definition of the Stieltjes integral, except that we insist on using the value of the function f at the left ends of the intervals in the sum. Thus, for the n-th interval $(t\frac{n-1}{k}, t\frac{n}{k}]$, we use the value $f(t\frac{n-1}{k}, B(t\frac{n-1}{k}))$ at the left end of the interval, and multiply it by the increment in the SBM over the interval, and then sum these products over all the intervals. This choice is very critical in the definition of Ito integral and has very important implications.

The question is: where do we encounter this type of an integral? We illustrate with an example.

Example 10.9   Suppose $P(t)$ is the price of a stock at time t and $f(t, x)$ is a given function of t and x. We want to conduct k trades at equally spaced times over $[0, t]$. Thus the nth trade takes place at time $t_n = t\frac{n-1}{k}$, $1 \le n \le k$. The nth trade consists of buying $f_n = f(t_n, P_n)$ number of shares at time t_n at price $P_n = P(t_n)$ and selling them at price $P_{n+1} = P(t_{n+1})$ at time t_{n+1}. Thus there is both selling and buying going on at times t_n for $2 \le n \le k - 1$, while there is only buying at time t_1 and only selling at time t_k. There are no transaction costs. Thus the net capital gain from the nth trade is $f_n(P_{n+1} - P_n)$, and the total capital gain over the interval $[0, t]$ from these k trades is

$$\sum_{n=1}^{k} f_n(P_{n+1} - P_n) = \sum_{n=1}^{k} f(t\frac{n-1}{k}, P(t\frac{n-1}{k}))(P(t\frac{n}{k}) - P(t\frac{n-1}{k})).$$

In the limit, as the number of trades goes to infinity, we see that the limiting capital gain is given by

$$\lim_{k\to\infty} \sum_{n=1}^{k} f(t\frac{n-1}{k}, P(t\frac{n-1}{k}))(P(t\frac{n}{k}) - P(t\frac{n-1}{k})) = \int_0^t f(u, P(u))dP(u).$$

If the price process is a BM (which is clearly not a good assumption, since a BM can go negative), the limiting capital gain is thus given by the Ito integral of f with respect to a BM.

It is now clear why we use the value of f at the left end of the interval $(t_n, t_{n+1}]$: number of shares to be bought or sold at time t_n cannot depend on price of the stock at time strictly after t_n, because it is not known at time t_n. This is why Ito integrals appear often in practice, where actions at time t can depend upon information available up to and including time t, but not on the information available in the future.

∎

We now study several important properties of the Ito integral.

Theorem 10.21 Linearity of Ito Integral. *For given real numbers a and b,*

$$\int_0^t (af(u, B(u)) + bg(u, B(u)))dB(u)$$

$$= a\int_0^t f(u, B(u))dB(u) + b\int_0^t g(u, B(u))dB(u). \qquad (10.47)$$

Proof: Follows directly from the definition. ∎

Note that the Ito integral of Equation 10.46 is a random variable, and hence it makes sense to compute its moments. The next theorem gives the first two moments.

Theorem 10.22 Moments of the Ito Integral.

$$\mathsf{E}\left(\int_0^t f(u, B(u))dB(u)\right) = 0, \qquad (10.48)$$

$$\mathsf{E}\left[\left(\int_0^t f(u, B(u))dB(u)\right)^2\right] = \int_0^t \mathsf{E}(f^2(u, B(u)))du. \qquad (10.49)$$

Proof: An SBM has independent increments, hence $f_n = f(t\frac{n-1}{k}, B(t\frac{n-1}{k}))$ is independent of $\Delta_n = B(t\frac{n}{k}) - B(t\frac{n-1}{k})$. (We suppress the dependence of f_n and Δ_n on k to simplify notation.) This independence implies that

$$\mathsf{E}(f_n\Delta_n) = \mathsf{E}(f_n)\mathsf{E}(\Delta_n) = 0.$$

Thus we have

$$E\left(\int_0^t f(u, B(u))dB(u)\right) = E\left(\lim_{k\to\infty}\sum_{n=1}^{k} f_n\Delta_n\right)$$

$$= \lim_{k\to\infty}\sum_{n=1}^{k} E(f_n)E(\Delta_n)$$

$$= 0.$$

The interchange of the limit and the expected value is allowed since the convergence is in mean square. This yields Equation 10.48.

To obtain Equation 10.49, we use

$$E\left[\left(\sum_{n=1}^{k} f_n\Delta_n\right)^2\right] = E\left(\sum_{n=1}^{k} f_n\Delta_n \sum_{m=1}^{k} f_m\Delta_m\right)$$

$$= \sum_{n=1}^{k}\sum_{m=1}^{k} E(f_n f_m \Delta_n \Delta_m)$$

$$= \sum_{n=1}^{k} E(f_n^2\Delta_n^2)$$

$$= \sum_{n=1}^{k} E(f_n^2)E(\Delta_n^2)$$

$$= \sum_{n=1}^{k} E(f^2(t\frac{n-1}{k}, B(t\frac{n-1}{k})))\frac{t}{k}.$$

Here we have used the fact that $E(f_n f_m \Delta_n \Delta_m) = 0$ if $n \neq m$. Finally,

$$E\left[\left(\int_0^t f(u, B(u))dB(u)\right)^2\right] = \lim_{k\to\infty}\sum_{n=1}^{k} E\left(f^2\left(t\frac{n-1}{k}, B\left(t\frac{n-1}{k}\right)\right)\right)\frac{t}{k}$$

$$= \int_0^t E(f^2(u, B(u)))du.$$

This yields Equation 10.49. ∎

We illustrate the computation of Ito integrals with an example.

Example 10.10   Show that

$$\int_0^t dB(u) = B(t), \quad t \geq 0,$$

$$\int_0^t B(u)dB(u) = \frac{1}{2}(B^2(t) - t), \quad t \geq 0.$$

Let $B_n = B(t\frac{n}{k})$. Using $f(u, B(u)) = 1$ in Equation 10.46 we get

$$\int_0^t dB(u) = \lim_{k \to \infty} \sum_{n=1}^k (B_n - B_{n-1})$$

$$= \lim_{k \to \infty} B_k$$

$$= \lim_{k \to \infty} B\left(t\frac{k}{k}\right) = B(t).$$

To compute the next integral, we use the identity

$$(B_n - B_{n-1})^2 = B_n^2 - B_{n-1}^2 - 2B_{n-1}(B_n - B_{n-1}), \quad 1 \le n \le k.$$

Summing over all n from 1 to k we get

$$\sum_{n=1}^k (B_n - B_{n-1})^2 = B_k^2 - 2\sum_{n=1}^k B_{n-1}(B_n - B_{n-1}) = B^2(t) - 2\sum_{n=1}^k B_{n-1}(B_n - B_{n-1}).$$

Now let $k \to \infty$. The sum on the left-hand side reduces to the squared variation of the SBM over $[0, t]$, while the sum on the right-hand side reduces to the Ito integral $\int_0^t B(u)dB(u)$. Hence, using Theorem 10.20 we get

$$V_B^2(t) = t = B^2(t) - 2\int_0^t B(u)dB(u).$$

This gives

$$\int_0^t B(u)dB(u) = \frac{1}{2}(B^2(t) - t).$$

Note that there is an unexpected $t/2$ term in the integral! This is the contribution of the non-zero quadratic variation that is absent in the standard calculus, but plays an important part in the Ito calculus! ∎

We have seen the terms $B(t)$ and $B^2(t) - t$ as examples of Martingales; see Examples 10.2 and 10.4. Is this just a coincidence that the two Ito integrals turn out to be Martingales? The next theorem shows that this is a general property of the Ito integrals. We omit the proof, since it is too technical for this book.

Theorem 10.23   Ito Integral as a Martingale. *Let*

$$X(t) = \int_0^t f(u, B(u))dB(u), \quad t \ge 0.$$

Then $\{X(t), t \ge 0\}$ is a Martingale with respect to $\{B(t), t \ge 0\}$.

We develop three peculiar integrals below that will help us in the development of stochastic calculus in the next section. Following Equation 10.46 we have

$$\int_0^t dB(u)dB(u)$$

$$= \lim_{k \to \infty} \sum_{n=1}^{k} \left(B\left(t\frac{n}{k}\right) - B\left(t\frac{n-1}{k}\right) \right) \left(B\left(t\frac{n}{k}\right) - B\left(t\frac{n-1}{k}\right) \right)$$

$$= V_B^2(t) = t.$$

$$\int_0^t du dB(u) = \lim_{k \to \infty} \sum_{n=1}^{k} \frac{1}{k} \left(B\left(t\frac{n}{k}\right) - B\left(t\frac{n-1}{k}\right) \right)$$

$$= \lim_{k \to \infty} \frac{B(t)}{k} = 0.$$

$$\int_0^t du du = \lim_{k \to \infty} \sum_{n=1}^{k} \left(\frac{t}{k}\right)^2$$

$$= t^2 \lim_{k \to \infty} \frac{k}{k^2} = 0.$$

In differential notation, we write the above "integrals" as

$$dB(t)dB(t) = (dB(t))^2 = dt, \tag{10.50}$$

$$dt dB(t) = dB(t)dt = dt dt = 0. \tag{10.51}$$

The above differentials form the basis of the Ito stochastic calculus developed in the next section. We illustrate one use of it in the next example.

Example 10.11 Show that

$$\int_0^t B^2(u)dB(u) = \frac{1}{3}B^3(t) - \int_0^t B(u)du. \tag{10.52}$$

We follow the steps used in computing $\int B(u)dB(u)$ in Example 10.10. We begin with the algebraic identity:

$$B_n^3 - B_{n-1}^3 = 3B_{n-1}^2(B_n - B_{n-1}) + 3B_n(B_n - B_{n-1})^2 + (B_n - B_{n-1})^3. \tag{10.53}$$

Now define

$$A_k = \sum_{n=1}^{k} (B_n^3 - B_{n-1}^3),$$

$$C_k = \sum_{n=1}^{k} B_{n-1}^2(B_n - B_{n-1}),$$

$$D_k = \sum_{n=1}^{k} B_n(B_n - B_{n-1})^2,$$

$$E_k = \sum_{n=1}^{k} (B_n - B_{n-1})^3.$$

Summing both sides of Equation 10.53 from $n = 1$ to k, we get

$$A_k = 3C_k + 3D_k + E_k. \tag{10.54}$$

Now, we have

$$\lim_{k\to\infty} A_k = \lim_{k\to\infty} B_k^3 = B^3(t).$$

From the definition of Ito integral we get

$$\lim_{k\to\infty} C_k = \lim_{k\to\infty} \sum_{n=1}^{k} B_{n-1}^2 (B_n - B_{n-1}) = \int_0^t B^2(u)dB(u).$$

Using Equation 10.50 we get

$$\lim_{k\to\infty} D_k = \lim_{k\to\infty} \sum_{n=1}^{k} B_{n-1}(B_n - B_{n-1})^2 = \int_0^t B(u)(dB(u))^2 = \int_0^t B(u)du,$$

Finally, from Theorem 10.20

$$\lim_{k\to\infty} |E_k| = \lim_{k\to\infty} \left| \sum_{n=1}^{k} (B_n - B_{n-1})^3 \right| \le V_B^3(t) = 0.$$

Hence

$$\lim_{k\to\infty} E_k = 0.$$

Letting $k \to \infty$ in Equation 10.54 we get

$$B^3(t) = 3\int_0^t B^2(u)dB(u) + 3\int_0^t B(u)du.$$

This can be rearranged to get Equation 10.52.

Note that the integral on the right-hand side of Equation 10.52 is the usual Riemann integral. Furthermore, unlike the stochastic integrals in Example 10.10, the stochastic integral in Equation 10.52 depends on the entire sample path of the SBM over $[0, t]$, and not just its value at time t.

Clearly it is not easy to compute the stochastic integrals from scratch using the definition. In the next section we develop Ito's formula, which is the stochastic calculus equivalent of the fundamental theorem of regular calculus.

10.10 Stochastic Differential Equations and Ito's Formula

We now introduce a more general stochastic process $\{Z(t), t \ge 0\}$ called a diffusion process, which generalizes a BM(μ, σ) $\{X(t), t \ge 0\}$. Note that when $\{X(t), t \ge 0\}$ is a BM(μ, σ), if $X(t) = x$, the increment $dX(t) = X(t+dt) - X(t)$ over the interval $(t, t + dt)$ is a sum of two components: (1) a deterministic drift component given by μdt, and (2) a random diffusion component $\sigma dB(t)$ that is a N$(0, \sigma^2 dt)$ random variable. The generalization $\{Z(t), t \ge 0\}$ is obtained by letting drift μ and the variance parameter σ^2 depend on t and $Z(t)$. More specifically, when $Z(t) = z$, the increment $dZ(t) = Z(t+dt) - Z(t)$ in a diffusion process over the interval $(t, t+dt)$ is a sum of two components: (1) a deterministic drift component given by $\mu(t, z)dt$, and (2) a random diffusion component $\sigma(t, z)dB(t)$ that is a $N(0, \sigma^2(t, z)dt)$ random variable. We formally define a diffusion process next using the stochastic differential notation.

Definition 10.11   Diffusion Process. *Let $\mu(t, z)$ and $\sigma(t, z) > 0$ be continuous functions of t and z. Let $\{Z(t), t \geq 0\}$ be a stochastic process that satisfies*

$$dZ(t) = \mu(t, Z(t))dt + \sigma(t, Z(t))dB(t), \quad t \geq 0. \tag{10.55}$$

Then $\{Z(t), t \geq 0\}$ is called a diffusion process, $\mu(t, z)$ is called its drift function, and $\sigma^2(t, z)$ the variance parameter function.

One can equivalently write Equation 10.55 in integral form as

$$Z(t) = Z(0) + \int_0^t \mu(u, Z(u))du + \int_0^t \sigma(u, Z(u))dB(u), \quad t \geq 0. \tag{10.56}$$

Note that the first integral is a simple Riemann integral, while the second integral seems like the Ito integral defined in the previous section, with one difference. Definition 10.10 defines an Ito integral of a function of an SBM with respect to that SBM, but the integral in Equation 10.56 is an integral of a function of the Z process with respect an SBM. To make this precise, we give the extended definition below.

Definition 10.12   Ito Integral. *Let $\{Z(t), t \geq 0\}$ be a stochastic process that is defined on the same probability space as the SBM $\{B(t), t \geq 0\}$. Let $f(t, z)$ be a continuous real-valued function in t and z. Suppose*

$$\int_0^t E(f^2(u, Z(u)))du < \infty.$$

Then the Ito integral of $f(t, Z(t))$ with respect to $B(t)$ is defined as

$$\int_0^t f(u, Z(u))dB(u) \tag{10.57}$$

$$= \lim_{k \to \infty} \sum_{n=1}^k f\left(t\frac{n-1}{k}, Z\left(t\frac{n-1}{k}\right)\right)\left(B\left(t\frac{n}{k}\right) - B\left(t\frac{n-1}{k}\right)\right) \tag{10.58}$$

Here the limit is defined in the mean-squared sense.

We further assume that $\{Z(t), t \geq 0\}$ is non-anticipative, that is, $\{Z(u), 0 \leq u \leq t\}$ is independent of $\{B(t + u) - B(t), u \geq 0\}$ for all $t \geq 0$. Note that we had seen a similar property in Section 7.2.3 while discussing PASTA. One can show that under this assumption the stochastic integral defined above remains a Martingale with respect to $\{B(t), t \geq 0\}$.

Example 10.12 Suppose $Z(0) = 0$, $\mu(t, z) = 0$, and $\sigma(t, z) = 1$. Substituting in Equation 10.56 we see that

$$Z(t) = \int_0^t dB(u) = B(t).$$

Thus the SBM is a diffusion process. Next suppose $Z(0) = z$, $\mu(t, z) = \mu$, and $\sigma(t, z) = \sigma$. Substituting in Equation 10.56 we see that

$$Z(t) = z + \int_0^t \mu dt + \int_0^t \sigma dB(u) = z + \mu t + \sigma B(t).$$

From Theorem 10.1 we see that $\{Z(t), t \geq 0\}$ is a BM(μ, σ) with initial state z. Thus a BM is a diffusion process. ∎

Next we consider a process $\{W(t), t \geq 0\}$ defined by

$$W(t) = g(t, Z(t)),$$

where $Z(t)$ is as defined in Equation 10.55 and $g(t, z)$ is a given function. The next theorem derives an expression for the differential $dW(t) = W(t + dt) - W(t)$.

Theorem 10.24 Functions of a Diffusion Process. *Let $\{Z(t), t \geq 0\}$ satisfy Equation 10.55. Let $g(t, z)$ be a function that is continuously differentiable in t and twice continuously differentiable in z. The differential of the stochastic process $\{g(t, Z(t)), t \geq 0\}$ is given by*

$$dg(t, Z(t)) =$$
$$\left(g_t(t, Z(t)) + \mu(t, Z(t))g_z(t, Z(t)) + \frac{1}{2}\sigma^2(t, Z(t))g_{zz}(t, Z(t)) \right) dt$$
$$+ \sigma(t, Z(t))g_z(t, Z(t))dB(t). \tag{10.59}$$

Proof: Let $\Delta_h Z(t) = Z(t + h) - Z(t)$. Taylor expansion of $g(t + h, Z(t + h)) = g(t + h, Z(t) + \Delta_h Z(t))$ around $(t, Z(t))$ yields :

$$\begin{aligned} g(t + h, Z(t + h)) &= g(t, Z(t)) + hg_t(t, Z(t)) + \Delta_h Z(t)g_z(t, Z(t)) \\ &\quad + \frac{1}{2}(\Delta_h Z(t))^2 g_{zz}(t, Z(t)) + R_h \end{aligned}$$

where R_h is a remainder error term that goes to zero as $h \to 0$. Hence we have

$$\Delta_h g(t, Z(t)) = g(t + h, Z(t + h)) - g(t, Z(t))$$
$$= hg_t(t, Z(t)) + \Delta_h Z(t)g_z(t, Z(t)) + \frac{1}{2}(\Delta_h Z(t))^2 g_{zz}(t, Z(t)) + R_h.$$

The above equation yields the differential form,

$$dg(t, Z(t)) = g_t(t, Z(t))dt + g_z(t, Z(t))dZ(t) + \frac{1}{2}g_{zz}(t, Z(t))(dZ(t))^2.$$

Now $\{Z(t), t \geq 0\}$ satisfies Equation 10.55. Hence, using Equations 10.50 and 10.51, we get

$$(dZ(t))^2 = (\mu(t, Z(t))dt + \sigma(t, Z(t))dB(t))^2$$

$$= \quad \mu^2(t, Z(t))dtdt + 2\mu(t, Z(t))\sigma(t, Z(t))dtdB(t) + \sigma^2(t, Z(t))(dB(t))^2$$
$$= \quad \sigma^2(t, Z(t))dt.$$

Substituting in the previous equation we get

$$dg(t, Z(t)) = g_t(t, Z(t))dt$$

$$+(\mu(t, Z(t))dt + \sigma(t, Z(t))dB(t))g_z(t, Z(t)) + \frac{1}{2}\sigma^2(t, Z(t))g_{zz}(t, Z(t))dt$$

$$= \quad \left(g_t(t, Z(t)) + \mu(t, Z(t))g_z(t, Z(t)) + \frac{1}{2}\sigma^2(t, Z(t))g_{zz}(t, Z(t)) \right) dt$$

$$+\sigma(t, Z(t))g_z(t, Z(t))dB(t),$$

which is Equation 10.59. ∎

Equation 10.59 is known as Ito's formula. If $Z(t) = B(t)$, i.e., if $\mu(t, z) = 0$ and $\sigma(t, z) = 1$, Ito's formula reduces to

$$dg(t, B(t)) = \left(g_t(t, B(t)) + \frac{1}{2}g_{zz}(t, B(t)) \right) dt + g_z(t, B(t))dB(t). \quad (10.60)$$

The above formula can be used to compute stochastic integrals defined by Equation 10.46. Let

$$G(t, z) = \int_0^z g(t, y)dy.$$

Then applying the integral form of Equation 10.60 to $G(t, z)$ instead of $g(t, z)$ and rearranging the terms we get a useful form as

$$\int_0^t g(u, B(u))dB(u) = G(t, B(t)) - G(0, 0) - \int_0^t \left(G_t(u, B(u)) + \frac{1}{2}g_z(u, B(u)) \right) du.$$
$$(10.61)$$

The constant $G(0, 0)$ is included on the right hand side so that both sides are zero when $t = 0$. We illustrate the use of the above formula by an example below.

Example 10.13 Compute $\int_0^t B(u)dB(u)$ using Equation 10.61.

Let $g(t, z) = z$. Then $G(t, z) = z^2/2$, and $g_z(t, z) = 1$, and $G_t(t, x) = 0$. Hence, Equation 10.61 yields

$$\int_0^t B(u)dB(u) = \frac{B^2(t)}{2} - \int_0^t \frac{1}{2}du = \frac{1}{2}(B^2(t) - t).$$

This matches with our brute force calculation of this integral in Example 10.10. ∎

Ito's formula can be used to solve stochastic differential equations, which, in their simplest form, are equations of the following form:

$$dW(t) = a(t, W(t))dt + b(t, W(t))dB(t), \quad (10.62)$$

where $a(\cdot, \cdot)$ and $b(\cdot, \cdot)$ are given functions. We have seen this as Equation 10.55 in the definition of a diffusion process. Solving a stochastic differential equation involves constructing a stochastic process $\{W(t), t \geq 0\}$ on the same probability space as that of $\{B(t), t \geq 0\}$, so that its differential $dW(t)$ satisfies Equation 10.62. A simplest solution occurs if we can represent $W(t)$ as

$$W(t) = g(t, B(t)) \tag{10.63}$$

for some function $g(\cdot, \cdot)$. We can use Ito's formula to compute $dW(t)$ to get

$$dW(t) = \left(g_t(t, B(t)) + \frac{1}{2}g_{zz}(t, B(t)) \right) dt + g_z(t, B(t))dB(t). \tag{10.64}$$

Thus the $\{W(t), t \geq 0\}$ process defined by Equation 10.63 is a solution to the stochastic differential equation 10.62 if

$$g_t(t, z) + \frac{1}{2}g_{zz}(t, z) = a(t, g(t, z)), \tag{10.65}$$

$$g_z(t, z) = b(t, g(t, z)). \tag{10.66}$$

We explain this procedure by an example below.

Example 10.14 Geometric Brownian Motion. Solve the stochastic differential equation

$$dW(t) = \mu W(t)dt + \sigma W(t)dB(t).$$

Here we have $a(t, w) = \mu w$ and $\sigma(t, w) = \sigma w$. We look for a solution of the form $W(t) = g(t, B(t))$ that satisfies the above differential equation. From Equations 10.65 and 10.66, we see that the g function has to satisfy:

$$g_t(t, z) + \frac{1}{2}g_{zz}(t, z) = \mu g(t, z), \tag{10.67}$$

and

$$g_z(t, z) = \sigma g(t, z). \tag{10.68}$$

The last equation implies that

$$g(t, z) = f(t)\exp(\sigma z),$$

where $f(t)$ is a function of t to be determined. Substituting in Equation 10.67 we get

$$f'(t)\exp(\sigma z) + \frac{1}{2}f(t)\sigma^2\exp(\sigma z) = \mu f(t)\exp(\sigma z).$$

Canceling $\exp(\sigma z)$ from both sides, we get

$$f'(t) = (\mu - \frac{1}{2}\sigma^2)f(t),$$

which yields

$$f(t) = c\exp((\mu - \frac{1}{2}\sigma^2)t),$$

where c is an arbitrary constant. Putting these equations together, we get

$$g(t, z) = c\exp((\mu - \frac{1}{2}\sigma^2)t + \sigma z).$$

Thus the solution is

$$W(t) = g(t, B(t)) = c\exp((\mu - \frac{1}{2}\sigma^2)t + \sigma B(t)).$$

Substituting $t = 0$, and using $B(0) = 0$, we get $c = W(0)$. Thus the final solution is

$$W(t) = W(0)\exp\left((\mu - \frac{1}{2}\sigma^2)t + \sigma B(t)\right). \qquad (10.69)$$

$\{W(t), t \geq 0\}$ given by the above equation is called the geometric Brownian motion (GBM) and is a common model of stock returns used in financial literature. ∎

Obviously, the solution to Equation 10.62 need not be of the form given in Equation 10.63. In this case we try the next class of solutions given by

$$W(t) = g(t, Z(t)) \qquad (10.70)$$

where $\{Z(t), t \geq 0\}$ is a diffusion process with a drift function $\mu(t, z)$ and variance parameter function $\sigma^2(t, z)$. Since the μ and σ functions can be chosen arbitrarily, this gives us a rich class of solutions to try out. Following similar steps as before, we see that the g function must satisfy

$$g_t(t, z) + \mu(t, z)g_z(t, z) + \frac{1}{2}\sigma^2(t, z)g_{zz}(t, z) = a(t, g(t, z)), \qquad (10.71)$$

and

$$\sigma(t, z)g_z(t, z) = b(t, g(t, z)). \qquad (10.72)$$

Clearly, this solution is useful only if the diffusion process $\{Z(t), t \geq 0\}$ can be represented as a function of the SBM easily. We illustrate with an example.

Example 10.15 Solve the stochastic differential equation

$$dW(t) = -\mu W(t)dt + \sigma dB(t). \qquad (10.73)$$

In this case we have $a(t, w) = -\mu w$ and $b(t, w) = \sigma$. Let us first try a solution of the type $W(t) = g(t, B(t))$. That is, we try to find a g satisfying

$$g_t(t, z) + \frac{1}{2}g_{zz}(t, z) = -\mu g(t, z), \qquad (10.74)$$

and

$$g_z(t, z) = \sigma. \qquad (10.75)$$

The last equation yields

$$g(t, z) = f(t) + \sigma z,$$

for an arbitrary function $f(t)$. Substituting in Equation 10.74 we get

$$f'(t) = -\mu(f(t) + \sigma z)$$

which has a solution

$$f(t) = A\exp(-\mu t) - \sigma z,$$

where A is a constant. However, this implies

$$g(t, z) = A \exp(-\mu t),$$

which does not satisfy Equation 10.75. Thus there is no function $g(t, z)$ that satisfies both Equations 10.74 and 10.75. Thus there is no stochastic process of the form $W(t) = g(t, B(t))$ that solves Equation 10.73.

Hence we try a solution of the form $W(t) = g(t, Z(t))$ where $\{Z(t), t \geq 0\}$ is a diffusion process with drift function $\mu(t, z)$ and variance parameter function $\sigma^2(t, z)$, where we are free to choose these functions. Equations 10.71 and 10.72 reduce to

$$g_t(t, z) + \mu(t, z)g_z(t, z) + \frac{1}{2}\sigma^2(t, z)g_{zz}(t, z) = -\mu g(t, z), \qquad (10.76)$$

and

$$\sigma(t, z)g_z(t, z) = \sigma. \qquad (10.77)$$

We choose $\sigma(t, z) = \sigma \exp(\mu t)$, and $\mu(t, z) = 0$, i.e., the diffusion process $\{Z(t), t \geq 0\}$ satisfies

$$dZ(t) = \sigma \exp(\mu t)dB(t),$$

or

$$Z(t) = Z(0) + \sigma \int_0^t e^{\mu u}dB(u). \qquad (10.78)$$

Using the results of Conceptual Exercise 10.10 we get

$$Z(t) = Z(0) + \sigma e^{\mu t}B(t) - \mu\sigma \int_0^t e^{\mu u}B(u)du.$$

Then Equations 10.76 and 10.77 reduce to

$$g_t(t, z) + \frac{1}{2}\sigma^2 \exp(2\mu t)g_{zz}(t, z) = -\mu g(t, z),$$

and

$$\sigma \exp(\mu t)g_z(t, z) = \sigma.$$

By following the earlier procedure, we see that these two equations admit the solution

$$g(t, z) = \exp(-\mu t)(c + z),$$

where c is an arbitrary constant. Thus the solution to Equation 10.73 is given by

$$W(t) = \exp(-\mu t)(c + Z(t)).$$

Setting $t = 0$ we get $c + Z(0) = W(0)$. Using Equation 10.78, we get the final solution as

$$\begin{aligned} W(t) &= \exp(-\mu t)(W(0) + \sigma \int_0^t \exp(\mu u)dB(u)) \\ &= \exp(-\mu t)W(0) + \sigma B(t) - \mu\sigma \int_0^t \exp(-\mu(t - u))B(u)du. \end{aligned}$$

The process defined above is known as the Ornstein–Uhlenbeck process. This example illustrates the complexity of the solutions exhibited by even simple stochastic differential equations. ∎

Clearly, the solution to Equation 10.62 may not be of the form given in Equation 10.70 for any diffusion process. What do we do then? This case is beyond the scope of this book, and we refer the reader to advanced books on this topic.

10.11 Applications to Finance

Let $X(t)$ be the price of a stock at time t. One of the simplest financial derivatives based on the stock price is called the European call option. (American call options are discussed at the end of this section.) A broker sells this option at a price C to a buyer. It gives the buyer (called the owner) the right (but not the obligation) to buy one share of this stock at a pre-specified time T in the future (called the maturity time, or expiry date), at a pre-specified price K (called the strike price). Clearly, if the stock price $X(T)$ at time T is greater than K, it makes sense for the owner to exercise the option, since the owner can buy the stock at price K and immediately sell it at price $X(T)$ and realize a net profit of $X(T) - K$. On the other hand, if $X(T) \leq K$, it makes sense to let the option lapse. Thus, the payout to the owner of this contract is $\max(X(T) - K, 0)$ at time T. How much should the broker sell this option for, i.e., what should be the value of C? This is the famous option pricing problem.

As mentioned in Example 10.14, the geometric Brownian motion as defined in Equation 10.69 is commonly used as a model of stock market price evolution. The ratio $(X(t_2) - X(t_1))/X(t_1)$ $(0 \leq t_1 < t_2)$ is called the return over the period $[t_1, t_2]$. It can be seen that the geometric Brownian motion model of the stock price implies that the returns over non-overlapping intervals are independent. This is one of the most compelling reasons for the wide use of this model. The other reason is the analytical tractability of this process.

Before we can settle the question of evaluating the proper value of C, we need to know what else we can do with the money. We assume that we can either invest it in the stock itself, or put it in a risk-free savings account that yields a continuous fixed rate of return r. Thus one dollar invested in the stock at time 0 will be worth $\$X(t)/X(0)$ at time t, while one dollar invested in the savings account will be worth e^{rt} at time t.

One would naively argue that the value of C should be given by the expected discounted (discount factor r) value of the option payout at time T, i.e.,

$$C = e^{-rT}\mathsf{E}(\max(X(T) - K, 0)). \tag{10.79}$$

One can show (see any book on mathematical finance) that if $\mu \neq r$, the broker can make positive profit with probability one by judiciously investing the above proceeds of $\$C$ in the stock and the savings account. Such a possibility is called an arbitrage opportunity, and it cannot exist in a perfect market. Hence we need to evaluate C assuming that $\mu = r$. The next theorem gives the expression for C when $\mu = r$. This is the celebrated Black–Scholes' formula.

Theorem 10.25 Black–Scholes' Formula: European Call Option. *The value of the European call option with maturity date T and strike price K is given by:*

$$C = X(0)\Phi(d_1) - Ke^{-rT}\Phi(d_2), \tag{10.80}$$

where

$$d_1 = \frac{\ln(X(0)/K) + (r + \sigma^2/2)T}{\sigma\sqrt{T}}, \tag{10.81}$$

$$d_2 = \frac{\ln(X(0)/K) + (r - \sigma^2/2)T}{\sigma\sqrt{T}}, \tag{10.82}$$

and Φ is the cdf of a $N(0,1)$ random variable.

Proof: Equation 10.69 implies that, for a fixed t, the stock price $X(t)$ can be represented as

$$X(t) = \exp(\ln X(0) + (\mu - \frac{1}{2}\sigma^2)t + \sigma\sqrt{t}Z), \tag{10.83}$$

where $Z \sim N(0,1)$.

Define

$$X(T)^+ = \begin{cases} X(T) & \text{if } X(T) > K, \\ 0 & \text{if } X(T) \le K. \end{cases}$$

and

$$K^+ = \begin{cases} K & \text{if } X(T) > K, \\ 0 & \text{if } X(T) \le K. \end{cases}$$

Then

$$\max(X(T) - K, 0) = X(T)^+ - K^+.$$

The value of the option as given in Equation 10.79 reduces to

$$C = e^{-rT}[E(X(T)^+) - E(K^+)], \tag{10.84}$$

where we use $\mu = r$ to compute the expectations. Next we compute the two expectations above. Using Equation 10.83 we get

$$E(X(T)^+) =$$

$$= E(\exp((\ln X(0) + (r - \frac{1}{2}\sigma^2)T + \sigma\sqrt{T}Z)^+)$$

$$= E\left(\exp(\ln X(0) + (r - \frac{1}{2}\sigma^2)T + \sigma\sqrt{T}Z) \cdot 1_{\{\ln X(0) + (r - \frac{1}{2}\sigma^2)T + \sigma\sqrt{T}Z > \ln K\}}\right)$$

$$= \exp(\ln X(0) + (r - \frac{1}{2}\sigma^2)T)E(\exp(\sigma\sqrt{T}Z)1_{\{Z > c_1\}}),$$

where

$$c_1 = \frac{\ln(K/X(0)) - (r - \frac{1}{2}\sigma^2)T}{\sigma\sqrt{T}}.$$

Now we can show by direct integration that

$$E(\exp(aZ)1_{\{Z > b\}}) = e^{a^2/2}\Phi(a - b), \tag{10.85}$$

for positive a. Substituting in the previous equation, we get

$$E(X(T)^+) = X(0)\exp((r - \frac{1}{2}\sigma^2)T)\exp(\frac{1}{2}\sigma^2 T)\Phi(d_1) = X(0)e^{rT}\Phi(d_1),$$

where $d_1 = \sigma\sqrt{T} - c_1$ is as given in Equation 10.81. Next, we have

$$
\begin{aligned}
E(K^+) &= E(K1_{\{X(T) > K\}}) = KP(X(T) > K) \\
&= KP(\ln X(0) + (r - \frac{1}{2}\sigma^2)T + \sigma\sqrt{T}Z > \ln K) \\
&= KP(Z > c_1) = KP(Z \le d_2),
\end{aligned}
$$

where $d_2 = -c_1$ is as given in Equation 10.82. Substituting in Equation 10.84 we get Equation 10.80. ∎

One can also study the European put option in a similar way. It gives the buyer the right (but not the obligation) to sell one share of the stock at time T at price K. The payout of this option at time T is $\max(K - X(T), 0)$. We leave it to the reader to prove the following theorem.

Theorem 10.26 Black–Scholes' Formula: European Put Option. *The value of the European put option with maturity date T and strike price K is given by:*

$$C = Ke^{-rT}\Phi(-d_2) - X(0)\Phi(-d_1), \tag{10.86}$$

where d_1, d_2, and Φ are as in Theorem 10.25.

There are options called the American call and put options with the maturity time T and strike price K. These are the same as the European options, except that they may be exercised at any time $t \in [0, T]$. If an American call option is exercised at time t, then its payout is $\max(X(t) - K, 0)$. Similarly the payout from an American put option exercised at time t is $\max(K - X(t), 0)$. One can show that under the assumption that the stock price satisfies Equation 10.83, it is optimal to exercise an American call option at maturity. Thus its value is the same as that of the European option. Pricing the American put option is much harder for the finite maturity date. However, if the maturity date is infinity (this case is called the perpetual American put option) one can evaluate it analytically. We state the main result in the following theorem.

Theorem 10.27 American Perpetual Put Option. *Consider a perpetual American put option with strike price K, and assume that the stock price process is as given in Equation 10.83. It is optimal to exercise the option as soon as the stock price falls below*

$$a^* = \frac{2r}{2r + \sigma^2}K.$$

The optimal expected payout under this policy, starting with $X(0) = x > a^*$ is

$$(K - a^*)(a^*/x)^{\frac{2r}{\sigma^2}}.$$

Proof: See Conceptual Exercises 10.11 and 10.13. ∎

Note that the result of the above theorem does not depend upon μ. As we have seen before, this is because we need to do the calculations under the assumption $\mu = r$ in order to avoid arbitrage.

10.12 Computational Exercises

10.1 Let $\{B(t), t \geq 0\}$ be an SBM. Show that $\mathrm{Cov}(B(s), B(t)) = \min(s, t)$.

10.2 Let $\{X(t), t \geq 0\}$ be a BM(μ, σ). Compute $\mathrm{Cov}(X(t), X(t + s))$.

10.3 Let $\{X(t), t \geq 0\}$ be a BM(μ, σ). Compute the joint density of $[X(t_1), X(t_2)]$, where $0 < t_1 < t_2$.

10.4 Let $\{B(t), t \geq 0\}$ be an SBM. Let $0 < s < t$. Show that, given $B(t) = y$, $B(s)$ is a normal random variable with mean ys/t and variance $s(t - s)/t$.

10.5 Let $\{X(t), t \geq 0\}$ be a BM(μ, σ). Let $0 < s < t$. Compute the conditional density of $X(s)$ given $X(t) = y$.

10.6 Let $t \in (0, 1)$. Show that $B(t) - tB(1)$ has the same distribution as the conditional density of $B(t)$ given $B(1) = 0$.

10.7 Verify that the solution given in Equation 10.8 satisfies the partial differential equation given in Equation 10.6.

10.8 Prove Theorem 10.7. Show that the pdf of a N($\mu t, \sigma^2 t$) random variable satisfies Equation 10.9.

10.9 Derive Equation 10.11 from Equation 10.12.

10.10 Let $\{X(t), t \geq 0\}$ be a BM(μ, σ). If we stop the process at time t, we earn a discounted reward equal to $e^{-\alpha t} X(t)$. Suppose we use the following policy: stop the process as soon as it reaches a state a. Find the optimal value of a that maximizes the expected discounted reward of this policy.

10.11 Let $\{X(t), t \geq 0\}$ be BM(μ, σ) reflected at 0. Assume $X(0) \geq 0$, and $\mu < 0$. Show that the limiting distribution of $X(t)$, as $t \to \infty$, is $\exp(-2\mu/\sigma^2)$. Hint: Set $a = 0$ and let $b \to \infty$ in Theorem 10.17.

10.12 Let $\{X(t), t \geq 0\}$ be a BM(μ, σ). Find functions $a(t)$ and $b(t)$ such that $X^2(t) + a(t)X(t) + b(t)$ is a Martingale with respect to $\{X(t), t \geq 0\}$. Hint: Use Example 10.4.

10.13 Let $\{X(t), t \geq 0\}$ be a BM(μ, σ). Find the functions $a(t)$, $b(t)$, and $c(t)$ such that $X^3(t) + a(t)X^2(t) + b(t)X(t) + c(t)$ is a Martingale with respect to $\{X(t), t \geq 0\}$. Hint: Use Conceptual Exercise 10.3.

10.14 Let $m_2(x) = \mathsf{E}(T_{a,b}^2(x))$ where $T_{a,b}(x)$ is the first passage time as defined in Equation 10.25 in a BM(μ, σ). Show that $m_2(\cdot)$ satisfies the following differential equation

$$\frac{\sigma^2}{2} m_2''(x) + \mu m_2'(x) = -2m(x), \quad a < x < b$$

where $m(x) = \mathsf{E}(T_{a,b}(x))$. The boundary conditions are $m_2(a) = m_2(b) = 0$.

10.15 Let $\{X(t), t \geq 0\}$ be a BM(μ, σ) with $X(0) = 0$ and $\mu < 0$, and define $M = \max_{t \geq 0} X(t)$. Show that M is an exp($-2\mu/\sigma^2$) random variable. Hint: Argue that

$$\mathsf{P}(M < y) = \lim_{a \to -\infty} \mathsf{P}(X(T_{a,y}) = a),$$

and use Theorem 10.11.

10.16 Let $\{X(t), t \geq 0\}$ be a BM(μ, σ^2) with $X(0) = 0$ and $\mu > 0$, and define $L = \min_{t \geq 0} X(t)$. Use the hint in Computational Exercise 10.15 to compute the distribution of L.

10.17 Use the exponential Martingale of Example 10.5 to derive Equation 10.14. Hint: Set

$$\theta\mu + \theta^2\sigma^2/2 = s$$

and use Theorem 10.18.

10.18 Let $X(t)$ be the price of a stock at time t. Suppose $\{X(t), t \geq 0\}$ is the geometric Brownian motion defined by Equation 10.69. Thus a dollar invested in this stock at time u will be worth $X(t)/X(u)$ at time $t > u$. Consider a static investment strategy under which we invest in this stock fresh money at a rate $\$d$ at all times $t \geq 0$. Let $Y(t)$ be total value of the investment at time t, assuming that $Y(0) = 0$. Compute $\mathsf{E}(Y(t))$.

10.19 Consider the optimal policy derived in Example 10.8. Using the regenerative nature of $\{Y(t), t \geq 0\}$ defined there, compute the limiting mean and variance of $Y(t)$ as $t \to \infty$.

10.20 Let $Y(t)$ be the level of inventory at time t. Assume that $Y(0) = q > 0$. When the inventory reaches 0, an order of size q is placed from an outside source. Assume that the order arrives instantaneously, so that the inventory level jumps to q. Between two consecutive orders the Y process behaves like a BM(μ, σ) starting in state q, with $\mu < 0$. Suppose it costs h dollars to hold one unit of the inventory for one unit of time, and the restoration operation costs K. Find the optimal value of q that minimizes the long run expected cost per unit time.

10.21 Use Ito's formula to derive Equation 10.52.

10.22 Using Equation 10.61 show that

$$\int_0^t B^n(u)dB(u) = \frac{1}{n+1}B^{n+1}(u) - \frac{n}{2}\int_0^t B^{n-1}(u)du.$$

10.23 Using Equation 10.61 show that

$$\int_0^t e^{-\alpha u}B^n(u)dB(u) = \frac{1}{n+1}e^{-\alpha t}B^{n+1}(t) - \int_0^t \left(\frac{n}{2}e^{-\alpha u}B^{n-1}(u) - \alpha e^{-\alpha u}\frac{B^{n+1}(u)}{n+1}\right)du.$$

10.24 Solve the stochastic differential equation

$$dY(t) = -\alpha Y(t)dt + e^{-\alpha t}dB(t),$$

assuming a solution of the form $Y(t) = g(t, B(t))$.

10.25 Solve the stochastic differential equation

$$dY(t) = -\frac{Y(t)}{1+t}dt + \frac{1}{1+t}dB(t),$$

assuming a solution of the form $Y(t) = g(t, B(t))$.

10.26 For $i = 1, 2$, let $\{Y_i(t), t \geq 0\}$ be a BM(μ_i, σ_i), with $\mu_1 > \mu_2$. Suppose the Y_1 process is independent of the Y_2 process and $Y_1(0) = y_1 < Y_2(0) = y_2$, where y_1 and y_2 are given real numbers. Define

$$T = \min\{t \geq 0 : Y_1(t) = Y_2(t)\}.$$

Compute E(T).

10.27 Let $X(t)$ be the temperature (in degrees Fahrenheit) of a furnace at time t. We can control the temperature by turning the furnace on and off. When the furnace is turned on the temperature behaves like a Brownian motion with drift parameter 1 degree per minute, and variance parameter 1. When it is down, it behaves like a Brownian motion with drift parameter -2 degrees per minute, and variance parameter 4. The aim is to keep the temperature as close to c degrees as possible. This is achieved by turning the furnace off when the temperature rises to $c + u$ degrees (where u is a fixed positive number.) Once the furnace is off it is kept off until the temperature drops to $c - d$ degrees (where d is a fixed positive number). At that time the furnace is turned on and kept on until the temperature rises to $c + u$ degrees. This control policy continues forever.

1. Let α be the chatter rate of the furnace, defined as the number of times the furnace switches from off to on per hour in the long run. Compute α as a function of u and d.

2. What fraction of the time is the furnace on in the long run?

3. What is the expected temperature of the furnace in the long run?

4. What values of u and d would you choose if the aim is to keep the long run average temperature at c degrees, keep the chatter rate bounded above by 10 per hour, and keep the temperature range $(u + d)$ to a minimum?

10.28 Let $\{B(t), t \geq 0\}$ be a standard Brownian motion. Define $S_0 = 0$ and

$$S_{n+1} = \min\{t > S_n : B(t) \in \{B(S_n) - 1, B(S_n) + 1\}\}, \ n \geq 0.$$

Let

$$N(t) = \sup\{n \geq 0 : S_n \leq t\}.$$

1. Compute $\lim_{t \to \infty} \frac{N(t)}{t}$.
2. Compute $E(\int_0^{S_1} B^2(u)du | B(0) = 0)$.

10.29 Let $Y(t)$ be as defined in Example 10.5. Show that $E(Y(t)) = 1$.

10.30 The following stochastic differential equation is used often as a model of the evolution of short term interest rates $\{X(t), t \geq 0\}$:

$$dX(t) = \alpha(\beta - X(t))dt + \sigma dB(t),$$

where $\alpha > 0, \beta > 0$, and $\sigma > 0$ are given constants.

1. Show that

$$X(t) = X(0)e^{-\alpha t} + \beta(1 - e^{-\alpha t}) + \sigma e^{-\alpha t} \int_0^t e^{\alpha u} dB(u)$$

   is a solution to the above equation.
2. Compute $E(X(t))$ as a function of t. (Equation 10.48 may be useful here.)
3. Compute $\text{Var}(X(t))$ as a function of t. (Equations 10.48 and 10.49 may be useful here.)

10.13 Conceptual Exercises

10.1 Let T be a stopping time for the stochastic process $\{X(t), t \geq 0\}$. Prove or disprove the following statements:

1. The event $\{T > t\}$ is completely described by $\{X(u) : 0 \leq u \leq t\}$.
2. The event $\{T > t\}$ is completely described by $\{X(u) : u > t\}$.
3. The event $\{T < t\}$ is completely described by $\{X(u) : 0 \leq u < t\}$.

10.2 Let $\{X(t), t \geq 0\}$ be a BM(μ, σ) and define

$$T = \min\{t \geq 0 : \int_0^t X(u)du \geq 1\}.$$

Is T a stopping time for the BM?

10.3 Show that $B^3(t) - 3tB(t)$ is a Martingale with respect to $\{B(t), t \geq 0\}$.

10.4 Show that $B^4(t) - 6tB^2(t) + 3t^2$ is a Martingale with respect to $\{B(t), t \geq 0\}$.

10.5 Let $L = \int_0^t B(u)du$. Show that

$$\mathsf{E}(L) = 0, \quad \mathsf{E}(L^2) = t^3/3.$$

10.6 Let L be as in Conceptual Exercise 10.5. Show that

$$\mathsf{E}(\exp(\theta L)) = \exp(\theta^2 t^3/6).$$

This implies that $L \sim N(0, t^3/3)$.

10.7 Let $L = \int_0^t \exp(\theta B(u))du$. Show that

$$\mathsf{E}(L) = \frac{2}{\theta^2}(\exp(\theta^2 t/2) - 1).$$

10.8 Let L be as in Conceptual Exercise 10.7. Show that

$$\mathsf{E}(L^2) = \frac{2}{3\theta^4}\exp(2\theta^2 t) - \frac{8}{3\theta^4}\exp(\theta^2 t/2) - \frac{2}{\theta^4}.$$

10.9 Using Equation 10.61 show that, for a differentiable function f,

$$\int_0^t f(u)dB(u) = f(t)B(t) - \int_0^t B(u)f'(u)du.$$

10.10 Solve the stochastic differential equation:

$$dW(t) = a(t)W(t)dt + b(t)W(t)dB(t), \quad W(0) = 1,$$

where $a(\cdot)$ and $b(\cdot)$ are differentiable functions of t. Hint: Try a solution of the type $W(t) = g(t, Z(t))$, where $Z(t)$ satisfies $dZ(t) = b(t)dB(t)$.

10.11 Proof of Theorem 10.27. Consider the perpetual American put option with strike price K for the stock price process given in Theorem 10.27. Suppose $X(0) > K$, for otherwise the buyer would exercise the option right away with a payout of $K - X(0)$. Let $\Pi(a)$ be a policy that exercises the option as soon as the stock price falls below a given constant $0 < a < K$. Thus the payout under policy $\Pi(a)$ is

$$C(a, x) = \mathsf{E}(e^{-rT_a}(K - a)|X(0) = x), \quad x > K$$

where

$$T_a = \min\{t \geq 0 : X(t) = a\}$$

and the expected value is computed under the assumption that $\mu = r$. Show that

$$C(a, x) = (K - a)(a/x)^{\frac{2r}{\sigma^2}}, \quad x > a.$$

Hint: First show that T_a is the same as the first time a BM($r - \sigma^2/2, \sigma$), starting from $\ln(X(0))$, reaches $\ln(a)$. Then use Theorem 10.10 to compute the LST of T_a, and then use Computational Exercise 10.10 to compute the expectations.

10.12 Prove Theorem 10.26.

10.13 Proof of Theorem 10.27, continued. Starting with the result in the Conceptual Exercise 10.11 show that the value of a that maximizes $C(a, x)$ is given by the a^* of Theorem 10.27. Also, $C(a^*, x)$ is as given in the theorem.

Epilogue

Here ends our journey. Congratulations! Now is the time to look back to see what we have learned and to look ahead to see what uncharted territory lies ahead.

We started with a basic knowledge of probability and built classes of increasingly powerful stochastic processes, each class providing a stepping stone to the next. We started with Markov chains and ended with Markov regenerative processes. The theories of Markov chains and renewal processes emerged as two main cornerstones of the entire development. At the end we dipped our toes in diffusion processes, and saw what is involved once we step away from the constraints of discrete state-space and discrete time. However, we stayed within the confines of Markov processes.

We saw a large number of examples of modeling a given system by an appropriate class of processes and then performing the transient analysis, first passage time analysis, steady state analysis, cost/reward analysis, etc. Indeed we now have a rich bag of tools to tackle the uncertain world.

What lies ahead? There is far more to diffusion processes than what we have covered here. Also, we have not covered the class of stationary processes: these processes look the same from any point of time. We saw them when studying the Markov chains starting in their stationary distributions. Stationary processes play an important role in statistics and forecasting.

We have also ignored the topic of controlling a stochastic system. The models studied here are descriptive — they describe the behavior of a system. They do not tell us how to control it. Of course, a given control scheme can be analyzed using the descriptive models. However, these models will not show how to find the optimal control scheme. This direction of inquiry will lead us to Markov decision processes.

Each of these topics merits a whole new book by itself. We stop here by wishing our readers well in the future as they explore these new and exciting areas.

APPENDIX A

Probability of Events

"All business proceeds on beliefs, on judgement of probabilities, and not on certainties."
– Charles W. Eliot

This appendix contains the brief review of probability and analysis topics that we use in the book. Its main aim is to act as a resource for the main results. It is not meant to be a source for beginners to learn these topics.

Probability Model. A random phenomenon or a random experiment is mathematically described by a probability model $(\Omega, \mathcal{F}, \mathsf{P})$. Ω is called the sample space: the set of all possible outcomes of the random experiment. A subset of Ω is called an event. $\mathcal{F}$ is the set of all possible events of interest about the random experiment. P is a consistent description of the likelihood of the occurrence of the events in $\mathcal{F}$.

Properties of $\mathcal{F}$. $\mathcal{F}$ is called the $\sigma-$algebra of events and has the following properties:

1. $\Omega \in \mathcal{F}$,
2. $E \in \mathcal{F} \Rightarrow E^c \in \mathcal{F}$,
3. $E_1, E_2, \cdots \in \mathcal{F} \Rightarrow \cup_{i=1}^{\infty} E_i \in \mathcal{F}$.

Properties of $\mathsf{P}(\cdot)$. P is called the probability function or the probability measure. It is a function $\mathsf{P} : \mathcal{F} \to [0,1]$ with the following properties:

1. $\mathsf{P}(\Omega) = 1$,
2. If $E_1, E_2, \cdots$ are mutually exclusive events in $\mathcal{F}$

$$\mathsf{P}(\cup_{i=1}^{\infty} E_i) = \sum_{i=1}^{\infty} \mathsf{P}(E_i).$$

One can deduce the following important properties of the probability function:

$$\mathsf{P}(\phi) = 0, \quad \mathsf{P}(E^c) = 1 - \mathsf{P}(E), \quad \mathsf{P}(E \cup F) = \mathsf{P}(E) + \mathsf{P}(F) - \mathsf{P}(EF).$$

Here EF represents the intersection of the events (sets) E and F.

Conditional Probability. Suppose $E, F \in \mathcal{F}$, and $P(F) > 0$. Then $P(E|F)$, the conditional probability of E given F, is given by

$$P(E|F) = \frac{P(EF)}{P(F)}.$$

If $P(F) = 0$, the above conditional probability is undefined.

Independence of Events. Events E and F are called independent if

$$P(EF) = P(E)P(F).$$

The events $E_1, E_2, \cdots, E_n \in \mathcal{F}$ are called mutually independent if for any $A \subseteq \{1, 2, \cdots, n\}$

$$P(\cap_{i \in A} E_i) = \prod_{i \in A} P(E_i).$$

Law of Total Probability. Let $E_1, E_2, \cdots \in \mathcal{F}$ be mutually exclusive and exhaustive events. Then, for $F \in \mathcal{F}$,

$$P(F) = \sum_{i=1}^{\infty} P(F|E_i)P(E_i).$$

Bayes' Rule. One consequence of the law of total probability is the Bayes' rule:

$$P(E_i|F) = \frac{P(F|E_i)P(E_i)}{\sum_{i=1}^{\infty} P(F|E_i)P(E_i)}.$$

Limits of Sets. Let $E_1, E_2, \cdots \in \mathcal{F}$. We define

$$\limsup E_n = \cap_{m \geq 1} \cup_{n \geq m} E_n, \quad \liminf E_n = \cup_{m \geq 1} \cap_{n \geq m} E_n.$$

In words, $\limsup E_n$ is the event that the events E_n occur infinitely often, and $\liminf E_n$ is the event that one of the events E_n occurs eventually. We have the following inequalities, Fatou's lemma for sets:

$$P(\limsup E_n) \geq \limsup P(E_n),$$

$$P(\liminf E_n) \leq \liminf P(E_n).$$

If $\limsup E_n = \liminf E_n = E$, we say that the sequence of sets $\{E_1, E_2, \cdots\}$ has a limit, and write $\lim E_n = E$. Probability is a continuous set function, that is, if $\{E_1, E_2, \cdots\}$ has a limit,

$$\lim P(E_n) = P(\lim E_n).$$

The following result is called the Borel–Cantelli lemma:

$$\sum_{n=1}^{\infty} P(E_n) < \infty \Rightarrow P(\limsup E_n) = 0.$$

The second Borel–Cantelli lemma states: If $\{E_n, n \geq 1\}$ are independent events, then

$$\sum_{n=1}^{\infty} P(E_n) = \infty \Rightarrow P(\limsup E_n) = 1.$$

Univariate Random Variables

A real valued random variable is a function $X : \Omega \to \mathcal{R} = (-\infty, \infty)$ such that $\{\omega \in \Omega : X(\omega) \leq x\} \in \mathcal{F}$ for all $-\infty < x < \infty$. It is described by its cumulative distribution function (cdf)

$$F_X(x) = \mathsf{P}(X \leq x) = \mathsf{P}(\{\omega \in \Omega : X(\omega) \leq x\}), \quad -\infty < x < \infty.$$

The cdf F_X is a non-decreasing right-continuous function of x and is bounded above by 1 and below by 0. X is said to be non-defective if

$$F_X(\infty) = \lim_{x \to \infty} F_X(x) = 1,$$

and

$$F_X(-\infty) = \lim_{x \to -\infty} F_X(x) = 0.$$

Otherwise it is called defective.

Discrete Random Variables. X is called discrete if F_X is a piecewise constant function with jumps in a discrete set A of real numbers. When A is a set of integers X is called an integer-valued random variable. It can be described in an equivalent way by giving its probability mass function (pmf) defined as

$$p_X(x) = \mathsf{P}(X = x) = F_X(x) - F_X(x-), \quad x \in A.$$

A pmf satisfies

$$p_X(x) \geq 0, \quad \sum_{x \in A} p_X(x) = 1.$$

Continuous Random Variables. X is said to be a continuous random variable if F_X is absolutely continuous, that is, if there exists a function $f_X(\cdot)$, called the probability density function (pdf), such that

$$F_X(x) = \int_{-\infty}^{x} f_X(u)du, \quad x \in \mathcal{R}.$$

A pdf satisfies

$$f_X(x) \geq 0, \quad \int_{-\infty}^{\infty} f_X(u)du = 1.$$

It is possible for a random variable to have a discrete part and a continuous part. There is a third possibility: $F_X(x)$ can be continuous, but not absolutely continuous. Such random variables are called singular, and we will not encounter them in this book.

Expectation of a Random Variable. The expectation of a random variable X is defined as

$$E(X) = \begin{cases} \sum_{x \in A} x p_X(x) & \text{if } X \text{ is discrete} \\ \int_{-\infty}^{\infty} x f_X(x) dx & \text{if } X \text{ is continuous,} \end{cases}$$

provided the sums and integrals are well defined. When a random variable has both discrete and continuous components, we employ the Stieltjes integral notation and write

$$E(X) = \int_{-\infty}^{\infty} x dF_X(x).$$

For a non-negative random variable we can show that

$$E(X) = \int_0^{\infty} (1 - F_X(x)) dx.$$

Functions of Random Variables. Let g be a function such that $Y = g(X)$ is also a random variable. This is the case if $g : \mathcal{R} \to \mathcal{R}$ is a piecewise continuous function. Computing the expected value of $g(X)$ can be done without first computing the cdf of $Y = g(X)$ by using the following formula:

$$E(g(X)) = \int_{-\infty}^{\infty} g(x) dF_X(x).$$

Suppose X is non-negative, and $g(\cdot)$ is a non-negative differentiable function such that

$$\lim_{x \to \infty} g(x)(1 - F_X(x)) = 0.$$

Then

$$E(g(X)) = \int_0^{\infty} g'(x)(1 - F_X(x)) dx.$$

Expectations of special functions of X have special names:

1. $E(X^n)$: nth moment of X,
2. $E(X^{(n)}) = E(X(X-1) \cdots (X-n+1))$: nth factorial moment of X, used primarily for non-negative integer valued random variables,
3. $E((X - E(X))^n)$: nth central moment of X,
4. $E((X - E(X))^2)$ or $\text{Var}(X)$: variance of X,
5. $\sqrt{\text{Var}(X)}$: standard deviation of X,
6. $\frac{\sqrt{\text{Var}(X)}}{E(X)}$: coefficient of variation of X,
7. $E(e^{-sX})$: Laplace–Stieltjes transform of X, used primarily for non-negative random variables,

8. $E(z^X)$: generating function of X, used primarily for non-negative integer valued random variables,

9. $E(e^{i\theta X})$: characteristic function of X, ($\theta \in \mathcal{R}$).

Properties of the expectation:

$$E(aX + b) = aE(X) + b,$$

$$Var(aX + b) = a^2 Var(X).$$

Common Discrete Random Variables. The facts about the commonly occurring integer valued random variables are given in the tables below.

Random Variable, X	Symbol	Parameter Range	$p_X(k)$
Bernoulli	Ber(p)	$0 \le p \le 1$	$p^k(1-p)^{1-k}$, $k = 0, 1$
Binomial	Bin(n, p)	$n \ge 0, 0 \le p \le 1$	$\binom{n}{k}p^k(1-p)^{n-k}$, $0 \le k \le n$
Geometric	G(p)	$0 \le p \le 1$	$(1-p)^{k-1}p$, $k \ge 1$
Modified Geometric	MG(p)	$0 \le p \le 1$	$(1-p)^k p$, $k \ge 0$
Poisson	P(λ)	$\lambda \ge 0$	$e^{-\lambda}\frac{\lambda^k}{k!}$, $k \ge 0$

X	Mean	Variance	Generating Function
Ber(p)	p	$p(1-p)$	$pz + 1 - p$
Bin(n, p)	np	$np(1-p)$	$(pz + 1 - p)^n$
G(p)	$1/p$	$(1-p)/p^2$	$\frac{pz}{1-(1-p)z}$
MG(p)	$(1-p)/p$	$(1-p)/p^2$	$\frac{p}{1-(1-p)z}$
P(λ)	λ	λ	$e^{-\lambda(1-z)}$

Common Continuous Random Variables. The facts about the commonly occurring real valued continuous random variables are given in the tables below.

Random Variable, X	Symbol	Parameter Range	$f_X(x)$
Uniform	U(a, b)	$-\infty < a < b < \infty$	$\frac{1}{b-a}$, $a \le x \le b$
Exponential	exp(λ)	$\lambda > 0$	$\lambda e^{-\lambda x}$, $x \ge 0$
Erlang	Erl(k, λ)	$\lambda > 0, k = 1, 2, \cdots$	$\lambda e^{-\lambda x} \frac{(\lambda x)^{k-1}}{(k-1)!}$, $x \ge 0$
Normal	N(μ, σ^2)	$-\infty < \mu < \infty, \sigma^2 > 0$	$\frac{1}{\sqrt{2\pi\sigma^2}} \exp\left(-\frac{1}{2}\left(\frac{x-\mu}{\sigma}\right)^2\right)$

X	Mean	Variance	LST(X)
U(a, b)	$\frac{a+b}{2}$	$\frac{(b-a)^2}{12}$	$\frac{e^{-sa}-e^{-sb}}{s(b-a)}$
exp(λ)	$\frac{1}{\lambda}$	$\frac{1}{\lambda^2}$	$\frac{\lambda}{\lambda+s}$
Erl(k, λ)	$\frac{k}{\lambda}$	$\frac{k}{\lambda^2}$	$\left(\frac{\lambda}{\lambda+s}\right)^k$
N(μ, σ^2)	μ	σ^2	$\exp\left(-\mu s + \frac{\sigma^2 s^2}{2}\right)$

Multivariate Random Variables

Let $X_i : \Omega \to \mathcal{R}$ $(i = 1, 2, \cdots, n)$ be n real valued random variables. Then $X = (X_1, X_2, \cdots, X_n)$ is called a multivariate random variable or a multivariate random vector. When $n = 2$ it is called a bivariate random variable. It is described by its joint cdf

$$F_X(x) = \mathsf{P}(X_i \leq x_i,\ 1 \leq i \leq n), \quad x = (x_1, x_2, \cdots, x_n) \in \mathcal{R}^n.$$

The marginal cdf $F_{X_i}(x_i)$ of X_i is given by

$$F_{X_i}(x_i) = F_X(\infty, \cdots, \infty, x_i, \infty, \cdots, \infty).$$

X is called discrete if each X_i $(1 \leq i \leq n)$ is a discrete random variable taking values in a discrete set $A_i \subset \mathcal{R}$. A discrete X can be described in an equivalent way by giving its joint pmf

$$p_X(x) = \mathsf{P}(X_i = x_i,\ 1 \leq i \leq n), \quad x_i \in A_i.$$

The marginal pmf $p_{X_i}(x_i)$ of X_i is given by

$$p_{X_i}(x_i) = \sum_{x_j \in A_j : j \neq i} p_X(x_1, \cdots, x_{i-1}, x_i, x_{i+1}, \cdots, x_n).$$

X is said to be a jointly continuous random variable if F_X is absolutely continuous, that is, if there exists a function $f_X(x_1, x_2, \cdots, x_n)$, called the joint pdf, such that

$$F_X(x) = \int_{u \leq x} f_X(u_1, u_2, \cdots, u_n) du_1 du_2 \cdots du_n, \quad x \in \mathcal{R}^n.$$

It is possible that each X_i is a continuous random variable but X is not a jointly continuous random variable. The marginal pdf $f_{X_i}(x_i)$ of X_i from a jointly continuous X is given by

$$f_{X_i}(x_i) = \int_{x_j \in \mathcal{R} : j \neq i} f_X(u_1, \cdots, u_{i-1}, x_i, u_{i+1}, \cdots, u_n) du_1 \cdots du_{i-1} du_{i+1} \cdots du_n.$$

Independent Random Variables. The random variables $(X_1, X_2, \cdots, X_n)$ are said to be independent if

$$F_X(x) = \prod_{i=1}^{n} F_{X_i}(x_i), \quad x = (x_1, x_2, \cdots, x_n) \in \mathcal{R}^n.$$

The discrete random variables $(X_1, X_2, \cdots, X_n)$ are said to be independent if

$$p_X(x) = \prod_{i=1}^{n} p_{X_i}(x_i), \quad x_i \in A_i.$$

The jointly continuous random variables $(X_1, X_2, \cdots, X_n)$ are said to be independent if

$$f_X(x) = \prod_{i=1}^{n} f_{X_i}(x_i), \quad x_i \in \mathcal{R}.$$

The random variables $(X_1, X_2, \cdots, X_n)$ are said to independent and identically distributed (iid) if

$$F_{X_1}(x) = F_{X_2}(x) = \cdots = F_{X_n}(x), \quad x \in \mathcal{R}$$

or

$$p_{X_1}(x) = p_{X_2}(x) = \cdots = p_{X_n}(x), \quad x \in \mathcal{R}$$

in the discrete case, or

$$f_{X_1}(x) = f_{X_2}(x) = \cdots = f_{X_n}(x), \quad x \in \mathcal{R}$$

in the continuous case.

Sums of Random Variables. Let (X_1, X_2) be a discrete bivariate random variable. Then the pmf of $Z = X_1 + X_2$ is given by

$$p_Z(z) = \sum_{(x_1, x_2): x_1 + x_2 = z} p_X(x_1, x_2).$$

If X_1 and X_2 are independent, we have

$$p_Z(z) = \sum_{x_1} p_{X_1}(x_1) p_{X_2}(z - x_1).$$

This is called a discrete convolution. As a special case, if X_1 and X_2 are non-negative integer valued random variables, we have

$$p_Z(n) = \sum_{i=0}^{n} p_{X_1}(i) p_{X_2}(n - i), \quad n \geq 0.$$

Let (X_1, X_2) be a jointly continuous bivariate random variable. Then the pdf of $Z = X_1 + X_2$ is given by

$$f_Z(z) = \int_{(x_1, x_2): x_1 + x_2 = z} f_X(x_1, x_2) dx_1 dx_2.$$

If X_1 and X_2 are independent, we have

$$f_Z(z) = \int_{x_1 \in \mathcal{R}} f_{X_1}(x_1) f_{X_2}(z - x_1) dx_1.$$

This is called the convolution of f_{X_1} and f_{X_2}. As a special case, if X_1 and X_2 are non-negative real valued random variables, we have

$$f_Z(z) = \int_{x=0}^{z} f_{X_1}(x) f_{X_2}(z - x) dx, \quad z \geq 0.$$

If X_1 and X_2 are non-negative real valued random variables (discrete, continuous, or mixed),

$$F_Z(z) = \int_{x=0}^{z} dF_{X_1}(x) F_{X_2}(z - x), \quad z \geq 0.$$

We call this the Stieltjes convolution of F_{X_1} and F_{X_2}. The following facts about the sums of independent random variables are useful:

1. Suppose $(X_1, X_2, \cdots, X_n)$ are iid $\mathrm{Ber}(p)$. Then $X_1 + X_2 + \cdots + X_n$ is $\mathrm{Bin}(n, p)$.
2. Suppose $(X_1, X_2, \cdots, X_n)$ are iid $\exp(\lambda)$. Then $X_1 + X_2 + \cdots + X_n$ is $\mathrm{Erl}(n, \lambda)$.
3. Suppose $X_i \sim \mathrm{P}(\lambda_i)$ $(1 \leq i \leq n)$ are independent. Then $X_1 + X_2 + \cdots + X_n$ is $\mathrm{P}(\lambda)$ where $\lambda = \lambda_1 + \lambda_2 + \cdots + \lambda_n$.
4. Suppose $X_i \sim \mathrm{N}(\mu_i, \sigma_i^2)$ $(1 \leq i \leq n)$ are independent. Then $X_1 + X_2 + \cdots + X_n$ is $\mathrm{N}(\mu, \sigma^2)$ where $\mu = \mu_1 + \mu_2 + \cdots + \mu_n$ and $\sigma^2 = \sigma_1^2 + \sigma_2^2 + \cdots + \sigma_n^2$.

Functions of Multivariate Random Vectors. Let $g : \mathcal{R}^n \to \mathcal{R}$ be a function such that $Z = g(X)$ is a real valued random variable. This is true if g is a piecewise continuous function, for example. The expectation of Z can be computed without computing its distribution as follows:

$$E(Z) = \int_{u \in \mathcal{R}^n} g(u) dF_X(u).$$

In particular we have

$$E\left(\sum_{i=1}^{n} X_i\right) = \sum_{i=1}^{n} E(X_i),$$

which holds even if the X_i's are dependent. If they are independent, we also have

$$E\left(\prod_{i=1}^{n} X_i\right) = \prod_{i=1}^{n} E(X_i).$$

We define the covariance of two random variables as

$$\mathrm{Cov}(X_1, X_2) = E(X_1 X_2) - E(X_1) E(X_2).$$

If X_1 and X_2 are independent, their covariance is zero. The converse is not true in general. We have

$$\mathrm{Var}\left(\sum_{i=1}^{n} X_i\right) = \sum_{i=1}^{n} \mathrm{Var}(X_i) + 2 \sum_{i=1}^{n} \sum_{j=i+1}^{n} \mathrm{Cov}(X_i, X_j).$$

In particular, when the X_i's are independent,

$$\mathrm{Var}\left(\sum_{i=1}^{n} X_i\right) = \sum_{i=1}^{n} \mathrm{Var}(X_i).$$

Conditional Distributions and Expectations. Let (X_1, X_2) be a discrete bivariate random variable. Then the conditional pmf of X_1 given $X_2 = x_2$ is given by

$$p_{X_1|X_2}(x_1|x_2) = \frac{p_X(x_1, x_2)}{p_{X_2}(x_2)},$$

and the conditional expected value of X_1 given $X_2 = x_2$ is given by

$$E(X_1|X_2 = x_2) = \sum_{x_1} x_1 p_{X_1|X_2}(x_1|x_2).$$

We have the following useful formula for computing expectations by conditioning:

$$E(X_1) = \sum_{x_2} E(X_1|X_2 = x_2) p_{X_2}(x_2).$$

Let (X_1, X_2) be a jointly continuous bivariate random variable. Then the conditional pdf of X_1 given $X_2 = x_2$ is given by

$$f_{X_1|X_2}(x_1|x_2) = \frac{f_X(x_1, x_2)}{f_{X_2}(x_2)},$$

and the conditional expected value of X_1 given $X_2 = x_2$ is given by

$$E(X_1|X_2 = x_2) = \int_{x_1} x_1 f_{X_1|X_2}(x_1|x_2) dx_1.$$

We have the following useful formula for computing expectations by conditioning:

$$E(X_1) = \int_{x_2} E(X_1|X_2 = x_2) f_{X_2}(x_2) dx_2.$$

In general we can define $E(X_1|X_2)$ to be a random variable that takes value $E(X_1|X_2 = x_2)$ with "probability $dF_{X_2}(x_2)$." With this interpretation we get

$$E(X_1) = E(E(X_1|X_2)) = \int_{x_2} E(X_1|X_2 = x_2) dF_{X_2}(x_2).$$

Order Statistics. Let $X_1, X_2, \cdots, X_n$ be iid random variables with common cdf $F(\cdot)$. Let $(Y_1, Y_2, \cdots, Y_n)$ be a permutation of $(X_1, X_2, \cdots, X_n)$ ordered in an ascending order, i.e., $Y_1 = \min(X_1, X_2, \cdots, X_n) \leq Y_2 \leq \cdots \leq Y_{n-1} \leq Y_n = \max(X_1, X_2, \cdots, X_n)$. Then $(Y_1, Y_2, \cdots, Y_n)$ is called the order statistics of $(X_1, X_2, \cdots, X_n)$. The marginal distribution of Y_k is given by

$$F_{Y_k}(t) = P(Y_k \leq t) = \sum_{j=k}^{n} \binom{n}{j} F(t)^j (1 - F(t))^{n-j}, \quad 1 \leq k \leq n.$$

In particular

$$F_{Y_n}(t) = F(t)^n, \quad F_{Y_1}(t) = 1 - (1 - F(t))^n.$$

In addition, when $(X_1, X_2, \cdots, X_n)$ are jointly continuous with common pdf $f(\cdot)$, the $(Y_1, Y_2, \cdots, Y_n)$ are also jointly continuous with joint pdf given by

$$f_Y(y_1, y_2, \cdots, y_n) = n! f(y_1) f(y_2) \cdots f(y_n), \quad y_1 \leq y_2 \leq \cdots \leq y_n.$$

The density is zero outside the above region.

Multivariate Normal Random Variable. Let μ be an n-dimensional column vector, and $\Sigma = [\Sigma_{i,j}]$ be an $n \times n$ positive definite matrix. A random vector $X = [X_1, X_2, \cdots, X_n]^\top$ is called a multivariate normal variable $N(\mu, \Sigma)$ if it has the joint pdf given by

$$f_X(x) = \frac{1}{\sqrt{(2\pi)^n \det(\Sigma)}} \exp\left(-\frac{1}{2}(x-\mu)\Sigma^{-1}(x-\mu)^\top\right), \quad x = (x_1, \ldots, x_n) \in \mathcal{R}^n.$$

In this case we have

$$\mathsf{E}(X_i) = \mu_i, \quad \mathrm{Cov}(X_i, X_j) = \Sigma_{i,j}, \quad \mathrm{Var}(X_i) = \Sigma_{i,i} = \sigma_i^2.$$

If Σ is diagonal, then $X_1, X_2, \cdots, X_n$ are independent normal random variables and $X_i \sim N(\mu_i, \sigma_i^2)$. If Σ is the identity matrix and $\mu = 0$, the n components of X are iid standard normal random variables. Now let $Z = AX + c$ where A is an $m \times n$ matrix of real numbers and c is an m-dimensional vector of reals. The Z is a $N(\mu_Z, \Sigma_Z)$ random variable where

$$\mu_Z = A\mu + c, \quad \Sigma_Z = A\Sigma A^\top.$$

Now suppose the n-dimensional multivariate vector $X \sim N(\mu, \Sigma)$ is partitioned as $[X^{(1)}, X^{(2)}]^\top$ where $X^{(i)}$ is $n^{(i)}$ dimensional ($n^{(1)} + n^{(2)} = n$). Partition the μ vector and the Σ matrix in a commensurate fashion as

$$\mu = \begin{bmatrix} \mu^{(1)} \\ \mu^{(2)} \end{bmatrix}, \quad \Sigma = \begin{bmatrix} \Sigma^{(1,1)} & \Sigma^{(1,2)} \\ \Sigma^{(2,1)} & \Sigma^{(2,2)} \end{bmatrix}.$$

Then the marginal distribution of $X^{(1)}$ is $N(\mu^{(1)}, \Sigma^{(1,1)})$, and the distribution of $X^{(1)}$ given $X^{(2)} = a$ is multivariate normal $N(\bar{\mu}, \overline{\Sigma})$ where

$$\bar{\mu} = \mu^{(1)} + \Sigma^{(1,2)}(\Sigma^{(2,2)})^{-1}(a - \mu^{(2)}),$$

and covariance matrix

$$\overline{\Sigma} = \Sigma^{(1,1)} - \Sigma^{(1,2)}(\Sigma^{(2,2)})^{-1}\Sigma^{(1,2)}.$$

Generating Functions

Let X be a non-negative integer valued random variable with pmf $\{p_k, k \geq 0\}$. The generating function (GF) of X (or of its pmf) is defined as

$$g_X(z) = \mathsf{E}(z^X) = \sum_{k=0}^{\infty} z^k p_k, \quad |z| \leq 1.$$

The GFs for common random variables are given in Appendix B. Important and useful properties of the GF are enumerated below:

1. $g_X(1) = \mathsf{P}(X < \infty)$.

2. A random variable or its pmf is uniquely determined by its GF.

3. The pmf can be obtained from its GF as follows:

$$p_k = \frac{1}{k!} \frac{d^k}{dz^k} g_X(z) \bigg|_{z=0}, \quad k \geq 0.$$

4. The factorial moments of X are given by

$$\mathsf{E}(X(X-1)\cdots(X-k+1)) = \frac{d^k}{dz^k} g_X(z) \bigg|_{z=1}, \quad k \geq 1.$$

5. Let X_1 and X_2 be independent random variables. Then

$$g_{X_1+X_2}(z) = g_{X_1}(z) g_{X_2}(z).$$

6. Let $\{X_n, n \geq 1\}$ be a given sequence of random variables and let X be such that

$$\lim_{n \to \infty} \mathsf{P}(X_n = k) = \mathsf{P}(X = k), \quad k \geq 0.$$

Then

$$\lim_{n \to \infty} g_{X_n}(z) = g_X(z), \quad |z| \leq 1.$$

The converse also holds.

Let $\{p_k, k \geq 0\}$ be a sequence of real numbers, not necessarily a pmf. Define its GF as

$$p(z) = \sum_{k=0}^{\infty} z^k p_k.$$

Let R be its radius of convergence. We list some important and useful properties of the GF:

1. Let $q(z)$ be a GF of a sequence $\{q_k, k \geq 0\}$. If $p(z) = q(z)$ for $|z| < r \leq R$, then $p_k = q_k$ for all $k \geq 0$.
2. Let $r_k = ap_k + bq_k$, $k \geq 0$, where a and b are constants. Then
$$r(z) = ap(z) + bq(z).$$
3. Generating function of $\{kp_k\}$ is $zp'(z)$.
4. Let $q_0 = p_0$, $q_k = p_k - p_{k-1}$ for $k \geq 1$. Then
$$q(z) = (1 - z)p(z).$$
5. $q_k = \sum_{r=0}^{k} p_r$ for $k \geq 0$. Then
$$q(z) = p(z)/(1 - z).$$
6. $\lim_{k \to \infty} \frac{1}{k+1} \sum_{r=0}^{k} p_r = \lim_{z \to 1}(1 - z)p(z)$ if the limit on either side exists.
7. $\lim_{k \to \infty} p_k = \lim_{z \to 1}(1 - z)p(z)$ if the limit on the left-hand side exists.

Suppose the GF of $\{p_k, k \geq 0\}$ is given by

$$p(z) = \frac{P(z)}{Q(z)},$$

where $P(z)$ is a polynomial of degree r, and $Q(z)$ is a polynomial of degree $s > r$. Suppose $Q(z)$ has distinct roots so that we can write

$$Q(z) = c \prod_{i=1}^{s}(1 - z\alpha_i),$$

where $\{\alpha_i, 1 \leq i \leq s\}$ are distinct. Then

$$p_k = \sum_{i=1}^{s} c_i \alpha_i^k,$$

where

$$c_i = -\alpha_i \frac{P(1/\alpha_i)}{Q'(1/\alpha_i)}, \quad 1 \leq i \leq s.$$

Laplace–Stieltjes Transforms

Let X be a non-negative real valued random variable with cdf $F_X(\cdot)$. The Laplace–Stieltjes transform (LST) of X (or of its cdf) is defined as

$$\phi_X(s) = \mathsf{E}(e^{-sX}) = \int_{x=0}^{\infty} e^{-sx} dF_X(x), \quad \mathrm{Re}(s) \geq 0.$$

A jump of size $F_X(0)$ at $x = 0$ is included in the above integral. The LSTs for common random variables are given in Appendix B. Important and useful properties of the LST are enumerated below:

1. $\phi_X(0) = \mathsf{P}(X < \infty)$.

2. A random variable or its cdf is uniquely determined by its LST.

3. The moments of X are given by

$$\mathsf{E}(X^k) = (-1)^k \frac{d^k}{ds^k} \phi_X(s) \bigg|_{s=0}, \quad k \geq 1.$$

4. Let X_1 and X_2 be independent random variables. Then

$$\phi_{X_1+X_2}(s) = \phi_{X_1}(s)\phi_{X_2}(s).$$

5. Let $\{X_n, n \geq 1\}$ be a given sequence of random variables and let X be such that

$$\lim_{n \to \infty} \mathsf{P}(X_n \leq x) = \mathsf{P}(X \leq x) = F(x), \quad k \geq 0,$$

at all points of continuity of F. Then

$$\lim_{n \to \infty} \phi_{X_n}(s) = \phi_X(s), \quad \mathrm{Re}(s) > 0.$$

The converse also holds.

Let $F : [0, \infty) \to (-\infty, \infty)$, not necessarily a cdf. Define its LST as

$$LST(F) = \tilde{F}(s) = \int_{x=0}^{\infty} e^{-sx} dF(x),$$

if the integral exists for some complex s with $\mathrm{Re}(s) > 0$. It is assumed $F(0-) = 0$ so that there is a jump of size $F(0+)$ at $x = 0$. We list some important and useful properties of the LST:

1. Let a and b be given constants. Then

$$LST(aF + bG) = a\tilde{F}(s) + b\tilde{G}(s).$$

2. $H(t) = \int_0^t F(t - u)dG(u),\ t \geq 0 \Leftrightarrow \tilde{H}(s) = \tilde{F}(s)\tilde{G}(s).$

3. Assuming the limits exist

$$\lim_{t \to \infty} F(t) = \lim_{s \to 0} \tilde{F}(s),$$

$$\lim_{t \to 0} F(t) = \lim_{s \to \infty} \tilde{F}(s),$$

$$\lim_{t \to \infty} \frac{F(t)}{t} = \lim_{s \to 0} s\tilde{F}(s).$$

Laplace Transforms

Let $f : [0, \infty) \to (-\infty, \infty)$. Define its Laplace transform (LT) as

$$LT(f) = f^*(s) = \int_{x=0}^{\infty} e^{-sx} f(x)dx,$$

if the integral exists for some complex s with $\text{Re}(s) > 0$. We list some important and useful properties of the GF:

1. $LT(af + bg) = af^*(s) + bg^*(s)$.
2. $\tilde{F}(s) = sF^*(s)$.
3. $h(t) = \int_0^t f(t-u)g(u)du,\ t \geq 0 \Leftrightarrow h^*(s) = f^*(s)g^*(s)$.
4. $LT(f'(t)) = sf^*(s) - f(0)$.
5. $LT(e^{-at}f(t)) = f^*(s+a)$.
6. $LT(t^n f(t)) = (-1)^n \frac{d^n}{ds^n} f^*(s)$.

Table of Laplace Transforms.

$f(t)$	$f^*(s)$
1	$1/s$
t	$1/s^2$
t^n	$n!/s^{n+1}$
e^{-at}	$1/(s+a)$
$e^{-at}t^{n-1}/(n-1)!$	$1/(s+a)^n$
$(e^{-at} - e^{-bt})/(b-a)$	$1/(s+a)(s+b)$

Modes of Convergence

Let the random variables $X_1, X_2, X_3, \cdots$ be defined on a common probability space $(\Omega, \mathcal{F}, \mathsf{P})$. Let X be another random variable defined on the same probability space.

1. Almost Sure Convergence.

   X_n is said to converge almost surely to X, written $X_n \to X$ $(a.s.)$, if

   $$\mathsf{P}(\lim_{n \to \infty} X_n = X) = \mathsf{P}(\{\omega \in \Omega : \lim_{n \to \infty} X_n(\omega) = X(\omega)\}) = 1.$$

   This mode of convergence is also called convergence with probability 1, or almost everywhere convergence, or sample-path convergence. If X_n converges to X almost surely and f is a continuous function, then $f(X_n)$ converges to $f(X)$ almost surely.

2. Convergence in Probability.

   X_n is said to converge to X in probability, written $X_n \to X$ (p), if

   $$\lim_{n \to \infty} \mathsf{P}(|X_n - X| > \epsilon) = 0$$

   for any $\epsilon > 0$. If X_n converges to X in probability and f is a continuous function, then $f(X_n)$ converges to $f(X)$ in probability.

3. Convergence in Distribution.

   X_n is said to converge to X in distribution, written $X_n \to X$ (d), if

   $$\lim_{n \to \infty} F_{X_n}(x) = F_X(x)$$

   at all points of continuity of F_X. Convergence in distribution is also called weak convergence. If X_n converges to X in distribution and f is a continuous function, then $f(X_n)$ converges to $f(X)$ in distribution.

4. Convergence in Mean.

   X_n is said to converge to X in mean, written $X_n \to X$ (m), if

   $$\lim_{n \to \infty} \mathsf{E}(|X_n - X|) = 0.$$

5. Convergence in Mean Square.

   X_n is said to converge to X in mean square, written $X_n \to X$ $(m.s.)$, if

   $$\lim_{n \to \infty} \mathsf{E}((X_n - X)^2) = 0.$$

The various modes of convergence are related as follows: Almost sure convergence implies convergence in probability which implies convergence in distribution. Mean square convergence implies convergence in probability. Convergence in distribution or with probability one does not imply convergence in mean. We need an additional condition called uniform integrability defined as follows: A sequence of random variables $\{X_n, n \geq 0\}$ is called uniformly integrable if for a given $\epsilon > 0$, there exists a $K < \infty$ such that

$$\mathsf{E}(|X_n|; |X_n| > K) = \int_{\{x:|x|>K\}} x dF_{X_n}(dx) < \epsilon, \quad n \geq 1.$$

Then we have the following result: If X_n converges to X in distribution (or almost surely), and $\{X_n, n \geq 0\}$ is uniformly integrable, the X_n converges to X in mean.

The three main convergence results associated with a sequence $\{X_n, n \geq 1\}$ of iid random variables with common finite mean τ and variance σ^2 are given below. Let $\bar{X}_n = (X_1 + X_2 + \cdots + X_n)/n$.

1. Weak Law of Large Numbers.

$$\overline{X}_n \to \tau \ (p).$$

2. Strong Law of Large Numbers.

$$\overline{X}_n \to \tau \ (a.s.).$$

3. Central Limit Theorem. Assume $\sigma^2 < \infty$. Then

$$\frac{\overline{X}_n - \tau}{\sigma/\sqrt{n}} \to N(0, 1) \ (d).$$

Results from Analysis

We frequently need to interchange the operations of sums, integrals, limits, etc. Such interchanges are in general not valid unless certain conditions are satisfied. In this section we collect some useful sufficient conditions which enable us to do such interchanges.

1. Monotone Convergence Theorem for Sums. Let $\pi(i) \geq 0$ for all $i \geq 0$ and $\{g_n(i), n \geq 1\}$ be a non-decreasing (in n) non-negative sequence for each $i \geq 0$. Then

$$\lim_{n \to \infty} \left(\sum_{i=0}^{\infty} g_n(i)\pi(i) \right) = \sum_{i=0}^{\infty} \left(\lim_{n \to \infty} g_n(i) \right) \pi(i).$$

2. Suppose $\{g_n(i), n \geq 1\}$ is a non-negative sequence for each $i \geq 0$ and $\pi(i) \geq 0$ for all $i \geq 0$. Then

$$\sum_{n=0}^{\infty} \sum_{i=0}^{\infty} g_n(i)\pi(i) = \sum_{i=0}^{\infty} \sum_{n=0}^{\infty} g_n(i)\pi(i).$$

3. Fatou's Lemma for Sums. Suppose $\{g_n(i), n \geq 1\}$ is a non-negative sequence for each $i \geq 0$ and $\pi(i) \geq 0$ for all $i \geq 0$. Then

$$\sum_{i=0}^{\infty} \left(\liminf_{n \geq 1} g_n(i) \right) \pi(i) \leq \liminf_{n \geq 1} \left(\sum_{i=0}^{\infty} g_n(i)\pi(i) \right).$$

4. Bounded Convergence Theorem for Sums. Suppose $\pi(i) \geq 0$ for all $i \geq 0$, and there exists a $\{g(i), i \geq 0\}$ such that $\sum g(i)\pi(i) < \infty$ and $|g_n(i)| \leq g(i)$ for all $n \geq 1$ and $i \geq 0$. Then

$$\lim_{n \to \infty} \sum_{i=0}^{\infty} g_n(i)\pi(i) = \sum_{i=0}^{\infty} \left(\lim_{n \to \infty} g_n(i) \right) \pi(i).$$

5. Let $\{\pi_n(i), n \geq 1\}$ be a non-negative sequence for each $i \geq 0$ and let

$$\pi(i) = \lim_{n \to \infty} \pi_n(i), \quad i \geq 0.$$

Suppose

$$\lim_{n \to \infty} \sum_{i=0}^{\infty} \pi_n(i) = \sum_{i=0}^{\infty} \pi(i) < \infty.$$

Suppose $\{g(i), i \geq 0\}$ is a non-negative bounded function with $0 \leq g(i) \leq c$ for all $i \geq 0$ for some $c < \infty$. Then

$$\lim_{n \to \infty} \sum_{i=0}^{\infty} g(i)\pi_n(i) = \sum_{i=0}^{\infty} g(i) \left(\lim_{n \to \infty} \pi_n(i) \right).$$

6. Monotone Convergence Theorem for Integrals. Let $\pi(x) \geq 0$ for all $x \in \mathcal{R}$ and $\{g_n(x), n \geq 1\}$ be a non-decreasing (in n) non-negative sequence for each $x \in \mathcal{R}$. Then

$$\lim_{n \to \infty} \left(\int_{x \in \mathcal{R}} g_n(x)\pi(x)dx \right) = \int_{x \in \mathcal{R}} \left(\lim_{n \to \infty} g_n(x) \right) \pi(x)dx.$$

7. Suppose $\{g_n(x), n \geq 1\}$ is a non-negative sequence for each $x \in \mathcal{R}$ and $\pi(x) \geq 0$ for all $x \in \mathcal{R}$. Then

$$\sum_{n=0}^{\infty} \int_{x \in \mathcal{R}} g_n(x)\pi(x)dx = \int_{x \in \mathcal{R}} \sum_{n=0}^{\infty} g_n(x)\pi(x)dx.$$

8. Fatou's Lemma for Integrals. Suppose $\{g_n(x), n \geq 1\}$ is a non-negative sequence for each $x \in \mathcal{R}$ and $\pi(x) \geq 0$ for all $x \in \mathcal{R}$. Then

$$\int_{x \in \mathcal{R}} \left(\liminf_{n \geq 1} g_n(x) \right) \pi(x)dx \leq \liminf_{n \geq 1} \left(\int_{x \in \mathcal{R}} g_n(x)\pi(x)dx \right).$$

9. Bounded Convergence Theorem for Integrals. Let $\pi(x) \geq 0$ for all $x \in \mathcal{R}$ and there exists a function $g : \mathcal{R} \to \mathcal{R}$ such that $\int g(x)\pi(x)dx < \infty$, and $|g_n(x)| \leq g(x)$ for each $x \in \mathcal{R}$ and all $n \geq 1$. Then

$$\lim_{n \to \infty} \left(\int_{x \in \mathcal{R}} g_n(x)\pi(x)dx \right) = \int_{x \in \mathcal{R}} \left(\lim_{n \to \infty} g_n(x) \right) \pi(x)dx.$$

10. Let $\{\pi_n(x), n \geq 1\}$ be a non-negative sequence for each $x \in \mathcal{R}$ and let

$$\pi(x) = \lim_{n \to \infty} \pi_n(x), \quad x \in \mathcal{R}.$$

Suppose

$$\lim_{n \to \infty} \int_{x \in \mathcal{R}} \pi_n(x)dx = \int_{x \in \mathcal{R}} \pi(x)dx < \infty.$$

Suppose $g : \mathcal{R} \to [0, \infty)$ is a non-negative bounded function with $0 \leq g(x) \leq c$ for all $x \in \mathcal{R}$ for some $c < \infty$. Then

$$\lim_{n \to \infty} \left(\int_{x \in \mathcal{R}} g(x)\pi_n(x)dx \right) = \int_{x \in \mathcal{R}} g(x) \left(\lim_{n \to \infty} \pi_n(x) \right) dx.$$

Difference and Differential Equations

A sequence $\{x_k, k \geq 0\}$ is said to satisfy an nth order difference equation with constant coefficients $\{a_i, 0 \leq i \leq n-1\}$ if

$$\sum_{i=0}^{n-1} a_i x_{k+i} + x_{k+n} = r_k, \quad k \geq 0,$$

where $\{r_k, k \geq 0\}$ is a given sequence of real numbers. If the righthand side r_k is zero for all k, the equation is called homogeneous, else it is called non-homogeneous. The polynomial

$$\sum_{i=0}^{n-1} a_i \alpha^i + \alpha^n = 0,$$

is called the characteristic polynomial corresponding to the difference equation. Let $\{\alpha(i), 1 \leq i \leq d\}$ be the d distinct roots of the characteristic polynomial, and let $m(i)$ be the multiplicity of $\alpha(i)$. ($\sum_{i=1}^{d} m(i) = n$.) Then any solution to the homogeneous equation

$$\sum_{i=0}^{n-1} a_i x_{k+i} + x_{k+n} = 0, \quad k \geq 0,$$

is of the form

$$x_n = \sum_{i=1}^{d} \sum_{j=0}^{m(i)-1} c_{ij} n^j \alpha(i)^n, \quad n \geq 0.$$

The constants $c_{i,j}$ are to be determined by using the initial conditions. Now consider the non-homogeneous equation

$$\sum_{i=0}^{n-1} a_i x_{k+i} + x_{k+n} = r_k, \quad k \geq 0.$$

Any solution of this equation is of the form

$$x_n = x_n^p + \sum_{i=1}^{d} \sum_{j=0}^{m(i)-1} c_{i,j} n^j \alpha(i)^n$$

549

where $\{x_n^p, n \geq 0\}$ is any one solution (called the particular solution) of the non-homogeneous equation. The constants $c_{i,j}$ are to be determined by using the initial conditions.

A function $\{x(t), t \geq 0\}$ is said to satisfy an nth order differential equation with constant coefficients $\{a_i, 0 \leq i \leq n - 1\}$ if

$$\sum_{i=0}^{n-1} a_i \frac{d^i}{dt^i} x(t) + \frac{d^n}{dt^n} x(t) = r(t), \quad t \geq 0.$$

If the righthand side $r(t)$ is zero for all $t \geq 0$, the equation is called homogeneous, else it is called non-homogeneous. The polynomial

$$\sum_{i=0}^{n-1} a_i \alpha^i + \alpha^n = 0,$$

is called the characteristic polynomial corresponding to the differential equation. Let $\{\alpha(i), 1 \leq i \leq d\}$ be the d distinct roots of the characteristic polynomial, and let $m(i)$ be the multiplicity of $\alpha(i)$. ($\sum_{i=1}^{d} m(i) = n$.) Then any solution to the homogeneous equation

$$\sum_{i=0}^{n-1} a_i \frac{d^i}{dt^i} x(t) + \frac{d^n}{dt^n} x(t) = 0, \quad t \geq 0$$

is of the form

$$x(t) = \sum_{i=1}^{d} \sum_{j=0}^{m(i)-1} c_{i,j} t^j e^{\alpha(i)t}.$$

The constants $c_{i,j}$ are to be determined by using the initial conditions. Now consider the non-homogeneous equation

$$\sum_{i=0}^{n-1} a_i \frac{d^i}{dt^i} x(t) + \frac{d^n}{dt^n} x(t) = r(t), \quad t \geq 0.$$

Any solution of this equation is of the form

$$x(t) = x^p(t) + \sum_{i=1}^{d} \sum_{j=0}^{m(i)-1} c_{i,j} t^j e^{\alpha(i)t}$$

where $\{x^p(t), t \geq 0\}$ is any one solution (called the particular solution) of the non-homogeneous equation. The constants $c_{i,j}$ are to be determined by using the initial conditions.

Answers to Selected Problems

Chapter 2

MODELING EXERCISES

2.1 State space $= \{0, 1, 2, \cdots\}$.

$$p_{i,0} = p_{i+1} / \sum_{j=i+1}^{\infty} p_j, \quad p_{i,i+1} = 1 - p_{i,0}, \quad i \geq 0.$$

2.3 State space $= \{b, b+1, b+2, \cdots\}$.

$$X_{n+1} = \begin{cases} X_n + k - 1 & \text{w.p. } X_n/(w + b + n(k-1)) \\ X_n & \text{w.p. } 1 - X_n/(w + b + n(k-1)) \end{cases}$$

Thus $\{X_n, n \geq 0\}$ is a DTMC, but it is not time-homogeneous.

2.5 $X_n = 1$ if the weather is sunny on day n, 0 otherwise. $Y_n = (X_{n-1}, X_n)$, $n \geq 1$. $\{Y_n, n \geq 1\}$ is a DTMC on $S = \{(0,0), (0,1), (1,0), (1,1)\}$ with the following transition probability matrix:

$$P = \begin{bmatrix} .6 & 0.4 & 0 & 0 \\ 0 & 0 & .25 & .75 \\ .5 & .5 & 0 & 0 \\ 0 & 0 & .2 & .8 \end{bmatrix}.$$

2.7 $\{X_n, n \geq 0\}$ is a space homogeneous random walk on $S = \{..., -2, -1, 0, 1, 2, ...\}$ with

$$p_i = p_1(1 - p_2), \quad q_i = p_2(1 - p_1), \quad r_i = 1 - p_i - q_i.$$

2.9 State space $= S = \{1, 2, 3, 4, 5, 6\}$. Transition probabilities:

$$p_{i,j} = 1/6, \quad i \neq j, \quad p_{ii} = 0.$$

2.11 B_n (G_n) = the bar the boy (girl) is in on the nth night. $\{(B_n, G_n), n \geq 0\}$ is a DTMC on $S = \{(1, 1), (1, 2), (2, 1), (2, 2)\}$ with the following transition probability matrix:

$$P = \begin{bmatrix} 1 & 0 & 0 & 0 \\ a(1-d) & ad & (1-a)(1-d) & (1-a)d \\ (1-b)c & (1-b)(1-c) & bc & b(1-c) \\ 0 & 0 & 0 & 1 \end{bmatrix}.$$

The story ends in bar k if the DTMC gets absorbed in state (k, k), for $k = 1, 2$.

2.13 $p_{i,i} = p_i$, $p_{i,i+1} = 1 - p_i$, $1 \le i \le k - 1$, $\quad p_{k,k} = p_k$, $p_{k,1} = 1 - p_k$.

2.15 The state space is $S = \{(1, 2), (2, 3), (3, 1)\}$. The transition probability matrix is

$$P = \begin{bmatrix} 0 & b_{2,1} & b_{1,2} \\ b_{2,3} & 0 & b_{3,2} \\ b_{1,3} & b_{3,1} & 0 \end{bmatrix}.$$

2.17 $\{X_n, n \ge 0\}$ is a DTMC on $S = \{0, 1, \cdots, M\}$ with

$$p_{i,j} = \alpha_j, \quad 0 \le i < j \le M,$$

$$p_{i,i} = \sum_{k=0}^{i} \alpha_k, \quad 0 \le i \le M.$$

2.19 $\{X_n, n \ge 0\}$ is a DTMC since $\{Y_n, n \ge 0\}$ are iid and $X_{n+1} = Y_n + \text{Bin}(X_n, 1 - p)$.

2.21. $X_{n+1} = \min(\text{Bin}(X_n, 1 - p) + Y_{n+1}, B)$. Hence $\{X_n, n \ge 0\}$ is a DTMC with transition probabilities:

$$p_{i,j} = \sum_{k=0}^{i} \binom{i}{k} p^k (1-p)^{i-k} \alpha_{k+j-i}, \quad 0 \le i \le B, \, 0 \le i < B,$$

and

$$p_{i,B} = 1 - \sum_{j=0}^{B-1} p_{i,j},$$

where we use the convention that $\alpha_k = 0$ if $k < 0$.

2.23 Since $\{D_n, n \ge 0\}$ are iid, and

$$X_{n+1} = \begin{cases} X_n - D_n & \text{if } X_n - D_n \ge s, \\ S & \text{if } X_n - D_n < s, \end{cases}$$

$\{X_n, n \ge 0\}$ is a DTMC on state space $\{s, s + 1, ..., S - 1, S\}$. The transition probabilities are

$$p_{i,j} = \alpha_{i-j}, \quad s \le j \le i \le S, j \ne S,$$

$$p_{i,S} = \sum_{k=i-s+1}^{\infty} \alpha_k, \quad s \le i < S, j = S$$

$$p_{S,S} = \alpha_0 + \sum_{k=S-s+1}^{\infty} \alpha_k.$$

2.25 $\{X_n, n \geq 0\}$ is a DTMC general simple random walk on $\{0, 1, 2, ...\}$ with transition probabilities

$$p_{0,0} = 1, \quad p_{i,i+1} = \beta_i \alpha_2, \quad p_{i,i-1} = \beta_k \alpha_0, \quad p_{i,i} = \beta_i \alpha_1 + 1 - \beta_i.$$

2.27 $\{X_n, n \geq 0\}$ is a DTMC with state space $S = \{rr, dr, dd\}$ and transition probability matrix

$$P = \begin{bmatrix} 0 & 1 & 0 \\ 0 & .5 & .5 \\ 0 & 0 & 1 \end{bmatrix}.$$

2.29. $\{X_n, n \geq 0\}$ is a branching process with common progeny distribution

$$P(Y_{i,n} = 0) = 1 - \alpha; \quad P(Y_{i,n} = 20) = \alpha.$$

The number of recipients in the $(n+1)$st generation are given by

$$X_{n+1} = \sum_{i=1}^{X_n} Y_{i,n}.$$

2.31 $X_{n+1} = \max(X_n - 1 + Y_n, Y_n)$. This is the same as the DTMC in Example 2.16.

COMPUTATIONAL EXERCISES

2.1 1. $E(X_{20}|X_0 = 8) = 5.0346$, $E(X_{20}|X_0 = 5) = 5.0000$, $E(X_{20}|X_0 = 3) = 4.9769$.

2.5 Bin$(k, .3187)$.

2.7 $E(X_n) = (1 - d)^n(p\alpha + 1 - p)^n$, $E(X_n^2) = (1 - d)^{2n}(p\alpha^2 + 1 - p)^n$, where $\alpha = (1 + u)/(1 - d)$.

2.9 $E(X_5) = 14.9942$, $E(X_{10}) = 14.5887$.

2.13 $E(X_1) = 9.9972$, $E(X_{10}) = 19.0538$.

2.19 $p_{11}^{(n)} = .4 + 0.5236(-0.2236)^n + 0.0764(0.2236)^n$, $n \geq 0$.

2.21

$$P = \begin{bmatrix} 0 & 1 & 0 \\ 0 & .5 & .5 \\ 0 & 0 & 1 \end{bmatrix}, \quad P^n = XD^nX^{-1} = \begin{bmatrix} 0 & 2^{1-n} & 1 - 2^{1-n} \\ 0 & 2^{-n} & 1 - 2^{-n} \\ 0 & 0 & 1 \end{bmatrix}, \quad n \geq 1.$$

2.23

$$E(X_n|X_0 = i) = i\mu^n$$

$$\text{Var}(X_n|X_0 = i) = \begin{cases} in\sigma^2 & \text{if } \mu = 1 \\ i\sigma^2\mu^{n-1}\frac{\mu^n-1}{\mu-1} & \text{if } \mu \neq 1. \end{cases}$$

2.25 $E(X_n) = i$, $\text{Var}(X_n) = (1 - a^n)i(N - i)$, where $a = (N - 1)/N$.

CONCEPTUAL EXERCISES

2.1 First show that

$$P(X_{n+2} = k, X_{n+1} = j|X_n = i, X_{n-1}, \cdots, X_0) = P(X_2 = k, X_1 = j|X_0 = i).$$

Sum over all $j \in A$ and $k \in B$ to get the desired result.

2.3 (a) False (b) True (c) False.

2.5 No.

2.7 $P(N_i = k|X_0 = i) = (p_{i,i})^{k-1}(1 - p_{i,i})$, $k \geq 1$.

2.9 Use induction.

2.11 $\{|X_n|, n \geq 0\}$ is a random walk on $\{0, 1, 2, ...\}$ with $p_{0,1} = 1$, and, for $i \geq 1$,

$$p_{i,i+1} = \frac{p^{i+1} + q^{i+1}}{p^i + q^i} = 1 - p_{i,i-1}.$$

Chapter 3

COMPUTATIONAL EXERCISES

3.1 1. .87, 2. 5.7895, 3. 16.5559.

3.3 14.5771.

3.5 2.6241.

3.7 18.

3.9 2(e-1) = 3.4366.

3.11 $\frac{p^{r-1}(1-q^m)}{1-(1-p^{r-1})(1-q^{m-1})}$

3.13 7.

3.15 $1/p^2q^2$.

3.17 If $p = q$ the formula reduces to $i(N - i)$.

3.25 $.5\frac{(2-p_2^r)p_1^r}{1-(1-p_1^r)(1-p_2^r)}$.

3.27 .2.

3.29 $M + 1$.

3.33 $1/(a + b - 2ab)$.

3.35 $(N - 1)/(2pq)$.

CONCEPTUAL PROBLEMS

3.3 $\tilde{v}_i = v_i$, $i > 0$, $\tilde{v}_0 = \sum_{j=1}^{\infty} p_{i,j} v_j$.

3.7 For $B \subset A$, and $i \geq 1$, let $u(i, B)$ be the conditional probability that the process visits all states in A before visiting state 0, given that currently it is in state i and it has visited the states in set B so far. Then

$$u(i, B) = \sum_{j \in A-B} p_{i,j} u(j, B \cup \{j\}) + \sum_{j \neq 0, j \notin A-B} p_{i,j} u(j, B).$$

3.9

$$w_i = \delta_{i,j} p_{i,0} + \sum_{k=1}^{\infty} p_{i,k} w_k.$$

Chapter 4

COMPUTATIONAL EXERCISES

4.1 All rows of P^n and $\frac{M^{(n)}}{n+1}$ converge to $[.132 \; .319 \; .549]$.

4.3 All rows of P^n and $\frac{M^{(n)}}{n+1}$ converge to $[1/(N + 1), 1/(N + 1), \cdots, 1/(N + 1)]$.

4.5 $p > q \rightarrow$ Transient, $p = q \rightarrow$ Null recurrent, $p < q \rightarrow$ Positive recurrent.

4.7 Results of special case 2 of Example 4.24 continue to hold.

4.9 Communicating class: $\{A, B, C\}$ All states are aperiodic, positive recurrent.

4.13 (a)

1. Communicating Classes $\{1\}, \{2\}, \{3\}$, transient, aperiodic.
2. Communicating Class $\{4\}$, positive recurrent, aperiodic.

(b)

1. Class 1: $\{1, 2\}$, positive recurrent, aperiodic.
2. Class 2: $\{3\}$, transient, aperiodic.
3. Class 3: $\{4\}$, positive recurrent, aperiodic.

4.15 (i) Positive recurrent, (ii) Null recurrent, (iii) Positive recurrent.

4.17 Positive recurrent if $\sum_{n=0}^{\infty} np_n < \infty$, null recurrent if $\sum_{n=0}^{\infty} np_n = \infty$.

4.19 Use Pakes's lemma.

4.21 (a) Limiting distribution: $[.25 \quad .25 \quad .25 \quad .25]$. (b) Limiting occupancy distribution: $[\frac{1}{6} \quad \frac{1}{3} \quad \frac{1}{3} \quad \frac{1}{6}]$.

4.23 Limiting probability that the bus is full $= 0.6822$.

4.25 $\dfrac{m_i}{\sum_{j=1}^{k} m_j}$, where $m_i = 1/(1 - p_i)$.

4.27 (i) $\pi_j = \dfrac{2\alpha - 1}{\alpha} \left(\dfrac{1 - \alpha}{\alpha} \right)^j, \quad j \geq 0$.

4.29 $1/r$.

4.31 $\pi_n = \dfrac{\binom{N}{n}^2}{\sum_{j=0}^{N} \binom{N}{j}^2}, \quad 0 \leq n \leq N$.

4.33 $\pi_n = \sum_{i=n+1}^{\infty} \alpha_i / \tau, \quad n \geq 0$.

4.35 $\phi(z) = \prod_{n=0}^{\infty} \psi((1 - p)^n z + 1 - (1 - p)^n)$.

4.37 $\pi_0 \alpha_0 + \pi_{B-1} \alpha_2$, where π_j is the long run probability that the buffer has j bytes in it.

4.39 $\tau_1 / (\tau_1 + \tau_2)$.

4.41 $[.25, .5, .25]$.

4.43

$(a) \quad \begin{bmatrix} 4/11 & 7/11 & 0 & 0 \\ 4/11 & 7/11 & 0 & 0 \\ 4/11 & 7/11 & 0 & 0 \\ 4/11 & 7/11 & 0 & 0 \end{bmatrix},$

$$(b) \begin{bmatrix} 90/172 & 27/172 & 55/172 & 0 & 0 & 0 \\ 90/172 & 27/172 & 55/172 & 0 & 0 & 0 \\ 90/172 & 27/172 & 55/172 & 0 & 0 & 0 \\ 0 & 0 & 0 & 1 & 0 & 0 \\ 1170/3956 & 351/3956 & 715/3956 & 10/23 & 0 & 0 \\ 540/3956 & 162/3956 & 330/3956 & 17/23 & 0 & 0 \end{bmatrix}.$$

4.45 $(C_1 r + C_2(1-r))/T$, where $r = \sum_{i=1}^{K-1} p_i$, $T = \sum_{i=1}^{K-1} ip_i + K(1-r)$.

4.47 $d(1-2p) + cr/(1-r)$ where $r = (p/(1-p))^2$.

4.51 3.005 dollars per day.

4.53 Optimum $B = 12$, maximum probability = .9818.

4.57 Model 2 ranking: 1-2-3-4-5; Model 3 ranking: 2-1-3-4-5.

4.59

1. X_n = the number of items in the warehouse at the beginning of day n, after the demand for that day has been satisfied. Then

$$X_{n+1} = \min\{d, \max\{X_n + 1 - D_{n+1}, 0\}\}.$$

$\{X_n, n \geq 0\}$ is a DTMC with tr. pr. matrix

$$P = \begin{bmatrix} \beta_0 & \alpha_0 & 0 & 0 & \cdots & 0 \\ \beta_1 & \alpha_1 & \alpha_0 & 0 & \cdots & 0 \\ \beta_2 & \alpha_2 & \alpha_1 & \alpha_0 & \cdots & 0 \\ \vdots & \vdots & \vdots & \vdots & \vdots & \ddots \\ \beta_d & \alpha_{d-1} & \alpha_{d-2} & \alpha_{d-3} & \cdots & \alpha_1 + \alpha_0 \end{bmatrix}. \qquad (I.1)$$

2. Assume $\alpha_0 > 0$, and $\alpha_0 + \alpha_1 < 1$. The DTMC is irreducible and aperiodic.
3. $\pi_i = \rho^i(1-\rho)/(1-\rho^{d+1})$, $i \geq 0$, where $\rho = \alpha_0/\alpha_2$.
4. $L = \sum_{i=0}^{d} i\pi_i$.
5. $\alpha_1 \sum_{i=1}^{d} i\pi_i + \alpha_2(\pi_1 + \sum_{i=2}^{d} \pi_i(i + i - 1)/2)$.
6. π_d.

4.61

1. $\{X_n^i, n \geq 0\}$ is a DTMC with state space is $\{0, 1, 2, \cdots\}$ and

$$p_{m,m+1} = ap_i(1-d_i), \ m \geq 1, \quad p_{m,m-1} = (1-ap_i)d_i, \ m \geq 1,$$

$$p_{0,1} = ap_i, \quad p_{0,0} = 1 - ap_i, \quad p_{m,m} = 1 - p_{m,m+1} - p_{m,m-1}, \ m \geq 1.$$

It is positive recurrent when $ap_i < d_i$.
2. Not independent.

3. $p_1 = \frac{d_1\sqrt{d_1} - d_2\sqrt{d_2} + a\sqrt{d_2}}{\sqrt{d_1} + \sqrt{d_2}}$.

4.63

1. $X_{n+1} = \text{Bin}(X_n, 1 - p) + K, \quad n \geq 0$.

2. $m_n = \frac{K}{p}(1 - (1 - p)^n), \quad n \geq 0$.

3. $\frac{\alpha^2(r-c)p^2 - h\alpha(1-p)p}{(1-\alpha)(1-\alpha(1-p))}$.

4. $\frac{K}{p}((r - c)p - h(1 - p))$.

5. $\max(c - \sqrt{hc}, 0)$.

CONCEPTUAL EXERCISES

4.1 (i) Reflexive, symmetric, transitive.
(iii) Reflexive, symmetric, not transitive.

4.17 The results follows from the fact that $c(i)$ is the expected cost incurred at time n if $X_n = i$.

4.21 Global balance equations are obtained by summing the local equation over all j.

4.29 Suppose the DTMC earns one dollar every time it undergoes a transition from state i to j, and 0 otherwise. Then use Conceptual Exercise 4.18.

4.31 $g(i) = r_i + \sum_{j=1}^{\infty} p_{i,j} g(j), \quad i > 0$.

4.33

$$g(0) = 0, \quad g(s) = p_{s,t} + \sum_{j=1}^{\infty} p_{s,j} g(j),$$

$$g(i) = \sum_{j=1}^{\infty} p_{i,j} g(j), \quad i \neq s.$$

Chapter 5

COMPUTATIONAL EXERCISES

5.1 P(Length of the shortest path $> x$) $= \exp(-\lambda_3 x) \left[\frac{\lambda_2}{\lambda_2 - \lambda_1} e^{-\lambda_1 x} + \frac{\lambda_1}{\lambda_1 - \lambda_2} e^{-\lambda_2 x}\right]$.

5.3 P(Length of the longest path $\leq x$) $= \left(1 - \frac{\lambda_2}{\lambda_2 - \lambda_1} e^{-\lambda_1 x} - \frac{\lambda_1}{\lambda_1 - \lambda_2} e^{-\lambda_2 x}\right)(1 - e^{-\lambda_3 x})$.

5.5 $(1 - e^{-\lambda x})^n$.

5.7 $\frac{T}{1-e^{-\lambda T}} - \frac{1}{\lambda}$.

5.9 $\frac{1}{\lambda+\mu} 2p(1 + 3q^2)$ where $p = \mu/(\lambda + \mu)$ and $q = 1 - p$.

5.11 Process job 1 first if $C_1\lambda_1 > C_2\lambda_2$, otherwise process job 2 first.

5.17 $k = \min\{n \geq 1 : e^{-\lambda T} \sum_{i=0}^{n-1} \frac{(\lambda T)^i}{i!} \geq \alpha\}$.

5.25 $\frac{1}{2}(1 - e^{-2\lambda t})$.

5.27 $\max(e^{-\lambda t}, e^{-\lambda \tau})$.

5.33 (i) e^{-1}, (ii) $e^{-1}/2, 1 - 2.5e^{-1}$

5.35 6.

5.37 $N(t) \sim P(\Lambda(t))$ where $\Lambda(t) = \int_0^t \lambda(u)du = \begin{cases} c(t - k) & \text{if } 2k \leq t < 2k + 1 \\ c(k + 1) & \text{if } 2k + 1 \leq t < 2k + 2. \end{cases}$

5.39 $R(t) \sim P(\int_0^t \lambda(u)(1 - G(t - u))du)$.

5.41 $E(Z(t)) = \lambda\tau$ and $\text{Var}(Z(t)) = \lambda s^2 t$, where

$$\tau = \frac{\lambda_d}{\lambda_d + \lambda_w}\tau_d + \frac{\lambda_w}{\lambda_d + \lambda_w}\tau_w,$$

$$s^2 = \frac{\lambda_d}{\lambda_d + \lambda_w}(\tau_d^2 + \sigma_d^2) + \frac{\lambda_w}{\lambda_d + \lambda_w}(\tau_w^2 + \sigma_w^2).$$

5.43 2. Let $R(t) = \int_0^t r(u)du$. $E(C(t)) = c(R(t) + R(t)^2/2)$.
3. $E(C(t)) = 3ct^2/2..$

5.45 Mean $= 200 \cdot 350(1.08)^n(1 + e^{-n}(1 - e^{-1}))$,
Variance $= 200 \cdot 350^2(1.08)^{2n}(1 + e^{-n}(1 - e^{-1}))$.

CONCEPTUAL EXERCISES

5.1 Let $H(x) = P(X > x)$. Definition of hazard rate implies $H'(x) = -r(x)H(x)$. The solution is Equation 5.4.

5.9 A_i is a modified geometric ($\frac{\lambda_i}{\lambda_1+\lambda_2}$) random variable.

5.13 (i) No, (ii) No, (iii) No.

5.21 It is not an NPP since it does not have independent increments property.

5.23 $e^{-\lambda\pi x^2}$.

5.25 $e^{-(\Lambda(s+t)-\Lambda(s))}$.

Chapter 6

MODELING EXERCISES

6.1 State space = $\{0, 1, ..., k\}$.

$$q_{i,i+1} = (k - i)\lambda, \ \ 0 \leq i \leq k - 1,$$
$$q_{i,i-1} = i\mu, \ \ 1 \leq i \leq k.$$

6.3 State space = $\{0, 1, 2, 12, 21\}$. The state represents the queue of failed machines.

$$Q = \begin{bmatrix} -(\mu_1 + \mu_2) & \mu_1 & \mu_2 & 0 & 0 \\ \lambda_1 & -(\lambda_1 + \mu_2) & 0 & \mu_2 & 0 \\ \lambda_2 & -(\lambda_2 + \mu_1) & 0 & 0 & \mu_1 \\ 0 & 0 & \lambda_1 & -\lambda_1 & 0 \\ 0 & \lambda_2 & 0 & 0 & -\lambda_2 \end{bmatrix}.$$

6.5 Pure death process on $\{0, 1, ..., k\}$ with

$$\mu_i = \mu M, \ \ 1 \leq i \leq k.$$

6.7 The state space = $\{0, 1A, 1B, 2, 3, 4, ...\}$. State 1A (1B) = one customer in the system and he is being served by server A (B). State $i = i$ customers in the system.

$$q_{0,1A} = \lambda\alpha, \ \ q_{0,1B} = \lambda(1 - \alpha),$$
$$q_{1A,0} = \mu_1, \ \ q_{1A,2} = \lambda,$$
$$q_{1B,0} = \mu_2, \ \ q_{1B,2} = \lambda,$$
$$q_{2,1A} = \mu_2, \ \ q_{2,1B} = \mu_1, \ \ q_{2,3} = \lambda,$$
$$q_{i,i+1} = \lambda, \ \ q_{i,i-1} = \mu_1 + \mu_2, \ \ i \geq 3.$$

6.9 $q_{0,i} = \lambda_i, \ \ q_{i,0} = \mu_i, \ \ 1 \leq i \leq n.$

6.11 $\lambda_i = \begin{cases} \lambda_1 + \lambda_2 & \text{if } 0 \leq i < s \\ \lambda_1 & \text{if } i \geq s, \end{cases}$, $\mu_i = \min(i, s)\mu, \ \ i \geq 0.$

6.13 $q_{(0,0),(1,j)} = \lambda\alpha_j, \ \ q_{(1,j),(0,0)} = \mu_j, \ \ j = 1, 2,$
$$q_{(i,j),(i+1,j)} = \lambda, \ \ i \geq 1, \ \ j = 1, 2,$$
$$q_{(i,j),(i-1,k)} = \mu_j\alpha_k, \ \ i \geq 2, \ \ j, k = 1, 2.$$

6.15 $q_{i,i+1} = .4i\lambda, \ \ q_{i,i+2} = .3i\lambda, \ \ q_{i,i-1} = .3i\mu, \ \ i \geq 0.$

6.17 State space = $\{0, 1, 2, 3, 4, 5\}, 0 = $ failed, $i = i$ CPUs working.

$$q_{i,0} = i\mu(1 - c), \ \ q_{i,i-1} = i\mu c, \ \ 2 \leq i \leq 5, \ \ q_{1,0} = \mu.$$

6.19 Let $X_i(t)$ be the number of customers of type i in the system at time t. $\{(X_1(t), X_2(t)), t \geq 0\}$ is a CTMC on $S = \{(i,j) : i \geq 0, \ 0 \leq j \leq s\}$ with transition rates

$$q_{(i,j),(i+1,j)} = \lambda_1, \ (i,j) \in S, \ q_{(i,j),(i,j+1)} = \lambda_2, \ (i,j) \in S, j < s,$$

$$q_{(i,j),(i-1,j)} = \min(i, s-j)\mu_1, \ q_{(i,j),(i,j-1)} = j\mu_2, \ (i,j) \in S.$$

6.21 $Y(t)$ = the number of packets in the buffer, $Z(t)$ = the number of tokens in the token pool at time t. $X(t) = M - Z(t) + Y(t)$. $\{X(t), t \geq 0\}$ is a birth and death process with birth rates $\lambda_i = \lambda, \ i \geq 0$ and death rates $\mu_0 = 0, \mu_i = \mu, \ i \geq 1$.

6.23 $q_{2,1} = \mu, \ q_{i,i-1} = 2\mu, \ i \geq 3,$

$q_{i,i+1} = 2\lambda, \ q_{i,iT} = 2\theta, \ i \geq 2,$

$q_{1T,1} = \mu, \ q_{iT,(i-1)T} = \mu, \ i \geq 2,$

$q_{iT,TiT} = \theta, \ i \geq 1.$

6.25 $\{X_k(t), t \geq 0\} \ k = 1,2$ are two independent birth and death processes on $\{0, 1, \cdots, \}$ with birth parameters $\lambda_i = \lambda p_k$ for $i \geq 0$, and death parameters μ_k for $i \geq 1$.

6.27 $q_{0,k} = \lambda_k, \ q_{k,0} = \mu_k \ 1 \leq k \leq K.$

6.29 State space = $\{0, 1, 2\}$. Transition rates:

$$q_{0,1} = 2\lambda, \ q_{1,0} = \mu, \ q_{1,2} = 2\lambda, q_{2,0} = 2\mu.$$

6.31 $\lambda_i = \lambda_1 + \lambda_2$ for $0 \leq i \leq K - 1, \lambda_i = \lambda_1$ for $i \geq K, \mu_i = \mu$ for $i \geq 1$.

6.33 Denote a space as E if it is empty, B if it is occupied by a car in service, and W if it is occupied by a car that is waiting to begin service or has finished service. The state space is $S = \{1 = EEE, 2 = BEE, 3 = BBE, 4 = EBE, 5 = BBW, 6 = BWE, 7 = EBW, 8 = BWW\}$. The rate matrix is:

$$Q = \begin{bmatrix} -\lambda & \lambda & 0 & 0 & 0 & 0 & 0 & 0 \\ \mu & -(\lambda+\mu) & \lambda & 0 & 0 & 0 & 0 & 0 \\ 0 & 0 & -(\lambda+2\mu) & \mu & \lambda & \mu & 0 & 0 \\ \mu & 0 & 0 & -(\lambda+\mu) & 0 & 0 & \lambda & 0 \\ 0 & 0 & 0 & 0 & -2\mu & 0 & \mu & \mu \\ \mu & 0 & 0 & 0 & 0 & -(\lambda+\mu) & 0 & \lambda \\ 0 & \mu & 0 & 0 & 0 & 0 & -\mu & 0 \\ 0 & \mu & 0 & 0 & 0 & 0 & 0 & -\mu \end{bmatrix}.$$

6.35 State space $\{0, 1, 2, \cdots, N\}$. Rates

$$q_{i,i+1} = \beta i(N - i), \ 0 \leq i \leq N.$$

6.37 $q_{(i,j),(i-1,j+1)} = \beta ij, \ 0 \le i \le N, \ 0 \le j \le N - i,$
$q_{(i,j),(i,j-1)} = (\gamma + \nu)j, \ 0 \le i \le N, \ 0 \le j \le N - i.$

6.39 State-space $\{0, 1, 2, 3, \cdots\}$. Rates

$$q_{i,i+1} = \lambda, \ i \ge 0; \quad q_{i,i-1} = i\theta + \mu, \ i \ge 1.$$

6.41 State space $S = \{(i, j) : 0 \le i \le B, 0 \le j \le \min(B - b, B - i)\}$. Rates

$$q_{(i,j)=(i+1,j)} = \lambda_1, \ q_{(i,j),(i-1,j)} = \mu_1, \ q_{(i,j),(i,j+1)} = \lambda_2, \ q_{(i,j),(i,j-1)} = \mu_2,$$

for $(i, j) \in S$.

COMPUTATIONAL EXERCISES

6.1 $\frac{\alpha}{\alpha+\beta}(1 - e^{-(\alpha+\beta)t})$.

6.3 Let $\alpha_i(t) = \frac{\lambda_i}{\lambda_i+\mu_i} + \frac{\mu_i}{\lambda_i+\mu_i}e^{-(\lambda_i+\mu_i)t}$. Then

$$P(X(t) = 0|X(0) = 2) = (1 - \alpha_1(t))(1 - \alpha_2(t)),$$

$$P(X(t) = 1|X(0) = 2) = \alpha_1(t)(1 - \alpha_2(t)) + \alpha_2(t)(1 - \alpha_1(t)),$$

$$P(X(t) = 2|X(0) = 2) = \alpha_1(t)\alpha_2(t).$$

6.5 $P(X(t) = j|X(0) = i) = \binom{i}{j}e^{-j\mu t}(1 - e^{-\mu t})^{i-j}$.

6.7 $\frac{2\alpha\beta}{\alpha+\beta}t + \frac{\alpha(\alpha-\beta)}{(\alpha+\beta)^2}(1 - e^{-(\alpha+\beta)t})$.

6.9 $\frac{\lambda(e^{-16\mu} - e^{-(8\lambda+24\mu)})}{(\lambda+\mu)(1-e^{-(8\lambda+24\mu)})}$.

6.11 Stable iff $\lambda < 5\mu$.

$$\rho_i = \begin{cases} (\frac{\lambda}{\mu})^i & \text{for } 0 \le i \le 5 \\ (\frac{1}{2^{i-5}}\frac{\lambda}{\mu})^i & \text{for } 6 \le i \le 8 \\ (\frac{1}{8 \cdot 3^{i-8}}\frac{\lambda}{\mu})^i & \text{for } 9 \le i \le 12 \\ (\frac{1}{648 \cdot 4^{i-12}}\frac{\lambda}{\mu})^i & \text{for } 13 \le i \le 15 \\ (\frac{1}{41472 \cdot 5^{i-15}}\frac{\lambda}{\mu})^i & \text{for } i \ge 16. \end{cases}$$

$$p_i = \rho_i p_0, \ i \ge 0.$$

6.13 $p_0 = \frac{1}{1+\sum_{i=1}^{n}\frac{\lambda_i}{\mu_i}}, \quad p_i = \frac{\frac{\lambda_i}{\mu_i}}{1+\sum_{i=1}^{n}\frac{\lambda_i}{\mu_i}}, \quad i = 1, 2, ..., n.$

6.15 $G(z) = \left(\frac{1-p}{1-pz}\right)^{\frac{\lambda}{\mu p}}$.

6.19 Long run probability that the kth space is occupied $= \rho\frac{\sum_{i=0}^{k-1}\rho^i/i!}{\sum_{i=0}^{k}\rho^i/i!} - \rho\frac{\sum_{i=0}^{k-2}\rho^i/i!}{\sum_{i=0}^{k-1}\rho^i/i!}$.

Long run fraction of the customers lost $= \frac{\rho^K/K!}{\sum_{i=0}^{K}\rho^i/i!}$.

6.21 The system is stable if $\alpha > 0$. State R = computer under repair. State j = computer is functioning and there are j jobs in the system. The limiting distribution:

$$p_R = \frac{\theta}{\theta + \alpha}$$
$$p_j = \frac{\alpha}{\theta + \alpha}(1 - b)b^j, \quad j \geq 0$$

where

$$b = \frac{1 + \frac{\lambda}{\mu} + \frac{\theta}{\mu} - \sqrt{(1 + \frac{\lambda}{\mu} + \frac{\theta}{\mu})^2 - 4\frac{\lambda}{\mu}}}{2}.$$

Long run fraction of job that are completed successfully $= \frac{\mu}{\lambda}(1 - p_R - p_0)$.

6.23 $p_0 = (1 + \sum_{k=1}^{K} \frac{\lambda_k}{\mu_k})^{-1}$, $p_k = \frac{\lambda_k}{\mu_k}p_0$, $1 \leq k \leq K$.

6.25 Let $\alpha_i = \lambda_i/\mu, i = 1, 2$

$$\rho_i = (\alpha_1 + \alpha_2)^i, \quad 0 \leq i \leq K,$$
$$\rho_i = (\alpha_1 + \alpha_2)^K \alpha_1^{i-K}, \quad i > K.$$

The system is stable if $\lambda_1 < \mu$. The limiting distribution is given by

$$p_j = \rho_j / \sum_{i=0}^{\infty} \rho_i, \quad j \geq 0.$$

6.27 Condition of stability: $\lambda < \mu$. Expected number in the system $= \frac{3\rho}{4(1-\rho)} + \rho/4$.

6.29 Idle time fraction $= \frac{2\lambda\mu^2(\mu+\theta+2\lambda)}{8\lambda^4+4\lambda^3\theta+8\lambda^3\mu+6\lambda^2\mu\theta+4\lambda\mu^2\theta+2\lambda\mu^3+\mu^3\theta+4\lambda^2\mu^2}$.

6.31 $p_i = 1/N$, $1 \leq i \leq N$.

6.35 $\frac{2\lambda+\mu}{\lambda(4\lambda+\mu)}$.

6.39 $\frac{1}{\lambda}\frac{1-(\lambda/\mu)^{K+1}}{1-\lambda/\mu}$.

6.43 Optimal $\lambda = .065$.

6.45 $\frac{kr(\alpha+\lambda)}{\alpha(\alpha+\lambda+\mu)}$.

6.47 $\frac{r}{\mu}\frac{1-c^5}{1-c}$.

6.49 $T = .74$ approximately.

6.51 Optimal $p = \frac{\sqrt{1+\rho}}{\sqrt{1+\rho}+\sqrt{\rho}}$. Fraction joining $= 1/\sqrt{1+\rho}$.

6.53 Use solution to Computational Exercise 6.25. The rate of revenue is $\lambda_1 c_1 + \lambda_2 c_2 \sum_{j=0}^{K-1} p_j$.

6.55 $p(\phi) = [1 + 3\frac{\lambda}{\mu} + 3(\frac{\lambda}{\mu})^2 + (\frac{\lambda}{\mu})^3 + \frac{\theta}{\mu}]^{-1}$.

6.57 Yes. $p_i \propto d_i/q_i$ where d_i is the number of neighbors of node i.

6.61 $\prod_{j=1}^{K} \frac{j\gamma}{j\gamma+j(N-K)\beta}$.

6.63 $p_{i,j} = \frac{\gamma+\nu}{\gamma+\nu+\beta i} p_{i,j-1} + \frac{\beta i}{(\gamma+\nu)+\beta i} p_{i-1,j+1}$
for $k \leq i \leq N-1, k \leq i+j \leq N, j \geq 1$ with boundary conditions $p_{k,j} = \alpha^j, \geq 0$.

6.65 $T = 251$ days.

6.67 Let $[p_i]$ be the limiting distribution of a birth and death process with rates given in Equations 6.13 and 6.14. Let

$$A = \sum_{i=1}^{\infty} \lambda_i^b p_i + \sum_{i=-\infty}^{-1} \lambda_i^s p_i,$$

$$B = \sum_{i=-\infty}^{\infty} \lambda_i^b p_i.$$

Answer $= A/B$.

6.69 Let

$$p_0 = 1, \quad \rho_i = \frac{\lambda^i}{\mu(\mu+\theta)\cdots(\mu+(i-1)\theta)}, \quad i \geq 1.$$

$$p_j = \rho_j / \sum_{j=0}^{\infty} \rho_i, \quad j \geq 0.$$

6.71 $\sum_{j=1}^{J}(\lambda_j/\theta_j) \sum_{k=1}^{I} \mu_k \beta_{k,j} \alpha_{k,j}$.

6.73 $p_{i,i}(t) = e^{-\lambda t \alpha^i}, \quad p_{i,j}(t) = \alpha^{j-1}(1-\alpha)(1-e^{-\lambda t \alpha^j})/(1-\alpha^j), \quad j > i$.

6.75 2. $p_i = \frac{1}{K+\rho}, \quad 0 \leq i \leq K-1, \quad p_K = \frac{\rho}{K+\rho}$. 3. $L_q = \sum_{i=1}^{K-1} i p_i = \frac{K(K-1)}{2(K+\rho)}$.
4. $L = L_q + K p_K$.

6.79 1. $X(t)$ be the number of passengers at the bus depot at time t. $\{X(t), t \geq 0\}$ is a CTMC with rates

$$q_{i,i+1} = \lambda, \quad i \geq 0, q_{i,0} = \mu, \quad i \geq k.$$

2. $\rho = \lambda/(\lambda+\mu), p_i = p_0, \quad 1 \leq i \leq k-1, p_i = \rho^{i-k+1} p_0, \quad i \geq k, p_0 = (k + \frac{1}{1-\rho})^{-1}$. 3. $(k-1)/\lambda + 1/\mu$.

CONCEPTUAL EXERCISES

6.1 Let B be the submatrix of Q obtained by deleting row and column for the state N. Then the matrix $M = [M_{i,j}]$ satisfies $BM = -I$.

6.3 Let $m_k = E(T|X(0) = k)$. Then

$$\sum_{n \in S} q_{kn} m_n = -1, \quad k \neq i, \quad \sum_{n \in S - \{j\}} q_{in} m_n = -1.$$

6.9 Let $\{Y(t), t \geq 0\}$ have transition rates $q'_{i,j} = q_{i,j}/r(i)$.

6.11

$$E(e^{-sT}|X(0) = i, S_1 = y, X(y) = j) = \begin{cases} e^{-sx/r_i} & \text{if } y > x/r_i \\ e^{-sy}\phi_j(s, x - yr_i) & \text{if } y \leq x/r_i. \end{cases}$$

The result follows by unconditioning.

Chapter 7

MODELING EXERCISES

7.1 $X(t)$ = number of waiting customers at time t. $\{X(t), t \geq 0\}$ is the queue length process in an $M/M/1$ queue with arrival rate λ and service rate μ.

7.3 $\lambda_i = \lambda$, $i \geq 0$, $\mu_i = \mu, 1 \leq i \leq 3$, $2\mu, 4 \leq i \leq 9$, $3\mu, i \geq 10$.

7.5 $\lambda_i = \lambda$, $i \geq 0$, $\mu_i = \mu_1, 1 \leq i \leq 3$, $\mu_2, i \geq 4$.

7.7 Arrival process is P(λ) where $\lambda = \sum_{i=1}^{k} \lambda_i$. Service times are iid with common cdf $\sum_{i=1}^{k} (\lambda_i/\lambda)(1 - e^{-\mu_i x})$.

7.9 Interarrival times are deterministic (equal to 1), and the service times are iid $\exp(\mu)$.

7.13 $\{X(t), t \geq 0\}$ not a CTMC. $\{X_n, n \geq 0\}$ is a DTMC of the type given in Computational Exercise 4.24.

7.15 $\frac{\lambda(\lambda+s)}{(\lambda+s)(\lambda+\mu+s)-\lambda\mu}$.

7.17 $X_1 + X_2$ is an $M/M/1$ queue with arrival rate λ and service rate $\mu_1 + \mu_2$.

COMPUTATIONAL EXERCISES

7.3 $W^q = \frac{1}{\mu} \frac{\rho}{1-\rho}$.

7.5 $\frac{i}{1-\rho}$.

7.9 P_K from Equation 7.19.

7.11 $L/(\lambda(1 - p_K))$.

7.13 $\lambda a(1 - p_K) - cL$.

7.23 $1 - \rho$.

7.27 $M/M/1$ queue with $\rho = \lambda/\alpha\mu$.

7.31 See answer to Computational Exercise 6.25. An arriving customer of either type, and an entering customer of type 1, sees j people with probability p_j, $j \geq 0$. An entering customer of type 2 see j people with probability $p_j / \sum_{k=0}^{K-1} p_k$, $0 \leq j \leq K - 1$.

7.33 1. $\sum_{i=1}^{N} \frac{\lambda}{(1-p)\mu_i - \lambda}$,   2. $\prod_{i=1}^{N} \left(1 - \frac{\lambda}{(1-p)\mu_i}\right)$.

7.37 $\phi_1(n) = \begin{cases} (1/\mu)^n & \text{if } 1 \leq n \leq 5, \\ (1/\mu)^5(1/2\mu)^{n-5} & \text{if } n \geq 6, \end{cases}$

$\phi_2(n) = \begin{cases} (1/\mu)^n & \text{if } 1 \leq n \leq 2, \\ (1/\mu)^2(1/2\mu)^{n-2} & \text{if } 3 \leq n \leq 10, \\ (1/\mu)^2(1/2\mu)^8(1/3\mu)^{n-10} & \text{if } n \geq 11. \end{cases}$

The network is always stable.

$$L = \sum_{n=0}^{\infty} n \frac{\lambda^n}{n!} \sum_{n_1+n_2=0}^{n} \phi_1(n_1)\phi_2(n_2).$$

7.39 $a_1 = \lambda_1 + a_N p_N$, $a_i = \lambda_i + a_{i-1}p_{i-1}$, $2 \leq i \leq N$.
Solution:

$$a_N = \left(\sum_{i=1}^{N} \lambda_i \prod_{j=i}^{N-1} p_j\right) / \left(1 - \prod_{j=1}^{N} p_j\right).$$

Other a_i's are obtained by symmetry.

7.45 Condition of stability: $\rho = \sum_{i=1}^{k} \frac{\lambda_i}{\mu_i} < 1$. Expected number in steady state $L = \rho + \frac{\lambda}{1-\rho} \sum_{i=1}^{k} \frac{\lambda_i}{\mu_i^2}$.

7.47 .9562.

7.53 Condition of stability: $\lambda < 2\mu$. The second system has smaller L.

7.59 $X(t)$ and X_n have the same limiting distribution, which differs from that of $\bar{X}_n$.

7.63 Stability condition: $\frac{\rho}{1+\rho} < \theta$. The solution to Equation 7.47:

$$\alpha = \frac{1}{2}\left((2\lambda + \mu + \theta) - \sqrt{(2\lambda + \mu + \theta)^2 - 4(\theta + \lambda)}\right).$$

Use Theorem 7.17.

7.67 1093 servers. .0013 minutes.

7.69 992, 10158.

7.73 Stability condition: $\rho = \theta/(d\mu) < 1$. Let $\alpha \in (0,1)$ be the solution to

$$\alpha = \frac{e^{-\mu(1-\alpha)d}\theta}{1 - e^{-\mu(1-\alpha)d}(1-\theta)}.$$

The limiting queue length distribution is $p_0 = 1 - \rho$, $p_j = \rho\alpha^{j-1}(1-\alpha)$, $j \geq 0$.

7.75 1. $a_i = \mu_1\Pi_{k=1}^{i-1}\alpha_k$, $\rho_i = \lambda_i/\mu_i$, $2 \leq i \leq n$. Condition of stability: $\rho_i < 1$ for $2 \leq i \leq n$. $L_i = \frac{\rho_i}{1-\rho_i}$, $i \geq 2$.

Let $a_1 = \mu_1$ and $\alpha_n = 0$. The long run net profit $= \sum_{i=1}^{n} d_i a_i(1-\alpha_i) - \sum_{i=2}^{n} \frac{h_i\rho_i}{1-\rho_i}$.

7.77 3. $L_1 = \rho_1/(1-\rho_1)$, $L_2 = \rho_2/((1-\rho_1)(1-\rho_1-\rho_2))$, where $\rho_i = \lambda_i/\mu$.

Chapter 8

COMPUTATIONAL EXERCISES

8.1 $\{S_n, n \geq 0\}$ is a renewal sequence in the $M/M/1/K$ and $G/M/1/K$ system, but not in the $M/G/1/K$ system.

8.3 Recurrent.

8.5 $P\{N(t) = k\} = e^{-\lambda t}\frac{(\lambda t)^{2k}}{(2k)!} + e^{-\lambda t}\frac{(\lambda t)^{2k+1}}{(2k+1)!}$.

8.7 $P(N(n) = k) = \binom{n}{k}(1-\alpha)^k\alpha^{n-k}$.

8.9 $p_k(n) = .2p_{k-1}(n) + .3p_{k-1}(n-1) + .5p_{k-1}(n-2)$, $k \geq 1, n \geq 2$.

8.11 1. 6/17, 2. 4/17, 3. 2/17.

8.13 $\tau = m_0 = 1 + \frac{1-\alpha}{1-\beta}$, $s^2 = 2 + \alpha - \beta + 2/(1-\beta)$, $N(n) \sim N(t/\tau, (s^2-\tau^2)t/\tau^3)$.

8.15 $M(t) = \frac{\lambda_1\lambda_2}{(1-r)\lambda_1+r\lambda_2}t + \frac{r(1-r)(\lambda_1-\lambda_2)^2}{((1-r)\lambda_1+r\lambda_2)^2}(1 - e^{-((1-r)\lambda_1+r\lambda_2)t})$.

8.17 $M(t) = \frac{\lambda\mu}{\lambda+\mu}t - \frac{\lambda\mu}{(\lambda+\mu)^2}(1 - e^{-(\lambda+\mu)t})$.

8.19 $M^*(t) = pM(t)$.

8.21 $H(t) = D(t) + \int_0^t H(t-u)dG(u)$ where $D(t) = \begin{cases} 1 - G(t) & t \leq x \\ 0 & \text{otherwise} \end{cases}$.

$\lim_{t\to\infty} H(t) = \frac{1}{\tau}\int_0^x(1 - G(t))dt$.

8.23 $p(t) = G(t) - \int_0^t p(t-x)dG(x)$. Not a renewal equation. For a PP(λ), $p(t) = \frac{1}{2}(1 - e^{-2\lambda t})$.

8.25 $\tilde{M}_k(s) = k! \left[\frac{\tilde{G}(s)}{1-\tilde{G}(s)}\right]^k$.

8.27 $H(t) = D(t) + \int_0^t H(t-u)\,dG(u)$ where $D(t) = \int_t^\infty u^k\,dG(u)$. $\lim_{t\to\infty} E(C(t)^k) = \frac{E(X^{k+1})}{\tau}$.

8.31 $\frac{\tau}{1-\rho}$.

8.33 μ_i/μ.

8.37 $a\min(1, \frac{C_2}{C_1}[\sqrt{1+\frac{2C_1}{C_2}} - 1])$.

8.39 1. SMP, 2. Not an SMP, 3. Not an SMP.

8.41 Let $X(t) = -1$ if the machine is under repair at time t. If the machine is up at time t, let $X(t)$ be the cumulative damage at time t (since the last repair). Then $\{X(t), t \geq 0\}$ is an SMP with state space $\{-1, 0, 1, 2, \cdots, K\}$. The non-zero entries of the kernel $G(x) = [G_{i,j}(x)]$ are given by

$$G_{-1,0}(x) = A(x),$$

$$G_{i,j}(x) = (1 - \exp(-\lambda x))\alpha_{j-i}, \quad 0 \leq i < j \leq K,$$

$$G_{i,0}(x) = (1 - \exp(-\lambda x))\sum_{j=K+1-i}^{\infty} \alpha_j, \quad 0 \leq i \leq K.$$

8.45 $M/G/1/1:\ \frac{\lambda\tau}{1+\lambda\tau}, G/M/1/1:\ \tilde{G}(\mu)$.

8.47 Let $\lambda_4 = 0$, $\alpha_i = (\lambda_i^2 + 2\lambda_i\mu_i)/(\lambda_i+\mu_i)^2$, $\rho_j := \prod_{i=1}^{j-1}\alpha_i$, $\pi_j = \rho_j/\sum_{i=1}^4 \rho_i$, $\tau_j = (\lambda_i + 2\mu_i)/(\lambda_i + \mu_i)^2$, $i = 1, 2, 3, 4$. Then

$$p_j = \rho_j\tau_j/\sum_{i=1}^4 \rho_i\tau_i.$$

8.51 3. Space 6 is occupied for 76.24% of the time in steady state.

8.55 1. Yes only if F is exponential. 2. No, unless F is exponential. 3. Yes, 4. Yes.

8.57 2. $\frac{1}{1+\lambda a e^{-\lambda b}}$

8.59 1. $(r\lambda \int_0^T (1 - G(u))du - K)/T$. 2. $\min(\sqrt{2aK/(\lambda r)}, a)$.

8.61 1. 1.26 per hour. 2. 1.28 per hour.

8.63 1. .0125. 2. $265/day.

8.65 Condition on the time of the second event, with cdf G. Then

$$H(t) = e^{-\lambda t} + \int_{u=0}^{t} H(t-u)dG(u).$$

Solution: $H(t) = \frac{1}{2}(1 + e^{-2\lambda t}), \; t \geq 0.$

8.67 2. .2172, 3. 94.58 days, 4. 335.31 \$/year, 5. 771.81 \$/year.

CONCEPTUAL EXERCISES

8.1 Yes if $X_0 = i$ with probability 1 for some fixed i, and no otherwise.

8.3 N_1 is an RP, N_2 is a delayed RP.

8.13 N_1 is a RP, N_0 is not a RP.

Chapter 9

MODELING EXERCISES

9.1 Let $p_{i,j} = \prod_{r=i+1}^{j+1} \frac{\lambda_r}{\lambda_r + \mu_r}, \; j \geq i-1, A_i(x) = 1 - e^{-(\lambda_i + \mu_i)x}$. Then $G_{i,j}(x) = p_{i,j} \cdot A_i * A_{i+1} * \cdots * A_j * A_{j+1}(x).$

9.3 $G_{i,j}(x)$ is as given in Example 9.5 for $0 \leq j \leq i+1 < K$. For $i = K, 0 \leq j \leq K, G_{K,j} = G_{K-1,j}$.

9.5 $G_{i,j}(x) = \int_0^x \binom{N-i}{j+1-i}(1 - e^{-\mu t})^{j+1-i}e^{-(N-j-1)\mu t}dG(t), \; N > j \geq i-1 \geq 0.$
$G_{0j}(x) = \int_0^x N\mu e^{-N\mu y}G_{1j}(x-y)dy, \; N > j \geq 0.$

9.7 $A = [h_{i,j}], \; 0 \leq i, j < k, B = [h_{i,j}], \; 0 \leq i < k \leq j \leq N, C = [h_{i,j}], \; 0 \leq j < k \leq i \leq N$. Then $G_{i,j}^n = [A^{n-2}BC]_{i,j}, \; n \geq 2.$

9.9 Use $S_n = n$ and $X_n = X(S_n+)$.

9.11 Use S_n time of departure of the nth server, and $X_n = X(S_n+)$.

COMPUTATIONAL EXERCISES

9.1

$$F_{i,j}(t) = \begin{cases} H_i * H_{i+1} * \cdots * H_{j-2} * H_{j-1}(t) & 1 \leq i < j \leq N \\ H_i * H_{i+1} * \cdots * H_N * H_1 * \cdots * H_{j-2} * H_{j-1}(t) & 1 \leq j \leq i \leq N \end{cases}$$

$$M_{i,j}(s) = \frac{\tilde{F}_{i,j}(s)}{1 - \tilde{F}_{i,i}(s)}.$$

9.3 $p_1 = \frac{2(1-\tilde{A}(\mu))}{2\mu\tau+\tilde{A}(\mu)}$.

9.5 (1). $\frac{\rho^{j+1}}{(j+1)!}\left(\frac{e^{-\rho}}{1-e^{-\rho}}\right)$, (2). $(1-e^{-\rho})/\rho$.

9.7 Long run fraction of the time the MRGP spends in state j =
$\frac{1}{\lambda}\left[1 - e^{-\lambda}\sum_{k=0}^{r-j}(\lambda^k/k!)\right]$.

9.11 $\lim_{n\to\infty} E(X_n) = \frac{\lambda(1-\alpha)\tau}{\alpha}$, $\lim_{t\to\infty} E(X(t)) = \frac{\lambda(1-\alpha)\tau}{\alpha} + \frac{\lambda s^2}{2\tau}$.

9.13

$$H(t) = \begin{bmatrix} \frac{\beta\theta_0}{\alpha+\beta} & \frac{\alpha\theta_1}{\alpha+\beta} \\ \frac{\beta\theta_0}{\alpha+\beta} & \frac{\alpha\theta_1}{\alpha+\beta} \end{bmatrix} t + \begin{bmatrix} \frac{\alpha\theta_0}{(\alpha+\beta)^2} & -\frac{\alpha\theta_1}{(\alpha+\beta)^2} \\ -\frac{\beta\theta_0}{(\alpha+\beta)^2} & \frac{\beta\theta_1}{(\alpha+\beta)^2} \end{bmatrix}(1-e^{-(\alpha+\beta)t}).$$

9.15 Let $A_i(x) = \int_0^x e^{-i\mu u}\,dA(u)$. Then $\{X(t), t \ge 0\}$ is an MRGP with embedded MRS $\{(X_n, S_n), n \ge 0\}$ with kernel G as given below.

$$G(x) = \begin{bmatrix} A_0(x) - A_1(x) & A_1(x) & 0 \\ A_0(x) - 2A_1(x) + A_2(x) & 2(A_1(x) - A_2(x)) & A_2(x) \\ A_0(x) - 2A_1(x) + A_2(x) & 2(A_1(x) - A_2(x)) & A_2(x) \end{bmatrix}.$$

The transition probability matrix of $\{X_n, n \ge 0\}$ is as given below. Let $A_i = A_i(\infty)$. Then

$$P = G(\infty) = \begin{bmatrix} 1 - A_1 & A_1 & 0 \\ 1 - 2A_1 + A_2 & 2(A_1 - A_2) & A_2 \\ 1 - 2A_1 + A_2 & 2(A_1 - A_2) & A_2 \end{bmatrix}.$$

$\tau_{i,j}$ = the expected time spent by the $\{X(t), t \ge 0\}$ process in state j over $(S_n, S_{n+1}]$ given that $X_{=i}$. Let a be the expected inter-arrival time, and $B_i = (1 - A_i)/(i\mu)$, $i = 1, 2$. Then $\tau = [\tau_{i,j}] =$

$$\begin{bmatrix} a - B_1 & B_1 & 0 \\ a + B_2 - 2B_1 & 2(B_1 - B_2) & B_2 \\ a + B_2 - 2B_1 & 2(B_1 - B_2) & B_2 \end{bmatrix}.$$

The limiting distribution of $X(t)$ as $t \to \infty$ is given by:

$$p_j = \lim_{t\to\infty} P(X(t) = j) = \frac{1}{a}\left(\sum_{i=0}^{2} \pi_i \tau_{i,j}\right),$$

where π is a solution to $\pi = \pi P$.

9.17 The long run cost rate us

$$\frac{w\rho}{1-\alpha} + h(1-\rho),$$

where ρ and α are from the solution to Computational Exercise 7.73.

Chapter 10

COMPUTATIONAL EXERCISES

10.1 For $0 < s < t$, $\mathsf{E}(B(s)B(t)) =$
$\mathsf{E}(B(s)(B(t) - B(s) + B(s))) = \mathsf{E}(B(s)(B(t) - B(s))) + \mathsf{E}(B(s)^2) = s$.

10.3
$$f(x_1, x_2) = \frac{1}{2\pi\sigma^2\sqrt{t_1(t_2 - t_1)}} \times$$

$$\exp\left(-\frac{1}{2\sigma^2 t_1(t_2 - t_1)}[(x_1 - \mu_1 t_1)^2 t_2 - 2(x_1 - \mu_1 t_1)(x_2 - \mu_2 t_2) + (x_2 - \mu_2 t_2)^2 t_1]\right).$$

10.5 $N(sy/t, \sigma^2 s(t - s)/t)$.

10.13 $a(t) = -3\mu t$, $b(t) = 3(\mu t)^2 - 3\sigma^2 t$, $c(t) = 3\sigma^2\mu t - (\mu t)^3$.

10.19 Limiting mean = 0, and limiting variance = $b^2/6$.

10.25 $Y(t) = (B(t) + Y(0))/(1 + t)$.

10.27 1. $\frac{40}{u+d}$. 2. 1/3 3. $\frac{c-d+c+u}{2} + \frac{(2e^{u+d}-1)e^{-(u+d)}}{6(u+d)}$ 4. $u = 2.0015$, $d = 1.9983$.

10.29 $X(t) \sim N(\mu t, \sigma^2 t)$. Hence $\mathsf{E}(e^{\theta X(t)}) = \exp((\theta\mu + \frac{1}{2}\theta^2\sigma^2)t)$. Hence $\mathsf{E}(Y(t)) = 1$.

CONCEPTUAL EXERCISES

10.1 1. True, 2. False, 3. True.

10.10 $Y(t) = Y(0)e^{C(t)+X(t)} = Y(0)e^{C(t)+b(t)B(t)-\int_0^t B(u)b'(u)du}$,
where $C(t) = \int_0^t (a(u) - b^2(u)/2)du$.

References

Assmussen, S. (1987). *Applied Probability and Queues,* Wiley, NY.

Bartholomew, D. J. (1982). *Stochastic Models for Social Processes,* Wiley, NY.

Bartholomew, D. J. (1991). *Statistical Techniques for Manpower Planning,* Wiley, NY.

Bartlett, M.S. (1978). *An Introduction to Stochastic Processes,* Cambridge University Press, Cambridge, UK.

Beard, R. E., T. Pentikeinen, and E. Pesonen (1984). *Risk Theory: The Stochastic Basis of Insurance*, Chapman & Hall, London.

Bertsekas, D. and R. Gallager (1987). *Data Networks*, Prentice Hall, Englewood Cliffs, NJ.

Billingsley, P. (1968). *Convergence of Probability Measures,* Wiley, NY.

Borodin, A. and P. Salminen (1996). *Handbook of Brownian Motion - Facts and Formulae,* Birkhauser, Boston.

Burghes, D. N. and A. D. Wood (1980). *Mathematical Models in Social, Management, and Life Sciences,* Halstead Press, NY.

Buzzacott, J. A. and J. G. Shanthikumar (1993). *Stochastic Models of Manufacturing Systems,* Prentice Hall, Englewood Cliffs, NJ.

Chaudhry, M. I. and J. G. C. Templeton (1983). *A First Course in Bulk Queues,* Wiley, NY.

Chung, K. I. (1967). *Markov Chains with Stationary Transition Probabilities,* Springer-Verlag, NY.

Cinlar, E. (1975). *Introduction to Stochastic Processes,* Prentice Hall, Englewood Cliffs, NJ.

Cooper, R. E. (1981). *Introduction to Queueing Theory,* North Holland, NY.

Cox, D. R. (1962). *Renewal Theory,* Metheun, London.

Cox, D. R. (1980). *Point Processes,* Chapman & Hall, London.

Cox, D. R. and H. D. Miller (1965). *The Theory of Stochastic Processes,* Chapman & Hall, London.

Cox, D. R. and W. L. Smith (1961). *Queues*, Metheun, London.

Daigle, J. N. (2005). *Queueing Theory with Applications to Packet Telecommunications*, Springer-Verlag, NY.

Durbin, R., S. Eddy, A. Krogh, and G. Mitchison (2001). *Biological Sequence Analysis: Probabilistic Models of Proteins and Nucleic Acids,* Cambridge University Press, Cambridge, UK.

Durett, R. (2001). *Essentials of Stochastic Processes,* Springer-Verlag, NY.

Durett, R. (1996). *Stochastic Calculus: A Practical Introduction,* CRC Press, Boca Raton, FL.

Elmaghraby, S. E. (1977). *Activity Networks: Project Planning and Control by Network Models,* Wiley, NY.

El-Taha, M. and S. Stidham (1998). *Sample Path Analysis of Queueing Systems,* Springer-Verlag, NY.

Feller, W. (1971). *An Introduction to Probability Theory and Its Applications,* Vols. I and II, Wiley, NY.

Fuller, L. E. (1962). *Basic Matrix Theory*, Prentice Hall, Englewood Cliffs, NJ.

Gallager, R. G. (1996). *Discrete Stochastic Processes,* Kluwer Academic Publishers, The Netherlands.

Gantmacher, F. R. (1960). *Matrix Theory,* Vols. I and II, Chelsea Publishing, NY.

Gautam, N. (2012) *Analysis of Queues: Methods and Applications*, Taylor and Francis, Boca Raton, FL.

Gelenbe, E. and G. Pujolle (1987). *Introduction to Queueing Networks,* Wiley, NY.

Goel, S. N. and N. Richter-Dyn (1974). *Stochastic Models in Biology,* Academic Press, NY.

Golub, V. K. and C. F. van Loan (1983). *Matrix Computations,* Johns Hopkins University Press, Baltimore, MD.

Gradshteyn, I. S. and I. M. Ryzhik (1990). *Table of Integrals, Series, and Products,* Academic Press, NY.

Griffiths, A. J. F., J. H. Miller, D. T. Suzuki, R. C. Lewontin, and W. M. Gelbert (1993). *An Introduction to Genetic Analysis,* W. H. Freeman and Co., NY.

Grinold, R. C. and K. T. Marshall (1977). *Manpower Planning Models,* North-Holland, NY.

Gross, D. and C. M. Harris (1985). *Fundamentals of Queueing Theory*, Wiley, NY.

Heyman, D. P. and M. J. Sobel (1982). *Stochastic Models in Operations Research,* Vols. I and II, McGraw-Hill, NY.

Hoel, P. G., S. C. Port, and C. J. Stone (1972). *Introduction to Stochastic Process,* Houghton-Mifflin, Boston, MA.

Horn, R. A. and C. R. Johnson (1990). *Matrix Analysis,* Cambridge University Press, Cambridge, UK.

Hull, J. C. (1997). *Options, Futures, and Other Derivatives,* Prentice Hall, Englewood Cliffs, NJ.

Jagers, P. (1975). *Branching Processes with Biological Applications,* Wiley, NY.

Jaiswal, N. K. (1968). *Priority Queues,* Academic Press, NY.

Karatzas, I. and S. E. Shreve (1991). *Brownian Motion and Stochastic Calculus,* Springer-Verlag, NY.

Karlin, S. and H. Taylor (1975). *A First Course in Stochastic Processes,* Academic Press, NY.

Karlin, S. and H. Taylor (1981). *A Second Course in Stochastic Processes,* Academic Press, NY.

Kelly, F. P. (1979). *Reversibility and Stochastic Networks,* Wiley, NY.

Kemeny, J. G. and L. J. Snell (1960). *Finite Markov Chains,* Van Nostrand, Princeton, NJ.

Kemeny, J. G., L. J. Snell, and A. W. Knapp (1960). *Denumerable Markov Chains,* Van Nostrand, Princeton, NJ.

Kempthorne, O. (1969). *An Introduction to Genetic Statistics,* The Iowa University Press, IA.

Kingman, J. F. C. (1993). *Poisson Processes,* Oxford University Press, Oxford, UK.

Kingman, J. F. C. (1972). *Regenerative Phenomena,* Wiley, London.

Kleinrock, L. (1976). *Queueing Systems, Vol. I: Theory*, Wiley, NY.

Kleinrock, L. (1976). *Queueing Systems, Vol. II: Computer Applications*, Wiley, NY.

Kohlas, J. (1982). *Stochastic Methods of Operations Research,* Cambridge University Press, Cambridge, UK.

Koski, T. (2001). *Hidden Markov Models of Bioinformatics,* Kluwer Academic Publishers, The Netherlands.

Kreyszig, E. (1973). *Advanced Engineering Mathematics,* Wiley, NY.

Langville, A. N. and C. D. Meyer (2011). *Google's PageRank and Beyond: The Science of Search Engine Rankings,* Princeton University Press.

Malliaris, A. G. (1982). *Stochastic Methods in Economics and Finance,* North-Holland.

Marsden, J. E. (1974). *Elementary Classical Analysis,* W. H. Freeman and Co., San Francisco.

Moran, P. A. P. (1962). *Statistical Processes of Evolutionary Theory,* Oxford University Press, UK.

Nelson, R. (1995). *Probability, Stochastic Processes and Queueing Theory,* Springer-Verlag, NY.

Neuts, M. F. (1981). *Matrix Geometric Solutions in Stochastic Models: An Algorithmic Approach,* Johns Hopkins University Press, Baltimore, MD.

Neuts, M. F. (1989). *Structured Stochastic Matrices of M/G/1 Type and Their Applications,* Marcel Dekker, NY.

Prabhu, N. U. (1985). *Queues and Inventories,* Wiley, NY.

Prabhu, N. U. (1980). *Stochastic Storage Processes: Queues Insurance Risk and Dams,* Springer-Verlag, NY.

Oksendal, B. (1995). *Stochastic Differential Equations,* Springer-Verlag, NY.

Resnick, S. I. (1992) *Adventures in Stochastic Processes,* Birkhauser, Boston.

Rolski, T., H. Schmidli, V. Schmidt, and J. Teugels (1999). *Stochastic Processes for Insurance and Finance,* Wiley, NY.

Ross, S. M. (1983). *Stochastic Processes,* Wiley, NY.

Ross, S. M. (2002). *An Elementary Introduction to Mathematical Finance: Options and Other Topics,* Cambridge University Press, Cambridge, UK.

Saaty, T. L. (1961). *Elements of Queueing Theory,* McGraw-Hill, NY.

Saur, C. H. and K. M. Chandy (1981). *Computer Systems Performance Modeling,* Prentice Hall, Englewood Cliffs, NJ.

Seneta, E. (1981). *Non-negative Matrices and Markov Chains,* Springer-Verlag, NY.

Serfozo, R. (1999). *Introduction to Stochastic Networks,* Springer-Verlag, NY.

Shreve, S. E. (2000). *Stochastic Calculus for Finance I: The Binomial Asset Pricing Model,* Springer-Verlag, NY.

Shreve, S. E. (2000). *Stochastic Calculus for Finance II: Continuous-Time Models,* Springer-Verlag, NY.

Spitzer, F. (1964). *Principles of Random Walk,* Van Nostrand, NJ.

Steele, J. M. (2001). *Stochastic Calculus and Financial Applications,* Springer-Verlag, NY.

Stoyan, D. (1983). *Comparison Methods for Queues and Other Stochastic Models,* Wiley, NY.

Syski, R. (1992). *Passage Times for Markov Chains,* IOS Press, The Netherlands.

Takacs, L. (1962). *Introduction to the Theory of Queues,* Oxford University Press, UK.

Tijms, H. C. (1986). *Stochastic Modeling and Analysis: A Computational Approach,* Wiley, NY.

Trivedi, K. S. (1982). *Probability and Statistics with Reliability, Queueing, and Computer Science Applications,* Prentice Hall, Englewood Cliffs, NJ.

Vajda, S. (1978). *Mathematics of Manpower Planning,* Wiley, NY.

Varga, R. S. (1962). *Matrix Iterative Analysis,* Prentice Hall, Englewood Cliffs, NJ.

Walrand, J. (1988). *An Introductions to Queueing Networks,* Prentice Hall, Englewood Cliffs, NJ.

Whittle, P. (1986). *Systems in Stochastic Equilibrium,* Wiley, NY.

Wilmott, P., S. Howison, and J. Dewynne (1995). *The Mathematics of Financial Derivatives: A Student Introduction,* Cambridge University Press, Cambridge, UK.

Wolff, R. W. (1989). *Stochastic Modeling and the Theory of Queues,* Prentice Hall, Englewood Cliffs, NJ.

INDEX